SUZUKI
OUTBOARD
1988-99 REPAIR MANUAL
2–225 HORSEPOWER, 1–4 CYLINDER

CEO	Rick Van Dalen
President	Dean F. Morgantini, S.A.E.
Vice President–Finance	Barry L. Beck
Vice President–Sales	Glenn D. Potere
Executive Editor	Kevin M. G. Maher, A.S.E.
Manager–Consumer Automotive	Richard Schwartz, A.S.E.
Manager–Marine/Recreation	James R. Marotta, A.S.E.
Manager–Electronic Fulfillment	Will Kessler, A.S.E., S.A.E.
Production Specialists	Brian Hollingsworth, Melinda Possinger
Project Managers	Thomas A. Mellon, A.S.E., S.A.E., Richard J. Rivele, Christine L. Sheeky, S.A.E., Todd W. Stidham, A.S.E., Ron Webb
Schematics Editors	Christopher G. Ritchie, A.S.E., S.A.E., S.T.S., Stephanie A. Spunt
Editor	Scott A. Freeman

Manufactured in USA
© 2000 W. G. Nichols
1020 Andrew Drive
West Chester, PA 19380
ISBN 0-89330-050-0
Library of Congress Catalog Card No. 98-074901
1234567890 9876543210

www.chiltonsonline.com

Contents

1 GENERAL INFORMATION AND BOATING SAFETY
- 1-2 HOW TO USE THIS MANUAL
- 1-4 BOATING SAFETY
- 1-12 SAFETY IN SERVICE

2 TOOLS AND EQUIPMENT
- 2-2 TOOLS AND EQUIPMENT
- 2-5 TOOLS
- 2-14 FASTENERS, MEASUREMENTS AND CONVERSIONS

3 MAINTENANCE
- 3-2 ENGINE MAINTENANCE
- 3-8 BOAT MAINTENANCE
- 3-12 TUNE-UP
- 3-35 WINTER STORAGE CHECKLIST
- 3-35 SPRING COMMISSIONING CHECKLIST

4 FUEL SYSTEM
- 4-2 FUEL AND COMBUSTION
- 4-3 FUEL SYSTEM
- 4-7 TROUBLESHOOTING
- 4-11 CARBURETOR SERVICE
- 4-25 REED VALVE SERVICE
- 4-27 FUEL PUMP SERVICE
- 4-32 ELECTRONIC FUEL INJECTION

5 IGNITION AND ELECTRICAL SYSTEMS
- 5-2 UNDERSTANDING AND TROUBLESHOOTING ELECTRICAL SYSTEMS
- 5-7 BREAKER POINTS IGNITION (MAGNETO IGNITION)
- 5-11 CAPACITOR DISCHARGE IGNITION (CDI) SYSTEM
- 5-38 ELECTRONIC IGNITION
- 5-39 CHARGING CIRCUIT
- 5-45 STARTING CIRCUIT
- 5-52 IGNITION AND ELECTRICAL WIRING DIAGRAMS

6 OIL INJECTION
- 6-2 OIL INJECTION SYSTEM
- 6-11 COOLING SYSTEM
- 6-14 OIL INJECTION WARNING SYSTEMS
- 6-17 OVERHEAT WARNING SYSTEM

Contents

7 POWERHEAD
- **7-2** ENGINE MECHANICAL
- **7-32** POWERHEAD RECONDITIONING

8 LOWER UNIT
- **8-2** LOWER UNIT
- **8-6** LOWER UNIT OVERHAUL
- **8-51** JET DRIVE

9 TRIM AND TILT
- **9-2** MANUAL TILT
- **9-2** GAS ASSISTED TILT
- **9-3** POWER TILT
- **9-6** POWER TRIM/TILT

10 REMOTE CONTROL
- **10-2** REMOTE CONTROL BOX
- **10-7** TILLER HANDLE

11 HAND REWIND STARTER
- **11-2** HAND REWIND STARTER
- **11-2** OVERHEAD TYPE STARTER
- **11-10** BENDIX TYPE STARTER

GLOSSARY
- **11-13** GLOSSARY

MASTER INDEX
- **11-17** INDEX

See last page for information on additional titles

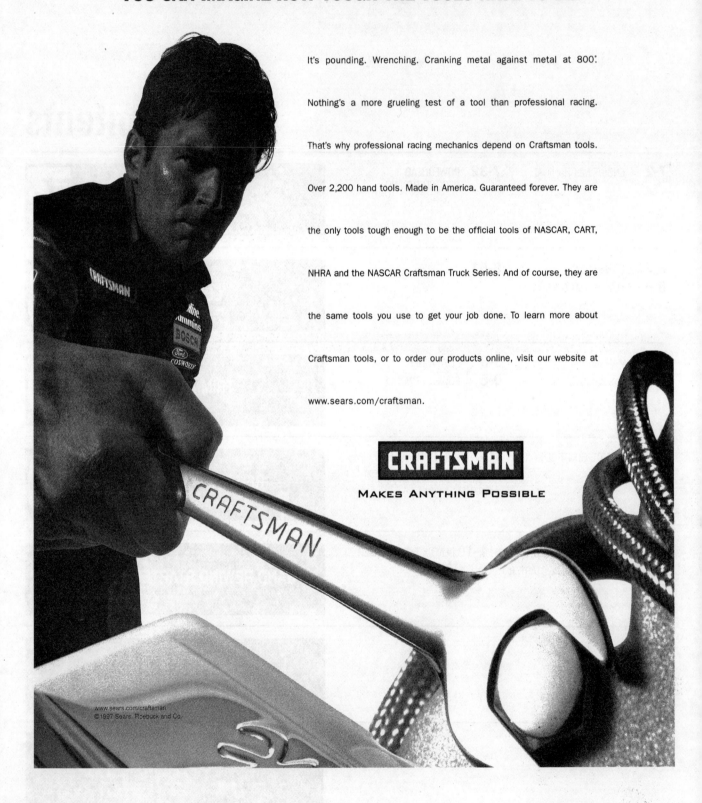

SAFETY NOTICE

Proper service and repair procedures are vital to the safe, reliable operation of all marine engines, as well as the personal safety of those performing repairs. This manual outlines procedures for servicing and repairing stern drives using safe, effective methods. The procedures contain many NOTES, CAUTIONS and WARNINGS which should be followed, along with standard procedures, to eliminate the possibility of personal injury or improper service which could damage the vessel or compromise its safety.

It is important to note that repair procedures and techniques, tools and parts for servicing marine engines, as well as the skill and experience of the individual performing the work, vary widely. It is not possible to anticipate all of the conceivable ways or conditions under which these engines may be serviced, or to provide cautions as to all possible hazards that may result. Standard and accepted safety precautions and equipment should be used during cutting, grinding, chiseling, prying, or any other process that can cause material removal or projectiles.

Some procedures require the use of tools specially designed for a specific purpose. Before substituting another tool or procedure, you must be completely satisfied that neither your personal safety, nor the performance of the marine engine, will be compromised.

Although information in this manual is based on industry sources and is complete as possible at the time of publication, the possibility exists that some vehicle manufacturers made later changes which could not be included here. While striving for total accuracy, Seloc® cannot assume responsibility for any errors, changes or omissions that may occur in the compilation of this data.

PART NUMBERS

Part numbers listed in this reference are not recommendations by Seloc® for any product brand name. They are references that can be used with interchange manuals and aftermarket supplier catalogs to locate each brand supplier's discrete part number.

SPECIAL TOOLS

Special tools are recommended by the marine manufacturer to perform a specific task. Use has been kept to a minimum, but, where absolutely necessary, they are referred to in the text by the part number of the tool manufacturer. These tools can be purchased, under the appropriate part number, from your local dealer or regional distributor, or an equivalent tool can be purchased locally from a tool supplier or parts outlet. Before substituting any tool for the one recommended, read the SAFETY NOTICE at the top of this page.

ALL RIGHTS RESERVED

No part of this publication may be reproduced, transmitted or stored in any form or by any means, electronic or mechanical, including photocopy, recording, or by information storage or retrieval system, without prior written permission from the publisher.

ACKNOWLEDGMENTS

Seloc® expresses appreciation to the following companies who supported the production of this manual:

- Suzuki Motor Corporation—Brea, CA
- Brown's Point Marina—Keyport, NJ
- Chester River Marine Services—Chestertown, MD
- Marine Mechanics Institute—Orlando, FL

Thanks to Mike Benfer and Robert Strickler of the Marine Mechanics Institute for providing the outboard engines and miscellaneous parts to get this project going, to Dennis Hogans and Jeremy Dolan of Chester River Marine Services for allowing us full access to their dealership for a portion of our photoshoot and finally to George Jansen of Brown's Point Marina and Ed Reed of Dorset Marine for their overall guidance in the finer points of outboard repair.

Seloc® would like to express thanks to all of the fine companies who participate in the production of our books. Hand tools supplied by Craftsman are used during all phases of teardown and photography. Many of the fine specialty tools used in our procedures were provided courtesy of Lisle Corporation. All American Manufacturing Inc. has provided the powerhead and lower unit stands used to hold and support our teardown subjects. Much of our shop's electronic testing equipment was supplied by Universal Enterprises Inc. (UEI). Rapiar® supplied the specialized ignition testing equipment used in the ignition system and tune-up sections of this manual.

MARINE TECHNICIAN TRAINING

INDUSTRY SUPPORTED PROGRAMS
OUTBOARD, STERNDRIVE & PERSONAL WATERCRAFT

- Dyno Testing • Boat & Trailer Rigging • Electrical & Fuel System Diagnostics
- Powerhead, Lower Unit & Drive Rebuilds • Powertrim & Tilt Rebuilds
- Instrument & Accessories Installation

TRAIN IN SUNNY FLORIDA!

For information regarding housing, financial aid and employment opportunities in the marine industry, contact us today:

CALL TOLL FREE
1-800-528-7995

An Accredited Institution

SM
Name
Address
City State Zip
Phone

FINANCIAL ASSISTANCE AVAILABLE FOR THOSE WHO QUALIFY!

MEMBER NMMA

MARINE MECHANICS INSTITUTE
A Division of CTI
9751 Delegates Drive • Orlando, Florida 32837
2844 W. Deer Valley Rd. • Phoenix, AZ 85027

HOW TO USE THIS MANUAL 1-2
CAN YOU DO IT? 1-2
WHERE TO BEGIN 1-2
AVOIDING TROUBLE 1-2
MAINTENANCE OR REPAIR? 1-2
DIRECTIONS AND LOCATIONS 1-2
PROFESSIONAL HELP 1-2
PURCHASING PARTS 1-3
AVOIDING THE MOST COMMON
  MISTAKES 1-3
BOATING SAFETY 1-4
REGULATIONS FOR YOUR BOAT 1-4
    DOCUMENTING OF VESSELS 1-4
    REGISTRATION OF BOATS 1-4
    NUMBERING OF VESSELS 1-4
    SALES AND TRANSFERS 1-4
    HULL IDENTIFICATION
      NUMBER 1-4
    LENGTH OF BOATS 1-4
    CAPACITY INFORMATION 1-4
    CERTIFICATE OF COMPLIANCE 1-4
    VENTILATION 1-5
    VENTILATION SYSTEMS 1-5
REQUIRED SAFETY EQUIPMENT 1-5
    TYPES OF FIRES 1-5
    FIRE EXTINGUISHERS 1-5
    WARNING SYSTEM 1-7
    PERSONAL FLOTATION
      DEVICES 1-7
    SOUND PRODUCING DEVICES 1-9
    VISUAL DISTRESS SIGNALS 1-9
EQUIPMENT NOT REQUIRED BUT
  RECOMMENDED 1-10
    SECOND MEANS OF
      PROPULSION 1-10
    BAILING DEVICES 1-10
    FIRST AID KIT 1-10
    ANCHORS 1-10
    VHF-FM RADIO 1-11
    TOOLS AND SPARE PARTS 1-11
COURTESY MARINE
  EXAMINATIONS 1-11
SAFETY IN SERVICE 1-12
DO'S 1-12
DON'TS 1-12

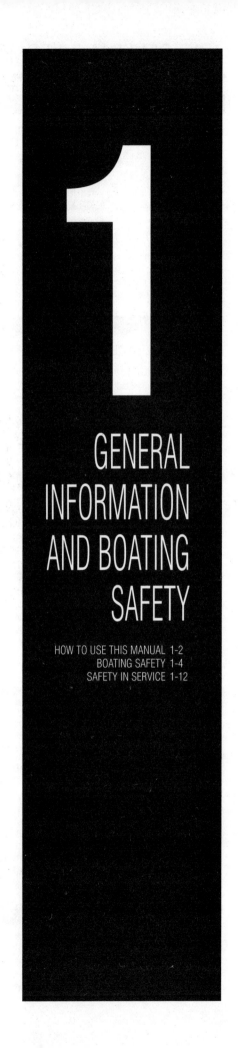

1

GENERAL INFORMATION AND BOATING SAFETY

HOW TO USE THIS MANUAL 1-2
BOATING SAFETY 1-4
SAFETY IN SERVICE 1-12

1-2 GENERAL INFORMATION AND BOATING SAFETY

HOW TO USE THIS MANUAL

This manual is designed to be a handy reference guide to maintaining and repairing your Suzuki 2-stroke outboard. We strongly believe that regardless of how many or how few years experience you may have, there is something new waiting here for you.

This manual covers the topics that a factory service manual (designed for factory trained mechanics) and a manufacturer owner's manual (designed more by lawyers these days) covers. It will take you through the basics of maintaining and repairing your outboard, step-by-step, to help you understand what the factory trained mechanics already know by heart. By using the information in this manual, any boat owner should be able to make better informed decisions about what they need to do to maintain and enjoy their outboard.

Even if you never plan on touching a wrench (and if so, we hope that you will change your mind), this manual will still help you understand what a mechanic needs to do in order to maintain your engine.

Can You Do It?

If you are not the type who is prone to taking a wrench to something, NEVER FEAR. The procedures in this manual cover topics at a level virtually anyone will be able to handle. And just the fact that you purchased this manual shows your interest in better understanding your outboard.

You may find that maintaining your outboard yourself is preferable in most cases. From a monetary standpoint, it could also be beneficial. The money spent on hauling your boat to a marina and paying a tech to service the engine could buy you fuel for a whole weekend's boating. If you are unsure of your own mechanical abilities, at the very least you should fully understand what a marine mechanic does to your boat. You may decide that anything other than maintenance and adjustments should be performed by a mechanic (and that's your call), but know that every time you board your boat, you are placing faith in the mechanic's work and trusting him or her with your well-being, and maybe your life.

It should also be noted that in most areas a factory trained mechanic will command a hefty hourly rate for off site service. This hourly rate is charged from the time they leave their shop to the time they return home. The cost savings in doing the job yourself should be readily apparent at this point.

Where to Begin

Before spending any money on parts, and before removing any nuts or bolts, read through the entire procedure or topic. This will give you the overall view of what tools and supplies will be required to perform the procedure or what questions need to be answered before purchasing parts. So read ahead and plan ahead. Each operation should be approached logically and all procedures thoroughly understood before attempting any work.

Avoiding Trouble

Some procedures in this manual may require you to "label and disconnect . . ." a group of lines, hoses or wires. Don't be lulled into thinking you can remember where everything goes — you won't. If you reconnect or install a part incorrectly, things may operate poorly, if at all. If you hook up electrical wiring incorrectly, you may instantly learn a very, very expensive lesson.

A piece of masking tape, for example, placed on a hose and another on its fitting will allow you to assign your own label such as the letter "A", or a short name. As long as you remember your own code, the lines can be reconnected by matching letters or names. Do remember that tape will dissolve when saturated in fluids. If a component is to be washed or cleaned, use another method of identification. A permanent felt-tipped marker can be very handy for marking metal parts; but remember that fluids will remove permanent marker.

SAFETY is the most important thing to remember when performing maintenance or repairs. Be sure to read the information on safety in this manual.

Maintenance or Repair?

Proper maintenance is the key to long and trouble-free engine life, and the work can yield its own rewards. A properly maintained engine performs better than one that is neglected. As a conscientious boat owner, set aside a Saturday morning, at least once a month, to perform a thorough check of items which could cause problems. Keep your own personal log to jot down which services you performed, how much the parts cost you, the date, and the amount of hours on the engine at the time. Keep all receipts for parts purchased, so that they may be referred to in case of related problems or to determine operating expenses. As a do-it-yourselfer, these receipts are the only proof you have that the required maintenance was performed. In the event of a warranty problem, these receipts will be invaluable.

It's necessary to mention the difference between maintenance and repair. Maintenance includes routine inspections, adjustments, and replacement of parts that show signs of normal wear. Maintenance compensates for wear or deterioration. Repair implies that something has broken or is not working. A need for repair is often caused by lack of maintenance.

For example: draining and refilling the engine oil is maintenance recommended by all manufacturers at specific intervals. Failure to do this can allow internal corrosion or damage and impair the operation of the engine, requiring expensive repairs. While no maintenance program can prevent items from breaking or wearing out, a general rule can be stated: MAINTENANCE IS CHEAPER THAN REPAIR.

Directions and Locations

♦ See Figure 1

Two basic rules should be mentioned here. First, whenever the Port side of the engine (or boat) is referred to, it is meant to specify the left side of the engine when you are sitting at the helm. Conversely, the Starboard means your right side. The Bow is the front of the boat and the Stern is the rear.

Most screws and bolts are removed by turning counterclockwise, and tightened by turning clockwise. An easy way to remember this is: righty-tighty; lefty-loosey. Corny, but effective. And if you are really dense (and we have all been so at one time or another), buy a ratchet that is marked ON and OFF, or mark your own.

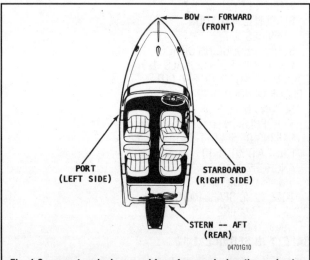

Fig. 1 Common terminology used for reference designation on boats of all size. These terms are used though out the manual

Professional Help

Occasionally, there are some things when working on an outboard that are beyond the capabilities or tools of the average Do-It-Yourselfer (DIYer). This shouldn't include most of the topics of this manual, but you will have to be the judge. Some engines require special tools or a selection of special parts, even for basic maintenance.

Talk to other boaters who use the same model of engine and speak with a trusted marina to find if there is a particular system or component on your engine that is difficult to maintain. For example, although the technique of valve adjustment on some engines may be easily understood and even performed by a DIYer, it might require a handy assortment of shims in various sizes and a few hours of disassembly to get to that point. Not having the assortment of shims handy might mean multiple trips back and forth to the parts store, and this might not be worth your time.

GENERAL INFORMATION AND BOATING SAFETY 1-3

You will have to decide for yourself where basic maintenance ends and where professional service should begin. Take your time and do your research first (starting with the information in this manual) and then make your own decision. If you really don't feel comfortable with attempting a procedure, DON'T DO IT. If you've gotten into something that may be over your head, don't panic. Tuck your tail between your legs and call a marine mechanic. Marinas and independent shops will be able to finish a job for you. Your ego may be damaged, but your boat will be properly restored to its full running order. So, as long as you approach jobs slowly and carefully, you really have nothing to lose and everything to gain by doing it yourself.

Purchasing Parts

▸ See Figures 2 and 3

When purchasing parts there are two things to consider. The first is quality and the second is to be sure to get the correct part for your engine. To get quality parts, always deal directly with a reputable retailer. To get the proper parts always refer to the information tag on your engine prior to calling the parts counter. An incorrect part can adversely affect your engine performance and fuel economy, and will cost you more money and aggravation in the end.

Just remember, a tow back to shore will cost plenty. That charge is per hour from the time the towboat leaves their home port, to the time they return to their home port. Get the picture. . . .$$$?

So who should you call for parts? Well, there are many sources for the parts you will need. Where you shop for parts will be determined by what kind of parts you need, how much you want to pay, and the types of stores in your neighborhood.

Your marina can supply you with many of the common parts you require. Using a marina for as your parts supplier may be handy because of location (just walk right down the dock) or because the marina specializes in your particular brand of engine. In addition, it is always a good idea to get to know the marina staff (especially the marine mechanic).

The marine parts jobber, who is usually listed in the yellow pages or whose name can be obtained from the marina, is another excellent source for parts. In addition to supplying local marinas, they also do a sizeable business in over-the-counter parts sales for the do-it-yourselfer.

Almost every community has one or more convenient marine chain stores. These stores often offer the best retail prices and the convenience of one-stop shopping for all your needs. Since they cater to the do-it-yourselfer, these stores are almost always open weeknights, Saturdays, and Sundays, when the jobbers are usually closed.

The lowest prices for parts are most often found in discount stores or the auto department of mass merchandisers. Parts sold here are name and private brand parts bought in huge quantities, so they can offer a competitive price. Private brand parts are made by major manufacturers and sold to large chains under a store label.

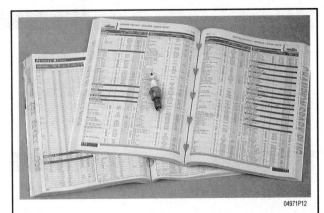

Fig. 3 Parts catalogs, giving application and part number information, are provided by manufacturers for most replacement parts

Fig. 2 By far the most important asset in purchasing parts is a knowledgeable and enthusiastic parts person

Avoiding the Most Common Mistakes

There are 3 common mistakes in mechanical work:

1. Incorrect order of assembly, disassembly or adjustment. When taking something apart or putting it together, performing steps in the wrong order usually just costs you extra time; however, it CAN break something. Read the entire procedure before beginning disassembly. Perform everything in the order in which the instructions say you should, even if you can't immediately see a reason for it. When you're taking apart something that is very intricate, you might want to draw a picture of how it looks when assembled at one point in order to make sure you get everything back in its proper position. When making adjustments, perform them in the proper order; often, one adjustment affects another, and you cannot expect satisfactory results unless each adjustment is made only when it cannot be changed by another.

2. Overtorquing (or undertorquing). While it is more common for overtorquing to cause damage, undertorquing may allow a fastener to vibrate loose causing serious damage. Especially when dealing with aluminum parts, pay attention to torque specifications and utilize a torque wrench in assembly. If a torque figure is not available, remember that if you are using the right tool to perform the job, you will probably not have to strain yourself to get a fastener tight enough. The pitch of most threads is so slight that the tension you put on the wrench will be multiplied many times in actual force on what you are tightening.

3. Crossthreading. This occurs when a part such as a bolt is screwed into a nut or casting at the wrong angle and forced. Crossthreading is more likely to occur if access is difficult. It helps to clean and lubricate fasteners, then to start threading with the part to be installed positioned straight in. Always start a fastener, etc. with your fingers. If you encounter resistance, unscrew the part and start over again at a different angle until it can be inserted and turned several times without much effort. Keep in mind that some parts may have tapered threads, so that gentle turning will automatically bring the part you're threading to the proper angle, but only if you don't force it or resist a change in angle. Don't put a wrench on the part until it has been tightened a couple of turns by hand. If you suddenly encounter resistance, and the part has not seated fully, don't force it. Pull it back out to make sure it's clean and threading properly.

1-4 GENERAL INFORMATION AND BOATING SAFETY

BOATING SAFETY

In 1971 Congress ordered the U.S. Coast Guard to improve recreational boating safety. In response, the Coast Guard drew up a set of regulations.

Beside these federal regulations, there are state and local laws you must follow. These sometimes exceed the Coast Guard requirements. This section discusses only the federal laws. State and local laws are available from your local Coast Guard. As with other laws, "Ignorance of the boating laws is no excuse." The rules fall into two groups: regulations for your boat and required safety equipment on your boat.

Regulations For Your Boat

Most boats on waters within Federal jurisdiction must be registered or documented. These waters are those that provide a means of transportation between two or more states or to the sea. They also include the territorial waters of the United States.

DOCUMENTING OF VESSELS

A vessel of five or more net tons may be documented as a yacht. In this process, papers are issued by the U.S. Coast Guard as they are for large ships. Documentation is a form of national registration. The boat must be used solely for pleasure. Its owner must be a U.S. citizen, a partnership of U.S. citizens, or a corporation controlled by U.S. citizens. The captain and other officers must also be U.S. citizens. The crew need not be.

If you document your yacht, you have the legal authority to fly the yacht ensign. You also may record bills of sale, mortgages, and other papers of title with federal authorities. Doing so gives legal notice that such instruments exist. Documentation also permits preferred status for mortgages. This gives you additional security and aids financing and transfer of title. You must carry the original documentation papers aboard your vessel. Copies will not suffice.

REGISTRATION OF BOATS

If your boat is not documented, registration in the state of its principal use is probably required. If you use it mainly on an ocean, a gulf, or other similar water, register it in the state where you moor it.

If you use your boat solely for racing, it may be exempt from the requirement in your state. States may also exclude dinghies. Some require registration of documented vessels and non-power driven boats.

All states, except Alaska, register boats. In Alaska, the U.S. Coast Guard issues the registration numbers. If you move your vessel to a new state of principal use, a valid registration certificate is good for 60 days. You must have the registration certificate (certificate of number) aboard your vessel when it is in use. A copy will not suffice. You may be cited if you do not have the original on board.

NUMBERING OF VESSELS

A registration number is on your registration certificate. You must paint or permanently attach this number to both sides of the forward half of your boat. Do not display any other number there.

The registration number must be clearly visible. It must not be placed on the obscured underside of a flared bow. If you can't place the number on the bow, place it on the forward half of the hull. If that doesn't work, put it on the superstructure. Put the number for an inflatable boat on a bracket or fixture. Then, firmly attach it to the forward half of the boat. The letters and numbers must be plain block characters and must read from left to right. Use a space or a hyphen to separate the prefix and suffix letters from the numerals. The color of the characters must contrast with that of the background, and they must be at least three inches high.

In some states your registration is good for only one year. In others, it is good for as long as three years. Renew your registration before it expires. At that time you will receive a new decal or decals. Place them as required by state law. You should remove old decals before putting on the new ones. Some states require that you show only the current decal or decals. If your vessel is moored, it must have a current decal even if it is not in use.

If your vessel is lost, destroyed, abandoned, stolen, or transferred, you must inform the issuing authority. If you lose your certificate of number or your address changes, notify the issuing authority as soon as possible.

SALES AND TRANSFERS

Your registration number is not transferable to another boat. The number stays with the boat unless its state of principal use is changed.

HULL IDENTIFICATION NUMBER

A Hull Identification Number (HIN) is like the Vehicle Identification Number (VIN) on your car. Boats built between November 1, 1972 and July 31, 1984 have old format HINs. Since August 1, 1984 a new format has been used.

Your boat's HIN must appear in two places. If it has a transom, the primary number is on its starboard side within two inches of its top. If it does not have a transom or if it was not practical to use the transom, the number is on the starboard side. In this case, it must be within one foot of the stern and within two inches of the top of the hull side. On pontoon boats, it is on the aft crossbeam within one foot of the starboard hull attachment. Your boat also has a duplicate number in an unexposed location. This is on the boat's interior or under a fitting or item of hardware.

LENGTH OF BOATS

For some purposes, boats are classed by length. Required equipment, for example, differs with boat size. Manufacturers may measure a boat's length in several ways. Officially, though, your boat is measured along a straight line from its bow to its stern. This line is parallel to its keel.

The length does not include bowsprits, boomkins, or pulpits. Nor does it include rudders, brackets, outboard motors, outdrives, diving platforms, or other attachments.

CAPACITY INFORMATION

▶ See Figure 4

Manufacturers must put capacity plates on most recreational boats less than 20 feet long. Sailboats, canoes, kayaks, and inflatable boats are usually exempt. Outboard boats must display the maximum permitted horsepower of their engines. The plates must also show the allowable maximum weights of the people on board. And they must show the allowable maximum combined weights of people, engines, and gear. Inboards and stern drives need not show the weight of their engines on their capacity plates. The capacity plate must appear where it is clearly visible to the operator when underway. This information serves to remind you of the capacity of your boat under normal circumstances. You should ask yourself, "Is my boat loaded above its recommended capacity" and, "Is my boat overloaded for the present sea and wind conditions?" If you are stopped by a legal authority, you may be cited if you are overloaded.

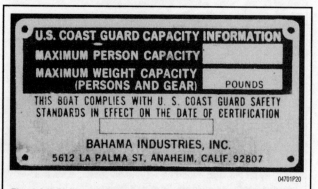

Fig. 4 A U.S. Coast Guard certification plate indicates the amount of occupants and gear appropriate for safe operation of the vessel

CERTIFICATE OF COMPLIANCE

Manufacturers are required to put compliance plates on motorboats greater than 20 feet in length. The plates must say, "This boat," or "This equipment

GENERAL INFORMATION AND BOATING SAFETY

complies with the U. S. Coast Guard Safety Standards in effect on the date of certification." Letters and numbers can be no less than one-eighth of an inch high. At the manufacturer's option, the capacity and compliance plates may be combined.

VENTILATION

A cup of gasoline spilled in the bilge has the potential explosive power of 15 sticks of dynamite. This statement, commonly quoted over 20 years ago, may be an exaggeration, however, it illustrates a fact. Gasoline fumes in the bilge of a boat are highly explosive and a serious danger. They are heavier than air and will stay in the bilge until they are vented out.

Because of this danger, Coast Guard regulations require ventilation on many power boats. There are several ways to supply fresh air to engine and gasoline tank compartments and to remove dangerous vapors. Whatever the choice, it must meet Coast Guard standards.

➡ The following is not intended to be a complete discussion of the regulations. It is limited to the majority of recreational vessels. Contact your local Coast Guard office for further information.

General Precautions

Ventilation systems will not remove raw gasoline that leaks from tanks or fuel lines. If you smell gasoline fumes, you need immediate repairs. The best device for sensing gasoline fumes is your nose. Use it! If you smell gasoline in an engine compartment or elsewhere, don't start your engine. The smaller the compartment, the less gasoline it takes to make an explosive mixture.

Ventilation for Open Boats

In open boats, gasoline vapors are dispersed by the air that moves through them. So they are exempt from ventilation requirements.

To be "open," a boat must meet certain conditions. Engine and fuel tank compartments and long narrow compartments that join them must be open to the atmosphere." This means they must have at least 15 square inches of open area for each cubic foot of net compartment volume. The open area must be in direct contact with the atmosphere. There must also be no long, unventilated spaces open to engine and fuel tank compartments into which flames could extend.

Ventilation for All Other Boats

Powered and natural ventilation are required in an enclosed compartment with a permanently installed gasoline engine that has a cranking motor. A compartment is exempt if its engine is open to the atmosphere. Diesel powered boats are also exempt.

VENTILATION SYSTEMS

There are two types of ventilation systems. One is "natural ventilation." In it, air circulates through closed spaces due to the boat's motion. The other type is "powered ventilation." In it, air is circulated by a motor driven fan or fans.

Natural Ventilation System Requirements

A natural ventilation system has an air supply from outside the boat. The air supply may also be from a ventilated compartment or a compartment open to the atmosphere. Intake openings are required. In addition, intake ducts may be required to direct the air to appropriate compartments.

The system must also have an exhaust duct that starts in the lower third of the compartment. The exhaust opening must be into another ventilated compartment or into the atmosphere. Each supply opening and supply duct, if there is one, must be above the usual level of water in the bilge. Exhaust openings and ducts must also be above the bilge water. Openings and ducts must be at least three square inches in area or two inches in diameter. Openings should be placed so exhaust gasses do not enter the fresh air intake. Exhaust fumes must not enter cabins or other enclosed, non-ventilated spaces. The carbon monoxide gas in them is deadly.

Intake and exhaust openings must be covered by cowls or similar devices. These registers keep out rain water and water from breaking seas. Most often, intake registers face forward and exhaust openings aft. This aids the flow of air when the boat is moving or at anchor since most boats face into the wind when anchored.

Power Ventilation System Requirements

♦ See Figure 5

Powered ventilation systems must meet the standards of a natural system. They must also have one or more exhaust blowers. The blower duct can serve as the exhaust duct for natural ventilation if fan blades do not obstruct the air flow when not powered. Openings in engine compartment, for carburetion are in addition to ventilation system requirements.

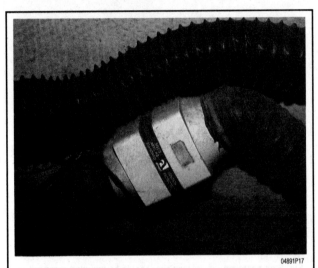

Fig. 5 Typical blower and duct system to vent fumes from the engine compartment

Required Safety Equipment

Coast Guard regulations require that your boat have certain equipment aboard. These requirements are minimums. Exceed them whenever you can.

TYPES OF FIRES

There are four common classes of fires:
- Class A—fires are in ordinary combustible materials such as paper or wood.
- Class B—fires involve gasoline, oil and grease.
- Class C—fires are electrical.
- Class D—fires involve ferrous metals

One of the greatest risks to boaters is fire. This is why it is so important to carry the correct number and type of extinguishers onboard.

The best fire extinguisher for most boats is a Class B extinguisher. Never use water on Class B or Class C fires, as water spreads these types of fires. You should never use water on a Class C fire as it may cause you to be electrocuted.

FIRE EXTINGUISHERS

♦ See Figure 6

If your boat meets one or more of the following conditions, you must have at least one fire extinguisher aboard. The conditions are:
- Inboard or stern drive engines
- Closed compartments under seats where portable fuel tanks can be stored
- Double bottoms not sealed together or not completely filled with flotation materials
- Closed living spaces
- Closed stowage compartments in which combustible or flammable materials are stored

1-6 GENERAL INFORMATION AND BOATING SAFETY

Fig. 6 An approved fire extinguisher should be mounted close to the operator for emergency use

- Permanently installed fuel tanks
- Boat is 26 feet or more in length.

Contents of Extinguishers

Fire extinguishers use a variety of materials. Those used on boats usually contain dry chemicals, Halon, or Carbon Dioxide (CO3). Dry chemical extinguishers contain chemical powders such as Sodium Bicarbonate—baking soda.

Carbon dioxide is a colorless and odorless gas when released from an extinguisher. It is not poisonous but caution must be used in entering compartments filled with it. It will not support life and keeps oxygen from reaching your lungs. A fire-killing concentration of Carbon Dioxide is lethal. If you are in a compartment with a high concentration of CO3, you will have no difficulty breathing. But the air does not contain enough oxygen to support life. Unconsciousness or death can result.

HALON EXTINGUISHERS

Some fire extinguishers and `built-in' or `fixed' automatic fire extinguishing systems contain a gas called Halon. Like carbon dioxide it is colorless and odorless and will not support life. Some Halons may be toxic if inhaled.

To be accepted to the Coast Guard, a fixed Halon system must have an indicator light at the vessel's helm. A green light shows the system is ready. Red means it is being discharged or has been discharged. Warning horns are available to let you know the system has been activated. If your fixed Halon system discharges, ventilate the space thoroughly before you enter it. There are no residues from Halon but it will not support life.

Although Halon has excellent fire fighting properties, it is thought to deplete the earth's ozone layer and has not been manufactured since January 1, 1994. Halon extinguishers can be refilled from existing stocks of the gas until they are used up, but high federal excise taxes are being charged for the service. If you discontinue using your Halon extinguisher, take it to a recovery station rather than releasing the gas into the atmosphere. Compounds such as FE 241, designed to replace Halon, are now available.

Fire Extinguisher Approval

Fire extinguishers must be Coast Guard approved. Look for the approval number on the nameplate. Approved extinguishers have the following on their labels: "Marine Type USCG Approved, Size . . . , Type . . . , 162.208/," etc. In addition, to be acceptable by the Coast Guard, an extinguisher must be in serviceable condition and mounted in its bracket. An extinguisher not properly mounted in its bracket will not be considered serviceable during a Coast Guard inspection.

Care and Treatment

Make certain your extinguishers are in their stowage brackets and are not damaged. Replace cracked or broken hoses. Nozzles should be free of obstructions. Sometimes, wasps and other insects nest inside nozzles and make them inoperable. Check your extinguishers frequently. If they have pressure gauges, is the pressure within acceptable limits? Do the locking pins and sealing wires show they have not been used since recharging?

Don't try an extinguisher to test it. Its valves will not reseat properly and the remaining gas will leak out. When this happens, the extinguisher is useless.

Weigh and tag carbon dioxide and Halon extinguishers twice a year. If their weight loss exceeds 10 percent of the weight of the charge, recharge them. Check to see that they have not been used. They should have been inspected by a qualified person within the past six months, and they should have tags showing all inspection and service dates. The problem is that they can be partially discharged while appearing to be fully charged.

Some Halon extinguishers have pressure gauges the same as dry chemical extinguishers. Don't rely too heavily on the gauge. The extinguisher can be partially discharged and still show a good gauge reading. Weighing a Halon extinguisher is the only accurate way to assess its contents.

If your dry chemical extinguisher has a pressure indicator, check it frequently. Check the nozzle to see if there is powder in it. If there is, recharge it. Occasionally invert your dry chemical extinguisher and hit the base with the palm of your hand. The chemical in these extinguishers packs and cakes due to the boat's vibration and pounding. There is a difference of opinion about whether hitting the base helps, but it can't hurt. It is known that caking of the chemical powder is a major cause of failure of dry chemical extinguishers. Carry spares in excess of the minimum requirement. If you have guests aboard, make certain they know where the extinguishers are and how to use them.

Using a Fire Extinguisher

A fire extinguisher usually has a device to keep it from being discharged accidentally. This is a metal or plastic pin or loop. If you need to use your extinguisher, take it from its bracket. Remove the pin or the loop and point the nozzle at the base of the flames. Now, squeeze the handle, and discharge the extinguisher's contents while sweeping from side to side. Recharge a used extinguisher as soon as possible.

If you are using a Halon or carbon dioxide extinguisher, keep your hands away from the discharge. The rapidly expanding gas will freeze them. If your fire extinguisher has a horn, hold it by its handle.

Legal Requirements for Extinguishers

You must carry fire extinguishers as defined by Coast Guard regulations. They must be firmly mounted in their brackets and immediately accessible.

A motorboat less than 26 feet long must have at least one approved hand-portable, Type B-1 extinguisher. If the boat has an approved fixed fire extinguishing system, you are not required to have the Type B-1 extinguisher. Also, if your boat is less than 26 feet long, is propelled by an outboard motor, or motors, and does not have any of the first six conditions described at the beginning of this section, it is not required to have an extinguisher. Even so, it's a good idea to have one, especially if a nearby boat catches fire, or if a fire occurs at a fuel dock.

A motorboat 26 feet to under 40 feet long, must have at least two Type B-1 approved hand-portable extinguishers. It can, instead, have at least one Coast Guard approved Type B-2. If you have an approved fire extinguishing system, only one Type B-1 is required.

A motorboat 40 to 65 feet long must have at least three Type B-1 approved portable extinguishers . It may have, instead, at least one Type B-1 plus a Type B-2. If there is an approved fixed fire extinguishing system, two Type B-1 or one Type B-2 is required.

GENERAL INFORMATION AND BOATING SAFETY 1-7

WARNING SYSTEM

Various devices are available to alert you to danger. These include fire, smoke, gasoline fumes, and carbon monoxide detectors. If your boat has a galley, it should have a smoke detector. Where possible, use wired detectors. Household batteries often corrode rapidly on a boat.

You can't see, smell, nor taste carbon monoxide gas, but it is lethal. As little as one part in 10,000 parts of air can bring on a headache. The symptoms of carbon monoxide poisoning—headaches, dizziness, and nausea—are like sea sickness. By the time you realize what is happening to you, it may be too late to take action. If you have enclosed living spaces on your boat, protect yourself with a detector. There are many ways in which carbon monoxide can enter your boat.

PERSONAL FLOTATION DEVICES

Personal Flotation Devices (PFDs) are commonly called life preservers or life jackets. You can get them in a variety of types and sizes. They vary with their intended uses. To be acceptable, they must be Coast Guard approved.

Type I PFDs

A Type I life jacket is also called an offshore life jacket. Type I life jackets will turn most unconscious people from facedown to a vertical or slightly backward position. The adult size gives a minimum of 22 pounds of buoyancy. The child size has at least 11 pounds. Type I jackets provide more protection to their wearers than any other type of life jacket. Type I life jackets are bulkier and less comfortable than other types. Furthermore, there are only two sizes, one for children and one for adults.

Type I life jackets will keep their wearers afloat for extended periods in rough water. They are recommended for offshore cruising where a delayed rescue is probable.

Type II PFDs

♦ See Figure 7

A Type II life jacket is also called a near-shore buoyant vest. It is an approved, wearable device. Type II life jackets will turn some unconscious people from facedown to vertical or slightly backward positions. The adult size gives at least 15.5 pounds of buoyancy. The medium child size has a minimum of 11 pounds. And the small child and infant sizes give seven pounds. A Type II life jacket is more comfortable than a Type I but it does not have as much buoyancy. It is not recommended for long hours in rough water. Because of this, Type IIs are recommended for inshore and inland cruising on calm water. Use them where there is a good chance of fast rescue.

Type III PFDs

Type III life jackets or marine buoyant devices are also known as flotation aids. Like Type IIs, they are designed for calm inland or close offshore water where there is a good chance of fast rescue. Their minimum buoyancy is 15.5 pounds. They will not turn their wearers face up.

Type III devices are usually worn where freedom of movement is necessary. Thus, they are used for water skiing, small boat sailing, and fishing among other activities. They are available as vests and flotation coats. Flotation coats are useful in cold weather. Type IIIs come in many sizes from small child through large adult.

Life jackets come in a variety of colors and patterns—red, blue, green, camouflage, and cartoon characters. From a safety standpoint, the best color is bright orange. It is easier to see in the water, especially if the water is rough.

Type IV PFDs

♦ See Figures 8 and 9

Type IV ring life buoys, buoyant cushions and horseshoe buoys are Coast Guard approved devices called throwables. They are made to be thrown to people in the water, and should not be worn. Type IV cushions are often used as

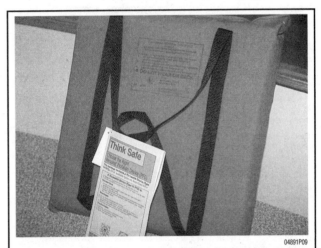
Fig. 8 Type IV buoyant cushions are made to be thrown to people in the water. If you can squeeze air out of the cushion, it is faulty and should be replaced

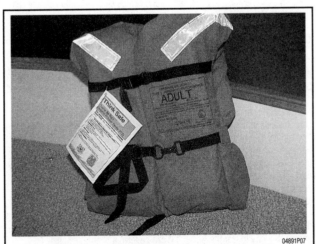
Fig. 7 Type II approved flotation devices are recommended for inshore and inland cruising on calm water. Use them where there is a good chance of fast rescue

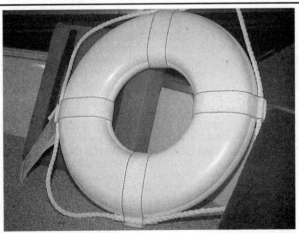
Fig. 9 Type IV throwables, such as this ring life buoy, are not designed as personal flotation devices for unconscious people, non-swimmers, or children

1-8 GENERAL INFORMATION AND BOATING SAFETY

seat cushions. Cushions are hard to hold onto in the water. Thus, they do not afford as much protection as wearable life jackets.

The straps on buoyant cushions are for you to hold onto either in the water or when throwing them. A cushion should never be worn on your back. It will turn you face down in the water.

Type IV throwables are not designed as personal flotation devices for unconscious people, non-swimmers, or children. Use them only in emergencies. They should not be used for, long periods in rough water.

Ring life buoys come in 18, 20, 24, and 30 inch diameter sizes. They have grab lines. You should attach about 60 feet of polypropylene line to the grab rope to aid in retrieving someone in the water. If you throw a ring, be careful not to hit the person. Ring buoys can knock people unconscious

Type V PFDs

Type V PFDs are of two kinds, special use devices and hybrids. Special use devices include boardsailing vests, deck suits, work vests, and others. They are approved only for the special uses or conditions indicated on their labels. Each is designed and intended for the particular application shown on its label. They do not meet legal requirements for general use aboard recreational boats.

Hybrid life jackets are inflatable devices with some built-in buoyancy provided by plastic foam or kapok. They can be inflated orally or by cylinders of compressed gas to give additional buoyancy. In some hybrids the gas is released manually. In others it is released automatically when the life jacket is immersed in water.

The inherent buoyancy of a hybrid may be insufficient to float a person unless it is inflated. The only way to find this out is for the user to try it in the water. Because of its limited buoyancy when deflated, a hybrid is recommended for use by anon-swimmer only if it is worn with enough inflation to float the wearer.

If they are to count against the legal requirement for the number of life jackets you must carry on your vessel, hybrids manufactured before February 8, 1995 must be worn whenever a boat is underway and the wearer is not below decks or in an enclosed space. To find out if your Type V hybrid must be worn to satisfy the legal requirement, read its label. If its use is restricted it will say, "REQUIRED TO BE WORN" in capital letters.

Hybrids cost more than other life jackets, but this factor must be weighed against the fact that they are more comfortable than Type I, II, or III life jackets. Because of their greater comfort, their owners are more likely to wear them than are the owners of Type I, II, or III life jackets.

The Coast Guard has determined that improved, less costly hybrids can save lives since they will be bought and used more frequently. For these reasons a new federal regulation was adopted effective February 8, 1995. The regulation increases both the deflated and inflated buoyancys of hybrids, makes them available in a greater variety of sizes and types, and reduces their costs by reducing production costs.

Even though it may not be required, the wearing of a hybrid or a life jacket is encouraged whenever a vessel is underway. Like life jackets, hybrids are now available in three types. To meet legal requirements, a Type I hybrid can be substituted for a Type I life jacket. Similarly Type II and III hybrids can be substituted for Type II and Type III life jackets. A Type I hybrid, when inflated, will turn most unconscious people from facedown to vertical or slightly backward positions just like a Type I life jacket. Type I and III hybrids function like Type II and III life jackets. If you purchase a new hybrid, it should have an owner's manual attached which describes its life jacket type and its deflated and inflated buoyancys. It warns you that it may have to be inflated to float you. The manual also tells you how to don the life jacket and how to inflate it. It also tells you how to change its inflation mechanism, recommended testing exercises, and inspection and maintenance procedures. The manual also tells you why you need a life jacket and why you should wear it. A new hybrid must be packaged with at least three gas cartridges. One of these may already be loaded into the inflation mechanism. Likewise, if it has an automatic inflation mechanism, it must be packaged with at least three of these water sensitive elements. One of these elements may be installed.

Legal Requirements

A Coast Guard approved life jacket must show the manufacturer's name and approval number. Most are marked as Type I, II, III, IV, or V. All of the newer hybrids are marked for type.

You are required to carry at least one wearable life jacket or hybrid for each person on board your recreational vessel. If your vessel is 16 feet or more in length and is not a canoe or a kayak, you must also have at least one Type IV on board. These requirements apply to all recreational vessels that are propelled or controlled by machinery, sails, oars, paddles, poles, or another vessel. Sailboards are not required to carry life jackets.

You can substitute an older Type V hybrid for any required Type I, II, or III life jacket provided that its approval label shows it is approved for the activity the vessel is engaged in, approved as a substitute for a life jacket of the type required on the vessel, used as required on the labels, and used in accordance with any requirements in its owner's manual, if the approval label makes reference to such a manual.

A water skier being towed is considered to be on board the vessel when judging compliance with legal requirements.

You are required to keep your Type I, II, or III life jackets or equivalent hybrids readily accessible, which means you must be able to reach out and get them when needed. All life jackets must be in good, serviceable condition.

General Considerations

The proper use of a life jacket requires the wearer to know how it will perform. You can gain this knowledge only through experience. Each person on your boat should be assigned a life jacket. Next, it should be fitted to the person who will wear it. Only then can you be sure that it will be ready for use in an emergency.

Boats can sink fast. There may be no time to look around for a life jacket. Fitting one on you in the water is almost impossible. This advice is good even if the water is calm, and you intend to boat near shore. Most drownings occur in inland waters within a few feet of safety. Most victims had life jackets, but they weren't wearing them.

Keeping life jackets in the plastic covers they came wrapped in and in a cabin assures that they will stay clean and unfaded. But this is no way to keep them when you are on the water. When you need a life jacket it must be readily accessible and adjusted to fit you. You can't spend time hunting for it or learning how to fit it.

There is no substitute for the experience of entering the water while wearing a life jacket. Children, especially, need practice. If possible, give, your guests this experience. Tell them they should keep their arms to their sides when jumping in to keep the life jacket from riding up. Let them jump in and see how the life jacket responds. Is it adjusted so it does not ride up? Is it the proper size? Are all straps snug? Are children's life jackets the right sizes for them? Are they adjusted properly? If a child's life jacket fits correctly, you can lift the child by the jacket's shoulder straps and the child's chin and ears will not slip through. Non-swimmers, children, handicapped persons, elderly persons and even pets should always wear life jackets when they are aboard. Many states require that everyone aboard wear them in hazardous waters.

Inspect your lifesaving equipment from time to time. Leave any questionable or unsatisfactory equipment on shore. An emergency is no time for you to conduct an inspection.

Indelibly mark your life jackets with your vessel's name, number, and calling port. This can be important in a search and rescue effort. It could help concentrate effort where it will do the most good.

Care of Life Jackets

Given reasonable care, life jackets last many years. Thoroughly dry them before putting them away. Stow them in dry, well ventilated places. Avoid the bottoms of lockers and deck storage boxes where moisture may collect. Air and dry them frequently.

Life jackets should not be tossed about or used as fenders or cushions. Many contain kapok or fibrous glass material enclosed in plastic bags. The bags can rupture and are then unserviceable. Squeeze your life jacket gently. Does air leak out? If so, water can leak in and it will no longer be safe to use. Cut it Up so no one will use it, and throw it away. The covers of some life jackets are made of nylon or polyester. These materials are plastics. Like many plastics, they break down after extended exposure to the ultraviolet light in sunlight. This process may be more rapid when the materials are dyed with bright dyes such as "neon" shades.

Ripped and badly faded fabric are clues that the covering of your life jacket is deteriorating. A simple test is to pinch the fabric between your thumbs and forefingers. Now try to tear the fabric. If it can be torn, it should definitely be destroyed and discarded. Compare the colors in protected places to those exposed to the sun. If the colors have faded, the materials have been weakened. A fabric covered life jacket should ordinarily last several boating seasons with normal use. A life jacket used every day in direct sunlight should probably be replaced more often.

GENERAL INFORMATION AND BOATING SAFETY   1-9

SOUND PRODUCING DEVICES

All boats are required to carry some means of making an efficient sound signal. Devices for making the whistle or horn noises required by the Navigation Rules must be capable of a four second blast. The blast should be audible for at least one-half mile. Athletic whistles are not acceptable on boats 12 meters or longer. Use caution with athletic whistles. When wet, some of them come apart and loose their "pea." When this happens, they are useless.

If your vessel is 12 meters long and less than 20 meters, you must have a power whistle (or power horn) and a bell on board. The bell must be in operating condition and have a minimum diameter of at least 200 mm (7.9 inches) at its mouth.

VISUAL DISTRESS SIGNALS

▶ See Figure 10

Visual Distress Signals (VDS) attract attention to your vessel if you need help. They also help to guide searchers in search and rescue situations. Be sure you have the right types, and learn how to use them properly.

It is illegal to fire flares improperly. In addition, they cost the Coast Guard and its Auxiliary many wasted hours in fruitless searches. If you signal a distress with flares and then someone helps you, please let the Coast Guard or the appropriate Search And Rescue Agency (SAR) know so the distress report will be canceled.

Recreational boats less than 16 feet long must carry visual distress signals on coastal waters at night. Coastal waters are:
- The ocean (territorial sea)
- The Great Lakes
- Bays or sounds that empty into oceans
- Rivers over two miles across at their mouths upstream to where they narrow to two miles.

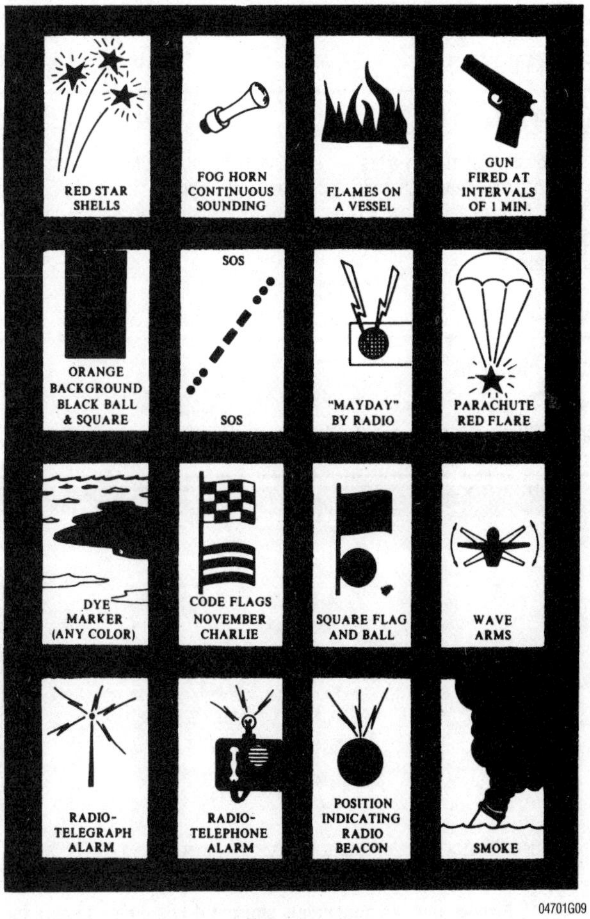

Fig. 10 Internationally accepted distress signals

Recreational boats 16 feet or longer must carry VDS at all times on coastal waters. The same requirement applies to boats carrying six or fewer passengers for hire. Open sailboats less than 26 feet long without engines are exempt in the daytime as are manually propelled boats. Also exempt are boats in organized races, regattas, parades, etc. Boats owned in the United States and operating on the high seas must be equipped with VDS.

A wide variety of signaling devices meet Coast Guard regulations. For pyrotechnic devices, a minimum of three must be carried. Any combination can be carried as long as it adds up to at least three signals for day use and at least three signals for night use. Three day/night signals meet both requirements. If possible, carry more than the legal requirement.

➡ **The American flag flying upside down is a commonly recognized distress signal. It is not recognized in the Coast Guard regulations, though. In an emergency, your efforts would probably be better used in more effective signaling methods.**

Types of VDS

VDS are divided into two groups; daytime and nighttime use. Each of these groups is subdivided into pyrotechnic and non-pyrotechnic devices.

DAYTIME NON-PYROTECHNIC SIGNALS

A bright orange flag with a black square over a black circle is the simplest VDS. It is usable, of course, only in daylight. It has the advantage of being a continuous signal. A mirror can be used to good advantage on sunny days. It can attract the attention of other boaters and of aircraft from great distances. Mirrors are available with holes in their centers to aid in "aiming." In the absence of a mirror, any shiny object can be used. When another boat is in sight, an effective VDS is to extend your arms from your sides and move them up and down. Do it slowly. If you do it too fast the other people may think you are just being friendly. This simple gesture is seldom misunderstood, and requires no equipment.

DAYTIME PYROTECHNIC DEVICES

Orange smoke is a useful daytime signal. Hand-held or floating smoke flares are very effective in attracting attention from aircraft. Smoke flares don't last long, and are not very effective in high wind or poor visibility. As with other pyrotechnic devices, use them only when you know there is a possibility that someone will see the display.

To be usable, smoke flares must be kept dry. Keep them in airtight containers and store them in dry places. If the "striker" is damp, dry it out before trying to ignite the device. Some pyrotechnic devices require a forceful "strike" to ignite them.

All hand-held pyrotechnic devices may produce hot ashes or slag when burning. Hold them over the side of your boat in such a way that they do not burn your hand or drip into your boat.

Nighttime Non-Pyrotechnic Signals

An electric distress light is available. This light automatically flashes the international morse code SOS distress signal (••• •••). Flashed four to six times a minute, it is an unmistakable distress signal. It must show that it is approved by the Coast Guard. Be sure the batteries are fresh. Dated batteries give assurance that they are current.

Under the Inland Navigation Rules, a high intensity white light flashing 50-70 times per minute is a distress signal. Therefore, use strobe lights on inland waters only for distress signals.

Nighttime Pyrotechnic Devices

▶ See Figure 11

Aerial and hand-held flares can be used at night or in the daytime. Obviously, they are more effective at night.

Currently, the serviceable life of a pyrotechnic device is rated at 42 months from its date of manufacture. Pyrotechnic devices are expensive. Look at their dates before you buy them. Buy them with as much time remaining as possible.

Like smoke flares, aerial and hand-held flares may fail to work if they have been damaged or abused. They will not function if they are or have been wet. Store them in dry, airtight containers in dry places. But store them where they are readily accessible.

Aerial VDSs, depending on their type and the conditions they are used in, may not go very high. Again, use them only when there is a good chance they will be seen.

1-10 GENERAL INFORMATION AND BOATING SAFETY

Fig. 11 Moisture protected flares should be carried onboard any vessel for use as a distress signal

A serious disadvantage of aerial flares is that they burn for only a short time. Most burn for less than 10 seconds. Most parachute flares burn for less than 45 seconds. If you use a VDS in an emergency, do so carefully. Hold hand-held flares over the side of the boat when in use. Never use a road hazard flare on a boat, it can easily start a fire. Marine type flares are carefully designed to lessen risk, but they still must be used carefully.

Aerial flares should be given the same respect as firearms since they are firearms! Never point them at another person. Don't allow children to play with them or around them. When you fire one, face away from the wind. Aim it downwind and upward at an angle of about 60 degrees to the horizon. If there is a strong wind, aim it somewhat more vertically. Never fire it straight up. Before you discharge a flare pistol, check for overhead obstructions. These might be damaged by the flare. They might deflect the flare to where it will cause damage.

Disposal of VDS

Keep outdated flares when you get new ones. They do not meet legal requirements, but you might need them sometime, and they may work. It is illegal to fire a VDS on federal navigable waters unless an emergency exists. Many states have similar laws.

Emergency Position Indicating Radio Beacon (EPIRB)

There is no requirement for recreational boats to have EPIRBs. Some commercial and fishing vessels, though, must have them if they operate beyond the three mile limit. Vessels carrying six or fewer passengers for hire must have EPIRBs under some circumstances when operating beyond the three mile limit. If you boat in a remote area or offshore, you should have an EPIRB. An EPIRB is a small (about 6 to 20 inches high), battery-powered, radio transmitting buoy-like device. It is a radio transmitter and requires a license or an endorsement on your radio station license by the Federal Communications Commission (FCC). EPIRBs are activated by being immersed in water or by a manual switch.

Equipment Not Required But Recommended

Although not required by law, there are other pieces of equipment that are good to have onboard.

SECOND MEANS OF PROPULSION

▶ See Figure 12

All boats less than 16 feet long should carry a second means of propulsion. A paddle or oar can come in handy at times. For most small boats, a spare trolling or outboard motor is an excellent idea. If you carry a spare motor, it should have its own fuel tank and starting power. If you use an electric trolling motor, it should have its own battery.

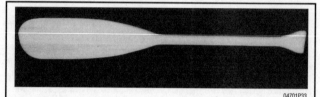

Fig. 12 A typical wooden oar should be kept onboard as an auxiliary means of propulsion. It can also function as a grab hook for someone fallen overboard

BAILING DEVICES

All boats should carry at least one effective manual bailing device in addition to any installed electric bilge pump. This can be a bucket, can, scoop, hand operated pump, etc. If your battery "goes dead" it will not operate your electric pump.

FIRST AID KIT

▶ See Figure 13

All boats should carry a first aid kit. It should contain adhesive bandages, gauze, adhesive tape, antiseptic, aspirin, etc. Check your first aid kit from time to time. Replace anything that is outdated. It is to your advantage to know how to use your first aid kit. Another good idea would be to take a Red Cross first aid course.

ANCHORS

▶ See Figure 14

All boats should have anchors. Choose one of suitable size for your boat. Better still, have two anchors of different sizes. Use the smaller one in calm

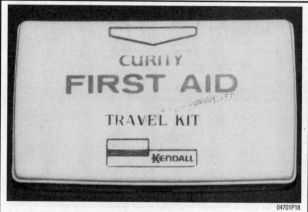

Fig. 13 Always carry an adequately stocked first aid kit on board for the safety of the crew and guests

GENERAL INFORMATION AND BOATING SAFETY

Fig. 14 Choose an anchor of sufficient weight to secure the boat without dragging. In some cases separate anchors may be needed for different situations

water or when anchoring for a short time to fish or eat. Use the larger one when the water is rougher or for overnight anchoring.

Carry enough anchor line of suitable size for your boat and the waters in which you will operate. If your engine fails you, the first thing you usually should do is lower your anchor. This is good advice in shallow water where you may be driven aground by the wind or water. It is also good advice in windy weather or rough water. The anchor will usually hold your bow into the waves.

VHF-FM RADIO

Your best means of summoning help in an emergency or in case of a breakdown is a VHF-FM radio. You can use it to get advice or assistance from the Coast Guard. In the event of a serious illness or injury aboard your boat, the Coast Guard can have emergency medical equipment meet you ashore.

TOOLS AND SPARE PARTS

▶ See Figures 15 and 16

Carry a few tools and some spare parts, and learn how to make minor repairs. Many search and rescue cases are caused by minor breakdowns that boat operators could have repaired. If your engine is an inboard or stern drive, carry spare belts and water pump impellers and the tools to change them.

Courtesy Marine Examinations

One of the roles of the Coast Guard Auxiliary is to promote recreational boating safety. This is why they conduct thousands of Courtesy Marine Examinations each year. The auxiliarists who do these examinations are well-trained and knowledgeable in the field.

These examinations are free and done only at the consent of boat owners. To pass the examination, a vessel must satisfy federal equipment requirements and certain additional requirements of the coast guard auxiliary. If your vessel does not pass the Courtesy Marine Examination, no report of the failure is made. Instead, you will be told what you need to correct the deficiencies. The examiner will return at your convenience to redo the examination.

If your vessel qualifies, you will be awarded a safety decal. The decal does not carry any special privileges, it simply attests to your interest in safe boating.

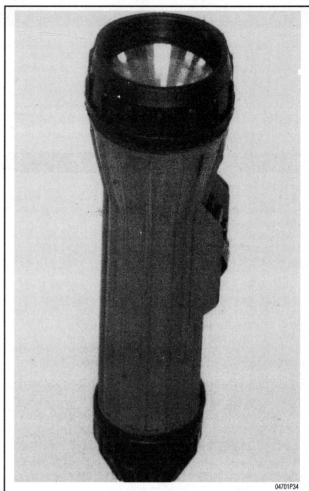

Fig. 15 A flashlight with a fresh set of batteries is handy when repairs are needed at night. It can also double as a signaling device

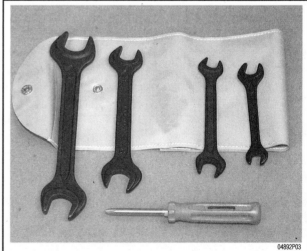

Fig. 16 A few wrenches, a screwdriver and maybe a pair of pliers can be very helpful to make emergency repairs

SAFETY IN SERVICE

It is virtually impossible to anticipate all of the hazards involved with maintenance and service, but care and common sense will prevent most accidents.

The rules of safety for mechanics range from "don't smoke around gasoline," to "use the proper tool(s) for the job." The trick to avoiding injuries is to develop safe work habits and to take every possible precaution. Whenever you are working on your boat, pay attention to what you are doing. The more you pay attention to details and what is going on around you, the less likely you will be to hurt yourself or damage your boat.

Do's

- Do keep a fire extinguisher and first aid kit handy.
- Do wear safety glasses or goggles when cutting, drilling, grinding or prying, even if you have 20–20 vision. If you wear glasses for the sake of vision, wear safety goggles over your regular glasses.
- Do shield your eyes whenever you work around the battery. Batteries contain sulfuric acid. In case of contact with the eyes or skin, flush the area with water or a mixture of water and baking soda, then seek immediate medical attention.
- Do use adequate ventilation when working with any chemicals or hazardous materials.
- Do disconnect the negative battery cable when working on the electrical system. The secondary ignition system contains EXTREMELY HIGH VOLTAGE. In some cases it can even exceed 50,000 volts.
- Do follow manufacturer's directions whenever working with potentially hazardous materials. Most chemicals and fluids are poisonous if taken internally.
- Do properly maintain your tools. Loose hammerheads, mushroomed punches and chisels, frayed or poorly grounded electrical cords, excessively worn screwdrivers, spread wrenches (open end), cracked sockets, or slipping ratchets can cause accidents.
- Likewise, keep your tools clean; a greasy wrench can slip off a bolt head, ruining the bolt and often harming your knuckles in the process.
- Do use the proper size and type of tool for the job at hand. Do select a wrench or socket that fits the nut or bolt. The wrench or socket should sit straight, not cocked.
- Do, when possible, pull on a wrench handle rather than push on it, and adjust your stance to prevent a fall.
- Do be sure that adjustable wrenches are tightly closed on the nut or bolt and pulled so that the force is on the side of the fixed jaw. Better yet, avoid the use of an adjustable if you have a fixed wrench that will fit.
- Do strike squarely with a hammer; avoid glancing blows. But, we REALLY hope you won't be using a hammer much in basic maintenance.
- Do use common sense whenever you work on your boat or motor. If a situation arises that doesn't seem right, sit back and have a second look. It may save an embarrassing moment or potential damage to your beloved boat.

Don'ts

- Don't run the engine in an enclosed area or anywhere else without proper ventilation—EVER! Carbon monoxide is poisonous; it takes a long time to leave the human body and you can build up a deadly supply of it in your system by simply breathing in a little every day. You may not realize you are slowly poisoning yourself.
- Don't work around moving parts while wearing loose clothing. Short sleeves are much safer than long, loose sleeves. Hard-toed shoes with neoprene soles protect your toes and give a better grip on slippery surfaces. Jewelry, watches, large belt buckles, or body adornment of any kind is not safe working around any vehicle. Long hair should be tied back under a hat.
- Don't use pockets for toolboxes. A fall or bump can drive a screwdriver deep into your body. Even a rag hanging from your back pocket can wrap around a spinning shaft.
- Don't smoke when working around gasoline, cleaning solvent or other flammable material.
- Don't smoke when working around the battery. When the battery is being charged, it gives off explosive hydrogen gas. Actually, you shouldn't smoke anyway. Save the cigarette money and put it into your boat!
- Don't use gasoline to wash your hands; there are excellent soaps available. Gasoline contains dangerous additives which can enter the body through a cut or through your pores. Gasoline also removes all the natural oils from the skin so that bone dry hands will suck up oil and grease.
- Don't use screwdrivers for anything other than driving screws! A screwdriver used as an prying tool can snap when you least expect it, causing injuries. At the very least, you'll ruin a good screwdriver.

TOOLS AND EQUIPMENT 2-2
SAFETY TOOLS 2-2
    WORK GLOVES 2-2
    EYE AND EAR PROTECTION 2-2
    WORK CLOTHES 2-3
CHEMICALS 2-3
    LUBRICANTS & PENETRANTS 2-3
    SEALANTS 2-4
    CLEANERS 2-4
TOOLS 2-5
HAND TOOLS 2-5
    SOCKET SETS 2-5
    WRENCHES 2-8
    PLIERS 2-9
    SCREWDRIVERS 2-9
    HAMMERS 2-9
    OTHER COMMON TOOLS 2-10
    SPECIAL TOOLS 2-10
    ELECTRONIC TOOLS 2-10
    GAUGES 2-11
MEASURING TOOLS 2-12
    MICROMETERS & CALIPERS 2-12
    DIAL INDICATORS 2-13
    TELESCOPING GAUGES 2-13
    DEPTH GAUGES 2-13
FASTENERS, MEASUREMENTS AND CONVERSIONS 2-14
BOLTS, NUTS AND OTHER THREADED RETAINERS 2-14
TORQUE 2-15
STANDARD AND METRIC MEASUREMENTS 2-15
SPECIFICATIONS CHARTS
    CONVERSION FACTORS 2-16

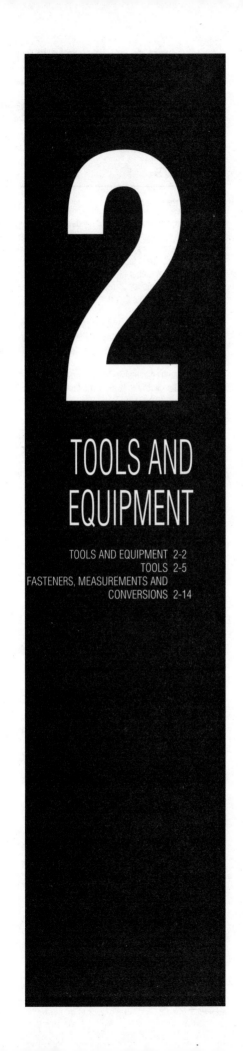

2
TOOLS AND EQUIPMENT

TOOLS AND EQUIPMENT 2-2
TOOLS 2-5
FASTENERS, MEASUREMENTS AND
    CONVERSIONS 2-14

2-2 TOOLS AND EQUIPMENT

TOOLS AND EQUIPMENT

Safety Tools

WORK GLOVES

▶ See Figures 1 and 2

Unless you think scars on your hands are cool, enjoy pain and like wearing bandages, get a good pair of work gloves. Canvas or leather are the best. And yes, we realize that there are some jobs involving small parts that can't be done while wearing work gloves. These jobs are not the ones usually associated with hand injuries.

A good pair of rubber gloves (such as those usually associated with dish washing) or vinyl gloves is also a great idea. There are some liquids such as solvents and penetrants that don't belong on your skin. Avoid burns and rashes. Wear these gloves.

And lastly, an option. If you're tired of being greasy and dirty all the time, go to the drug store and buy a box of disposable latex gloves like medical professionals wear. You can handle greasy parts, perform small tasks, wash parts, etc. all without getting dirty! These gloves take a surprising amount of abuse without tearing and aren't expensive. Note however, that it has been reported that some people are allergic to the latex or the powder used inside some gloves, so pay attention to what you buy.

EYE AND EAR PROTECTION

▶ See Figures 3 and 4

Don't begin any job without a good pair of work goggles or impact resistant glasses! When doing any kind of work, it's all too easy to avoid eye injury through this simple precaution. And don't just buy eye protection and leave it on the shelf. Wear it all the time! Things have a habit of breaking, chipping, splashing, spraying, splintering and flying around. And, for some reason, your eye is always in the way!

Fig. 2 Latex gloves come in handy when you are doing those messy jobs

If you wear vision correcting glasses as a matter of routine, get a pair made with polycarbonate lenses. These lenses are impact resistant and are available at any optometrist.

Often overlooked is hearing protection. Power equipment is noisy! Loud noises damage your ears. It's as simple as that! The simplest and cheapest form of ear protection is a pair of noise-reducing ear plugs. Cheap insurance for your ears. And, they may even come with their own, cute little carrying case.

More substantial, more protection and more money is a good pair of noise

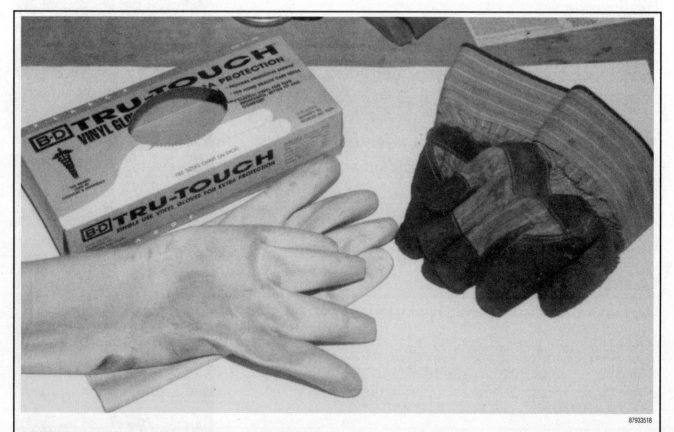

Fig. 1 Three different types of work gloves. The box contains latex gloves

TOOLS AND EQUIPMENT 2-3

Fig. 3 Don't begin any job without a good pair of work goggles or impact resistant glasses. Also good noise reducing earmuffs are cheap insurance to protect your hearing

Fig. 4 Things have a habit of breaking, chipping, splashing, spraying, splintering and flying around. And, for some reason, your eye is always in the way

reducing earmuffs. They protect from all but the loudest sounds. Hopefully those are sounds that you'll never encounter since they're usually associated with disasters.

WORK CLOTHES

Everyone has "work clothes." Usually these consist of old jeans and a shirt that has seen better days. That's fine. In addition, a denim work apron is a nice accessory. It's rugged, can hold some spare bolts, and you don't feel bad wiping your hands or tools on it. That's what it's for.

When working in cold weather, a one-piece, thermal work outfit is invaluable. Most are rated to below zero (Fahrenheit) temperatures and are ruggedly constructed. Just look at what the marine mechanics are wearing and that should give you a clue as to what type of clothing is good.

Chemicals

There is a whole range of chemicals that you'll find handy for maintenance work. The most common types are, lubricants, penetrants and sealers. Keep these handy onboard. There are also many chemicals that are used for detailing or cleaning.

When a particular chemical is not being used, keep it capped, upright and in a safe place. These substances may be flammable, may be irritants or might even be caustic and should always be stored properly, used properly and handled with care. Always read and follow all label directions and be sure to wear hand and eye protection!

LUBRICANTS & PENETRANTS

▶ See Figure 5

Anti-seize is used to coat certain fasteners prior to installation. This can be especially helpful when two dissimilar metals are in contact (to help prevent corrosion that might lock the fastener in place). This is a good practice on a lot of different fasteners, BUT, NOT on any fastener which might vibrate loose causing a problem. If anti-seize is used on a fastener, it should be checked periodically for proper tightness.

Lithium grease, chassis lube, silicone grease or a synthetic brake caliper grease can all be used pretty much interchangeably. All can be used for coating rust-prone fasteners and for facilitating the assembly of parts that are a tight fit. Silicone and synthetic greases are the most versatile.

➡Silicone dielectric grease is a non-conductor that is often used to coat the terminals of wiring connectors before fastening them. It may sound odd to coat metal portions of a terminal with something that won't conduct electricity, but here is it how it works. When the connector is fastened the metal-to-metal contact between the terminals will displace the grease (allowing the circuit to be completed). The grease that is displaced will then coat the non-contacted surface and the cavity around the terminals, SEALING them from atmospheric moisture that could cause corrosion.

Silicone spray is a good lubricant for hard-to-reach places and parts that shouldn't be gooped up with grease.

Penetrating oil may turn out to be one of your best friends when taking something apart that has corroded fasteners. Not only can they make a job easier, they can really help to avoid broken and stripped fasteners. The most familiar penetrating oils are Liquid Wrench® and WD-40®. A newer penetrant, PB Blaster® also works well. These products have hundreds of uses. For your purposes, they are vital!

Before disassembling any part (especially on an exhaust system), check the fasteners. If any appear rusted, soak them thoroughly with the penetrant and let them stand while you do something else (for particularly rusted or frozen parts you may need to soak them a few days in advance). This simple act can save you hours of tedious work trying to extract a broken bolt or stud.

Fig. 5 Antiseize, penetrating oil, lithium grease, electronic cleaner and silicone spray. These products have hundreds of uses and should be a part of your chemical tool collection

2-4 TOOLS AND EQUIPMENT

SEALANTS

♦ See Figures 6 and 7

Sealants are an indispensable part for certain tasks, especially if you are trying to avoid leaks. The purpose of sealants is to establish a leak-proof bond between or around assembled parts. Most sealers are used in conjunction with gaskets, but some are used instead of conventional gasket material.

The most common sealers are the non-hardening types such as Permatex®No.2 or its equivalents. These sealers are applied to the mating surfaces of each part to be joined, then a gasket is put in place and the parts are assembled.

➥A sometimes overlooked use for sealants like RTV is on the threads of vibration prone fasteners.

One very helpful type of non-hardening sealer is the "high tack" type. This type is a very sticky material that holds the gasket in place while the parts are being assembled. This stuff is really a good idea when you don't have enough hands or fingers to keep everything where it should be.

The stand-alone sealers are the Room Temperature Vulcanizing (RTV) silicone gasket makers. On some engines, this material is used instead of a gasket.

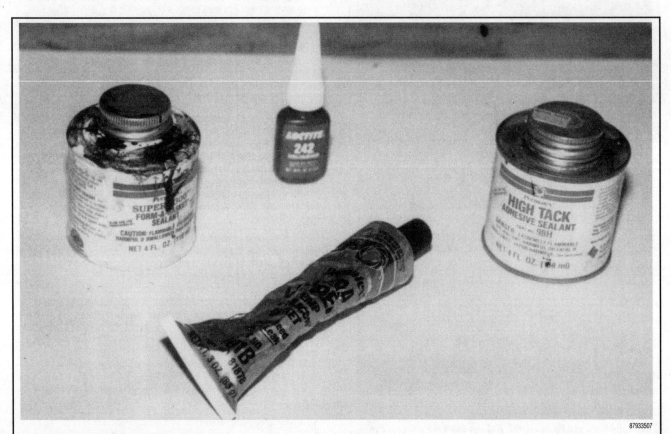

Fig. 6 Sealants are essential for preventing leaks

In those instances, a gasket may not be available or, because of the shape of the mating surfaces, a gasket shouldn't be used. This stuff, when used in conjunction with a conventional gasket, produces the surest bonds.

RTV does have its limitations though. When using this material, you will have a time limit. It starts to set-up within 15 minutes or so, so you have to assemble the parts without delay. In addition, when squeezing the material out of the tube, don't drop any glops into the engine. The stuff will form and set and travel around the oil gallery, possibly plugging up a passage. Also, most types are not fuel-proof. Check the tube for all cautions.

CLEANERS

♦ See Figures 8 and 9

There are two types of cleaners on the market today: parts cleaners and hand cleaners. The parts cleaners are for the parts; the hand cleaners are for you. They are not interchangeable.

There are many good, non-flammable, biodegradable parts cleaners on the market. These cleaning agents are safe for you, the parts and the environment. Therefore, there is no reason to use flammable, caustic or toxic substances to clean your parts or tools.

As far as hand cleaners go, the waterless types are the best. They have always been efficient at cleaning, but leave a pretty smelly odor. Recently

Fig. 7 On some engines, RTV is used instead of gasket material to seal components

TOOLS AND EQUIPMENT 2-5

Fig. 8 The new citrus hand cleaners not only work well, but they smell pretty good too. Choose one with pumice for added cleaning power

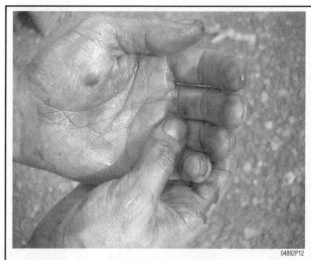

Fig. 9 The use of hand lotion seals your hands and keeps dirt and grease from sticking to your skin

though, just about all of them have eliminated the odor and added stuff that actually smells good. Make sure that you pick one that contains lanolin or some other moisture-replenishing additive. Cleaners not only remove grease and oil but also skin oil.

➡ Most women will tell you to use a hand lotion when you're all cleaned up. It's okay. Real men DO use hand lotion! Believe it or not, using hand lotion before your hands are dirty will actually make them easier to clean when you're finished with a dirty job. Lotion seals your hands, and keeps dirt and grease from sticking to your skin.

TOOLS

◆ See Figure 10

Tools; this subject could fill a completely separate manual. The first thing you will need to ask yourself, is just how involved do you plan to get. If you are serious about your maintenance you will want to gather a quality set of tools to make the job easier, and more enjoyable. BESIDES, TOOLS ARE FUN!!!

Almost every do-it-yourselfer loves to accumulate tools. Though most find a way to perform jobs with only a few common tools, they tend to buy more over time, as money allows. So gathering the tools necessary for maintenance does not have to be an expensive, overnight proposition.

When buying tools, the saying "You get what you pay for . . ." is absolutely true! Don't go cheap! Any hand tool that you buy should be drop forged and/or chrome vanadium. These two qualities tell you that the tool is strong enough for the job. With any tool, go with a name that you've heard of before, or, that is recommended buy your local professional retailer. Let's go over a list of tools that you'll need.

Most of the world uses the metric system. However, some American-built engines and aftermarket accessories use standard fasteners. So, accumulate your tools accordingly. Any good DIYer should have a decent set of both U.S. and metric measure tools.

➡ Don't be confused by terminology. Most advertising refers to "SAE and metric", or "standard and metric." Both are misnomers. The Society of Automotive Engineers (SAE) did not invent the English system of measurement; the English did. The SAE likes metrics just fine. Both English (U.S.) and metric measurements are SAE approved. Also, the current "standard" measurement IS metric. So, if it's not metric, it's U.S. measurement.

Hand Tools

SOCKET SETS

◆ See Figures 11 thru 17

Socket sets are the most basic hand tools necessary for repair and maintenance work. For our purposes, socket sets come in three drive sizes: ¼ inch, ⅜ inch and ½ inch. Drive size refers to the size of the drive lug on the ratchet, breaker bar or speed handle.

A ⅜ inch set is probably the most versatile set in any mechanic's tool box. It allows you to get into tight places that the larger drive ratchets can't and gives you a range of larger sockets that are still strong enough for heavy duty work. The socket set that you'll need should range in sizes from ⅜ inch through 1 inch for standard fasteners, and a 6mm through 19mm for metric fasteners.

You'll need a good ½ inch set since this size drive lug assures that you won't break a ratchet or socket on large or heavy fasteners. Also, torque wrenches with a torque scale high enough for larger fasteners are usually ½ inch drive.

¼ inch drive sets can be very handy in tight places. Though they usually duplicate functions of the ⅜ inch set, ¼ inch drive sets are easier to use for smaller bolts and nuts.

As for the sockets themselves, they come in standard and deep lengths as well as 6 or 12 point. 6 and 12 points refers to how many sides are in the

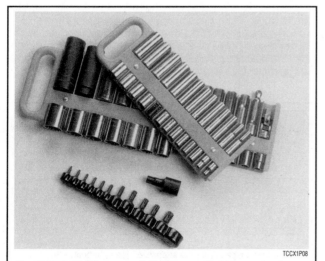

Fig. 10 Socket holders, especially the magnetic type, are handy items to keep tools in order

2-6 TOOLS AND EQUIPMENT

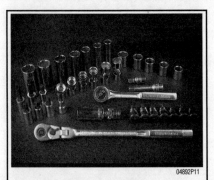
Fig. 11 A ⅜ inch socket set is probably the most versatile tool in any mechanic's tool box

Fig. 12 A swivel (U-joint) adapter (left), a ¼ inch-to-⅜ inch adapter (center) and a ⅜ inch-to-¼ inch adapter (right)

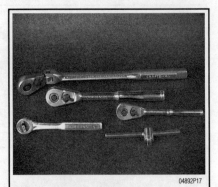
Fig. 13 Ratchets come in all sizes and configurations from rigid to swivel-headed

Fig. 14 Standard length sockets (top) are good for just about all jobs. However, some bolts may require deep sockets (bottom)

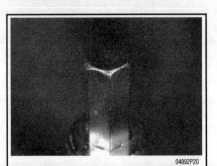
Fig. 15 Hex-head fasteners retain many components on modern powerheads. These fasteners require a socket with a hex shaped driver

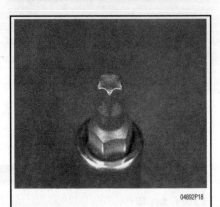
Fig. 16 Torx® drivers . . .

Fig. 17 . . . and tamper resistant drivers are required to remove special fasteners installed by the manufacturers

socket itself. Each has advantages. The 6 point socket is stronger and less prone to slipping which would strip a bolt head or nut. 12 point sockets are more common, usually less expensive and can operate better in tight places where the ratchet handle can't swing far.

Standard length sockets are good for just about all jobs, however, some stud-head bolts, hard-to-reach bolts, nuts on long studs, etc., require the deep sockets.

Most manufacturers use recessed hex-head fasteners to retain many of the engine parts. These fasteners require a socket with a hex shaped driver or a large sturdy hex key. To help prevent torn knuckles, we would recommend that you stick to the sockets on any tight fastener and leave the hex keys for lighter applications. Hex driver sockets are available individually or in sets just like conventional sockets.

More and more, manufacturers are using Torx® head fasteners, which were once known as tamper resistant fasteners (because many people did not have tools with the necessary odd driver shape). They are still used where the manufacturer would prefer only knowledgeable mechanics or advanced Do-It-Yourselfers (DIYers) to work.

Torque Wrenches

▶ See Figure 18

In most applications, a torque wrench can be used to assure proper installation of a fastener. Torque wrenches come in various designs and most stores will carry a variety to suit your needs. A torque wrench should be used any time you have a specific torque value for a fastener. Keep in mind that because there is no worldwide standardization of fasteners, the charts at the end of this section are a general guideline and should be used with caution. If you are using the right tool for the job, you should not have to strain to tighten a fastener.

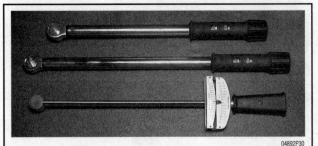
Fig. 18 Three types of torque wrenches. Top to bottom: a ⅜ inch drive beam type that reads in inch lbs., a ½ inch drive clicker type and a ½ inch drive beam type

TOOLS AND EQUIPMENT 2-7

BEAM TYPE

♦ See Figures 19 and 20

The beam type torque wrench is one of the most popular styles in use. If used properly, it can be the most accurate also. It consists of a pointer attached to the head that runs the length of the flexible beam (shaft) to a scale located near the handle. As the wrench is pulled, the beam bends and the pointer indicates the torque using the scale.

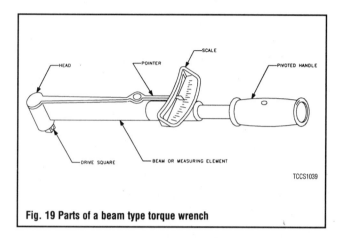

Fig. 19 Parts of a beam type torque wrench

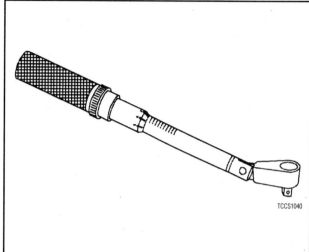

Fig. 21 A click type or breakaway torque wrench—note this one has a pivoting head

Fig. 20 A beam type torque wrench consists of a pointer attached to the head that runs the length of the flexible beam (shaft) to a scale located near the handle

Fig. 22 Setting the proper torque on a click type torque wrench involves turning the handle until the proper torque specification appears on the dial

CLICK (BREAKAWAY) TYPE

♦ See Figures 21 and 22

Another popular torque wrench design is the click type. The clicking mechanism makes achieving the proper torque easy and most use ratcheting head for ease of bolt installation. To use the click type wrench you pre-adjust it to a torque setting. Once the torque is reached, the wrench has a reflex signaling feature that causes a momentary breakaway of the torque wrench body, sending an impulse to the operator's hand.

Breaker Bars

♦ See Figure 23

Breaker bars are long handles with a drive lug. Their main purpose is to provide extra turning force when breaking loose tight bolts or nuts. They come in all drive sizes and lengths. Always take extra precautions and use proper technique when using a breaker bar.

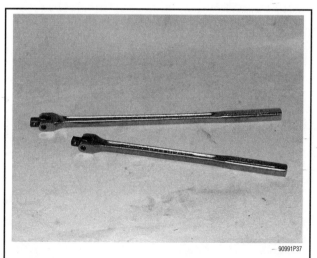

Fig. 23 Breaker bars are great for loosening large or stuck fasteners

2-8 TOOLS AND EQUIPMENT

WRENCHES

♦ See Figures 24, 25, 26, 27 and 28

Basically, there are 3 kinds of fixed wrenches: open end, box end, and combination.

Open end wrenches have 2-jawed openings at each end of the wrench. These wrenches are able to fit onto just about any nut or bolt. They are extremely versatile but have one major drawback. They can slip on a worn or rounded bolt head or nut, causing bleeding knuckles and a useless fastener.

Box-end wrenches have a 360° circular jaw at each end of the wrench. They come in both 6 and 12 point versions just like sockets and each type has the same advantages and disadvantages as sockets.

Combination wrenches have the best of both. They have a 2-jawed open end and a box end. These wrenches are probably the most versatile.

As for sizes, you'll probably need a range similar to that of the sockets, about ¼ inch through 1 inch for standard fasteners, or 6mm through 19mm for metric fasteners. As for numbers, you'll need 2 of each size, since, in many instances, one wrench holds the nut while the other turns the bolt. On most fasteners, the nut and bolt are the same size so having two wrenches of the same size comes in handy.

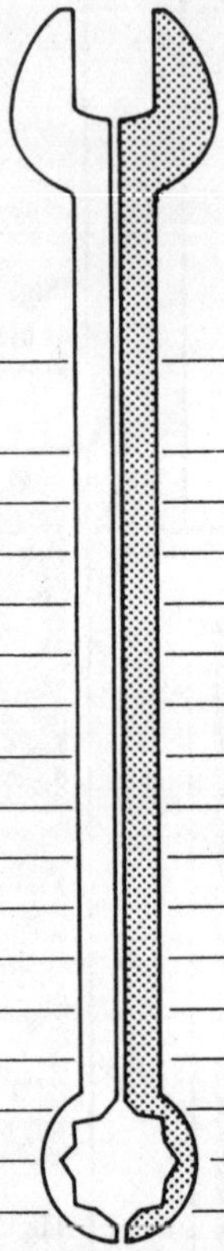

INCHES	DECIMAL	DECIMAL	MILLIMETERS
1/8"	.125	.118	3mm
3/16"	.187	.157	4mm
1/4"	.250	.236	6mm
5/16"	.312	.354	9mm
3/8"	.375	.394	10mm
7/16"	.437	.472	12mm
1/2"	.500	.512	13mm
9/16"	.562	.590	15mm
5/8"	.625	.630	16mm
11/16"	.687	.709	18mm
3/4"	.750	.748	19mm
13/16"	.812	.787	20mm
7/8"	.875	.866	22mm
15/16"	.937	.945	24mm
1"	1.00	.984	25mm

Fig. 24 Comparison of U.S. measure and metric wrench sizes

Fig. 25 Always use a backup wrench to prevent rounding flare nut fittings

Fig. 26 Note how the flare wrench sides are extended to grip the fitting tighter and prevent rounding

TOOLS AND EQUIPMENT   2-9

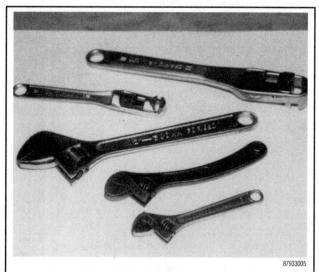

Fig. 27 Several types and sizes of adjustable wrenches

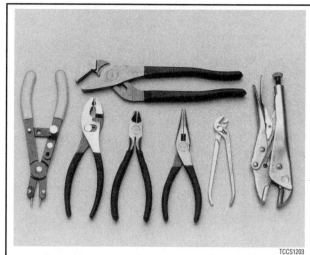

Fig. 29 Pliers and cutters come in many shapes and sizes. You should have an assortment on hand

In addition to standard pliers there are the slip-joint, multi-position pliers such as ChannelLock® pliers and locking pliers, such as Vise Grips®.

Slip joint pliers are extremely valuable in grasping oddly sized parts and fasteners. Just make sure that you don't use them instead of a wrench too often since they can easily round off a bolt head or nut.

Locking pliers are usually used for gripping bolts or studs that can't be removed conventionally. You can get locking pliers in square jawed, needle-nosed and pipe-jawed. Locking pliers can rank right up behind duct tape as the handy-man's best friend.

SCREWDRIVERS

You can't have too many screwdrivers. They come in 2 basic flavors, either standard or Phillips. Standard blades come in various sizes and thicknesses for all types of slotted fasteners. Phillips screwdrivers come in sizes with number designations from 1 on up, with the lower number designating the smaller size. Screwdrivers can be purchased separately or in sets.

HAMMERS

♦ See Figure 30

You always need a hammer for just about any kind of work. You need a ball-peen hammer for most metal work when using drivers and other like tools. A

Fig. 28 Occasionally you will find a nut which requires a particularly large or particularly small wrench. Rest assured that the proper wrench to fit is available at your local tool store

➡Although you will typically just need the sizes we specified, there are some exceptions. Occasionally you will find a nut which is larger. For these, you will need to buy ONE expensive wrench or a very large adjustable. Or you can always just convince the spouse that we are talking about safety here and buy a whole (read expensive) large wrench set.

One extremely valuable type of wrench is the adjustable wrench. An adjustable wrench has a fixed upper jaw and a moveable lower jaw. The lower jaw is moved by turning a threaded drum. The advantage of an adjustable wrench is its ability to be adjusted to just about any size fastener.

The main drawback of an adjustable wrench is the lower jaw's tendency to move slightly under heavy pressure. This can cause the wrench to slip if it is not facing the right way. Pulling on an adjustable wrench in the proper direction will cause the jaws to lock in place. Adjustable wrenches come in a large range of sizes, measured by the wrench length.

PLIERS

♦ See Figure 29

Pliers are simply mechanical fingers. They are, more than anything, an extension of your hand. At least 3 pair of pliers are an absolute necessity—standard, needle nose and channel lock.

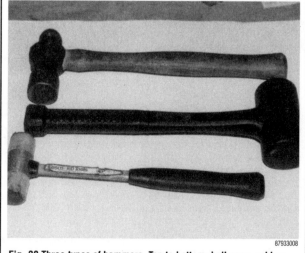

Fig. 30 Three types of hammers. Top to bottom: ball peen, rubber dead-blow, and plastic

TOOLS AND EQUIPMENT

plastic hammer comes in handy for hitting things safely. A soft-faced dead-blow hammer is used for hitting things safely and hard. Hammers are also VERY useful with non air-powered impact drivers.

OTHER COMMON TOOLS

There are a lot of other tools that every DIYer will eventually need (though not all for basic maintenance). They include:
- Funnels (for adding fluid)
- Chisels
- Punches
- Files
- Hacksaw
- Portable Bench Vise
- Tap and Die Set
- Flashlight
- Magnetic Bolt Retriever
- Gasket scraper
- Putty Knife
- Screw/Bolt Extractors
- Prybar

Hacksaws have just one use—cutting things off. You may wonder why you'd need one for something as simple as maintenance, but you never know. Among other things, guide studs to ease parts installation can be made from old bolts with their heads cut off.

A tap and die set might be something you've never needed, but you will eventually. It's a good rule, when everything is apart, to clean-up all threads, on bolts, screws and threaded holes. Also, you'll likely run across a situation in which stripped threads will be encountered. The tap and die set will handle that for you.

Gasket scrapers are just what you'd think, tools made for scraping old gasket material off of parts. You don't absolutely need one. Old gasket material can be removed with a putty knife or single edge razor blade. However, putty knives may not be sharp enough for some really stubborn gaskets and razor blades have a knack of breaking just when you don't want them to, inevitably slicing the nearest body part! As the old saying goes, "always use the proper tool for the job". If you're going to use a razor to scrape a gasket, be sure to always use a blade holder.

Putty knives really do have a use in a repair shop. Just because you remove all the bolts from a component sealed with a gasket doesn't mean it's going to come off. Most of the time, the gasket and sealer will hold it tightly. Lightly driving a putty knife at various points between the two parts will break the seal without damage to the parts.

A small — 8-10 inches (20–25 centimeters) long — prybar is extremely useful for removing stuck parts.

➡ **Never use a screwdriver as a prybar! Screwdrivers are not meant for prying. Screwdrivers, used for prying, can break, sending the broken shaft flying!**

Screw/bolt extractors are used for removing broken bolts or studs that have broke off flush with the surface of the part.

SPECIAL TOOLS

▶ See Figure 31

Almost every marine engine around today requires at least one special tool to perform a certain task. In most cases, these tools are specially designed to overcome some unique problem or to fit on some oddly sized component.

When manufacturers go through the trouble of making a special tool, it is usually necessary to use it to assure that the job will be done right. A special tool might be designed to make a job easier, or it might be used to keep you from damaging or breaking a part.

Don't worry, MOST basic maintenance procedures can either be performed without any special tools OR, because the tools must be used for such basic things, they are commonly available for a reasonable price. It is usually just the low production, highly specialized tools (like a super thin 7-point star-shaped

Fig. 31 Almost every marine engine requires at least one special tool to perform a certain task

socket capable of 150 ft. lbs. (203 Nm) of torque that is used only on the crankshaft nut of the limited production what-dya-callit engine) that tend to be outrageously expensive and hard to find. Luckily, you will probably never need such a tool.

Special tools can be as inexpensive and simple as an adjustable strap wrench or as complicated as an ignition tester. A few common specialty tools are listed here, but check with your dealer or with other boaters for help in determining if there are any special tools for YOUR particular engine. There is an added advantage in seeking advice from others, chances are they may have already found the special tool you will need, and know how to get it cheaper.

ELECTRONIC TOOLS

Battery Testers

The best way to test a non-sealed battery is using a hydrometer to check the specific gravity of the acid. Luckily, these are usually inexpensive and are available at most parts stores. Just be careful because the larger testers are usually designed for larger batteries and may require more acid than you will be able to draw from the battery cell. Smaller testers (usually a short, squeeze bulb type) will require less acid and should work on most batteries.

Electronic testers are available and are often necessary to tell if a sealed battery is usable. Luckily, many parts stores have them on hand and are willing to test your battery for you.

Battery Chargers

▶ See Figure 32

If you are a weekend boater and take your boat out every week, then you will most likely want to buy a battery charger to keep your battery fresh. There are many types available, from low amperage trickle chargers to electronically controlled battery maintenance tools which monitor the battery voltage to prevent over or undercharging. This last type is especially useful if you store your boat for any length of time (such as during the severe winter months found in many Northern climates).

Even if you use your boat on a regular basis, you will eventually need a battery charger. Remember that most batteries are shipped dry and in a partial charged state. Before a new battery can be put into service it must be filled and properly charged. Failure to properly charge a battery (which was shipped dry) before it is put into service will prevent it from ever reaching a fully charged state.

TOOLS AND EQUIPMENT 2-11

Fig. 32 The Battery Tender® is more than just a battery charger, when left connected, it keeps your battery fully charged

Digital Volt/Ohm Meter (DVOM)

♦ See Figure 33

Multimeters are an extremely useful tool for troubleshooting electrical problems. They can be purchased in either analog or digital form and have a price range to suit any budget. A multimeter is a voltmeter, ammeter and ohmmeter (along with other features) combined into one instrument. It is often used when testing solid state circuits because of its high input impedance (usually 10 megaohms or more). A brief description of the multimeter main test functions follows:

• Voltmeter—the voltmeter is used to measure voltage at any point in a circuit, or to measure the voltage drop across any part of a circuit. Voltmeters usually have various scales and a selector switch to allow the reading of different voltage ranges. The voltmeter has a positive and a negative lead. To avoid damage to the meter, always connect the negative lead to the negative (-) side of the circuit (to ground or nearest the ground side of the circuit) and connect the positive lead to the positive (+) side of the circuit (to the power source or the nearest power source). Note that the negative voltmeter lead will always be black and that the positive voltmeter will always be some color other than black (usually red).

• Ohmmeter—the ohmmeter is designed to read resistance (measured in ohms) in a circuit or component. Most ohmmeters will have a selector switch which permits the measurement of different ranges of resistance (usually the selector switch allows the multiplication of the meter reading by 10, 100, 1,000 and 10,000). Some ohmmeters are "auto-ranging" which means the meter itself will determine which scale to use. Since the meters are powered by an internal battery, the ohmmeter can be used like a self-powered test light. When the ohmmeter is connected, current from the ohmmeter flows through the circuit or component being tested. Since the ohmmeter's internal resistance and voltage are known values, the amount of current flow through the meter depends on the resistance of the circuit or component being tested. The ohmmeter can also be used to perform a continuity test for suspected open circuits. In using the meter for making continuity checks, do not be concerned with the actual resistance readings. Zero resistance, or any ohm reading, indicates continuity in the circuit. Infinite resistance indicates an opening in the circuit. A high resistance reading where there should be none indicates a problem in the circuit. Checks for short circuits are made in the same manner as checks for open circuits, except that the circuit must be isolated from both power and normal ground. Infinite resistance indicates no continuity, while zero resistance indicates a dead short.

✲✲ WARNING

Never use an ohmmeter to check the resistance of a component or wire while there is voltage applied to the circuit.

• Ammeter—an ammeter measures the amount of current flowing through a circuit in units called amperes or amps. At normal operating voltage, most circuits have a characteristic amount of amperes, called "current draw" which can be measured using an ammeter. By referring to a specified current draw rating, then measuring the amperes and comparing the two values, one can determine what is happening within the circuit to aid in diagnosis. An open circuit, for example, will not allow any current to flow, so the ammeter reading will be zero. A damaged component or circuit will have an increased current draw, so the reading will be high. The ammeter is always connected in series with the circuit being tested. All of the current that normally flows through the circuit must also flow through the ammeter; if there is any other path for the current to follow, the ammeter reading will not be accurate. The ammeter itself has very little resistance to current flow and, therefore, will not affect the circuit, but it will measure current draw only when the circuit is closed and electricity is flowing. Excessive current draw can blow fuses and drain the battery, while a reduced current draw can cause motors to run slowly, lights to dim and other components to not operate properly.

GAUGES

Compression Gauge

♦ See Figure 34

An important element in checking the overall condition of your engine is to check compression. This becomes increasingly more important on outboards with high hours. Compression gauges are available as screw-in types and hold-in types. The screw-in type is slower to use, but eliminates the possibility of a faulty reading due to escaping pressure. A compression reading will uncover many problems that can cause rough running. Normally, these are not the sort of problems that can be cured by a tune-up.

Vacuum Gauge

♦ See Figures 35 and 36

Vacuum gauges are handy for discovering air leaks, late ignition or valve timing, and a number of other problems.

Fig. 33 Multimeters are an extremely useful tool for troubleshooting electrical problems

2-12 TOOLS AND EQUIPMENT

Fig. 34 Cylinder compression test results are extremely valuable indicators of internal engine condition

Fig. 35 Vacuum gauges are useful for many diagnostic tasks including testing of some fuel pumps

Fig. 36 In a pinch, you can also use the vacuum gauge on a hand operated vacuum pump

Measuring Tools

Eventually, you are going to have to measure something. To do this, you will need at least a few precision tools in addition to the special tools mentioned earlier.

MICROMETERS & CALIPERS

Micrometers and calipers are devices used to make extremely precise measurements. The simple truth is that you really won't have the need for many of these items just for simple maintenance. You will probably want to have at least one precision tool such as an outside caliper to measure rotors or brake pads, but that should be sufficient to most basic maintenance procedures.

Should you decide on becoming more involved in boat engine mechanics, such as repair or rebuilding, then these tools will become very important. The success of any rebuild is dependent, to a great extent on the ability to check the size and fit of components as specified by the manufacturer. These measurements are made in thousandths and ten-thousandths of an inch.

Micrometers

▶ See Figure 37

A micrometer is an instrument made up of a precisely machined spindle which is rotated in a fixed nut, opening and closing the distance between the end of the spindle and a fixed anvil.

Outside micrometers can be used to check the thickness parts such shims or the outside diameter of components like the crankshaft journals. They are also used during many rebuild and repair procedures to measure the diameter of components such as the pistons. The most common type of micrometer reads in 1/1000 of an inch. Micrometers that use a vernier scale can estimate to 1/10 of an inch.

Inside micrometers are used to measure the distance between two parallel surfaces. For example, in powerhead rebuilding work, the inside mike measures cylinder bore wear and taper. Inside mikes are graduated the same way as outside mikes and are read the same way as well.

Remember that an inside mike must be absolutely perpendicular to the work being measured. When you measure with an inside mike, rock the mike gently from side to side and tip it back and forth slightly so that you span the widest part of the bore. Just to be on the safe side, take several readings. It takes a certain amount of experience to work any mike with confidence.

Metric micrometers are read in the same way as inch micrometers, except that the measurements are in millimeters. Each line on the main scale equals 1 mm. Each fifth line is stamped 5, 10, 15, and so on. Each line on the thimble scale equals 0.01 mm. It will take a little practice, but if you can read an inch mike, you can read a metric mike.

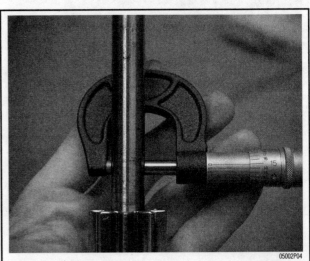

Fig. 37 Outside micrometers can be used to measure the thickness of shims or the outside diameter of a shaft

TOOLS AND EQUIPMENT 2-13

Calipers

♦ See Figures 38, 39 and 40

Inside and outside calipers are useful devices to have if you need to measure something quickly and precise measurement is not necessary. Simply take the reading and then hold the calipers on an accurate steel rule.

DIAL INDICATORS

♦ See Figure 41

A dial indicator is a gauge that utilizes a dial face and a needle to register measurements. There is a movable contact arm on the dial indicator. When the arms moves, the needle rotates on the dial. Dial indicators are calibrated to show readings in thousandths of an inch and typically, are used to measure end-play and runout on various parts.

Dial indicators are quite easy to use, although they are relatively expensive. A variety of mounting devices are available so that the indicator can be used in a number of situations. Make certain that the contact arm is always parallel to the movement of the work being measured.

Fig. 38 Calipers, such as this dial caliper, are the fast and easy way to make precise measurements

Fig. 39 Calipers can also be used to measure depth . . .

Fig. 40 . . . and inside diameter measurements, usually to 0.001 inch accuracy

Fig. 41 Here, a dial indicator is used to measure the axial clearance (end play) of a crankshaft during a powerhead rebuilding procedure

TELESCOPING GAUGES

♦ See Figure 42

A telescope gauge is used during rebuilding procedures (NOT usually basic maintenance) to measure the inside of bores. It can take the place of an inside mike for some of these jobs. Simply insert the gauge in the hole to be measured and lock the plungers after they have contacted the walls. Remove the tool and measure across the plungers with an outside micrometer.

DEPTH GAUGES

♦ See Figure 43

A depth gauge can be inserted into a bore or other small hole to determine exactly how deep it is. One common use for a depth gauge is measuring the distance the piston sits below the deck of the block at top dead center. Some outside calipers contain a built-in depth gauge so money can be saved by just buying one tool.

2-14 TOOLS AND EQUIPMENT

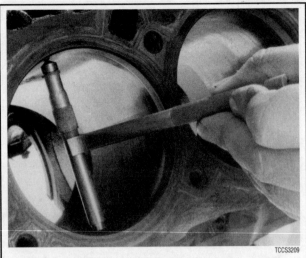

Fig. 42 Telescoping gauges are used during powerhead rebuilding procedures to measure the inside diameter of bores

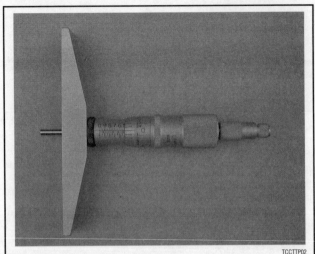

Fig. 43 Depth gauges are used to measure the depth of bore or other small holes

FASTENERS, MEASUREMENTS AND CONVERSIONS

Bolts, Nuts and Other Threaded Retainers

♦ See Figures 44, and 45

Although there are a great variety of fasteners found in the modern boat engine, the most commonly used retainer is the threaded fastener (nuts, bolts, screws, studs, etc). Most threaded retainers may be reused, provided that they are not damaged in use or during the repair.

➡Some retainers (such as stretch bolts or torque prevailing nuts) are designed to deform when tightened or in use and should not be reused.

Whenever possible, we will note any special retainers which should be replaced during a procedure. But you should always inspect the condition of a retainer when it is removed and you should replace any that show signs of damage. Check all threads for rust or corrosion which can increase the torque necessary to achieve the desired clamp load for which that fastener was originally selected. Additionally, be sure that the driver surface of the fastener has not

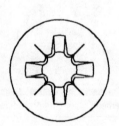

POZIDRIVE   PHILLIPS RECESS   TORX®   CLUTCH RECESS

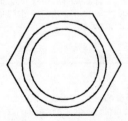
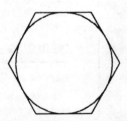
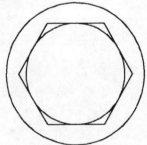

INDENTED HEXAGON   HEXAGON TRIMMED   HEXAGON WASHER HEAD

Fig. 44 Here are a few of the most common screw/bolt driver styles

TOOLS AND EQUIPMENT　2-15

Fig. 45 Thread gauges measure the threads-per-inch and the pitch of a bolt or stud's threads

been compromised by rounding or other damage. In some cases a driver surface may become only partially rounded, allowing the driver to catch in only one direction. In many of these occurrences, a fastener may be installed and tightened, but the driver would not be able to grip and loosen the fastener again. (This could lead to frustration down the line should that component ever need to be disassembled again).

If you must replace a fastener, whether due to design or damage, you must always be sure to use the proper replacement. In all cases, a retainer of the same design, material and strength should be used. Markings on the heads of most bolts will help determine the proper strength of the fastener. The same material, thread and pitch must be selected to assure proper installation and safe operation of the vehicle afterwards.

Thread gauges are available to help measure a bolt or stud's thread. Most part or hardware stores keep gauges available to help you select the proper size. In a pinch, you can use another nut or bolt for a thread gauge. If the bolt you are replacing is not too badly damaged, you can select a match by finding another bolt which will thread in its place. If you find a nut which threads properly onto the damaged bolt, then use that nut to help select the replacement bolt. If however, the bolt you are replacing is so badly damaged (broken or drilled out) that its threads cannot be used as a gauge, you might start by looking for another bolt (from the same assembly or a similar location) which will thread into the damaged bolt's mounting. If so, the other bolt can be used to select a nut; the nut can then be used to select the replacement bolt.

In all cases, be absolutely sure you have selected the proper replacement. Don't be shy, you can always ask the store clerk for help.

✱✱ WARNING

Be aware that when you find a bolt with damaged threads, you may also find the nut or drilled hole it was threaded into has also been damaged. If this is the case, you may have to drill and tap the hole, replace the nut or otherwise repair the threads. NEVER try to force a replacement bolt to fit into the damaged threads.

Torque

Torque is defined as the measurement of resistance to turning or rotating. It tends to twist a body about an axis of rotation. A common example of this would be tightening a threaded retainer such as a nut, bolt or screw. Measuring torque is one of the most common ways to help assure that a threaded retainer has been properly fastened.

When tightening a threaded fastener, torque is applied in three distinct areas, the head, the bearing surface and the clamp load. About 50 percent of the measured torque is used in overcoming bearing friction. This is the friction between the bearing surface of the bolt head, screw head or nut face and the base material or washer (the surface on which the fastener is rotating). Approximately 40 percent of the applied torque is used in overcoming thread friction. This leaves only about 10 percent of the applied torque to develop a useful clamp load (the force which holds a joint together). This means that friction can account for as much as 90 percent of the applied torque on a fastener.

Standard and Metric Measurements

Specifications are often used to help you determine the condition of various components, or to assist you in their installation. Some of the most common measurements include length (in. or cm/mm), torque (ft. lbs., inch lbs. or Nm) and pressure (psi, in. Hg, kPa or mm Hg).

In some cases, that value may not be conveniently measured with what is available in your toolbox. Luckily, many of the measuring devices which are available today will have two scales so Standard or Metric measurements may easily be taken. If any of the various measuring tools which are available to you do not contain the same scale as listed in your specifications, use the accompanying conversion factors to determine the proper value.

The conversion factor chart is used by taking the given specification and multiplying it by the necessary conversion factor. For instance, looking at the first line, if you have a measurement in inches such as "free-play should be 2 in." but your ruler reads only in millimeters, multiply 2 in. by the conversion factor of 25.4 to get the metric equivalent of 50.8mm. Likewise, if the specification was given only in a Metric measurement, for example in Newton Meters (Nm), then look at the center column first. If the measurement is 100 Nm, multiply it by the conversion factor of 0.738 to get 73.8 ft. lbs.

TOOLS AND EQUIPMENT

CONVERSION FACTORS

LENGTH–DISTANCE

Inches (in.)	x 25.4	= Millimeters (mm)	x .0394	= Inches
Feet (ft.)	x .305	= Meters (m)	x 3.281	= Feet
Miles	x 1.609	= Kilometers (km)	x .0621	= Miles

VOLUME

Cubic Inches (in3)	x 16.387	= Cubic Centimeters	x .061	= in3
IMP Pints (IMP pt.)	x .568	= Liters (L)	x 1.76	= IMP pt.
IMP Quarts (IMP qt.)	x 1.137	= Liters (L)	x .88	= IMP qt.
IMP Gallons (IMP gal.)	x 4.546	= Liters (L)	x .22	= IMP gal.
IMP Quarts (IMP qt.)	x 1.201	= US Quarts (US qt.)	x .833	= IMP qt.
IMP Gallons (IMP gal.)	x 1.201	= US Gallons (US gal.)	x .833	= IMP gal.
Fl. Ounces	x 29.573	= Milliliters	x .034	= Ounces
US Pints (US pt.)	x .473	= Liters (L)	x 2.113	= Pints
US Quarts (US qt.)	x .946	= Liters (L)	x 1.057	= Quarts
US Gallons (US gal.)	x 3.785	= Liters (L)	x .264	= Gallons

MASS–WEIGHT

Ounces (oz.)	x 28.35	= Grams (g)	x .035	= Ounces
Pounds (lb.)	x .454	= Kilograms (kg)	x 2.205	= Pounds

PRESSURE

Pounds Per Sq. In. (psi)	x 6.895	= Kilopascals (kPa)	x .145	= psi
Inches of Mercury (Hg)	x .4912	= psi	x 2.036	= Hg
Inches of Mercury (Hg)	x 3.377	= Kilopascals (kPa)	x .2961	= Hg
Inches of Water (H_2O)	x .07355	= Inches of Mercury	x 13.783	= H_2O
Inches of Water (H_2O)	x .03613	= psi	x 27.684	= H_2O
Inches of Water (H_2O)	x .248	= Kilopascals (kPa)	x 4.026	= H_2O

TORQUE

Pounds–Force Inches (in-lb)	x .113	= Newton Meters (N·m)	x 8.85	= in–lb
Pounds–Force Feet (ft-lb)	x 1.356	= Newton Meters (N·m)	x .738	= ft–lb

VELOCITY

Miles Per Hour (MPH)	x 1.609	= Kilometers Per Hour (KPH)	x .621	= MPH

POWER

Horsepower (Hp)	x .745	= Kilowatts	x 1.34	= Horsepower

FUEL CONSUMPTION*

Miles Per Gallon IMP (MPG)	x .354	= Kilometers Per Liter (Km/L)
Kilometers Per Liter (Km/L)	x 2.352	= IMP MPG
Miles Per Gallon US (MPG)	x .425	= Kilometers Per Liter (Km/L)
Kilometers Per Liter (Km/L)	x 2.352	= US MPG

*It is common to covert from miles per gallon (mpg) to liters/100 kilometers (1/100 km), where mpg (IMP) x 1/100 km = 282 and mpg (US) x 1/100 km = 235.

TEMPERATURE

Degree Fahrenheit (°F) = (°C x 1.8) + 32
Degree Celsius (°C) = (°F – 32) x .56

ENGINE MAINTENANCE 3-2
SERIAL NUMBER IDENTIFICATION 3-2
2-STROKE OIL 3-2
    OIL RECOMMENDATIONS 3-2
    FILLING 3-2
LOWER UNIT 3-3
    OIL RECOMMENDATIONS 3-3
    DRAINING & FILLING 3-3
FUEL FILTER 3-4
    RELIEVING FUEL SYSTEM
      PRESSURE 3-5
    REMOVAL & INSTALLATION 3-5
FUEL/WATER SEPARATOR 3-6
TRIM/TILT & PIVOT POINTS 3-6
    INSPECTION & LUBRICATION 3-6
PROPELLER 3-7
BOAT MAINTENANCE 3-8
INSIDE THE BOAT 3-8
FIBERGLASS HULLS 3-8
TRIM TABS, ANODES AND LEAD
  WIRES 3-8
BATTERY 3-9
    CLEANING 3-10
    CHECKING SPECIFIC GRAVITY 3-10
    BATTERY TERMINALS 3-11
    BATTERY & CHARGING SAFETY
      PRECAUTIONS 3-11
    BATTERY CHARGERS 3-11
    REPLACING BATTERY CABLES 3-12
TUNE-UP 3-12
INTRODUCTION 3-12
TUNE-UP SEQUENCE 3-12
COMPRESSION CHECK 3-12
    CHECKING COMPRESSION 3-12
    LOW COMPRESSION 3-13
SPARK PLUGS 3-13
    SPARK PLUG HEAT RANGE 3-13
    SPARK PLUG SERVICE 3-14
    REMOVAL & INSTALLATION 3-14
    READING SPARK PLUGS 3-14
    INSPECTION & GAPPING 3-15
SPARK PLUG WIRES 3-17
    TESTING 3-17
    REMOVAL & INSTALLATION 3-17
IGNITION SYSTEM 3-17
TIMING AND SYNCHRONIZATION 3-17
    TIMING 3-17
    SYNCHRONIZATION 3-17
    PREPARATION 3-17
DT2 AND DT2.2 3-18
    IGNITION TIMING 3-18
    IDLE SPEED 3-19
DT4 3-19
    IGNITION TIMING 3-19
    IDLE SPEED 3-19
DT6 AND DT8 3-19
    IGNITION TIMING 3-19
    IDLE SPEED 3-20
DT9.9 AND DT15 3-21
    IGNITION TIMING 3-21
    IDLE SPEED 3-21

DT20, DT25 AND DT30 3-22
    IGNITION TIMING 3-22
    THROTTLE LINKAGE
      ADJUSTMENT 3-22
    IDLE SPEED 3-22
DT35 AND DT40 3-23
    IGNITION TIMING 3-23
    IDLE SPEED 3-25
    THROTTLE LINKAGE 3-26
DT55 AND DT65 3-26
    IGNITION TIMING 3-26
    IDLE SPEED 3-26
    THROTTLE LINKAGE 3-27
DT75 AND DT85 3-29
    IGNITION TIMING 3-29
    CARBURETOR LINKAGE
      ADJUSTMENT 3-29
    IDLE SPEED 3-29
DT90 AND DT100 3-30
    IGNITION TIMING 3-30
    CARBURETOR LINKAGE
      ADJUSTMENT 3-30
    IDLE SPEED 3-30
DT115 AND DT140 3-30
    IGNITION TIMING 3-30
    CARBURETOR LINKAGE 3-31
    IDLE SPEED 3-32
DT150, DT175, DT200 3-32
    IGNITION TIMING 3-32
    CARBURETOR LINKAGE 3-32
    IDLE SPEED 3-32
WINTER STORAGE CHECKLIST 3-35
SPRING COMMISSIONING
  CHECKLIST 3-35
SPECIFICATIONS CHARTS
    CAPACITIES 3-2
    CARBURETOR IDLE AIR SCREW
      SPECIFICATION 3-33
    TUNE UP SPECIFICATIONS
      CHARTS 3-34
    GENERAL ENGINE
      SPECIFICATIONS 3-38
    SERIAL NUMBER
      IDENTIFICATION 3-43

3
MAINTENANCE

ENGINE MAINTENANCE 3-2
BOAT MAINTENANCE 3-8
TUNE-UP 3-12
WINTER STORAGE CHECKLIST 3-35
SPRING COMMISSIONING
CHECKLIST 3-35

3-2 MAINTENANCE

ENGINE MAINTENANCE

Serial Number Identification

Suzuki uses engine serial numbers and models numbers for identification purposes. These numbers are stamped on plates riveted to the port side transom bracket or to the starboard side of the support plate.

This information identifies the specific engine and will indicate to the owner or service technician if there are any unique parts or if any changes have been made to that particular model during its production run. The serial and models number should be used any time you order replacement parts.

For more information, refer to the "Serial Number Identification" and the "General Engine Specifications" charts at the end of this section.

2-Stroke Oil

OIL RECOMMENDATIONS

♦ See Figures 1 and 2

Use only Suzuki CCI oil or NMMA (National Marine Manufacturers Association) certified 2-stroke lubricants. These oils are proprietary lubricants designed to ensure optimal engine performance and to minimize combustion chamber deposits, avoid detonation and prolong spark plug life. If Suzuki CCI oil or a NMMA certified lubricant is unavailable, use only 2-stroke outboard oil.

➡**Remember, it is this oil, mixed with the gasoline that lubricates the internal parts of the engine. Lack of lubrication due to the wrong mix or improper type of oil can cause catastrophic powerhead failure.**

FILLING

There are two methods of adding 2-stroke oil to an outboard. The first is the pre-mix method used on outboards up to 6 horsepower and on some commercial models. The second is the automatic oil injection method which automatically injects the correct quantity of oil into the engine for all operating conditions.

Pre-Mix

Mixing the engine lubricant with gasoline before pouring it into the tank is by far the simplest method of lubrication for 2-stroke outboards. However, this method is the most messy and causes the most amount of harm to our environment.

The most important part of filling a pre-mix system is to determine the proper fuel/oil ratio. Most manufacturers use a 50:1 ratio (that is 50 parts of fuel to 1 part of oil) or a 100:1 ratio. Consult your owners manual to determine what the appropriate ratio should be for your engine.

The procedure itself is uncomplicated. Simply add the correct amount of lubricant to your fuel tank and then fill the tank with gasoline. The order in which you do this is important because as the gasoline is poured into the fuel tank it will mix with and agitate the oil for a complete blending.

If you are attempting to top off your tank, here is a general guideline to determine how much oil to add. For three gallons of fuel you would add 4 ounces of oil to obtain a 100:1 ratio; 8 ounces of oil to obtain a 50:1 ratio and 16 ounces of oil to obtain a 25:1 ratio.

Oil Injection

Most outboard manufacturers use a mechanically driven oil pump mounted on the engine block that is connected to the throttle by way of a linkage arm.

Fig. 1 2-Stroke outboard oils are proprietary lubricants designed to ensure optimal engine performance and to minimize combustion chamber deposits, avoid detonation and prolong spark plug life

Fig. 2 This scuffed piston is an example of what can happen when the proper 2-stroke oil is not used. The outboard required a complete overhaul

Capacities

Model	Injection Oil Quart (Liter)	Lower Unit Oz. (ml)	Fuel Tank Gal. (Liter)
DT2	PreMix	2.4(70)	0.3(1.2)
DT2.2	PreMix	2.4(70)	0.3(1.2)
DT4	PreMix	6.4(190)	0.7(2.6)
DT6	PreMix	11.5(240)	6.3(24)
DT8	PreMix	11.5(240)	6.3(24)
DT9.9	2.3(2.1)	5.7(170)	6.3(24)
DT15	2.3(2.1)	5.7(170)	6.3(24)
DT20	2.3(2.1)	10.1(300)	6.3(24)
DT25	2.1(2.0)	7.8(230)	6.3(24)
DT30	2.1(2.0)	7.8(230)	6.3(24)
DT35	2.1(2.0)	20.6(610)	6.3(24)
DT40	2.1(2.0)	20.6(610)	6.3(24)
DT55	3.2(3.0)	22(650)	6.3(24)
DT65	3.2(3.0)	22(650)	6.3(24)
DT75	2.4(2.3)	23.7(700)	6.3(24)
DT85	2.4(2.3)	23.7(700)	6.3(24)
DT90	4.8(4.5)	18.9(560)	6.3(24)
DT100	4.8(4.5)	18.9(560)	6.3(24)
DT115	6.3(6.0)	37.2(1100)	6.3(24)
DT140	6.3(6.0)	37.2(1100)	6.3(24)
DT150	9.5(9.0)	35.5(1050)	6.3(24)
DT175	9.5(9.0)	35.5(1050)	6.3(24)
DT200	9.5(9.0)	35.5(1050)	6.3(24)
DT225	9.0(8.5)	35.5(1050)	-

MAINTENANCE 3-3

The system is powered by the crankshaft which drives a gear in the pump, creating oil pressure. As the throttle lever is advanced to increase engine speed, the linkage arm also moves, opening a valve that allows more oil to flow into the oil pump.

Most mechanical-injection systems incorporate low-oil warning alarms that are also connected to an engine overheating sensor. Also, these systems may have a built-in speed limiter. This sub-system is designed to reduce engine speed automatically when oil problems occur. This important feature goes a long way toward preventing severe engine damage in the event of an oil injection problem.

The procedure for filling these systems is simple. On each powerhead there is an auxiliary oil reservoir which holds the 2-stroke oil. Simply fill the oil take to the proper capacity.

➡It is highly advisable to carry several spare bottles of 2-stroke oil with you onboard.

For more information on the oil injection system refer to the "Lubrication and Cooling" section of this manual.

Lower Unit

♦ See Figures 3 and 4

Regular maintenance and inspection of the lower unit is critical for proper operation and reliability. A lower unit can quickly fail if it becomes heavily contaminated with water, or excessively low on oil. The most common cause of a lower unit failure is water contamination.

Water in the lower unit is usually caused by fishing line, or other foreign material, becoming entangled around the propeller shaft and damaging the seal. If the line is not removed, it will eventually cut the propeller shaft seal and allow water to enter the lower unit. Fishing line has also been known to cut a groove in the propeller shaft if left neglected over time. This area should be checked frequently.

Fig. 3 This lower unit was destroyed because the bearing carrier was frozen due to lack of lubrication

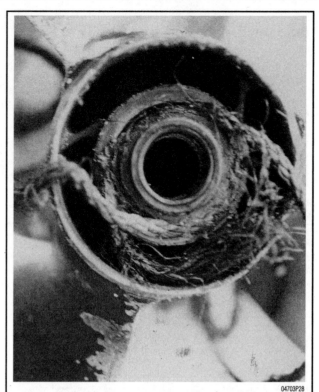

Fig. 4 Excellent view of rope and fishing line entangled behind the propeller. Entangled fishing line can actually cut through the seal, allowing water to enter the lower unit and lubricant to escape

OIL RECOMMENDATIONS

Use only Suzuki Outboard Motor Gear Oil or and equivalent high quality SAE 90 hypoid gear oil. These oils are proprietary lubricants designed to ensure optimal performance and to minimize corrosion in the lower unit.

➡Remember, it is this lower unit lubricant that prevents corrosion and lubricates the internal parts of the drive gears. Lack of lubrication due to water contamination or the improper type of oil can cause catastrophic lower unit failure.

DRAINING & FILLING

♦ See accompanying illustrations

> **CAUTION**
>
> The EPA warns that prolonged contact with used engine oil may cause a number of skin disorders, including cancer! You should make every effort to minimize your exposure to used engine oil. Protective gloves should be worn when changing the oil. Wash your hands and any other exposed skin areas as soon as possible after exposure to used engine oil. Soap and water, or waterless hand cleaner should be used.

1. Place a suitable container under the lower unit.
2. Loosen the oil level plug on the lower unit. This step is important! If the oil level plug cannot be loosened or removed, the complete lower unit lubricant service cannot be performed.

➡Never remove the vent or filler plugs when the lower unit is hot. Expanded lubricant will be released through the plug hole.

3. Remove the fill plug from the lower end of the gear housing followed by the oil level plug.
4. Allow the lubricant to completely drain from the lower unit.

3-4 MAINTENANCE

Step 2

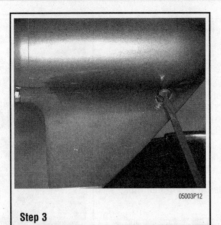
Step 3

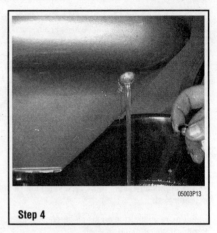
Step 4

➡ **If applicable, check the magnet end of the drain screw for metal particles. Some normal wear is to be expected, but if there are signs of metal chips or excessive metal particles, the gear case needs to be disassembled and inspected.**

5. Inspect the lubricant for the presence of a milky white substance, water or metallic particles. If any of these conditions are present, the lower unit should be serviced immediately.
6. Place the outboard in the proper position for filling the lower unit. The lower unit should not list to either port or starboard, and should be completely vertical.
7. On smaller outboards, insert the lubricant tube into the oil drain hole at the bottom of the lower unit, and squeeze lubricant until the excess begins to come out the oil level hole.
8. On larger outboards, oil should be injected, to fill the gear case through the drain plug.

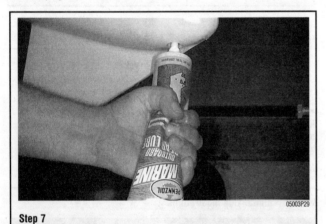
Step 7

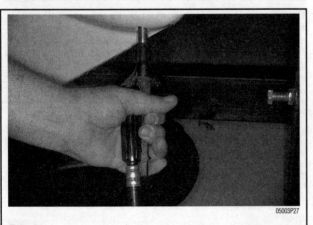
Step 8

9. Pump kits are available from marine manufacturers such as Rapiar®.
10. Using new gaskets, (washers) install the oil level and vent plugs (if applicable) first, then install the oil fill plug.
11. Place the used lubricant in a suitable container for transportation to an authorized recycling station.

Step 9

Fuel Filter

▶ **See Figures 5 and 6**

The fuel filter is designed to keep particles of dirt and debris from entering the carburetor(s) and clogging the internal passages. A small speck of dirt or sand can drastically affect the ability of the carburetor(s) to deliver the proper amount of air and fuel to enter the engine. If a filter becomes clogged, the flow of gasoline will be impeded. This could cause lean fuel mixtures, hesitation and stumbling, and idle problems.

Regular replacement of the fuel filter will decrease the risk of blocking the flow of fuel to the engine, which could leave you stranded on the water. Fuel filters are usually inexpensive, and replacement is a simple task. Change your fuel filter on a regular basis to avoid fuel delivery problems to the carburetor.

In addition to the fuel filter mounted on the engine, a filter is usually found inside or near the fuel tank (with the exception of DT2 and DT2.2). Because of the large variety of differences in both portable and fixed fuel tanks, it is impossible to give a detailed procedure for removal and installation. Most in-tank filters are simply a screen on the pick-up line inside the fuel tank. Filters of this type usually only need to be cleaned and returned to service. Fuel filters on the outside of the tank are typically of the inline type, and are replaced by simply removing the clamps, disconnecting the hoses, and installing a new filter. When installing the new filter, make sure the arrow on the filter points in the direction of fuel flow.

MAINTENANCE 3-5

Fig. 5 Typical fuel filter mounting location

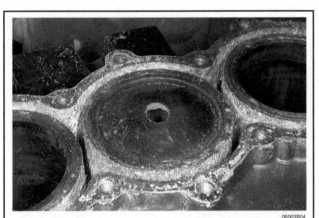

Fig. 6 A clogged fuel filter resulted in a lean fuel mixture at speed and caused the burn hole in the top of the piston. This powerhead required a complete overhaul

RELIEVING FUEL SYSTEM PRESSURE

On fuel injected engines, always relieve system pressure prior to disconnecting any fuel system component, fitting or fuel line.

✳✳ CAUTION

Exercise extreme caution whenever relieving fuel system pressure to avoid fuel spray and potential serious bodily injury. Please be advised that fuel under pressure may penetrate the skin or any part of the body it contacts.

To avoid the possibility of fire and personal injury, always disconnect the negative battery cable.

Always place a shop towel or cloth around the fitting or connection prior to loosening to absorb any excess fuel due to spillage. Ensure that all fuel spillage is removed from engine surfaces. Ensure that all fuel soaked clothes or towels in suitable waste container.

1. Remove the engine cover.
2. Place a wrench on both the service check bolt and fitting nut to prevent the fitting from twisting and breaking off.
3. Holding the service check bolt and fuel pressure check nut with both wrenches, place a shop towel or equivalent material over the service check bolt.
4. Loosen the service check bolt approximately one turn slowly to relieve the fuel pressure.

5. After relieving the fuel pressure, remove the service check bolt and replace the 6mm sealing washer with a new one. Tighten the service check bolt to 9 ft. lbs. (12 Nm).

REMOVAL & INSTALLATION

♦ See Figures 7 and 8

✳✳ CAUTION

Observe all applicable safety precautions when working around fuel. Whenever servicing the fuel system, always work in a well-ventilated area. Do not allow fuel spray or vapors to come in contact with a spark or open flame. Do not smoke while working around gasoline. Keep a dry chemical fire extinguisher near the work area. Always keep fuel in a container specifically designed for fuel storage; also, always properly seal fuel containers to avoid the possibility of fire or explosion.

1. Remove the engine cover.
2. Locate the fuel filter in the engine pan.
3. Lift the fuel filter from the engine pan, and place a pan or clean rag underneath it to absorb any spilled fuel.
4. Slide the hose retaining clips off the filter nipple with a pair of pliers and disconnect the hoses from the filter.
5. Reinstall the hoses on the filter nipples of the new filter. Make sure the embossed arrow on the filter points in the direction of fuel flow.
6. Slide the clips on each hose over the filter nipples.
7. Check the fuel filter installation for leakage by priming the fuel system with the fuel line primer bulb.
8. Once it is confirmed that there is no leakage from the connections, place the filter back to its proper position in the engine pan.
9. Replace the engine cover.

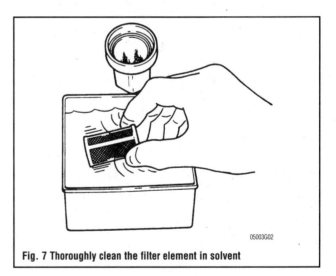

Fig. 7 Thoroughly clean the filter element in solvent

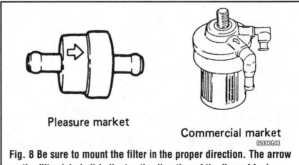

Fig. 8 Be sure to mount the filter in the proper direction. The arrow on the filter (circled) indicates the direction of the flow of fuel

3-6 MAINTENANCE

Fuel/Water Separator

♦ See Figures 9 and 10

In addition to the engine and inline fuel filters, there is usually another filter located in the fuel supply line. This is the fuel/water separator. It is used to remove water particles from the fuel prior to entering the engine or inline filter. Water can enter the fuel supply from a variety of sources and can lead to poor engine performance and ultimately, serious engine damage.

Because of the large variety of differences in both portable and fixed fuel tanks, it is impossible to give a single procedure to cover all applications.

Check with the boat manufacturer or the marina who rigged the boat to get the specifics of your particular fuel filtration system.

Trim/Tilt & Pivot Points

INSPECTION & LUBRICATION

♦ See Figures 11, 12 and 13

The steering head and other pivot points of the outboard-to-engine mounting components need periodic lubrication with marine grade grease to provide smooth operation and prevent corrosion. Usually, these pivot points are easily lubricated by simply attaching a grease gun to the fittings.

If the engine is used in salt water, the frequency of applying lubricant is usually doubled in comparison to operation in fresh water. Due to the very corrosive nature of salt water, an anti-seize thread compound should be used on all exposed fasteners outside of the cowling to reduce the chance of them seizing in place and breaking off when you try to remove them.

➡ Rinsing off the engine after each use is a very good habit to get into, not only does it help preserve the appearance of the engine, it virtually eliminates the corrosive effects of operating in salt water.

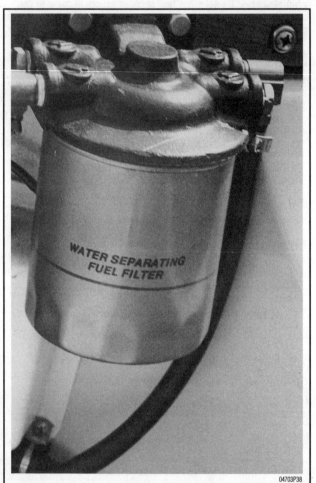

Fig. 9 A water separating fuel filter installed inside the boat on the transom

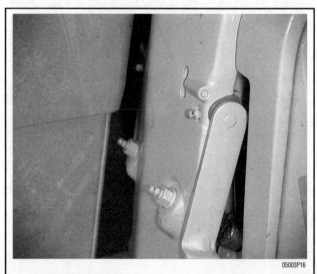

Fig. 11 The steering head . . .

Fig. 10 A typical water separating fuel filter assembly ready to be installed on the boat

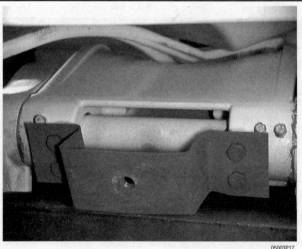

Fig. 12 . . . and steering tube both contain grease fittings which should be lubricated regularly

MAINTENANCE 3-7

Fig. 13 Due to the very corrosive nature of salt water, some sort of anti-seize type thread compound should be used on all exposed fasteners outside of the cowling to reduce the chance of them seizing in place

Propeller

♦ See Figures 14, 15 and 16

The propeller should be inspected regularly to be sure the blades are in good condition. If any of the blades become bent or nicked, this condition will set up vibrations in the motor. Remove and inspect the propeller. Use a file to trim nicks and burrs. Take care not to remove any more material than is absolutely necessary.

Also, check the rubber and splines inside the propeller hub for damage. If there is damage to either of these, take the propeller to your local marine dealer or a "prop shop". They can evaluate the damaged propeller and determine if it can be saved by rehubbing.

Additionally, the propeller should be removed each time the boat is hauled from the water at the end of an outing. Any material entangled behind the propeller should be removed before any damage to the shaft and seals can occur. This may seem like a waste of time, but the small amount of time involved in removing the propeller is returned many times by reduced maintenance and repair, including the replacement of expensive parts.

Fig. 15 A block of wood inserted between the propeller and the anti-cavitation plate will prevent the propeller from turning while the nut is being removed or installed

Fig. 14 An application of anti-seize on the propeller shaft splines will prevent the propeller from seizing on the shaft and facilitate easier removal for the next service

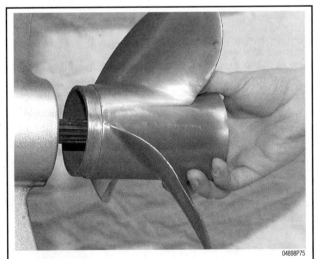

Fig. 16 Once the propeller nut and washer is removed, the propeller can be removed by sliding it off the shaft

3-8 MAINTENANCE

BOAT MAINTENANCE

Inside The Boat

♦ See Figure 17

The following points may be lubricated with an all purpose marine lubricant:
• Remote control cable ends next to the hand nut. DO NOT over-lubricate the cable
• Steering arm pivot socket
• Exposed shaft of the cable passing through the cable guide tube
• Steering link rod to steering cable

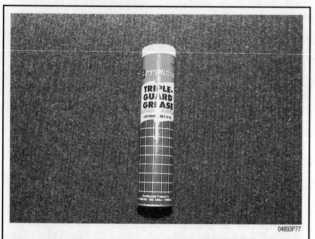

Fig. 17 Use only a good quality marine grade grease for lubrication

Fiberglass Hulls

♦ See Figures 18, 19 and 20

Fiberglass reinforced plastic hulls are tough, durable, and highly resistant to impact. However, like any other material they can be damaged. One of the advantages of this type of construction is the relative ease with which it may be repaired. Because of its break characteristics, and the simple techniques used in restoration, these hulls have gained popularity throughout the world. From the most congested urban marina, to isolated lakes in wilderness areas, to the severe cold of far off northern seas, and in sunny tropic remote rivers of primitive islands or continents, fiberglass boats can be found performing their daily task with a minimum of maintenance.

A fiberglass hull has almost no internal stresses. Therefore, when the hull is broken or stove-in, it retains its true form. It will not dent to take an out-of-shape set. When the hull sustains a severe blow, the impact will be either absorbed by deflection of the laminated panel or the blow will result in a definite, localized break. In addition to hull damage, bulkheads, stringers, and other stiffening structures attached to the hull may also be affected and therefore, should be checked. Repairs are usually confined to the general area of the rupture.

➥**The best way to care for a fiberglass hull is to wash it thoroughly, immediately after hauling the boat while the hull is still wet.**

A fouled bottom can seriously affect boat performance. This is one reason why racers, large and small, both powerboat and sail, are constantly giving attention to the condition of the hull below the waterline.

In areas where marine growth is prevalent, a coating of vinyl, anti-fouling bottom paint should be applied. If growth has developed on the bottom, it can be removed with a solution of Muriatic acid applied with a brush or swab and then rinsed with clear water. Always use rubber gloves when working with Muriatic acid and take extra care to keep it away from your face and hands. The fumes are toxic. Therefore, work in a well-ventilated area, or if outside, keep your face on the windward side of the work.

Barnacles have a nasty habit of making their home on the bottom of boats which have not been treated with anti-fouling paint. Actually they will not harm the fiberglass hull, but can develop into a major nuisance.

If barnacles or other crustaceans have attached themselves to the hull, extra work will be required to bring the bottom back to a satisfactory condition. First, if practical, put the boat into a body of fresh water and allow it to remain for a few days. A large percentage of the growth can be removed in this manner. If this remedy is not possible, wash the bottom thoroughly with a high-pressure fresh water source and use a scraper. Small particles of hard shell may still hold fast. These can be removed with sandpaper.

Trim Tabs, Anodes and Lead Wires

♦ See Figures 21 thru 28

Check the trim tabs and the anodes (zinc). Replace them, if necessary. The trim tab must make a good ground inside the lower unit. Therefore, the trim tab and the cavity must not be painted. In addition to trimming the boat, the trim tab acts as a zinc electrode to prevent electrolysis from acting on more expensive parts. It is normal for the tab to show signs of erosion. The tabs are inexpensive and should be replaced frequently.

Clean the exterior surface of the unit thoroughly. Inspect the finish for damage or corrosion. Clean any damaged or corroded areas, and then apply primer and matching paint.

Check the entire unit for loose, damaged, or missing parts.

An anode is attached across both clamp brackets. It also serves as protection for the coil of hydraulic hoses beneath the trim/tilt unit between the brackets.

Lead wires provide good electrical continuity between various brackets which might be isolated from the trim tab by a coating of lubricant between moving parts.

Fig. 18 In areas where marine growth is a problem, a coating of anti-foul bottom paint should be applied

Fig. 19 The best way to care for a fiberglass hull is to wash it thoroughly

Fig. 20 Fiberglass, vinyl and rubber care products, such as those available from Meguiar's are available to protect every part of your boat

MAINTENANCE 3-9

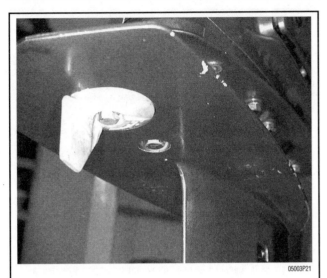

Fig. 21 What a trim tab should look like when it's in good condition

Fig. 22 Such extensive erosion of a trim tab compared with a new tab suggests an electrolysis problem or complete disregard for periodic maintenance

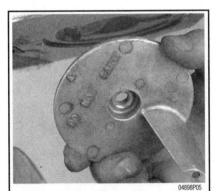

Fig. 23 Although many outboards use the trim tab as an anode . . .

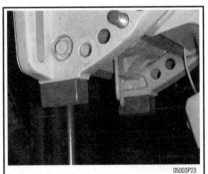

Fig. 24 . . . other types of anodes are also used throughout the outboard, like this one on the stern bracket . . .

Fig. 25 . . . and this one on the lower unit

Fig. 26 Anodes installed in the water jacket of a powerhead provide added protection against corrosion

Fig. 27 Most anodes are easily removed by loosening and removing their attaching fasteners

Fig. 28 One of the many lead wires used to connect bracketed parts. Lead wires are used as an assist in reducing corrosion

Battery

Difficulty in starting accounts for almost half of the service required on boats each year. A survey by Champion Spark Plug Company indicated that roughly one third of all boat owners experienced a "won't start" condition in a given year. When an engine won't start, most people blame the battery when, in fact, it may be that the battery has run down in a futile attempt to start an engine with other problems.

Maintaining your battery in peak condition may be though of as either tune-up or maintenance material. Most wise boaters will consider it to be both. A complete check up of the electrical system in your boat at the beginning of the boating season is a wise move. Continued regular maintenance of the battery will ensure trouble free starting on the water.

A complete battery service procedure is included in the "Maintenance" section of this manual. The following are a list of basic electrical system service procedures that should be performed as part of any tune-up.
- Check the battery for solid cable connections
- Check the battery and cables for signs of corrosion damage
- Check the battery case for damage or electrolyte leakage
- Check the electrolyte level in each cell

3-10 MAINTENANCE

- Check to be sure the battery is fastened securely in position
- Check the battery's state of charge and charge as necessary
- Check battery voltage while cranking the starter. Voltage should remain above 9.5 volts
- Clean the battery, terminals and cables
- Coat the battery terminals with dielectric grease or terminal protector

Batteries which are not maintained on a regular basis can fall victim to parasitic loads (small current drains which are constantly drawing current from the battery). Normal parasitic loads may drain a battery on boat that is in storage and not used frequently. Boats that have additional accessories with increased parasitic load may discharge a battery sooner. Storing a boat with the negative battery cable disconnected or battery switch turned off will minimize discharge due to parasitic loads.

CLEANING

Keep the battery clean, as a film of dirt can help discharge a battery that is not used for long periods. A solution of baking soda and water mixed into a paste may be used for cleaning, but be careful to flush this off with clear water.

➥Do not let any of the solution into the filler holes on non-sealed batteries. Baking soda neutralizes battery acid and will de-activate a battery cell.

CHECKING SPECIFIC GRAVITY

The electrolyte fluid (sulfuric acid solution) contained in the battery cells will tell you many things about the condition of the battery. Because the cell plates must be kept submerged below the fluid level in order to operate, maintaining the fluid level is extremely important. In addition, because the specific gravity of the acid is an indication of electrical charge, testing the fluid can be an aid in determining if the battery must be replaced. A battery in a boat with a properly operating charging system should require little maintenance, but careful, periodic inspection should reveal problems before they leave you stranded.

✳✳ CAUTION

Battery electrolyte contains sulfuric acid. If you should splash any on your skin or in your eyes, flush the affected area with plenty of clear water. If it lands in your eyes, get medical help immediately.

As stated earlier, the specific gravity of a battery's electrolyte level can be used as an indication of battery charge. At least once a year, check the specific gravity of the battery. It should be between 1.20 and 1.26 on the gravity scale. Most parts stores carry a variety of inexpensive battery testing hydrometers. These can be used on any non-sealed battery to test the specific gravity in each cell.

Conventional Battery

▶ See Figures 29 and 30

A hydrometer is required to check the specific gravity on all batteries that are not maintenance-free. The hydrometer has a squeeze bulb at one end and a nozzle at the other. Battery electrolyte is sucked into the hydrometer until the float or pointer is lifted from its seat. The specific gravity is then read by noting the position of the float/pointer. If gravity is low in one or more cells, the battery should be slowly charged and checked again to see if the gravity has come up. Generally, if after charging, the specific gravity of any two cells varies more than 50 points (0.50), the battery should be replaced, as it can no longer produce sufficient voltage to guarantee proper operation.

Check the battery electrolyte level at least once a month, or more often in hot weather or during periods of extended operation. Electrolyte level can be checked either through the case on translucent batteries or by removing the cell caps on opaque-case types. The electrolyte level in each cell should be kept filled to the split ring inside each cell, or the line marked on the outside of the case.

If the level is low, add only distilled water through the opening until the level is correct. Each cell is separate from the others, so each must be checked and filled individually. Distilled water should be used, because the chemicals and minerals found in most drinking water are harmful to the battery and could significantly shorten its life.

If water is added in freezing weather, the battery should be warmed to allow the water to mix with the electrolyte. Otherwise, the battery could freeze.

Fig. 29 On non-maintenance free batteries with translucent cases, the electrolyte level can be seen through the case; on other types (such as the one shown), the cell cap must be removed

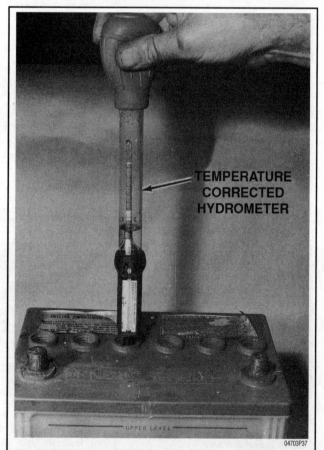

Fig. 30 The best way to determine the condition of a battery is to test the specific gravity of the electrolyte with a battery tester

Maintenance-Free Batteries

▶ See Figure 31

Although some maintenance-free batteries have removable cell caps for access to the electrolyte, the electrolyte condition and level is usually checked using the built-in hydrometer "eye". The exact type of eye varies between battery manufacturers, but most apply a sticker to the battery itself explaining the pos-

MAINTENANCE 3-11

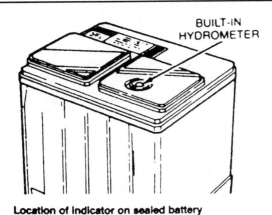
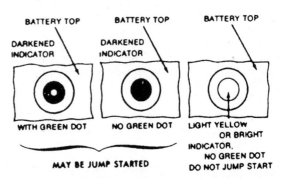

Fig. 31 A typical sealed (maintenance-free) battery with a built-in hydrometer—note that the hydrometer eye may vary between manufacturers; always refer to the battery's label

sible readings. When in doubt, refer to the battery manufacturer's instructions to interpret battery condition using the built-in hydrometer.

The readings from built-in hydrometers may vary, however a green eye usually indicates a properly charged battery with sufficient fluid level. A dark eye is normally an indicator of a battery with sufficient fluid, but one that may be low in charge. In addition, a light or yellow eye is usually an indication that electrolyte supply has dropped below the necessary level for battery (and hydrometer) operation. In this last case, sealed batteries with an insufficient electrolyte level must usually be discarded.

BATTERY TERMINALS

At least once a season, the battery terminals and cable clamps should be cleaned. Loosen the clamps and remove the cables, negative cable first. On batteries with top mounted posts, the use of a puller specially made for this purpose is recommended. These are inexpensive and available from most auto parts stores.

Clean the cable clamps and the battery terminal with a wire brush, until all corrosion, grease, etc., is removed and the metal is shiny. It is especially important to clean the inside of the clamp thoroughly (a wire brush is useful here), since a small deposit of foreign material or oxidation there will prevent a sound electrical connection and inhibit either starting or charging. It is also a good idea to apply some dielectric grease to the terminal, as this will aid in the prevention of corrosion.

After the clamps and terminals are clean, reinstall the cables, negative cable last; Do not hammer the clamps onto battery posts. Tighten the clamps securely, but do not distort them. Give the clamps and terminals a thin external coating of grease after installation, to retard corrosion.

Check the cables at the same time that the terminals are cleaned. If the insulation is cracked or broken, or if its end is frayed, that cable should be replaced with a new one of the same length and gauge.

BATTERY & CHARGING SAFETY PRECAUTIONS

Always follow these safety precautions when charging or handling a battery.
1. Wear eye protection when working around batteries. Batteries contain corrosive acid and produce explosive gas as a byproduct of their operation. Acid on the skin should be neutralized with a solution of baking soda and water made into a paste. In case acid contacts the eyes, flush with clear water and seek medical attention immediately.
2. Avoid flame or sparks that could ignite the hydrogen gas produced by the battery and cause an explosion. Connection and disconnection of cables to battery terminals is one of the most common causes of sparks.
3. Always turn a battery charger OFF, before connecting or disconnecting the leads. When connecting the leads, connect the positive lead first, then the negative lead, to avoid sparks.

4. When lifting a battery, use a battery carrier or lift at opposite corners of the base.
5. Ensure there is good ventilation in a room where the battery is being charged.
6. Do not attempt to charge or load-test a maintenance-free battery when the charge indicator dot is indicating insufficient electrolyte.
7. Disconnect the negative battery cable if the battery is to remain in the boat during the charging process.
8. Be sure the ignition switch is OFF before connecting or turning the charger ON. Sudden power surges can destroy electronic components.
9. Use proper adapters to connect charger leads to batteries with non-conventional terminals.

BATTERY CHARGERS

♦ See Figure 32

Before using any battery charger, consult the manufacturer's instructions for its use. Battery chargers are electrical devices that change Alternating Current (AC) to a lower voltage of Direct Current (DC) that can be used to charge a marine battery. There are two types of battery chargers—manual and automatic.

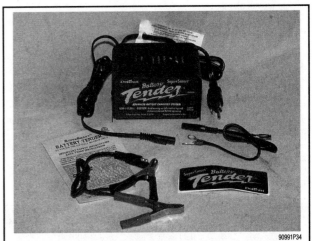

Fig. 32 Automatic battery chargers, like the Battery Tender® from Deltran, have an important advantage—they can stay connected to your battery for extended periods without the possibility of overcharging

3-12 MAINTENANCE

A manual battery charger must be physically disconnected when the battery has come to a full charge. If not, the battery can be overcharged, and possibly fail. Excess charging current at the end of the charging cycle will heat the electrolyte, resulting in loss of water and active material, substantially reducing battery life.

➥**As a rule, on manual chargers, when the ammeter on the charger registers half the rated amperage of the charger, the battery is fully charged. This can vary, and it is recommended to use a hydrometer to accurately measure state of charge.**

Automatic battery chargers have an important advantage—they can be left connected (for instance, overnight) without the possibility of overcharging the battery. Automatic chargers are equipped with a sensing device to allow the battery charge to taper off to near zero as the battery becomes fully charged. When charging a low or completely discharged battery, the meter will read close to full rated output. If only partially discharged, the initial reading may be less than full rated output, as the charger responds to the condition of the battery. As the battery continues to charge, the sensing device monitors the state of charge and reduces the charging rate. As the rate of charge tapers to zero amps, the charger will continue to supply a few milliamps of current—just enough to maintain a charged condition.

REPLACING BATTERY CABLES

Battery cables don't go bad very often, but like anything else, they can wear out. If the cables on your boat are cracked, frayed or broken, they should be replaced.

When working on any electrical component, it is always a good idea to disconnect the negative (-) battery cable. This will prevent potential damage to many sensitive electrical components

Always replace the battery cables with one of the same length, or you will increase resistance and possibly cause hard starting. Coat the battery posts with a light film of dielectric grease, or a battery terminal protectant spray once you've installed the new cables. If you replace the cables one at a time, you won't mix them up.

➥**Any time you disconnect the battery cables, it is recommended that you disconnect the negative (-) battery cable first. This will prevent you from accidentally grounding the positive (+) terminal when disconnecting it, thereby preventing damage to the electrical system.**

Before you disconnect the cable(s), first turn the ignition to the OFF position. This will prevent a draw on the battery which could cause arcing. When the battery cable(s) are reconnected (negative cable last), be sure to check all electrical accessories are all working correctly.

TUNE-UP

Introduction

A proper tune-up is the key to long and trouble-free engine life, and the work can yield its own rewards. Studies have shown that a properly tuned and maintained engine can achieve better fuel mileage than an out-of-tune engine. As a conscientious boater, set aside a Saturday morning, say once a month, to check or replace items which could cause major problems later. Keep your own personal log to jot down which services you performed, how much the parts cost you, the date, and the number of hours on the engine at the time. Keep all receipts for such items as engine oil and filters, so that they may be referred to in case of related problems or to determine operating expenses. As a do-it-yourselfer, these receipts are the only proof you have that the required maintenance was performed. In the event of a warranty problem, these receipts will be invaluable.

The efficiency, reliability, fuel economy and enjoyment available from engine performance are all directly dependent on having your outboard tuned properly. The importance of performing service work in the proper sequence cannot be over emphasized. Before making any adjustments, check the specifications. Never rely on memory when making critical adjustments.

Before beginning to tune any engine, ensure the engine has satisfactory compression. An engine with worn or broken piston rings, burned pistons, or scored cylinder walls, will not perform properly no matter how much time and expense is spent on the tune-up. Poor compression must be corrected or the tune-up will not give the desired results.

A practical maintenance program that is followed throughout the year, is one of the best methods of ensuring the engine will give satisfactory performance. As they say, you can spend a little time now or a lot of time later.

The extent of the engine tune-up is usually dependent on the time lapse since the last service. A complete tune-up of the entire engine would entail almost all of the work outlined in this manual. However, this is usually not necessary in most cases.

In this section, a logical sequence of tune-up steps will be presented in general terms. If additional information or detailed service work is required, refer to the section containing the appropriate instructions.

Each year higher compression ratios are built into modern outboard engines and the electrical systems become more complex. Therefore, the need for reliable, authoritative, and detailed instructions becomes more critical. The information in this section will fulfill that requirement.

Tune-Up Sequence

During a major tune-up, a definite sequence of service work should be followed to return the engine its maximum performance level. This type of work should not be confused with troubleshooting (attempting to locate a problem when the engine is not performing satisfactorily). In many cases, these two areas will overlap, because many times a minor or major tune-up will correct the malfunction and return the system to normal operation.

The following list is a suggested sequence of tasks to perform during a tune-up.
- Perform a compression check of each cylinder.
- Inspect the spark plugs to determine their condition. Test for adequate spark at the plug.
- Start the engine in a body of water and check the water flow through the engine.
- Check the gear oil in the lower unit.
- Check the carburetor adjustments and the need for an overhaul.
- Check the fuel pump for adequate performance and delivery.
- Make a general inspection of the ignition system.
- Test the starter motor and the solenoid, if so equipped.
- Check the internal wiring.
- Check the timing and synchronization.

Compression Check

Cylinder compression test results are extremely valuable indicators of internal engine condition. The best marine mechanics automatically check an engine's compression as the first step in a comprehensive tune-up. Obviously, it is useless to try to tune an engine with extremely low or erratic compression readings, since a simple tune-up will not cure the problem.

The pressure created in the combustion chamber may be measured with a gauge that remains at the highest reading it measures during the action of a one-way valve. This gauge is inserted into the spark plug hole. A compression test will uncover many mechanical problems that can cause rough running or poor performance.

If the powerhead shows any indication of overheating, such as discolored or scorched paint, inspect the cylinders visually through the transfer ports for possible scoring. It is possible for a cylinder with satisfactory compression to be scored slightly. Also, check the water pump. A faulty water pump may cause the overheating condition.

CHECKING COMPRESSION

♦ See Figures 33, 34 and 35

Prepare the engine for a compression test as follows:
1. Run the engine until it reaches operating temperature. If the test is performed on a cold engine, the readings will be considerably lower than normal, even if the engine is in perfect mechanical condition.
2. Label and disconnect the spark plug wires. Always grasp the molded cap and pull it loose with a twisting motion to prevent damage to the connection.
3. Clean all dirt and foreign material from around the spark plugs, and then remove all the plugs. Keep them in order by cylinder for later evaluation.
4. Ground the spark plug leads to the engine to render the ignition system inoperative while performing the compression check.

MAINTENANCE 3-13

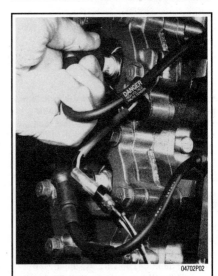

Fig. 33 Removing the high tension lead. Always use a twist and pull motion on the boot to prevent damage to the wire

Fig. 34 All spark plugs should be grounded while making compression tests. this action will prevent placing an extra load on the ignition coil

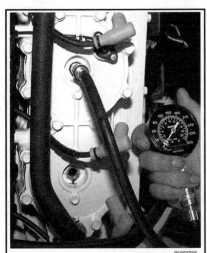

Fig. 35 Crank the engine with the starter through at least 4 complete strokes with the throttle at the wide-open position, to obtain the highest possible reading

5. Insert a compression gauge into the No. 1, top, spark plug opening.
6. Crank the engine with the starter through at least 4 complete strokes with the throttle at the wide-open position, to obtain the highest possible reading. Then record the reading.
7. Repeat the test and record the compression for each cylinder.
8. A variation between cylinders is far more important than the actual readings. A variation of more than 15 psi (103 kPa), between cylinders indicates the lower compression cylinder is defective. Not all engines will exhibit the same compression readings. In fact, two identical engines may not have the same compression. Generally, the rule of thumb is that the lowest cylinder should be within 25% of the highest (difference between the two readings).
9. If compression is low in one or more cylinders, the problem may be worn, broken, or sticking piston rings, scored pistons or worn cylinders.

LOW COMPRESSION

Compression readings that are generally low indicate worn, broken, or sticking piston rings, scored pistons or worn cylinders, and usually indicate an engine that has a lot of hours on it. Low compression in two adjacent cylinders (with normal compression in the other cylinders) indicates a blown head gasket between the low-reading cylinders. Other problems are possible (broken ring, hole burned in a piston), but a blown head gasket is most likely.

A conventional compression check will only show secondary compression readings and not primary crankcase compression. If there is an air leak in the crankcase, this will cause insufficient fuel to be brought into the crankcase and cylinder for normal operation. If it is a small leak, the powerhead will run poorly, because the fuel mixture will be too lean, and cylinder temperatures will be hotter than normal.

Air leaks are possible around any seal, O-ring, cylinder block mating surface, or gasket surface. Always replace O-rings, gaskets and seals when service work has been preformed. If the powerhead is running poorly, spray soapy water on the suspected sealing surface and look for bubbles to form, indicating an air leak. The base of the powerhead and the lower crankshaft seal are impossible to check in this manner, and will need to be checked by another method, a crankcase pressure test

To pressure test the crankcase, make up adapters to fit the carburetor mounting studs. Into one adapter fit an air fitting, which will accept a hand pump, which is used for testing the lower unit. With the powerhead on the bench, place some rubber gasket material over the exhaust, leaving the water passages open. Using the hand pump, pressurize the crankcase to 5 psi.

Spray soapy water around the lower crankcase seal area and other seals and gasket sealing surfaces looking for telltale bubbles. Also, if possible, pull a vacuum in the crankcase to check the seals in the opposite direction and watch for a pressure drop.

Spark Plugs

♦ See Figure 36

Spark plug life and efficiency depend upon the condition of the engine and the combustion chamber temperatures to which the plug is exposed. These temperatures are affected by many factors, such as compression ratio of the engine, air/fuel mixtures and the type of normally placed on your engine.

Factory installed plugs are, in a way, compromise plugs, since the factory has no way of knowing what typical loads your engine will see. However, most people never have reason to change their plugs

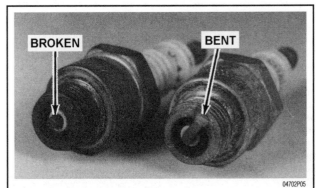

Fig. 36 Damaged spark plugs. Notice the broken electrode on the left plug. The electrode must be found and retrieved prior to returning the powerhead to service

SPARK PLUG HEAT RANGE

♦ See Figure 37

Spark plug heat range is the ability of the plug to dissipate heat. The longer the insulator (or the farther it extends into the engine), the hotter the plug will operate; the shorter the insulator,(the closer the electrode is to the block's cooling passages) the cooler it will operate. A plug that absorbs little heat and remains too cool will quickly accumulate deposits of oil and carbon since it is not hot enough to burn them off. This leads to plug fouling and consequently to misfiring. A plug that absorbs too much heat will have no deposits but, due to the excessive heat, the electrodes will burn away quickly and might possibly

3-14 MAINTENANCE

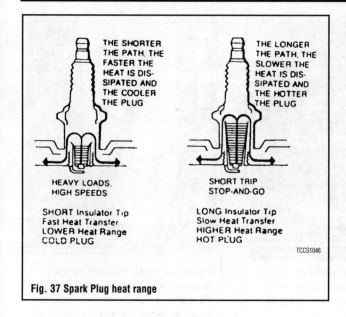

Fig. 37 Spark Plug heat range

lead to pre-ignition or other ignition problems. Pre-ignition takes place when plug tips get so hot that they glow sufficiently to ignite the air/fuel mixture before the actual spark occurs. This early ignition will usually cause a pinging during heavy loads.

SPARK PLUG SERVICE

➡New technologies in spark plug and ignition system design have pushed the recommended replacement interval to every 100 hours of operation (6 months). However, this depends on usage and conditions. This holds true unless internal engine wear or damage cause plug fouling. If you suspect this, you may wish to remove and inspect the plugs before the recommended time.

Spark plugs should only require replacement once a season. The electrode on a new spark plug has a sharp edge, but with use, this edge becomes rounded by wear, causing the plug gap to increase. As the gap increases, the plug's voltage requirement also increases. It requires a greater voltage to jump the wider gap and about two to three times as much voltage to fire a plug at high speeds than at idle.

Tools needed for spark plug replacement include: a ratchet, short extension, spark plug socket (there are two types; either 13/16 inch or 5/8 inch, depending upon the type of plug), a combination spark plug gauge and gapping tool, and a can of penetrating oil or anti-seize type grease for engines with aluminum heads.

When removing spark plugs, work on one at a time. Don't start by removing the plug wires all at once, because unless you number them, they may become mixed up. Take a minute before you begin and number the wires with tape.

REMOVAL & INSTALLATION

1. Disconnect the negative battery cable, and if the engine has been run recently, allow the engine to thoroughly cool. Attempting to remove plugs from a hot cylinder head could cause the plugs to seize and damage the threads in the cylinder head. Especially on aluminum heads!

2. Carefully twist the spark plug wire boot to loosen it, then pull the boot using a twisting motion and remove it from the plug. Be sure to pull on the boot and not on the wire, otherwise the connector located inside the boot may become separated.

➡A spark plug wire removal tool is recommended as it will make removal easier and help prevent damage to the boot and wire assembly.

3. Using compressed air (and safety glasses), blow debris from the spark plug well to assure that no harmful contaminants are allowed to enter the combustion chamber when the spark plug is removed. If compressed air is not available, use a rag or a brush to clean the area. Compressed air is available from both an air compressor or from compressed air in cans available at photography stores.

➡Remove the spark plugs when the engine is cold, if possible, to prevent damage to the threads. If plug removal is difficult, apply a few drops of penetrating oil to the area around the base of the plug, and allow it a few minutes to work.

4. Using a spark plug socket that is equipped with a rubber insert to properly hold the plug, turn the spark plug counterclockwise to loosen and remove the spark plug from the bore.

※※ WARNING

Avoid the use of a flexible extension on the socket. Use of a flexible extension may allow a shear force to be applied to the plug. A shear force could break the plug off in the cylinder head, leading to costly and frustrating repairs. In addition, be sure to support the ratchet with your other hand—this will also help prevent the socket from damaging the plug.

Evaluate each cylinder's performance by comparing the spark condition. Check each spark plug to be sure they are all of the same manufacturer and have the same heat range rating. Inspect the threads in the spark plug opening of the block, and clean the threads before installing the plug.

When purchasing new spark plugs, always ask the dealer if there has been a spark plug change for the engine being serviced.

Crank the engine through several revolutions to blow out any material which might have become dislodged during cleaning. Always use a new gasket (if applicable). The gasket must be fully compressed on clean seats to complete the heat transfer process and to provide a gas tight seal in the cylinder.

5. Inspect the spark plug boot for tears or damage. If a damaged boot is found, the spark plug boot and possible the entire wire will need replacement.

6. Apply a thin coating of anti-seize on the thread of the plug. This is extremely important on aluminum head engines.

7. Carefully thread the plug into the bore by hand. If resistance is felt before the plug completely bottomed, back the plug out and begin threading again.

※※ WARNING

Do not use the spark plug socket to thread the plugs. Always carefully thread the plug by hand or using an old plug wire to prevent the possibility of crossthreading and damaging the cylinder head bore.

8. Carefully tighten the spark plug. If the plug you are installing is equipped with a crush washer, seat the plug, then tighten to 10–15 ft. lbs. (14–20 Nm) or about ¼ turn to crush the washer. Whenever possible, spark plugs should be tightened to the factory torque specification.

9. Apply a small amount of silicone dielectric compound to the end of the spark plug lead or inside the spark plug boot to prevent sticking, then install the boot to the spark plug and push until it clicks into place. The click may be felt or heard. Gently pull back on the boot to assure proper contact.

READING SPARK PLUGS

▶ See Figures 38 thru 44

Your spark plugs are the single most valuable indicator of your engine's internal condition. Study your spark plugs carefully every time you remove them. Compare them to illustrations shown to identify the most common plug conditions.

MAINTENANCE 3-15

Fig. 38 A normally worn spark plug should have light tan or gray deposits on the firing tip (electrode)

Fig. 39 A carbon-fouled plug, identified by soft, sooty black deposits, may indicate an improperly tuned vehicle. Check the air cleaner, ignition components and the engine control system.

Fig. 40 A physically damaged spark plug may be evidence of severe detonation in that cylinder. Watch that cylinder carefully between services, as a continued detonation will not only damage the plug, but could also damage the engine

Fig. 41 An oil-fouled spark plug indicates an engine with worn piston rings and/or bad valve seals allowing excessive oil to enter the combustion chamber

Fig. 42 This spark plug has been left in the engine too long, as evidenced by the extreme gap—Plugs with such an extreme gap can cause misfiring and stumbling accompanied by a noticeable lack of power

Fig. 43 A bridged or almost bridged spark plug, identified by the build-up between the electrodes caused by excessive carbon or oil build-upon the plug

INSPECTION & GAPPING

Check spark plug gap before installation. The ground electrode (the L-shaped one connected to the body of the plug) must be parallel to the center electrode and the specified size wire gauge must pass between the electrodes with a slight drag.

Always check the gap on new plugs as they are not always set correctly at the factory. Do not use a flat feeler gauge when measuring the gap on a used plug, because the reading may be inaccurate. A round-wire type gapping tool is the best way to check the gap. The correct gauge should pass through the electrode gap with a slight drag. If you're in doubt, try a wire that is one size smaller and one larger. The smaller gauge should go through easily, while the larger one shouldn't go through at all.

Wire gapping tools usually have a bending tool attached. Use this tool to adjust the side electrode until the proper distance is obtained. Never attempt to

3-16 MAINTENANCE

Tracking Arc
High voltage arcs between a fouling deposit on the insulator tip and spark plug shell. This ignites the fuel/air mixture at some point along the insulator tip, retarding the ignition timing which causes a power and fuel loss.

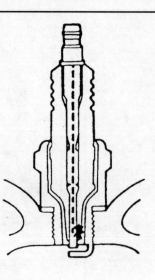

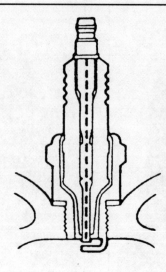

Wide Gap
Spark plug electrodes are worn so that the high voltage charge cannot arc across the electrodes. Improper gapping of electrodes on new or "cleaned" spark plugs could cause a similar condition. Fuel remains unburned and a power loss results.

Flashover
A damaged spark plug boot, along with dirt and moisture, could permit the high voltage charge to short over the insulator to the spark plug shell or the engine. A buttress insulator design helps prevent high voltage flashover.

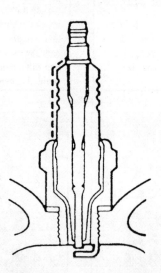

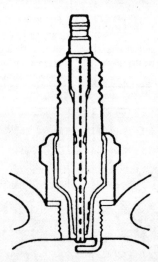

Fouled Spark Plug
Deposits that have formed on the insulator tip may become conductive and provide a "shunt" path to the shell. This prevents the high voltage from arcing between the electrodes. A power and fuel loss is the result.

Bridged Electrodes
Fouling deposits between the electrodes "ground out" the high voltage needed to fire the spark plug. The arc between the electrodes does not occur and the fuel air mixture is not ignited. This causes a power loss and exhausting of raw fuel.

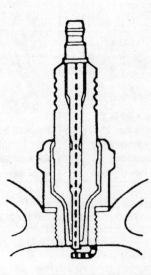

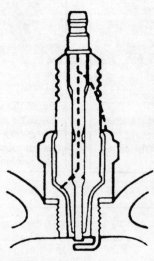

Cracked Insulator
A crack in the spark plug insulator could cause the high voltage charge to "ground out." Here, the spark does not jump the electrode gap and the fuel air mixture is not ignited. This causes a power loss and raw fuel is exhausted.

TCCS201A

Fig. 44 Typical spark plug problems showing damage which may indicate engine problems

MAINTENANCE 3-17

bend the center electrode. Also, be careful not to bend the side electrode too far or too often as it may weaken and break off within the engine, requiring removal of the cylinder head to retrieve it.

Spark Plug Wires

TESTING

At every tune-up/inspection, visually check the spark plug wires for burns, cuts, or breaks in the insulation. Check the boots on the coil and at the spark plug. Replace any wire that is damaged.

Once a year, usually when you change your spark plugs, check the resistance of the spark plug wires with an ohmmeter. Wires with excessive resistance will cause misfiring and may make the engine difficult to start. In addition worn wires will allow arcing and misfiring in humid conditions.

Remove the spark plug wire from the engine. Test the wires by connecting one lead of the ohmmeter to the coil end of the wire and the other lead to the spark plug end of the wire. Resistance should measure approximately 7000 ohms per foot of wire.

When installing a new set of spark plug wires, replace the wires one at a time so there will be no confusion. Coat the inside of the boots with dielectric grease to prevent sticking. Install the boot firmly over the spark plug until it clicks into place. The click may be felt or heard. Gently pull back on the boot to assure proper contact. Route the wire the same as the original and install it in a similar manner on the engine. Repeat the process for each wire.

REMOVAL & INSTALLATION

When installing a new set of spark plug wires, replace the wires one at a time so there will be no confusion. Coat the inside of the boots with dielectric grease to prevent sticking. Install the boot firmly over the spark plug until it clicks into place. The click may be felt or heard. Gently pull back on the boot to assure proper contact. Route the wire the same as the original and install it in a similar manner on the engine. Repeat the process for each wire.

Ignition System

The electronic CDI ignition system has become one of the most reliable components on the modern outboard engine. There is very little maintenance involved in the operation of the ignition and even less to repair if the component fails. Most systems are sealed and there is no option other than to replace the failed component.

It is very important to narrow down the ignition problem and replace the correct component rather than just replace parts hoping to solve the problem. Electronic components can be very expensive and are usually not returnable. Please refer to the "Ignition and Electrical" Section for more information on troubleshooting and repairing the CDI ignition system.

Timing And Synchronization

TIMING

Timing and synchronization on an outboard engine is extremely important to obtain maximum efficiency. The powerhead cannot perform properly and produce its designed horsepower output if the fuel and ignition systems have not been precisely adjusted.

All units covered in this manual except those equipped with the Integrated Circuit (IC) and Micro Link Ignition System, are equipped with a mechanical advance type Capacitor Discharge Ignition (CDI) system and use a series of link rods between the carburetor and the ignition base plate assembly. At the time the throttle is opened, the ignition base plate assembly is rotated by means of the link rod, thus advancing the timing.

On the IC and Micro Link equipped models, the microcomputer decides when to advance or retard the timing, based on input from various sensors. Therefore, there is no link rod between the magneto control lever and the stator assembly.

Many models have timing marks on the flywheel and CDI base. A timing light is normally used to check the ignition timing dynamically—with the powerhead operating. An alternate method is to check the static timing—with the powerhead not operating. This second method requires the use of a dial indicator gauge.

Various models have unique methods of checking ignition timing. These differences are explained in detail later in this section.

SYNCHRONIZATION

In simple terms, synchronization is timing the fuel system to the ignition. As the throttle is advanced to increase powerhead rpm, the carburetor and the ignition systems are both advanced equally and at the same rate.

Any time the fuel system or the ignition system on a powerhead is serviced to replace a faulty part or any adjustments are made for any reason, powerhead timing and synchronization must be carefully checked and verified. For this reason the timing and synchronizing procedures have been separated from all others and presented alone in this section.

Before making adjustments with the timing or synchronizing, the ignition system should be thoroughly checked and the fuel system verified to be in good working order.

On the breaker point ignitions, synchronization is automatic once the point gap and the piston travel or timing mark alignments are correct.

Models equipped with electronic ignitions are statically timed by aligning the timing marks on the throttle cam or throttle stopper with timing marks on the flywheel. Initial timing and timing advance are both set this way before using a timing light to check the timing.

Before making adjustments with the timing or synchronizing, the ignition system should be thoroughly checked and the fuel system verified to be in good working order.

PREPARATION

Timing and synchronizing the ignition and fuel systems on an outboard motor are critical adjustments. The following equipment is essential and is called out repeatedly in this section. This equipment must be used as described, unless otherwise instructed by the equipment manufacturer. Naturally, the equipment is removed following completion of the adjustments.

Suzuki also recommends the use of a test wheel instead of a normal propeller in order to put a load on the engine and propeller shaft. The use of the test wheel prevents the engine from excessive rpm.

The Synchronizing of the fuel systems on an outboard motor are critical adjustments. The following equipment is essential and is called out repeatedly in this section. This equipment must be used as described, unless otherwise instructed by the equipment manufacturer. Naturally, the equipment is removed following completion of the adjustments.

Dial Indicator

Top dead center (TDC) of the No. 1 (top) piston must be precisely known before the timing adjustment can be made. TDC can only be determined through installation of a dial indicator into the No. 1 spark plug opening.

Timing Light

During many procedures in this section, the timing mark on the flywheel must be aligned with a stationary timing mark on the engine while the powerhead is being cranked or is running. Only through use of a timing light connected to the No. 1 spark plug lead, can the timing mark on the flywheel be observed while the engine is operating.

Tachometer

A tachometer connected to the powerhead must be used to accurately determine engine speed during idle and high-speed adjustment. Engine speed readings range from 0 to 6,000 rpm in increments of 100 rpm. Choose a tachometers with solid state electronic circuits which eliminates the need for relays or batteries and contribute to their accuracy.

A tachometer is installed as standard equipment on most powerheads covered in this manual. Due to local conditions, it may be necessary to adjust the carburetor while the outboard unit is running in a test tank or with the boat in a body of water. For maximum performance, the idle rpm should be adjusted under actual operating conditions. Under such conditions it might be necessary to attach a tachometer closer to the powerhead than the one installed on the control panel.

3-18 MAINTENANCE

Flywheel Rotation

The instructions may call for rotating the flywheel until certain marks are aligned with the timing pointer. When the flywheel must be rotated, always move the flywheel in the indicated direction. If the flywheel should be rotated in the opposite direction, the water pump impeller vanes would be twisted.

Should the powerhead be started with the pump tangs bent back in the wrong direction, the tangs may not have time to bend in the correct direction before they are damaged. The least amount of damage to the water pump will affect cooling of the powerhead.

Test Tank

Since the engine must be operated at various times and engine speeds during some procedures, a test tank or moving the boat into a body of water, is necessary. If installing the engine in a test tank, outfit the engine with an appropriate test propeller.

✳✳ CAUTION

Water must circulate through the lower unit to the powerhead anytime the powerhead is operating to prevent damage to the water pump in the lower unit. Just five seconds without water will damage the water pump impeller.

➥Remember the powerhead will not start without the emergency tether in place behind the kill switch knob.

✳✳ CAUTION

Never operate the powerhead above a fast idle with a flush attachment connected to the lower unit. Operating the powerhead at a high rpm with no load on the propeller shaft could cause the powerhead to runaway causing extensive damage to the unit.

DT2 and DT2.2

IGNITION TIMING

♦ See Figures 45, 46, 47 and 48

1. Mount the engine in a test tank or move the boat to a body of water.
2. Remove the cowling and connect a tachometer to the powerhead.
3. Remove the flywheel.
4. Remove the spark plug.
5. Disconnect the magneto lead (usually white) from the connector and the stator lead (usually black) from the connector.

➥Before checking the ignition timing, make sure that the contact point faces are in good condition. Sand and make parallel the two faces by grinding with an oil stone if necessary and wipe the points clean with solvent. Apply a small dab of grease to the breaker shaft.

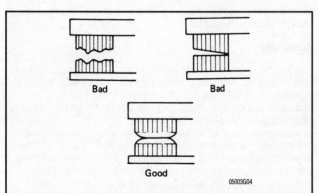
Fig. 45 Check the condition of the points before setting the point gap

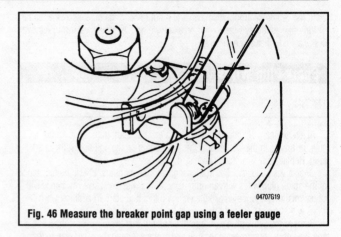
Fig. 46 Measure the breaker point gap using a feeler gauge

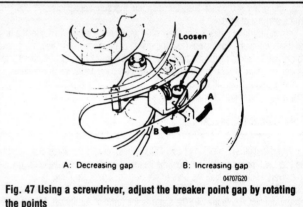
Fig. 47 Using a screwdriver, adjust the breaker point gap by rotating the points

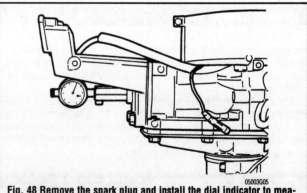
Fig. 48 Remove the spark plug and install the dial indicator to measure the piston travel

6. Check and adjust the breaker point gap to 0.012–0.016 in. (0.3–0.4 mm) by moving the breaker base plate.
7. Install a dial indicator with a special adapter (09931-00112) in the spark plug hole.
8. Rotate the flywheel clockwise until the piston has reached TDC then reset the indicator to zero.
9. Connect an ohmmeter between the magneto wire and a good engine ground. A timing tester (09900-27003) can also be used.
10. Gently turn the rotor clockwise (with the tester turned on) until the ohmmeter indicates continuity or the timing tester starts buzzing. Read the dial indicator, this reading is the piston travel and if the timing is set correctly, the indicator should read: 0.032 in. (0.804 mm).

If the reading is not within specification, retime the ignition system as follows:

11. Remove the flywheel magneto, loosen the screws securing the stator, and manually turn the stator clockwise to retard and counterclockwise to advance the timing by the amount necessary to meet specification.

MAINTENANCE 3-19

➡ **If the correct gap cannot be obtained by adjustment, the points should be replaced.**

IDLE SPEED

♦ See Figure 49

1. Mount the engine in a test tank or move the boat to a body of water.
2. Remove the cowling and connect a tachometer to the powerhead.
3. Start the engine and allow it to reach operating temperature.
4. Check idle speed and compare it with the specified idle speed in the "Tune-Up Specifications" chart.
5. If adjustment is necessary, rotate the idle adjustment screw on the carburetor until the powerhead idles at the required rpm.

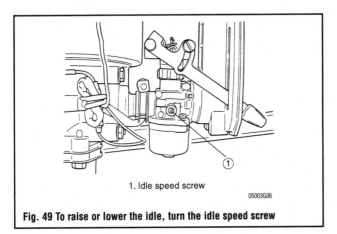

1. Idle speed screw

Fig. 49 To raise or lower the idle, turn the idle speed screw

DT4

IGNITION TIMING

♦ See Figure 50

The DT4 ignition system is a magneto CDI which provides high spark performance regardless of engine rpm. The electronic advance system provides optimum ignition timing for all conditions. Ignition timing is not adjustable. Each coil is not provided with any base plate for installation, but instead is constructed so that the coil itself can be mounted directly to the boss projected from the cylinder or crankcase. If the coils are installed in the correct position, the ignition timing will be within specification.

1. Mount the engine in a test tank or move the boat to a body of water.
2. Remove the cowling and connect a tachometer and a timing light to the powerhead.
3. Start the engine and allow it to warm to operating temperature. Place the engine in gear.

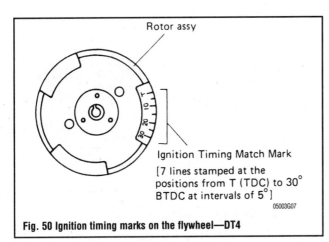

Ignition Timing Match Mark
[7 lines stamped at the positions from T (TDC) to 30° BTDC at intervals of 5°]

Fig. 50 Ignition timing marks on the flywheel—DT4

4. Aim the timing light at the timing window and the pointer on the recoil starter should line up with the timing mark on the rotor.
5. Compare ignition timing to specification in "Tuneup Specifications" chart.
6. Timing cannot be adjusted. If timing is incorrect, a fault has occurred in the CDI system and a test of the CDI unit needs to be performed.

IDLE SPEED

♦ See Figure 51

1. Mount the engine in a test tank or move the boat to a body of water.
2. Remove the cowling and connect a tachometer to the powerhead.
3. Start the engine and allow it to warm to operating temperature.
4. Turn the air screw in until it lightly seats and then back it out gradually. The engine will pick up speed correspondingly and then cease to rise. Set the air screw slightly before this point. See the "Idle Air Screw Specifications" chart for the base setting.
5. Shift the clutch into the forward position.
6. Run the throttle stop screw "B" in and out until the correct engine speed is reached. Idle speed specifications are located in the "Tune-Up Specifications" chart.

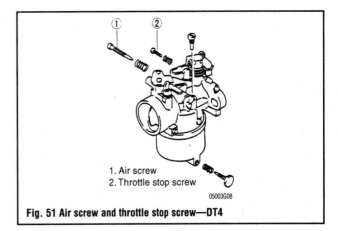

1. Air screw
2. Throttle stop screw

Fig. 51 Air screw and throttle stop screw—DT4

DT6 and DT8

IGNITION TIMING

♦ See Figures 52, 53, 54, 55 and 56

The DT6 and DT8 models use the Suzuki PEI simultaneous ignition system. The ignition timing is advanced in this system by moving the magneto stator according to the carburetor throttle opening.

The Suzuki PEI system is maintenance free because of the absence of breaker points. It produces a strong spark over a wide range of engine speeds from idle to wide open throttle.

The CDI unit which is integral with the ignition coil is compact and easy to handle.

The stator base moves according to the throttle opening to obtain the correct ignition timing. For this reason, brass is cast in the spigot joint of the oil seal housing and the stator base. Parts of the stator base include a coil which charges a capacitor of the CDI unit, a pulser coil which sends a signal to the CDI unit at ignition timing, and a lighting coil which generates lighting output of 12V and 80W.

1. Mount the engine in a test tank or the boat in a body of water.
2. Remove the cowling and connect a tachometer and timing light to the powerhead.
   • Setting the static ignition timing
3. Bring the face of the retainer stopper in line with the alignment mark of the magneto stator and fix the retainer stopper with bolts. When the end face of the stopper retainer is aligned with the boss of the cylinder center, ignition timing is -2°–+2°(no advance angle).
4. For the full-advance angle, adjust the length of the stator rod so that the throttle arm contacts the inlet case-side stopper. Now the ignition timing is 23–27°BTDC (full advance angle).

3-20 MAINTENANCE

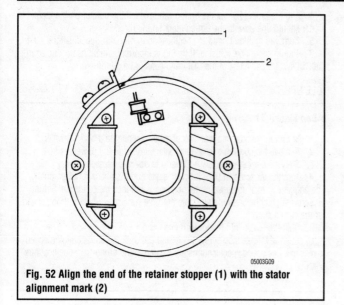

Fig. 52 Align the end of the retainer stopper (1) with the stator alignment mark (2)

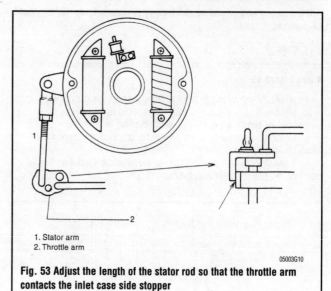

1. Stator arm
2. Throttle arm

Fig. 53 Adjust the length of the stator rod so that the throttle arm contacts the inlet case side stopper

Fig. 54 Correct position of the retainer stopper against the boss of the cylinder center

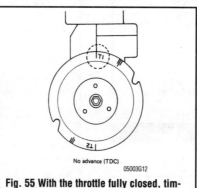

Fig. 55 With the throttle fully closed, timing mark should be like this (no advance TDC)—DT6 and DT8

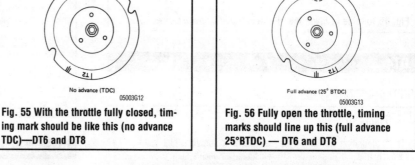

Fig. 56 Fully open the throttle, timing marks should line up this (full advance 25°BTDC) — DT6 and DT8

5. Checking the ignition timing (dynamic adjustment). To check ignition timing, warm up the engine for about 5 minutes. Then check if the cylinder center line is in line with the mark engraved beside the letter "T" on the flywheel with the throttle fully closed. If the marks in line, the engine piston is at TDC. Next fully open the throttle. If the cylinder center line is within the range bounded by the three mark lines engraved on the flywheel, the piston is within 2° of 25°BTDC.

IDLE SPEED

♦ See Figures 57 and 58

1. Mount the engine in a test tank or move the boat to a body of water.

2. Remove the cowling and connect a tachometer to the powerhead.
3. Start the engine and allow it to warm to operating temperature.
4. Turn the air screw in until it lightly seats and then back it out gradually. The engine will pick up speed correspondingly and then cease to rise. Set the air screw slightly before this point. See the "Idle Air Screw Specifications" chart for the base setting.
5. Shift the clutch into the forward position.
6. Run the throttle stop screw in and out until the correct engine speed is reached. Idle speed specifications are located in the "Tune-Up Specifications" chart.

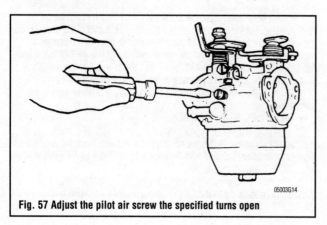

Fig. 57 Adjust the pilot air screw the specified turns open

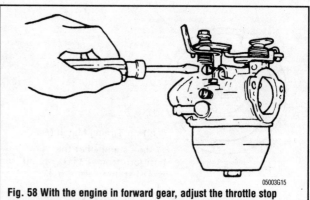

Fig. 58 With the engine in forward gear, adjust the throttle stop screw to the correct position

MAINTENANCE

DT9.9 and DT15

IGNITION TIMING

▶ See Figures 59, 60, 61, 62 and 63

The DT9.9 and DT15 are equipped with a single ignition coil for both cylinders and uses a mechanical advance system that changes the ignition timing by interlocking the stator to the carburetor opening. The CDI unit and the ignition coil are integrated to make the size compact and maintenance easier.

• The stator is made up of the ignition coil and a 12v 80w battery charge coil mounted on the aluminum base plate.
• The flywheel is constructed of cast iron and incorporates a tetra-pole. The electric start model has a ring gear attached to the outer edge while the manual start model does not. The flywheel has 4 engraved lines marked in 2° increments from 4°ATDC to 2°BTDC and 6 engraved lines marked in 2 °increments from 17°BTDC to 27°BTDC. These marks are used in conjunction with a timing light to inspect and adjust the ignition timing.

The static ignition timing adjustment is performed as follows:

• With the engine shut down, fully close the throttle. Contact the projection end surface of the magneto stator retainer to the inner side surface of the throttle cam. Thus adjust the setting marks of both parts so that they align with each other. Loosen the two cam set-screws and move the cam so that the stator-side and cam-side setting marks align.

➡On the DT9.9 make sure to tighten the cam stopper screw after adjusting.

• The maximum retard timing is 0°–4°ATDC at 650 rpm. Maximum advance timing is 16.5°–20.5°BTDC at 5000 rpm for the DT9.9 and 23°–27°BTDC at 5000 rpm for the DT15.

Dynamic ignition timing adjustment:
• Start and warm up the engine for about 5 minutes.
• Shift the clutch to neutral and keep the engine speed to 650 rpm.
• Ignition timing is indicated by the timing arrow attached to the side of the recoil starter pointing at any of the lines engraved on the outer surface of the flywheel.
• Using a timing light, make sure the timing arrow remains within a range of 0°–4°ATDC at 650 rpm. If not, the timing must be adjusted to align the timing marks.

IDLE SPEED

▶ See Figures 64 and 65

1. Mount the engine in a test tank or move the boat to a body of water.
2. Remove the cowling and connect a tachometer to the powerhead.

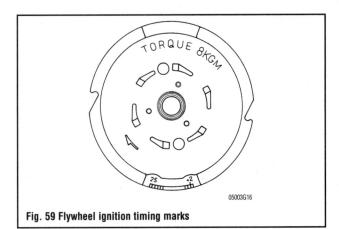
Fig. 59 Flywheel ignition timing marks

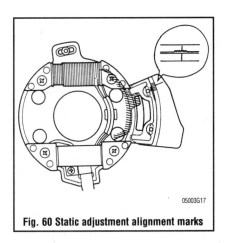
Fig. 60 Static adjustment alignment marks

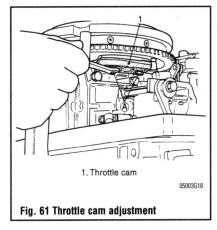
Fig. 61 Throttle cam adjustment

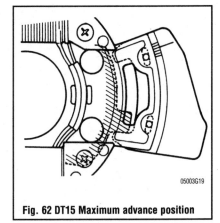
Fig. 62 DT15 Maximum advance position

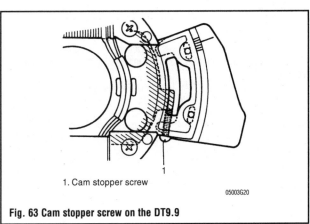
Fig. 63 Cam stopper screw on the DT9.9

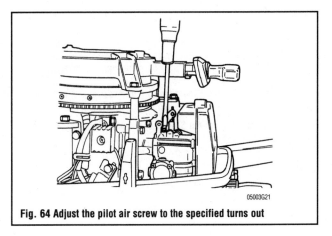
Fig. 64 Adjust the pilot air screw to the specified turns out

MAINTENANCE

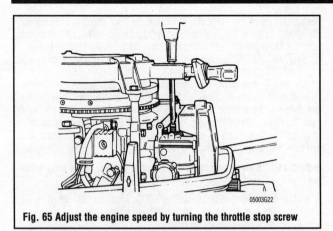

Fig. 65 Adjust the engine speed by turning the throttle stop screw

3. Turn the pilot air screw in until it lightly seats and then back it out the number of turns listed in the "Idle Air Screw Specifications" chart.
4. Start the engine and allow it to warm to operating temperature. Place the engine in gear.
5. Set the throttle lever to the full closed position.
6. Adjust the engine speed to the specification listed in the "Tune-Up Specifications" chart by turning the throttle stop screw.

➡ Make sure that the choke valve on the carburetor is in the full open position before adjusting the idle speed.

DT20, DT25 and DT30

IGNITION TIMING

This model uses the Suzuki IC (Integrated Circuit) ignition control to maintain precise spark timing for better power and acceleration.

A built in IC control unit monitors the degree of throttle opening and the engine rpm, it then determines the ideal spark timing. This not only improves acceleration, but by maintaining optimum carburetion and ignition synchronization the engine runs smoother and responds to throttle changes much quicker.

Since the ignition timing is controlled according to the opening of the throttle, it is not affected by any change in the engine speed.

For easier starting, every time the engine started, the ignition timing automatically advances to 5°BTDC for 15 seconds, after which time the ignition timing will return to the idle speed circuit and what ever position the "Idle Speed Adjustment Switch" is set at.

➡ Due the higher rpm of this setting, do not shift gears until the rpm's have returned to the Idle setting.

The idle speed ignition timing can be changed between 5 positions by means of the "Idle Speed Adjustment Switch" when the carburetor throttle valves are in the full closed position.

The ignition timing range of this switch is 9°ATDC when it is in the slow position and 1°ATDC in the fast position. Each position of the switch represents a 50 rpm change. When the throttle valve is in the full closed position, ignition timing is returned automatically to the timing of the "Idle Speed Adjustment Switch". This guarantees that the engine rpm will return to idle automatically.

To further assure the exact ignition timing, a gear counter coil measures flywheel position and relays this information to the CDI unit.

➡ The "Idle Speed Adjustment Switch" is the only means of adjusting the engine trolling speed.

THROTTLE LINKAGE ADJUSTMENT

▶ See Figure 66

1. Loosen the throttle lever adjusting screws on the top and 3rd carburetors. Then turn the plates counterclockwise to ensure full they're at the full closed throttle position. Hold the plates and tighten the screws.
2. Adjust the lever link rod to an initial length of 3.0 in. (75.5 mm) between the center holes in each connector.
3. At the full closed throttle position, the throttle lever arm must be against the stop boss on the block and a clearance of 0.02–0.06 in. (0.5–1.5 mm) must exist at this point. If the clearance is not correct, recheck the carburetor throttle valves and make sure that they are in the full closed position and recheck the throttle rod length. Readjust if necessary.
4. Check the synchronization of all three carburetors to make sure they fully open and close.

➡ Always adjust the throttle linkage after adjusting the oil pump linkage.

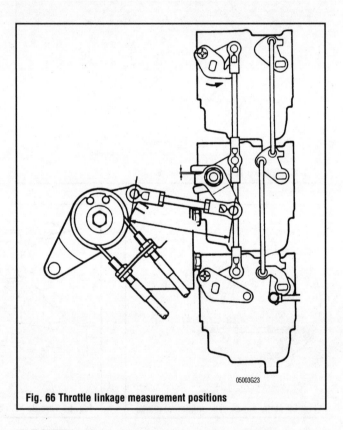

Fig. 66 Throttle linkage measurement positions

IDLE SPEED

1. Mount the engine in a test tank or move the boat to a body of water.
2. Remove the cowling and connect a tachometer to the powerhead.
3. Start the engine and allow it to warm to operating temperature. Place the engine in gear.
4. Check engine speed at idle. The powerhead should idle in at the rpm specified in the Tune-up Specifications chart.
5. Turn the pilot screw in until it lightly seats and then back it out the number of turns specified in the "Idle Air Screw Specifications" chart.
6. Place the engine in gear and check engine trolling speed in the same manner.

MAINTENANCE 3-23

7. Turn the idle speed adjusting knob to adjust the idle speed to the specification listed in the "Tune-Up Specifications chart.

➡ The engine trolling speed has been factory adjusted to the optimum speed. Trolling speed varies depending on boat type, weather conditions, propeller type and other variables. Adjustment can only be done with the engine in the water so there is back pressure against the exhaust system.

DT35 and DT40

IGNITION TIMING

♦ See Figures 67, 68 and 69

In these engines, a 2-cylinder simultaneous ignition CDI system has been adopted. The component parts of the ignition system are a magneto and CDI unit. The CDI unit contains the ignition coil. The ignition timing characteristics are made up of the advance angle of the magneto itself and the advance angle of stator sliding. An electronic advance system employing IC has been adopted in the advance angle of the magneto itself to assure highly precise ignition characteristics. The CDI unit also includes the over-rev limiter and oil warning circuit.

The total 27° ignition timing is the result of the combination of magneto and CDI unit that produces the 7° electrical advance angle shown in the illustration and the 20° advance angle when the magneto stator slides.

When the flywheel rotates, electromotive force is generated in the condenser charge coil and causes the current output from the positive side to flow from the condenser charge coil to the diode to the condenser to the ignition coil where it then charges the condenser charge coil. Next, when the timing coil and the flywheel magneto pole piece position become opposed, output is generated in the timing coil, resulting in the current flow from the timing coil to the diode to the SCR gate to ground, causing the SCR to change over from OFF to ON.

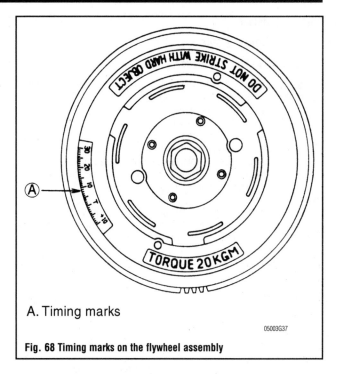

A. Timing marks

Fig. 68 Timing marks on the flywheel assembly

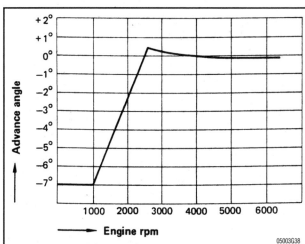

Fig. 69 The total ignition timing is the result of the combination of magneto and CDI unit that produces the electrical advance angle shown

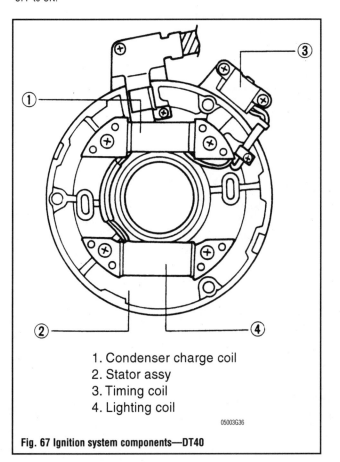

1. Condenser charge coil
2. Stator assy
3. Timing coil
4. Lighting coil

Fig. 67 Ignition system components—DT40

The electrical charge in the previously charged condenser flows from the condenser to the SCR to ground then to the ignition coil primary side and causes high voltage to be induced in the ignition coil secondary side causing the spark to jump across the spark plug gap. Since this is a simultaneous ignition system, both cylinders fire at the same time irregardless of which cylinder is firing.

This type of ignition system has flat wave characteristics up to 1,000 rpm. When engine speed rises above 1,000 rpm, the output of the arithmetic circuit built into the IC advances ahead of the output of the timing coil which starts the angle advance.

Up to 2,500 rpm, the advance angle increases in a designated ratio corresponding to the engine rpm. When above this rpm, the angle advance stops and exhibits a nearly flat wave characteristic. The advance angle of the magneto itself is 7° but in actuality, while mechanically sliding the magneto stator 20°, a total of 27° (ATDC 2°~ BTDC 25°) advance angle is realized.

MAINTENANCE

Static Adjustment

MAXIMUM ADVANCE SIDE

▶ See accompanying illustration

To adjust the ignition timing without starting the engine, proceed as follows while referring to the accompanying illustration:

1. Shift the clutch into forward gear.
2. Put the throttle in the wide full open position.
3. Keep this condition to align the mark (A) on the magneto stator (1) and the cylinder-to-crankcase fitting surface (C) with each other.
4. Keep this condition to loosen the screws (9) on the throttle cam (2) toward the direction indicated by the arrow mark (S) until contacting the crankcase side stopper (E), and fix the cam by tightening the screws (9) at this position.
5. Still keep this position to loosen the screw (10) on the carburetor, open the carburetor link toward the direction indicated by the arrow mark (T) fully hitting against the opponent, move the rotor (U) toward the direction indicated by the arrow mark (V) so as to contact the throttle cam (2) and tighten the screw (10) at this condition.

MAXIMUM RETARD SIDE

1. Place shift lever into the Forward 1st notch.
2. Align the mark (B) on the magneto stator and the cylinder-to-crankcase fitting surface (C) with each other. Turn the adjust bolt (5) so that the adjust bolt cap (4) contacts the crankcase-side stopper (D), and fix it by the nut (6).
3. Move the throttle limiter (1) toward the direction (W) until hitting against the undercover-side stopper (F) and check to see whether the carburetor can be fully opened at this condition.
4. If the carburetor cannot be fully opened, adjust the rod (7) for the proper length and fix it by the nut (8). The rod (7) should have a standard length of 4.25 in. (108 mm).
5. Adjust the throttle cable so that the stator can be fully opened and closed (DT40C/DT40CE).

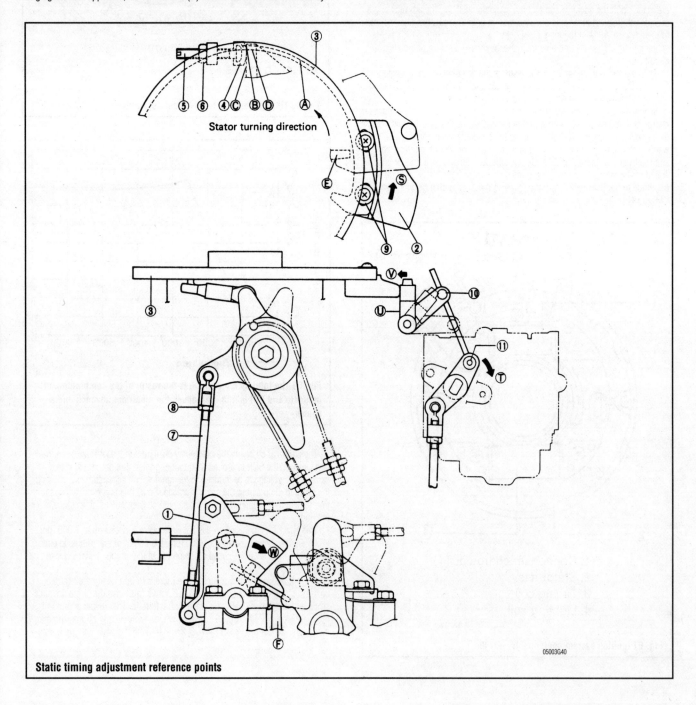

Static timing adjustment reference points

MAINTENANCE 3-25

Dynamic adjustment

❈❈ CAUTION

Before this adjustment, start the engine and warm it up for about 5 minutes. Before starting this adjustment, be sure to remove the throttle rod (11).

FULL ADVANCED IGNITION TIMING ADJUSTMENT

♦ See accompanying illustrations

1. Shift the clutch to forward.
2. To check the full-advanced ignition timing, keep the engine running at 1,000 rpm and move the throttle limiter (1) to the full advanced position (W).

❈❈ WARNING

This procedure is required to check the ignition timing while the engine is running at 5,000 rpm, but this method is very dangerous and here is an alternative method in which the ignition timing is advanced by hand with the engine running less than 1,000 rpm.

3. Keep the throttle limiter (1) in this position.
4. Check that the timing plate points to 18.5°BTDC, using a timing light.

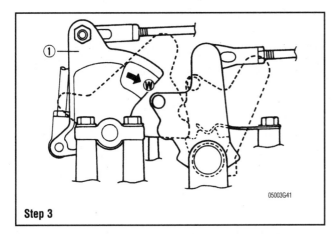
Step 3

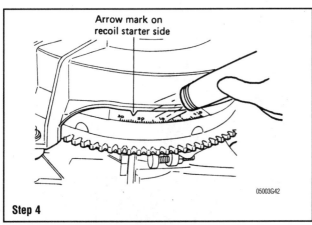
Step 4

➡When this method is used as a substitute, the ignition timing should be set at 18.5°BTDC at 1,000 rpm, but in actual operation, the ignition timing will be 18°BTDC at 5,000 rpm, showing a time lag of 0.5°. The actual ignition timing is 25°BTDC, this being the result of 7°of advance by the magneto and 18°of advance caused by sliding the stator.

5. If the ignition timing is off 18.5°BTDC, loosen the two screws (1) and move throttle cam (2) to right and left as shown in the illustration to adjust. After adjusting, tighten the screws (1) securely.
6. After adjustment, install the throttle rod to the original position.

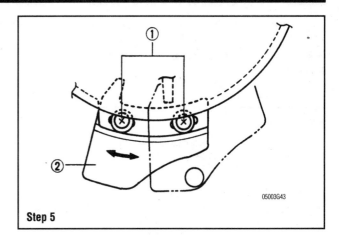
Step 5

FULL RETARDED IGNITION ADJUSTMENT

1. Shift the clutch to Neutral and keep the engine speed at 1,000 rpm.
2. Ignition timing is indicated by the timing arrow attached on the side of the recoil starter pointing at any of the lines engraved on the outer surface of the flywheel.
3. Using a timing light, make sure the timing arrow remains within a range of 2°plus or minus 1°ATDC at 1,000 rpm. If not, the ignition timing must be readjusted.
4. To retard the timing, loosen the lock nut and turn the adjusting screw counterclockwise. To advance the timing, turn the adjusting screw clockwise.
5. After the timing has been adjusted correctly, tighten the lock nut.

IDLE SPEED

♦ See accompanying illustration

❈❈ CAUTION

Before this adjustment is made, be sure to check the ignition timing and adjust it as needed. Start and thoroughly warm up the engine for about 5-minutes.

1. Using the access holes in the air silencer, insert a screwdriver to the carburetor, and turn in the pilot air screw all the way in until it lightly seats, and then back it out the number of turns specified in the "Idle Air Screw Specifications" chart.

❈❈ CAUTION

Be extra cautious to not tighten the air screw too much. It will damage the screw and seat.

2. Place the shift lever (or remote control) in the first notch of Forward gear.
3. Keeping the engine in this condition, turn the carburetor-side throttle

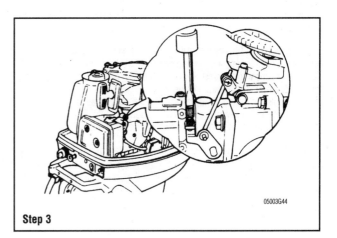
Step 3

MAINTENANCE

screw. This will allow you to adjust the engine speed to the specification located in the "Tune-Up Specifications" chart.

➡ **Make sure that the choke valve on the carburetor is in the full-open position.**

THROTTLE LINKAGE

♦ **See Figure 70**

The rod length can be determined by measuring the two throttle rods. The lengths should measure as follows:
- Dimension "A" 4.25 in. (108 mm)
- Dimension "B" 1.77 in. (45 mm)

Measure the rods between the connectors.

1. Turn the throttle grip (or remote control lever) to the full open position and turn the shaft lever "C" to the full open position until the stopper "D" contacts the protrusion on the carburetor. The lock the lever in position with the lock screw.

※※ CAUTION

After making the above adjustment, move the throttle grip to check for smooth throttle operation.

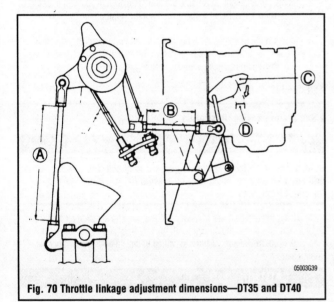

Fig. 70 Throttle linkage adjustment dimensions—DT35 and DT40

DT55 and DT65

IGNITION TIMING

♦ **See Figures 71, 72 and 73**

The DT55 and DT65 use the Suzuki IC (integrated circuit) ignition system. Ignition timing adjustment is not necessary on models equipped with these ignition systems with the exception of adjusting the throttle valve sensor. They are equipped with the following features:

- Engine Start Advance Mechanism. This feature ensures easy engine starting by automatically advancing the ignition advance to 10°BTDC for about 15 seconds, after which, the IC control circuit changes over to trolling ignition timing "A"
- Trolling Speed Adjusting Mechanism. The trolling ignition timing can be changed at 2° intervals from 0° to 6°ATDC by means of an idle speed adjusting switch. By changing over the trolling ignition timing, the trolling speed can be adjusted.
- All models after 1991 have had the Idle Speed Adjustment Switch removed and instead an ignition timing resistor has been installed. With this modification, the in gear idle timing with the throttle fully returned is kept at a constant 6°ATDC.

➡ **The engine rpm at trolling speed has been factory set at approximately 700 rpm. The trolling speed varies depending on boat type, weather conditions, propeller types and other variables. Adjust the trolling speed with the idle speed adjusting switch to obtain the desired engine speed.**

- Advance Stop Mechanism. When closing the throttle valve fully, an idle switch "1" is "ON" in conjunction with the carburetor and regardless of the engine rpm, the trolling ignition timing can be obtained. Therefore, by returning the throttle valve to its fully closed position during high speed travel, the boat's speed can be decreased suddenly.
- Acceleration Advance Mechanism. This device is available to increase engine rpm quickly during sudden acceleration. When an acceleration switch "2" is "ON" in cooperation with the carburetor, the ignition timing of the basic advance characteristic "2" is quickened to the ignition timing of the acceleration advance characteristic "C"
- Throttle Valve Switch and Cam. As the throttle valve moves, a cam fitted to the end of the throttle valve shaft moves accordingly to put the roller of the throttle valve switch in motion. By moving the roller, the switch is turned ON and OFF sending a signal to the CDI unit where the ignition timing is changed. When the throttle valve is fully closed, the idle switch "1" is "ON" and the acceleration switch is "OFF". Once the throttle valve is opened, the idle switch "1" is "OFF" and the acceleration switch "2" is "ON"

On the DT55 and DT65, the working angles at which the idle switch and acceleration switch turn "On" and "Off" differ from each other. Therefore, two different cams have been developed, one being for thew DT55 and the other for the DT65. They can be told apart by the different markings on them. The numerals on the cams denote the angle until the switch is "OFF" "C" from the vertical line "A" and an angle until the acceleration switch is "ON" "D" from the vertical line "A"

IDLE SPEED

1. On the carburetor, turn the pilot air screw all the way in until it lightly seats and then back it out the number of turns specified in the "Idle Air Screw Specifications" chart.
2. Place the remote control lever forward gear (idle).
3. Turn the idle speed adjusting switch, and adjust the idle speed to the specification listed in the "Tune-Up Specifications" chart.

➡ **Make sure the choke valve is in the fully open position.**

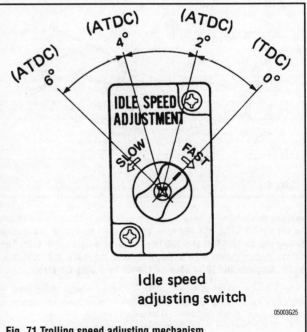

Fig. 71 Trolling speed adjusting mechanism

MAINTENANCE 3-27

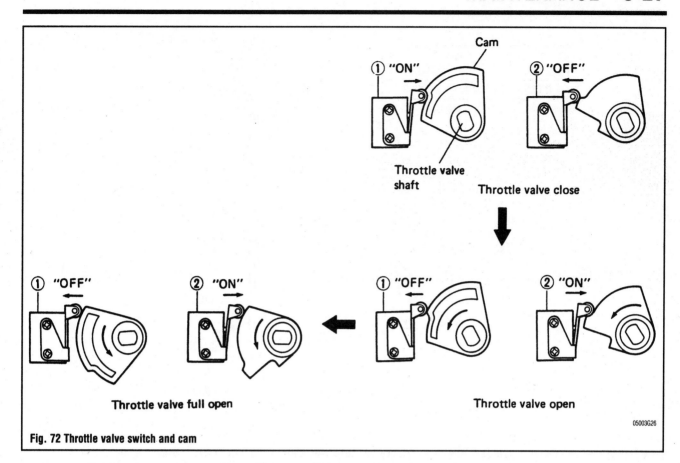

Fig. 72 Throttle valve switch and cam

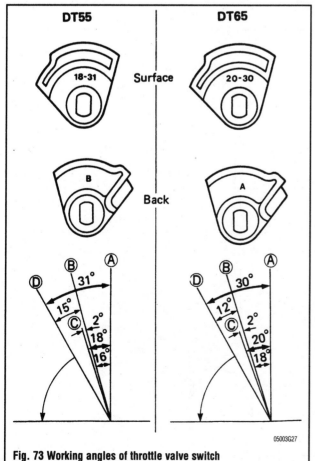

Fig. 73 Working angles of throttle valve switch

THROTTLE LINKAGE

▶ See Figures 74 and 75

Full-close adjustment of the throttle valve. DT55 (serial number 501001–502859) and DT65 (serial number 501001–502959)

1. Remove the throttle lever rod (1) from the throttle control (2) lever.
2. Loosen the set screw (4) on the lever (3) of the #1 and #3 carburetors.
3. Tighten the set screw (4) with the throttle valves of the #1 and #3 carburetors fully closed.
4. Check operation by moving the lever (5) of the #2 carburetor to make sure the individual throttle valves of each carburetor operate together.

➡ If they do not work together uniformly, make the above adjustments again.

Full close adjustment of the throttle valve. DT55 (serial number 502860 to present) and DT65 502960 to present).

5. Remove the throttle lever rod (1) from the throttle control lever (2).
6. Loosen the set screw (4) of the lever (3) of #1 and #3 carburetors. In this case, the throttle valve is set to its full-closed position by the action of a return spring.
7. Move the lever (5) of the second carburetor, a few times (more than 30°) as shown by the arrow (A) to eliminate any play in the throttle rod between the carburetors. All the throttle valves should be closed evenly.
8. Apply a thread locker to the loosened set screw (4) and tighten.
9. Check operation by moving the lever (5). All throttle valve should move at the same time.

Adjustment of the throttle lever rod

10. Adjust the dimension (B) of the throttle lever rod (1) to the following length and attach the control lever:
    - Standard dimension (B): DT55–4.5 in. (114 mm); DT65–4.3 in. (108 mm)
11. Move the control lever in the direction of the arrow (C) and adjust the length of the rod (1) with the connector (7) so that the control lever comes in contact with the stopper (6) at a position where the throttle valve fully opens or a position of 1° to 2° this side from the full-open position.

3-28 MAINTENANCE

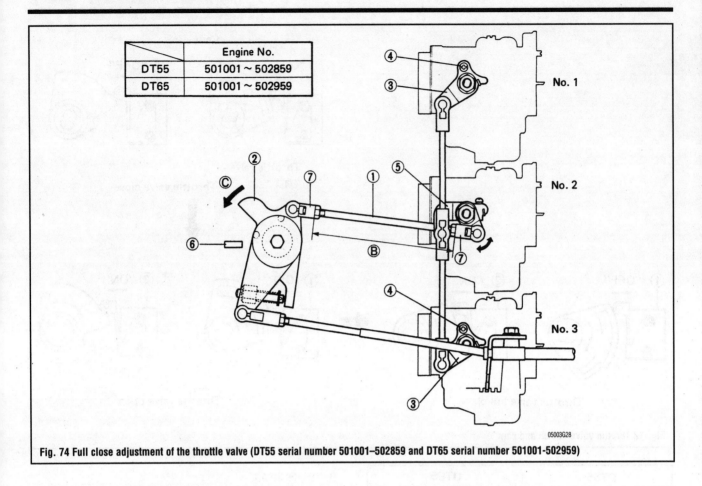

Fig. 74 Full close adjustment of the throttle valve (DT55 serial number 501001–502859 and DT65 serial number 501001-502959)

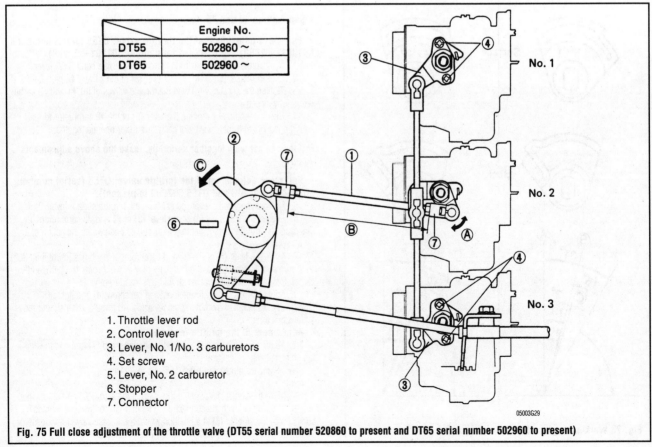

1. Throttle lever rod
2. Control lever
3. Lever, No. 1/No. 3 carburetors
4. Set screw
5. Lever, No. 2 carburetor
6. Stopper
7. Connector

Fig. 75 Full close adjustment of the throttle valve (DT55 serial number 520860 to present and DT65 serial number 502960 to present)

MAINTENANCE 3-29

⚠ CAUTION

If there is a gap between the control lever (2) and the stopper (6) when the throttle valve has opened fully, the throttle rod, throttle valve or carburetor(s) may be damaged and may not operate correctly at full throttle operation.

12. If all adjustments are correct, tighten the throttle rod lock nuts securely.

DT75 and DT85

IGNITION TIMING

▶ See Figure 76

Starting in 1988, the Suzuki digital IC ignition system was adopted. This system eliminates a direct mechanical linkage between the engine end the ignition system. Instead, sensors relay information detailing throttle position sensor, gear counter (engine speed) and engine temperature to the ignition module which processes this information and then determines the optimal ignition timing.

No adjustment is necessary on this system.

The DT75 and DT85 use the Suzuki IC (integrated circuit) ignition system. Ignition timing adjustment is not necessary on models equipped with these ignition systems with the exception of adjusting the throttle valve sensor. They are equipped with the following features:

- Engine Start Advance Mechanism. This feature ensures easy engine starting by automatically advancing the ignition advance to 5°BTDC for about 15 seconds, after which, the IC control circuit changes over to trolling ignition timing "A"
- Trolling Speed Adjusting Mechanism. The trolling ignition timing can be changed from 7°ATDC in the slow position to 1°BTDC by means of an idle speed adjusting switch. Each position on the switch represents approximately 50 rpm change. By changing over the trolling ignition timing, the trolling speed can be adjusted.

All models from 1991 have had the Idle Speed Adjustment Switch removed and instead an ignition timing resistor has been installed. With this modification, the in gear idle timing with the throttle fully returned is kept at a constant 2°–6°ATDC and the in-gear idle speed is now adjusted by the throttle stop screw on the #3 carburetor.

➡ **The engine rpm at trolling speed has been factory set at approximately 700 rpm. The trolling speed varies depending on boat type, weather conditions, propeller types and other variables. Adjust the trolling speed with the idle speed adjusting switch to obtain the desired engine speed.**

- Advance Stop Mechanism. When closing the throttle valve fully, an idle switch is "ON" in conjunction with the carburetor and regardless of the engine rpm, the trolling ignition timing can be obtained. Therefore, by returning the throttle valve to its fully closed position during high speed travel, the boat's speed can be decreased suddenly.

CARBURETOR LINKAGE ADJUSTMENT

▶ See Figure 77

Fully closed adjustment of the throttle valve.

1. Remove the throttle lever rod (1) from the throttle control lever.
2. Ensure that the throttle stop screw (on the #3 carburetor) if fully backed out.
3. Loosen the screws (4) of the adjustable levers on the #1 and #3 carburetors. The return springs will close the throttle valves fully.
4. Flick the lever (5) of the #2 carburetor 2 or 3 times, as shown by the arrow (A), which will ensure that all three throttle valves are closed evenly.
5. Tighten the lever screws (4) on the #1 and #3 carburetors and apply thread lock compound.
6. Finally, check the operation by flicking the lever (5), to see if the three carburetor throttle valves are balanced and synchronized with each other.

Adjustment of the throttle lever rod.

1. Adjust the dimension (B) of the throttle lever (1) to the correct length. For the DT75: 6.1 in. (155 mm) and the DT85: 5.7 in. (145 mm). Attach the control lever.
2. Move the control lever (2) in the direction of the arrow (C) and adjust the length of the rod (1) by screwing the connector (7) accordingly. The cam on the control lever should touch the stopper (6) when the throttle valves are fully open, or within 1°–2° of being fully open.

⚠ CAUTION

If there is a gap between the control lever (2) and stopper (6) at full throttle, damage may result to the throttle rod, throttle valves and carburetors.

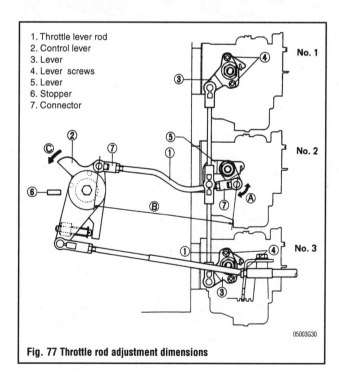

1. Throttle lever rod
2. Control lever
3. Lever
4. Lever screws
5. Lever
6. Stopper
7. Connector

Fig. 77 Throttle rod adjustment dimensions

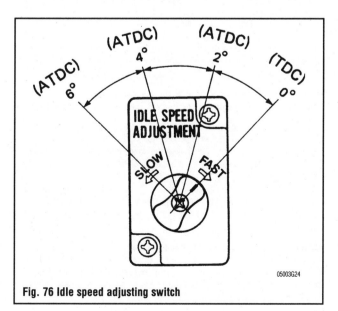

Fig. 76 Idle speed adjusting switch

IDLE SPEED

▶ See Figure 78

Adjust the in-gear idle speed in the following way.
1. Warm up the engine for approximately 5 minutes

3-30 MAINTENANCE

Fig. 78 Idle adjustment screw

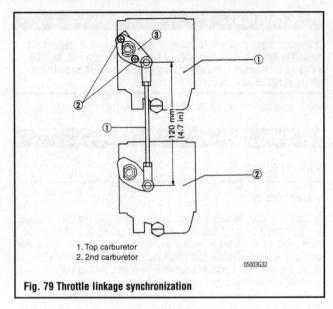

1. Top carburetor
2. 2nd carburetor

Fig. 79 Throttle linkage synchronization

2. Turn the pilot air screws in fully (clockwise) until lightly seated, then back them out (counterclockwise) the number of turns specified in the "Idle Air Screw Specifications" chart.
3. Engage forward gear
4. Set the pilot air screws as specified, then maintain a stable idle by turning the idle adjustment screw (A) clockwise to increase idle speed or counterclockwise to decrease idle speed. Idle speed specifications are located in the "Tune-Up Specifications" chart.

➡ If in-gear idle speed is not in the specified range, or if the engine will not maintain idle rpm, it is possible that there is an abnormality in either the fuel or ignition systems. If the fuel and ignition systems are working correctly, it is possible that the throttle valve sensor may need adjustment. For this adjustment refer to the applicable section in "Electrical and Ignition".

DT90 and DT100

IGNITION TIMING

The DT90 and DT100 are all equipped with the Micro Link Ignition system. This system uses a microcomputer to maximize combustion control and thus improve engine performance. The system uses various sensors and switches to monitor engine rpm, throttle valve opening, shift lever position and operator selected idle speed. The computer constantly evaluates this information and provides the optimal ignition spark timing for the current engine running condition.

No adjustment is necessary on these models. If there is a problem with ignition timing, it is most likely caused by a faulty CDI module. Refer to the appropriate section in "Electrical and Ignition" for CDI troubleshooting.

CARBURETOR LINKAGE ADJUSTMENT

▶ See Figure 79

1. Check the length of the throttle linkage rod (1) to the carburetors.
2. Loosen the throttle lever adjusting screws (2) on the top carburetor
3. Lightly push the throttle lever (3) clockwise until the throttle valves are completely closed. Then tighten the adjusting screws (2)
4. Actuate the throttle linkage and check if both throttle valves are synchronized in the completely closed position. If the throttle valves are not synchronized, perform the adjustment again

IDLE SPEED

1. Warm up the engine for approximately five minutes.
2. On the carburetor, turn the pilot screw all the way in until it lightly seats and then back it out the number of turns specified in the "Idle Air Screw Specifications" chart.

➡ Do not overtighten the pilot screw or you may damage the screw. Just lightly seat it and then turn it out.

3. Place the remote control lever into forward gear, first notch.
4. Turn the idle speed adjusting switch to position 5 (slow). This adjusts the engine speed to a range of 600–650 rpm. The engine must maintain this speed for 3 minutes.
5. If the engine speed is less than 600 rpm or it will not maintain trolling speed for three minutes:
• Adjust the top carburetor throttle stop screw to maintain the idle speed specified in the "Tune-Up Specifications" chart.
• Recheck the throttle valve sensor resistance value and readjust if necessary. See "Electrical and Ignition"

DT115 and DT140

IGNITION TIMING

The DT115 and DT140 models are equipped with the Suzuki digital IC ignition system. This system eliminates a direct mechanical linkage between the engine end the ignition system. Instead, sensors relay information detailing throttle position sensor, gear counter (engine speed) and engine temperature to the ignition module which processes this information and then determines the optimal ignition timing.

Ignition timing adjustment is not necessary on models equipped with these ignition systems with the exception of adjusting the throttle valve sensor. They are equipped with the following features:
• Engine Start Advance Mechanism. This feature ensures easy engine starting by automatically advancing the ignition advance to 7°BTDC for about 15 seconds, after which, the time the ignition timing will return to the idle speed circuit and what ever position the "Idle Speed Adjustment Switch" is set at.

➡ Due to the higher rpm created by the automatic starting device, do not shift gears until the engine speed has returned to normal idle speed.

• Trolling Speed Adjusting Mechanism. The trolling ignition timing can be changed from 7°ATDC in the slow position to 1°BTDC by means of an idle speed adjusting switch. Each position on the switch represents approximately 50 rpm change. By changing over the trolling ignition timing, the trolling speed can be adjusted.

Within idle speed range, the timing is not affected by any change in engine speed, up to 900 rpm. To further assure exact ignition timing, a gear counter coil electrically measures the flywheel position and sends this information to the CDI module.

➡ All models from 1991 have had the "Idle Speed Adjustment Switch" removed and instead an ignition timing resistor has been installed. With this modification, the in gear idle timing with the throttle fully returned is kept at a constant 6°ATDC and the in-gear idle speed is now adjusted by the throttle stop screw on the #4 carburetor.

MAINTENANCE 3-31

➡ The engine rpm at trolling speed has been factory set at approximately 700 rpm. The trolling speed varies depending on boat type, weather conditions, propeller types and other variables. Adjust the trolling speed with the idle speed adjusting switch to obtain the desired engine speed.

• Advance Stop Mechanism. When closing the throttle valve fully, an idle switch is "ON" in conjunction with the carburetor and regardless of the engine rpm, the trolling ignition timing can be obtained. Therefore, by returning the throttle valve to its fully closed position during high speed travel, the boat's speed can be decreased suddenly.

CARBURETOR LINKAGE

♦ See Figures 80 and 81

1. Check the throttle linkage rod (1) to the carburetors.
2. Loosen the throttle lever adjusting screws (2) on the #1, #2, #3 carburetors.
3. Lightly push the throttle lever (3) clockwise until the throttle valves are completely closed. Tighten the adjusting screws (2) after applying a thread locking agent.
4. Move the throttle linkage back and forth and check if all the throttle valves are synchronized in the completely closed position.
5. Adjust the link rod (1) to an initial length of 6.3 in. (160 mm) measured between the centers of each connector; the connectors must be at the same angle after adjusting.
6. Install the link rod (1) onto the anchor pins (3) on the carburetor lever (4) and the throttle control lever (5).
7. Push the throttle control lever (5) counterclockwise until the throttle valves are completely opened. Then, the clearance between the throttle control lever (5) and the stopper (6) on the crankcase must be zero.
8. Readjust the connectors to achieve this clearance.
9. Check for freedom of movement in the linkage.

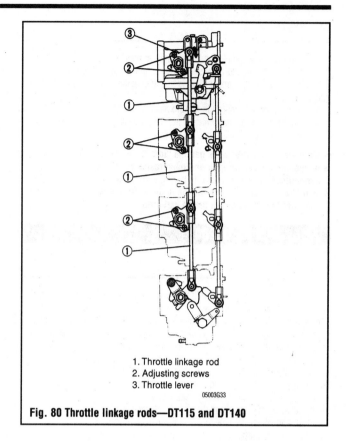

1. Throttle linkage rod
2. Adjusting screws
3. Throttle lever

Fig. 80 Throttle linkage rods—DT115 and DT140

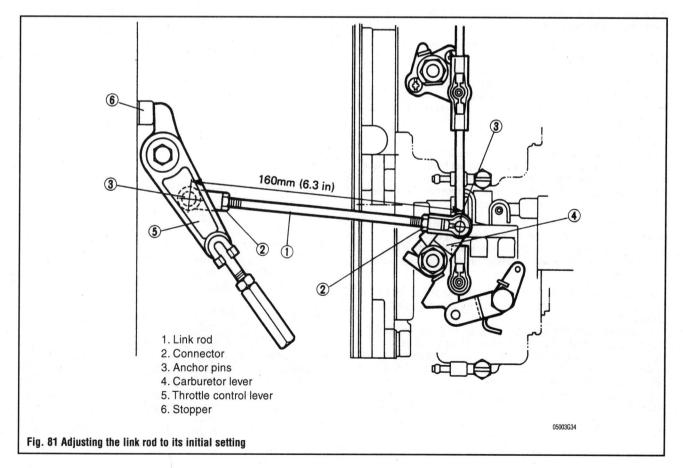

1. Link rod
2. Connector
3. Anchor pins
4. Carburetor lever
5. Throttle control lever
6. Stopper

Fig. 81 Adjusting the link rod to its initial setting

MAINTENANCE

IDLE SPEED

1. Warm up the engine for approximately five minutes.
2. On the carburetor, turn the pilot screw in all the way until it lightly seats, and then back it out the number of turns specified in the "Idle Air Screw Specifications" chart.
3. Place the remote control in forward gear, first notch.
4. Turn the idle adjusting switch to maintain the minimum idle speed specified in the "Tune-Up Specifications" chart.

➡ All models from 1991 have had the "Idle Speed Adjustment Switch" removed and instead an ignition timing resistor has been installed. With this modification, the in gear idle timing with the throttle fully returned is kept at a constant 6°ATDC and the in-gear idle speed is now adjusted by the throttle stop screw on the #4 carburetor.

DT150, DT175, DT200

IGNITION TIMING

The V6 models are equipped with the Suzuki digital IC ignition system. This system eliminates a direct mechanical linkage between the engine end the ignition system. Instead, sensors relay information detailing throttle position sensor, gear counter (engine speed) and engine temperature to the ignition module which processes this information and then determines the optimal ignition timing.

Ignition timing adjustment is not necessary on models equipped with these ignition systems with the exception of adjusting the throttle valve sensor. They are equipped with the following features:

- Engine Start Advance Mechanism. This feature ensures easy engine starting by automatically advancing the ignition advance to 5°BTDC for about 15 seconds, after which, the time the ignition timing will return to the idle speed circuit and what ever position the "Idle Speed Adjustment Switch" is set at.

➡ Due to the higher rpm created by the automatic starting device, do not shift gears until the engine speed has returned to normal idle speed.

- Trolling Speed Adjusting Mechanism. The trolling ignition timing can be changed from 6.5°ATDC in the slow position to 0.5°BTDC by means of an idle speed adjusting switch. Each position on the switch represents approximately 50 rpm change. By changing over the trolling ignition timing, the trolling speed can be adjusted. Within idle speed range, the timing is not affected by any change in engine speed, up to 900 rpm. To further assure exact ignition timing, a gear counter coil electrically measures the flywheel position and sends this information to the CDI module.

➡ All models from 1991 have had the "Idle Speed Adjustment Switch" removed and instead an ignition timing resistor has been installed. With this modification, the in gear idle timing with the throttle fully returned is kept at a constant 5°ATDC and the in-gear idle speed is now adjusted by the throttle stop screw on the #3 carburetor.

➡ The engine rpm at trolling speed has been factory set at approximately 700 rpm. The trolling speed varies depending on boat type, weather conditions, propeller types and other variables. Adjust the trolling speed with the idle speed adjusting switch to obtain the desired engine speed.

- Accelerator and Idle Return Switches. These switches are mounted in a single sealed unit on the port side of the #3 carburetor and are activated by throttle position.
- Accelerator switch. This switch prevents the ignition timing from lagging behind on quick acceleration by automatically giving the ignition 5 more degrees of advance as soon as it is activated. This switch is activated at 10 degrees of throttle valve opening. When the engine is accelerated slowly, the switch is activated at approximately 1,800 rpm.
- Idle return switch. This switch is "ON" when the throttle valve position is between 0°(full closed) and 2°open. When the switch is on, the basic advance curve is cancelled and ignition timing is returned automatically to the timing of the "Idle Speed Adjustment Switch". This guarantees that the engine speed will return to idle automatically. If this switch does not function properly, the engine rpm will take longer to return to idle speed.

CARBURETOR LINKAGE

▶ See Figure 82

1. Check the throttle linkage rod "1" to the carburetors.
2. Loosen the throttle lever adjusting screws "2" on the top and center carburetors.
3. Lightly push the throttle lever "3" clockwise until the throttle valves are completely closed. Then tighten the adjusting screws "2".
4. Move the throttle linkage and check to make sure all the throttle valves are synchronized and in the completely closed position.

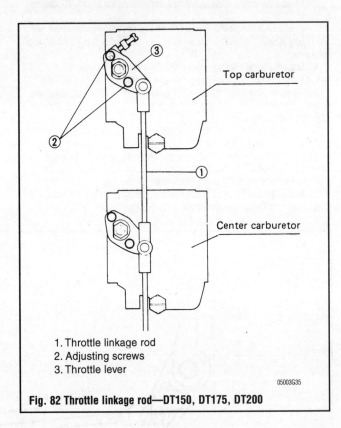

1. Throttle linkage rod
2. Adjusting screws
3. Throttle lever

Fig. 82 Throttle linkage rod—DT150, DT175, DT200

IDLE SPEED

1. Warm up the engine for approximately five minutes.
2. On the carburetor, turn the pilot screw in all the way until it lightly seats, and then back it out the number of turns specified in the "Idle Air Screw Specifications" chart.
3. Place the remote control in forward gear, first notch.
4. Turn the idle adjusting switch to maintain the minimum idle speed specified in the "Tune-Up Specifications" chart.

➡ All models from 1991 have had the "Idle Speed Adjustment Switch" removed and instead an ignition timing resistor has been installed. With this modification, the in gear idle timing with the throttle fully returned is kept at a constant 5°ATDC and the in-gear idle speed is now adjusted by the throttle stop screw on the #3 carburetor.

MAINTENANCE 3-33

Carburetor Idle Air Screw Specification

Model	Year	Type	Turns Out From Lightly Seated
DT 2	1988-96		1.25-1.75
DT2.2	1997-98		0.375-0.875
DT 4			1-1.50
DT 6		S-type	0.875-1.375
		L and UL type	1-1.50
DT 8	1988-91		1.75-2.25
	1992-97		0.50-1.0
DT 9.9			1.125-1.625
DT 15			1.50-2
DT 20			1.75-2.25
DT 25 (2-cyl)			1.25-1.75
DT 25 (3-cyl)	1989	MC	1.25-1.75
	1988-90	Except MC	1.5-2.0
	1991-99		1.0-1.5
DT 30	1989	MC	1.25-1.75
	1988-90	Except MC	1.5-2.0
	1991-97		1.0-1.5
DT 35			1.5-2.0
DT 40	1988-91		1.5-2.0
	1992-98		0.875-1.375
DT 55	1988-89		1.25-1.75
	1990-97		1.0-1.5
DT 65	1988-89		1.25-1.75
	1990-on		1.0-1.5
DT75	1988-90		1.75-2.25
	1991-94		1.5-2.0
	1995-97		1.375-1.875
DT85	1988-90		1.625-2.125
	1991		0.75-1.25
	1992		0.50-1.0
	1993-97		1.375-1.875
DT 90			1.125-1.625
DT100	1989-91		1.125-1.625
	1992-99		1.375-1.875
DT 115	1988		1.25-1.75
	1989		0.875
	1990-91		1.0-1.5
	1992-95		0.625-1.125
	1996		0.75-1.25
DT 140	1988		1.0-1.5
	1989-91		1.125-1.625
	1992-96		0.625-1.125
DT 150	1988		1-1.50
	1989-94		1.25-1.75
DT 175	1988		1.5-2.0
	1989-92		1.25-1.75
DT 200	1988		1.25-1.75
	1989-93		1.25-1.75

Tuneup Specifications Chart

Model		Spark Plug NGK	Spark Plug Champion	Spark Plug Gap Inch(mm)	Ignition Timing ° BTDC	Idle Speed RPM (Neutral)
DT2	1988-89	B4H	L81, L88A	.024-.028 (.6-.7)	15 @ 4500	800-900
	1990	B4H	L81, L88A	.024-.028 (.6-.7)	19 @ 4500	800-900
	1991	B5HS	L81, L88A	.024-.028 (.6-.7)	19 @ 4500	800-900
	1992-96	B5HS	L81, L88A	.024-.028 (.6-.7)	17 - 21 @ 4500	800-900
DT2.2	1997	BR5HS	L81, L88A	.024-.028 (.6-.7)	23 - 27 @ 5000	1000-1100
DT4	1988-89	BP6HS	RL12Y, RL87Y, L66Y	.024-.028 (.6-.7)	25 @ 5000	850-900
	1990-98	BP5HS	L81, L88A	.024-.028 (.6-.7)	24 @ 5000	850-900
DT6		BR6HS-10	RL12Y, RL87Y, L66Y	.035-.039 (.9-1.0)	23-27 @ 5000	600-650
DT8		BPR6HS	RL12Y, RL87Y, L66Y	.031-.035 (.08-.09)	23-27 @ 5000	600-650
DT9.9	1988	B6HS	L9J, QL7J, RL7J	.035-.039 (.9-1.0)	23-27 @ 5000	600-650
DT15	1988	B7HS	L5, L7J	.035-.039 (.9-1.0)	23-27 @ 5000	600-650
	1989	B7HS-10	N/A	.035-.039 (.9-1.0)	27 @ 2000	600-650
	1991-97	BR7HS-10	N/A	.035-.039 (.9-1.0)	28 @ 2000	650-700
DT20		BR7HS	L5, L7J	.035-.039 (.9-1.0)	25 @ 5000	600-650
DT25	1990	BR7HS-10	N/A	.035-.039 (.9-1.0)	25 @ 5000	650-700
	1991	BR7HS	L5, L7J	.035-.039 (.9-1.0)	23-27 @ 5000	600-650
	1992-99	BR7HS-10	N/A	.035-.039 (.9-1.0)	23-27 @ 5000	600-650
DT30	1989-90	BR7HS-10	N/A	.035-.039 (.9-1.0)	25 @ 5000	650-700
	1991-97	BR7HS-10	N/A	.035-.039 (.9-1.0)	23-27 @ 5000	650-700
DT35		B8HS	L4J, RL4J, L78, RL78	.031-.035 (.08-.09)	23-27 @ 5000	650-700
DT40		B8HS	L4J, RL4J, L78, RL79	.031-.035 (.08-.09)	23-27 @ 5000	650-700
DT55	1988-89	B8HS-10	L4J, RL4J, L78, RL78	.035-.039 (.9-1.0)	23 @ 5000	650-700
	1990	B8HS-10	L4J, RL4J, L78, RL78	.035-.039 (.9-1.0)	17 @ 5000	650-700
	1991-97	B8HS-10	L4J, RL4J, L78, RL78	.035-.039 (.9-1.0)	15-19 @ 5000	750-800
DT65	1988-90	B8HS-10	L4J, RL4J, L78, RL78	.031-.035 (.8-.9)	25 @ 5000	650-700
	1991-97	B8HS-10	L4J, RL4J, L78, RL78	.031-.035 (.8-.9)	23-27 @ WOT	650-700
DT75	1988-94	B8HS-10	L4J, RL4J, L78, RL78	.031-.035 (.8-.9)	16-20 @ WOT	600-700
	1995-97	BR8HS-10	N/A	.035-.039 (.9-1.0)	18 @ WOT	700-800
DT85	1988-91	B8HS	L4J, RL4J, L78, RL78	.031-.035 (.8-.9)	18 @ WOT	600-700
	1992-94	B8HS-10	L4J, RL4J, L78, RL78	.031-.035 (.8-.9)	16-20 @ WOT	600-700
	1995-99	BR8HS-10	N/A	.035-.039 (.9-1.0)	18 @ WOT	700-800
DT90	1989-91	BR8HS-10	N/A	.035-.039 (.9-1.0)	23 @ 5000	650-700
	1992-99	BR8HCS-10	N/A	.035-.039 (.9-1.0)	21-25 @ WOT	650-700
DT100	1989-91	BR8HS-10	N/A	.035-.039 (.9-1.0)	23 @ 5000	650-700
	1992-1999	BR8HCS-10	N/A	.035-.039 (.9-1.0)	21-25 @ WOT	650-700
DT 115	1988	B8HS	L4J, RL4J, L78, RL78	.031-.035 (.8-.9)	23 @ 5000	600-700
	1989-90	B8HS	L4J, RL4J, L78, RL78	.031-.035 (.8-.9)	20 @ 5000	600-700
	1991-95	B8HS	L4J, RL4J, L78, RL78	.035-.039 (.9-1.0)	22-26 @ WOT	600-700
	1996	BR8HCS-10	N/A	.035-.039 (.9-1.0)	18-22 @ WOT	650-750
	1996-97 EFI	BR8HS-10	L4J, RL4J, L78, RL78	.035-.039 (.9-1.0)	26 @ 5870-6130	600-700
DT 140	1988	B8HS	L4J, RL4J, L78, RL78	.031-.035 (.8-.9)	23 @ 5000	600-700
	1989-90	B8HS	L4J, RL4J, L78, RL78	.031-.035 (.8-.9)	20 @ 5000	600-700
	1991-95	B8HS	L4J, RL4J, L78, RL78	.035-.039 (.9-1.0)	22-26 @ WOT	600-700
	1996	BR8HCS-10	N/A	.035-.039 (.9-1.0)	18-22 @ WOT	650-750
	1996-97 EFI	BR8HS-10	L4J, RL4J, L78, RL78	.035-.039 (.9-1.0)	26 @ 5870-6130	600-700
DT 150		B8HS-10	L4J, RL4J, L78, RL78	.035-.039 (.9-1.0)	22 @ 5000	600-700
	1988	B8HS-10	L4J, RL4J, L78, RL78	.035-.039 (.9-1.0)	22 @ 5000	600-700
	1989-94	BR8HS-10	N/A	.035-.039 (.9-1.0)	20-24 @ 5000	600-700
	1995-99 EFI	BP8HS-10	N/A	.035-.039 (.9-1.0)	24 @ WOT	750-850
DT175	1988	B8HS-10	L4J, RL4J, L78, RL78	.035-.039 (.9-1.0)	22 @ 5000	600-700
	1989-92	BR8HS-10	N/A	.035-.039 (.9-1.0)	20-24 @ 5000	600-700
DT200	1988	B8HS-10	L4J, RL4J, L78, RL78	.035-.039 (.9-1.0)	22 @ 5000	600-700
	1989-92	BR8HS-10	N/A	.035-.039 (.9-1.0)	20-24 @ 5000	600-700
	1993-94 EFI	BR8HS-10	N/A	.035-.039 (.9-1.0)	24 @ WOT	650-750
	1995-99 EFI	BP8HS-10	N/A	.035-.039 (.9-1.0)	24 Non-Adjustable	750-850
DT 225		BR8HS-10	N/A	.035-.039 (.9-1.0)	13-17 @ WOT	600-700

MAINTENANCE 3-35

WINTER STORAGE CHECKLIST

Taking extra time to store the boat properly at the end of each season will increase the chances of satisfactory service at the next season. Remember, storage is the greatest enemy of an outboard motor. The unit should be run on a monthly basis. The boat steering and shifting mechanism should also be worked through complete cycles several times each month. If a small amount of time is spent in such maintenance, the reward will be satisfactory performance, increased longevity and greatly reduced maintenance expenses.

For many years there has been the widespread belief simply shutting off the fuel at the tank and then running the powerhead until it stops is the proper procedure before storing the engine for any length of time. Right? WRONG!

First, it is not possible to remove all fuel in the carburetor by operating the powerhead until it stops. Considerable fuel is trapped in the float chamber and other passages and in the line leading to the carburetor. The only guaranteed method of removing all fuel is to take the time to remove the carburetors, and drain the fuel.

Proper storage involves adequate protection of the unit from physical damage, rust, corrosion, and dirt. The following steps provide an adequate maintenance program for storing the unit at the end of a season.

1. Squirt a small quantity of engine oil into each spark plug hole and crank the engine over to distribute the oil around the engine internals. Reinstall the old spark plugs (you will install new spark plugs in the spring).

2. Drain all fuel from the carburetor float bowls. On fuel injected models, drain the fuel from the vapor separator.
3. Drain the fuel tank and the fuel lines Store the fuel tank in a cool dry area with the vent OPEN to allow air to circulate through the tank. Do not store the fuel tank on bare concrete. Place the tank to allow air to circulate around it.
4. Change the fuel filter.
5. Drain, and then fill the lower unit with new lower unit gear oil.
6. Lubricate the throttle and shift linkage and the steering pivot shaft.
7. Clean the outboard unit thoroughly. Coat the powerhead with a commercial corrosion and rust preventative spray. Install the cowling, and then apply a thin film of fresh engine oil to all painted surfaces.
8. Remove the propeller. Apply Perfect Seal® or a waterproof sealer to the propeller shaft splines, and then install the propeller back in position.
9. Be sure all drain holes in the gear housing are open and free of obstructions. Check to be sure the flush plug has been removed to allow all the water to drain. Trapped water could freeze, expand, and cause expensive castings to crack.
10. Always store the outboard unit off the boat with the lower unit below the powerhead to prevent any water from being trapped inside.
11. Be sure to consult your owners manual for any particular storage procedures applicable to your specific model.

SPRING COMMISSIONING CHECKLIST

▶ See Figures 83 thru 90

A spring tune-up is essential to getting the most out of your engine. If the engine has been properly winterized, it is usually no problem to get it in top running condition again in the springtime. If the engine has just been put in the garage and forgotten for the winter, then it is doubly important to do a complete tune up before putting the engine back into service. If you have ever been stranded out on the water because your engine has died, and you had to suffer the embarrassment of having to be towed back to the marina, now is the time to prevent that from occurring.

Satisfactory performance and maximum enjoyment can be realized if a little time is spent in preparing the outboard unit for service at the beginning of the season. Assuming the unit has been properly stored, a minimum amount of work is required to prepare the unit for use. The following steps outline an adequate and logical sequence of tasks to be performed before using the outboard the first time in a new season.

1. Lubricate the outboard according to the manufacturer's recommendations.
2. Perform a tune-up on the engine. This should include replacing the spark plugs and making a thorough check of the ignition system. The ignition

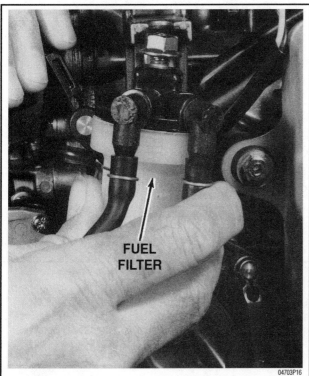

Fig. 83 Removing the fuel filter for inspection and possible replacement

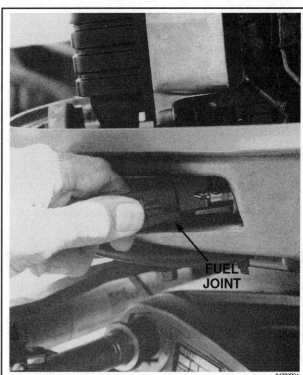

Fig. 84 Make a pre-season check of the fuel line coupling at the fuel joint to ensure a proper and clean connection

3-36 MAINTENANCE

Fig. 85 This popular and inexpensive flushing device should be included in every boat owner's maintenance kit

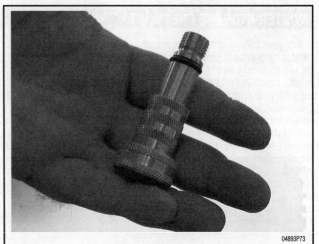

Fig. 86 Honda outboards come with a self contained flushing port on the lower unit that uses a special flush kit adapter available from your dealer

system check should include the ignition coils, stator assembly, condition of the wiring and the battery.

3. If a built-in fuel tank is installed, take time to check the tank and all fuel lines, fittings, couplings, valves, including the flexible tank fill and vent. Turn on the fuel supply valve at the tank. If the fuel was not drained at the end of the previous season, make a careful inspection for gum formation. If a six-gallon fuel tank is used, take the same action. When gasoline is allowed to stand for long periods of time, particularly in the presence of copper, gummy deposits form. This gum can clog the filters, lines, and passageways in the carburetor.

4. Replace the oil in the lower unit.
5. Replace the fuel filter.
6. Replace the engine oil and filter. Make sure to use only a quality four stroke engine oil and NEVER use two stroke oil in a four stroke engine.
7. Close all water drains. Check and replace any defective water hoses. Check to be sure the connections do not leak. Replace any spring-type hose clamps with band-type clamps, if they have lost their tension or if they have distorted the water hose.
8. The engine can be run with the lower unit in water to flush it. If this is not practical, a flush attachment may be used. This unit is attached to the water pick-up in the lower unit. Attach a garden hose, turn on the water, allow the water to flow into the engine for awhile, and then run the engine.

✷✷ CAUTION

Water must circulate through the lower unit to the powerhead anytime the powerhead is operating to prevent the engine from overheating and damage to the water pump in the lower unit. Just five seconds without water will damage the water pump impeller.

9. Check the exhaust outlet for water discharge. Check for leaks. Check operation of the thermostat.

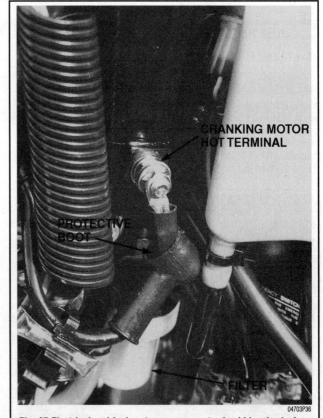

Fig. 87 Electrical and fuel system components should be checked on a regular basis

10. Check the electrolyte level in the battery and the voltage for a full charge. Clean and inspect the battery terminals and cable connections. Take time to check the polarity, if a new battery is being installed. Cover the cable connections with grease or special protective compound as a prevention to corrosion formation. Check all electrical wiring and grounding circuits.

MAINTENANCE   3-37

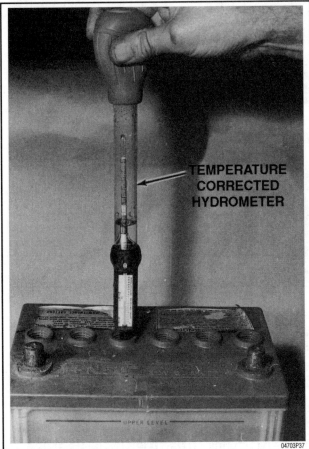

Fig. 88 Checking the condition of the battery electrolyte using a hydrometer

Fig. 89 A water separating fuel filter installed inside the boat on the transom

11. Check all electrical parts on the engine and lower portions of the hull. Rubber boots help keep electrical connections clean and reduce the possibility of arcing.

➡ Electric cranking motors and high tension wiring harnesses should be of a marine type that cannot cause an explosive mixture to ignite.

12. If a water separating filter is installed between the fuel tank and the powerhead fuel filter, replace the element at least once each season. This filter removes water and fuel system contaminants such as dirt, rust, and other solids, thus reducing potential problems.

13. As a last step in spring commissioning, perform a full engine tune-up.

✱✱ CAUTION

Before putting the boat in the water, take time to verify the drain plugs are installed. Countless number of boating excursions have had a very sad beginning because the boat was eased into the water only to have the boat begin to fill with the water.

Fig. 90 The thermostat is usually located in an accessible place for easy maintenance or replacement

General Engine Specifications

Year	Model (Horsepower)	Engine Type	Displace cu.in. (cc)	Bore and Stroke	Oil Injection System	Ignition System	Starting System	Cooling System
1999	DT 6	Inline 2-cylinder	10.1 (165)	1.97 x 1.65	Pre-Mix	Suzuki P.E.I.	Manual w/ manual choke	Impeller Pump
	DT 25	Inline 3-cylinder	33.1 (543)	2.44 x 2.36	Oil Injection	Digital I.C.	Electric/Manual w/ electric/manual choke	Impeller Pump / Thermostat Controlled
	DT 85	Inline 3-cylinder	73 (1197)	3.31 x 2.83	Oil Injection	Digital I.C.	Electric w/ electric choke	Impeller Pump / Thermostat Controlled
	DT 100	V-4 (70°)	86.6 (1419)	3.31 x 2.52	Oil Injection	MicroLink	Electric	Impeller Pump / Thermostat Controlled
	DT 115 EFI	Inline 4-cylinder	103.2 (1773)	3.31 x 3.15	Oil Injection	Digital I.C.	Electric	Impeller Pump / Thermostat Controlled
	DT 140 EFI	Inline 4-cylinder	103.2 (1773)	3.31 x 3.15	Oil Injection	Digital I.C.	Electric	Impeller Pump / Thermostat Controlled
	DT 150 EFI	V-6 (60°)	164.3 (2693)	3.31 x 3.19	Oil Injection	MicroLink	Electric	Impeller Pump / Thermostat Controlled
	DT 200 EFI	V-6 (60°)	164.3 (2693)	3.31 x 3.19	Oil Injection	MicroLink	Electric	Impeller Pump / Thermostat Controlled
	DT 225 EFI	V-6 (60°)	164.3 (2693)	3.31 x 3.19	Oil Injection	MicroLink	Electric	Impeller Pump / Thermostat Controlled
1998	DT 4	1-cylinder	5.5 (90)	1.97 x 1.81	Pre-Mix	Suzuki P.E.I.	Manual w/ manual choke	Impeller Pump
	DT 6	Inline 2-cylinder	10.1 (165)	1.97 x 1.65	Pre-Mix	Suzuki P.E.I.	Manual w/ manual choke	Impeller Pump
	DT 25	Inline 3-cylinder	33.1 (543)	2.44 x 2.36	Oil Injection	Digital I.C.	Electric/Manual w/ electric/manual choke	Impeller Pump / Thermostat Controlled
	DT 40	Inline 3-cylinder	42.5 (696)	3.11 x 2.80	Oil Injection	Suzuki P.E.I.	Electric/Manual w/ electric/manual choke	Impeller Pump / Thermostat Controlled
	DT 85	Inline 3-cylinder	73 (1197)	3.31 x 2.83	Oil Injection	Digital I.C.	Electric	Impeller Pump / Thermostat Controlled
	DT 100	V-4 (70°)	86.6 (1419)	3.31 x 2.52	Oil Injection	MicroLink	Electric	Impeller Pump / Thermostat Controlled
	DT 115 EFI	Inline 4-cylinder	103.2 (1773)	3.31 x 3.15	Oil Injection	Digital I.C.	Electric	Impeller Pump / Thermostat Controlled
	DT 140 EFI	Inline 4-cylinder	103.2 (1773)	3.31 x 3.15	Oil Injection	Digital I.C.	Electric	Impeller Pump / Thermostat Controlled
	DT 150 EFI	V-6 (60°)	164.3 (2693)	3.31 x 3.19	Oil Injection	MicroLink	Electric	Impeller Pump / Thermostat Controlled
	DT 200 EFI	V-6 (60°)	164.3 (2693)	3.31 x 3.19	Oil Injection	MicroLink	Electric	Impeller Pump / Thermostat Controlled
	DT 225 EFI	V-6 (60°)	164.3 (2693)	3.31 x 3.19	Oil Injection	MicroLink	Electric	Impeller Pump / Thermostat Controlled
1997	DT 2.2	1-cylinder	3.4 (55)	1.61 x 1.50	100:1 Pre-Mix	Suzuki P.E.I.	Manual w/ manual choke	Impeller Pump
	DT 4	1-cylinder	5.5 (90)	1.97 x 1.81	100:1 Pre-Mix	Suzuki P.E.I.	Manual w/ manual choke	Impeller Pump
	DT 6	Inline 2-cylinder	10.1 (165)	1.97 x 1.65	100:1 Pre-Mix	Suzuki P.E.I.	Manual w/ manual choke	Impeller Pump
	DT 8	Inline 2-cylinder	12.8 (211)	2.13 x 1.81	Oil Injection	Suzuki P.E.I.	Manual w/ manual choke	Impeller Pump
	DT 9.9	Inline 2-cylinder	12.8 (211)	2.13 x 1.81	Oil Injection	Digital I.C.	Manual w/ manual choke	Impeller Pump
	DT 15	Inline 2-cylinder	17.3 (284)	2.32 x 2.05	Oil Injection	Digital I.C.	Manual/Electric w/ electric/manual choke	Impeller Pump
	DT 25	Inline 3-cylinder	33.1 (543)	2.44 x 2.36	Oil Injection	Digital I.C.	Manual/Electric w/ Suzuki Start System	Impeller Pump / Thermostat Controlled
	DT 30	Inline 3-cylinder	33.1 (543)	2.44 x 2.36	Oil Injection	Digital I.C.	Manual/Electric w/ Suzuki Start System	Impeller Pump / Thermostat Controlled
	DT 40	Inline 2-cylinder	42.5 (696)	3.11 x 2.80	Oil Injection	Suzuki P.E.I.	Electric/Manual w/ manual/electric choke	Impeller Pump / Thermostat Controlled
	DT 55	Inline 3-cylinder	54.4 (891)	2.87 x 2.80	Oil Injection	Suzuki P.E.I.	Electric w/ electric choke	Impeller Pump / Thermostat Controlled
	DT 65	Inline 3-cylinder	54.4 (891)	2.87 x 2.80	Oil Injection	Suzuki P.E.I.	Electric w/ electric choke	Impeller Pump / Thermostat Controlled
	DT 75	Inline 3-cylinder	73 (1197)	3.31 x 2.83	Oil Injection	Digital I.C.	Electric w/ Suzuki EFI	Impeller Pump / Thermostat Controlled
	DT 85	Inline 3-cylinder	73 (1197)	3.31 x 2.83	Oil Injection	Digital I.C.	Electric w/ Suzuki EFI	Impeller Pump / Thermostat Controlled
	DT 90	V-4 (70°)	86.6 (1419)	3.31 x 2.52	Oil Injection	Microlink	Electric w/ Suzuki EFI	Impeller Pump / Thermostat Controlled
	DT 100	V-4 (70°)	86.6 (1419)	3.31 x 2.52	Oil Injection	Microlink	Electric w/ Suzuki EFI	Impeller Pump / Thermostat Controlled
	DT 115	Inline 4-cylinder	108.2 (1773)	3.31 x 3.15	Oil Injection	Digital I.C.	Manual/Electric w/ electric/manual choke	Impeller Pump / Thermostat Controlled
	DT 115 EFI	Inline 4-cylinder	108.2 (1773)	3.31 x 3.15	Oil Injection	Microlink	Manual/Electric w/ electric/manual choke	Impeller Pump / Thermostat Controlled
	DT 140 EFI	Inline 4-cylinder	108.2 (1773)	3.31 x 3.15	Oil Injection	Microlink	Manual/Electric w/ electric/manual choke	Impeller Pump / Thermostat Controlled
	DT 150 EFI	V-6 (60°)	164.3 (2693)	3.31 x 3.19	Oil Injection	Microlink	Electric/Manual w/ manual/electric choke	Impeller Pump / Thermostat Controlled
	DT 200 EFI	V-6 (60°)	164.3 (2693)	3.31 x 3.19	Oil Injection	Microlink	Electric w/ electric choke	Impeller Pump / Thermostat Controlled
	DT 225 EFI	V-6 (60°)	164.3 (2693)	3.31 x 3.19	Oil Injection	Microlink	Electric w/ electric choke	Impeller Pump / Thermostat Controlled
1996	DT 2	1-cylinder	3.05 (50)	1.61 x 1.49	100:1 Pre-Mix	Flywheel Magneto	Manual w/ manual choke	Impeller Pump
	DT 4	1-cylinder	5.5 (90)	1.97 x 1.81	100:1 Pre-Mix	Suzuki P.E.I.	Manual w/ manual choke	Impeller Pump
	DT 6	Inline 2-cylinder	10.1 (165)	1.97 x 1.65	100:1 Pre-Mix	Suzuki P.E.I.	Manual w/ manual choke	Impeller Pump
	DT 8	Inline 2-cylinder	12.8 (211)	2.13 x 1.81	Oil Injection	Suzuki P.E.I.	Manual w/ manual choke	Impeller Pump
	DT 9.9	Inline 2-cylinder	12.8 (211)	2.13 x 1.81	Oil Injection	Suzuki P.E.I.	Manual w/ manual choke	Impeller Pump
	DT 15	Inline 2-cylinder	17.3 (284)	2.32 x 2.05	Oil Injection	Digital I.C.	Manual/Electric w/ electric/manual choke	Impeller Pump
	DT 25	Inline 3-cylinder	33.1 (543)	2.44 x 2.36	Oil Injection	Digital I.C.	Manual/Electric w/ electric/manual choke	Impeller Pump / Thermostat Controlled
	DT 30	Inline 3-cylinder	33.1 (543)	2.44 x 2.36	Oil Injection	Digital I.C.	Manual/Electric w/ electric/manual choke	Impeller Pump / Thermostat Controlled
	DT 40	Inline 2-cylinder	42.5 (696)	3.11 x 2.80	Oil Injection	Suzuki P.E.I.	Electric/Manual w/ manual/electric choke	Impeller Pump / Thermostat Controlled
	DT 55	Inline 3-cylinder	54.4 (891)	2.87 x 2.80	Oil Injection	Suzuki P.E.I.	Electric w/ electric choke	Impeller Pump / Thermostat Controlled
	DT 65	Inline 3-cylinder	54.4 (891)	2.87 x 2.80	Oil Injection	Digital I.C.	Electric w/ electric choke	Impeller Pump / Thermostat Controlled
	DT 75	Inline 3-cylinder	73 (1197)	3.31 x 2.83	Oil Injection	Digital I.C.	Electric w/ electric choke	Impeller Pump / Thermostat Controlled
	DT 85	Inline 3-cylinder	73 (1197)	3.31 x 2.83	Oil Injection	Digital I.C.	Electric w/ electric choke	Impeller Pump / Thermostat Controlled

MAINTENANCE 3-39

General Engine Specifications

Year	Model (Horsepower)	Engine Type	Displace cu.in. (cc)	Bore and Stroke	Oil Injection System	Ignition System	Starting System	Cooling System
1996	DT 90	V-4 (70°)	86.6 (1419)	3.31 x 2.52	Oil Injection	Microlink	Electric w/ Suzuki Start System	Impeller Pump / Thermostat Controlled
	DT 100	V-4 (70°)	86.6 (1419)	3.31 x 2.52	Oil Injection	Microlink	Electric w/ Suzuki Start System	Impeller Pump / Thermostat Controlled
	DT 115	Inline 4-cylinder	108.2 (1773)	3.31 x 3.15	Oil Injection	Digital I.C.	Electric w/ electric choke	Impeller Pump / Thermostat Controlled
	DT 115 EFI	Inline 4-cylinder	108.2 (1773)	3.31 x 3.15	Oil Injection	Microlink	Electric w/ Suzuki Start System	Impeller Pump / Thermostat Controlled
	DT 140	Inline 4-cylinder	108.2 (1773)	3.31 x 3.15	Oil Injection	Microlink	Electric w/ Suzuki Start System	Impeller Pump / Thermostat Controlled
	DT 140 EFI	Inline 4-cylinder	108.2 (1773)	3.31 x 3.15	Oil Injection	Microlink	Electric w/ Suzuki EFI	Impeller Pump / Thermostat Controlled
	DT 150 EFI	V-6 (60°)	164.3 (2693)	3.31 x 3.19	Oil Injection	Microlink	Electric w/ Suzuki EFI	Impeller Pump / Thermostat Controlled
	DT 200 EFI	V-6 (60°)	164.3 (2693)	3.31 x 3.19	Oil Injection	Microlink	Electric w/ Suzuki EFI	Impeller Pump / Thermostat Controlled
	DT 225 EFI	V-6 (60°)	164.3 (2693)	3.31 x 3.19	Oil Injection	Microlink	Electric w/ Suzuki EFI	Impeller Pump / Thermostat Controlled
1995	DT 2	1-cylinder	3.05 (50)	1.61 x 1.49	100:1 Pre-Mix	Flywheel Magneto	Manual w/ manual choke	Impeller Pump
	DT 4	1-cylinder	5.5 (90)	1.97 x 1.81	100:1 Pre-Mix	Suzuki P.E.I.	Manual w/ manual choke	Impeller Pump
	DT 6	Inline 2-cylinder	10.1 (165)	1.97 x 1.65	100:1 Pre-Mix	Suzuki P.E.I.	Manual w/ electric choke	Impeller Pump
	DT 8	Inline 2-cylinder	12.8 (211)	2.13 x 1.81	Oil Injection	Suzuki P.E.I.	Manual w/ manual choke	Impeller Pump
	DT 9.9	Inline 2-cylinder	12.8 (211)	2.13 x 1.81	Oil Injection	Suzuki P.E.I.	Manual w/ manual choke	Impeller Pump
	DT 15	Inline 2-cylinder	17.3 (284)	2.32 x 2.05	Oil Injection	Digital I.C.	Manual/Electric w/ electric/manual choke	Impeller Pump
	DT 25	Inline 3-cylinder	33.1 (543)	2.44 x 2.36	Oil Injection	Digital I.C.	Manual/Electric w/ electric/manual choke	Impeller Pump
	DT 30	Inline 3-cylinder	33.1 (543)	2.44 x 2.36	Oil Injection	Digital I.C.	Manual/Electric w/ manual/electric choke	Impeller Pump
	DT 40	Inline 2-cylinder	42.5 (696)	3.11 x 2.80	Oil Injection	Suzuki P.E.I.	Electric/Manual w/ manual/electric choke	Impeller Pump
	DT 55	Inline 3-cylinder	54.4 (891)	2.87 x 2.80	Oil Injection	Microlink	Electric w/ electric choke	Impeller Pump / Thermostat Controlled
	DT 65	Inline 3-cylinder	54.4 (891)	2.87 x 2.80	Oil Injection	Microlink	Electric w/ electric choke	Impeller Pump / Thermostat Controlled
	DT 75	Inline 3-cylinder	73 (1197)	3.31 x 2.83	Oil Injection	Digital I.C.	Electric w/ electric choke	Impeller Pump / Thermostat Controlled
	DT 85	Inline 3-cylinder	73 (1197)	3.31 x 2.83	Oil Injection	Digital I.C.	Electric w/ electric choke	Impeller Pump / Thermostat Controlled
	DT 90	V-4 (70°)	86.6 (1419)	3.31 x 2.52	Oil Injection	Microlink	Electric w/ Suzuki Start System	Impeller Pump / Thermostat Controlled
	DT 100	V-4 (70°)	86.6 (1419)	3.31 x 2.52	Oil Injection	Microlink	Electric w/ Suzuki Start System	Impeller Pump / Thermostat Controlled
	DT 115	Inline 4-cylinder	108.2 (1773)	3.31 x 3.15	Oil Injection	Digital I.C.	Electric w/ electric choke	Impeller Pump / Thermostat Controlled
	DT 140	Inline 4-cylinder	108.2 (1773)	3.31 x 3.15	Oil Injection	Microlink	Electric w/ electric choke	Impeller Pump / Thermostat Controlled
	DT 150 EFI	V-6 (60°)	164.3 (2693)	3.31 x 3.19	Oil Injection	Microlink	Electric w/ Suzuki EFI	Impeller Pump / Thermostat Controlled
	DT 200 EFI	V-6 (60°)	164.3 (2693)	3.31 x 3.19	Oil Injection	Microlink	Electric w/ Suzuki EFI	Impeller Pump / Thermostat Controlled
	DT 225 EFI	V-6 (60°)	164.3 (2693)	3.31 x 3.19	Oil Injection	Microlink	Electric w/ Suzuki EFI	Impeller Pump / Thermostat Controlled
1994	DT 2	1-cylinder	3.05 (50)	1.61 x 1.49	100:1 Pre-Mix	Flywheel Magneto	Manual w/ manual choke	Impeller Pump
	DT 4	1-cylinder	5.5 (90)	1.97 x 1.81	100:1 Pre-Mix	Suzuki P.E.I.	Manual w/ manual choke	Impeller Pump
	DT 6	Inline 2-cylinder	10.1 (165)	1.97 x 1.65	100:1 Pre-Mix	Suzuki P.E.I.	Manual w/ manual choke	Impeller Pump
	DT 8	Inline 2-cylinder	12.8 (211)	2.13 x 1.81	Oil Injection	Suzuki P.E.I.	Manual w/ manual choke	Impeller Pump
	DT 9.9	Inline 2-cylinder	12.8 (211)	2.13 x 1.81	Oil Injection	Suzuki P.E.I.	Manual w/ manual choke	Impeller Pump
	DT 15	Inline 2-cylinder	17.3 (284)	2.32 x 2.05	Oil Injection	Digital I.C.	Manual w/ manual choke	Impeller Pump
	DT 25	Inline 3-cylinder	33.1 (543)	2.44 x 2.36	Oil Injection	Digital I.C.	Manual w/ manual choke	Impeller Pump
	DT 30	Inline 3-cylinder	33.1 (543)	2.44 x 2.36	Oil Injection	Digital I.C.	Manual w/ manual choke	Impeller Pump
	DT 40	Inline 2-cylinder	42.5 (696)	3.11 x 2.80	Oil Injection	Suzuki P.E.I.	Electric/Manual w/ manual/electric choke	Impeller Pump
	DT 55	Inline 3-cylinder	54.4 (891)	2.87 x 2.80	Oil Injection	Microlink	Electric w/ electric choke	Impeller Pump / Thermostat Controlled
	DT 65	Inline 3-cylinder	54.4 (891)	2.87 x 2.80	Oil Injection	Microlink	Electric w/ electric choke	Impeller Pump / Thermostat Controlled
	DT 75	Inline 3-cylinder	73 (1197)	3.31 x 2.83	Oil Injection	Digital I.C.	Electric w/ electric choke	Impeller Pump / Thermostat Controlled
	DT 85	Inline 3-cylinder	73 (1197)	3.31 x 2.83	Oil Injection	Digital I.C.	Electric w/ electric choke	Impeller Pump / Thermostat Controlled
	DT 90	V-4 (70°)	86.6 (1419)	3.31 x 2.52	Oil Injection	Microlink	Electric w/ Suzuki Start System	Impeller Pump / Thermostat Controlled
	DT 100	V-4 (70°)	86.6 (1419)	3.31 x 2.52	Oil Injection	Microlink	Electric w/ Suzuki Start System	Impeller Pump / Thermostat Controlled
	DT 115	Inline 4-cylinder	108.2 (1773)	3.31 x 3.15	Oil Injection	Digital I.C.	Electric w/ electric choke	Impeller Pump / Thermostat Controlled
	DT 140	Inline 4-cylinder	108.2 (1773)	3.31 x 3.15	Oil Injection	Microlink	Electric w/ electric choke	Impeller Pump / Thermostat Controlled
	DT 150	V-6 (60°)	164.3 (2693)	3.31 x 3.19	Oil Injection	Microlink	Electric w/ Suzuki EFI	Impeller Pump / Thermostat Controlled
	DT 200 EFI	V-6 (60°)	164.3 (2693)	3.31 x 3.19	Oil Injection	Microlink	Electric w/ Suzuki EFI	Impeller Pump / Thermostat Controlled
	DT 225 EFI	V-6 (60°)	164.3 (2693)	3.31 x 3.19	Oil Injection	Microlink	Electric w/ Suzuki EFI	Impeller Pump / Thermostat Controlled
1993	DT 2	1-cylinder	3.05 (50)	1.61 x 1.49	100:1 Pre-Mix	Flywheel Magneto	Manual w/ manual choke	Impeller Pump
	DT 4	1-cylinder	5.5 (90)	1.97 x 1.81	100:1 Pre-Mix	Suzuki P.E.I.	Manual w/ manual choke	Impeller Pump
	DT 6	Inline 2-cylinder	10.1 (165)	1.97 x 1.65	100:1 Pre-Mix	Suzuki P.E.I.	Manual w/ manual choke	Impeller Pump
	DT 8	Inline 2-cylinder	12.8 (211)	2.13 x 1.81	Oil Injection	Suzuki P.E.I.	Manual w/ manual choke	Impeller Pump
	DT 9.9	Inline 2-cylinder	12.8 (211)	2.13 x 1.81	Oil Injection	Suzuki P.E.I.	Manual w/ manual choke	Impeller Pump

3-40 MAINTENANCE

General Engine Specifications

Year	Model (Horsepower)	Engine Type	Displace cu.in. (cc)	Bore and Stroke	Oil Injection System	Ignition System	Starting System	Cooling System
1993	DT 15	Inline 2-cylinder	17.3 (284)	2.32 x 2.05	Oil Injection	Digital I.C.	Manual/Electric w/ electric/manual choke	Impeller Pump
	DT 25	Inline 3-cylinder	33.1 (543)	2.44 x 2.36	Oil Injection	Digital I.C.	Manual/Electric w/ electric/manual choke	Impeller Pump
	DT 30	Inline 3-cylinder	33.1 (543)	2.44 x 2.36	Oil Injection	Digital I.C.	Manual/Electric w/ electric/manual choke	Impeller Pump
	DT 40	Inline 2-cylinder	42.5 (696)	3.11 x 2.80	Oil Injection	Suzuki P.E.I.	Electric/Manual w/ manual/electric choke	Impeller Pump
	DT 55	Inline 3-cylinder	54.4 (891)	2.87 x 2.80	Oil Injection	Suzuki P.E.I.	Electric w/ electric choke	Impeller Pump / Thermostat Controlled
	DT 65	Inline 3-cylinder	54.4 (891)	2.87 x 2.80	Oil Injection	Suzuki P.E.I.	Electric w/ electric choke	Impeller Pump / Thermostat Controlled
	DT 75	Inline 3-cylinder	73 (1197)	3.31 x 2.83	Oil Injection	Digital I.C.	Electric w/ electric choke	Impeller Pump / Thermostat Controlled
	DT 85	Inline 3-cylinder	73 (1197)	3.31 x 2.83	Oil Injection	Digital I.C.	Electric w/ electric choke	Impeller Pump / Thermostat Controlled
	DT 90	V-4 (70°)	86.6 (1419)	3.31 x 2.52	Oil Injection	Microlink	Electric w/ Suzuki Start System	Impeller Pump / Thermostat Controlled
	DT 100	V-4 (70°)	86.6 (1419)	3.31 x 2.52	Oil Injection	Microlink	Electric w/ Suzuki Start System	Impeller Pump / Thermostat Controlled
	DT 115	Inline 4-cylinder	108.2 (1773)	3.31 x 3.15	Oil Injection	Digital I.C.	Electric w/ Suzuki Start System	Impeller Pump / Thermostat Controlled
	DT 140	Inline 4-cylinder	108.2 (1773)	3.31 x 3.15	Oil Injection	Microlink	Electric w/ Suzuki Start System	Impeller Pump / Thermostat Controlled
	DT 150	V-6 (60°)	164.3 (2693)	3.31 X 3.19	Oil Injection	Microlink	Electric w/ Suzuki EFI	Impeller Pump / Thermostat Controlled
	DT 200	V-6 (60°)	164.3 (2693)	3.31 x 3.19	Oil Injection	Microlink	Electric w/ Suzuki EFI	Impeller Pump / Thermostat Controlled
	DT 225 EFI	V-6 (60°)	164.3 (2693)	3.31 X 3.19	Oil Injection	Microlink	Electric w/ Suzuki EFI	Impeller Pump / Thermostat Controlled
1992	DT 2	1-cylinder	3.05 (50)	1.61 x 1.49	100:1 Pre-Mix	Flywheel Magneto	Manual w/ manual choke	Impeller Pump
	DT 4	1-cylinder	5.5 (90)	1.97 x 1.81	100:1 Pre-Mix	Suzuki P.E.I.	Manual w/ manual choke	Impeller Pump
	DT 6	Inline 2-cylinder	10.1 (165)	1.97 x 1.65	100:1 Pre-Mix	Suzuki P.E.I.	Manual w/ manual choke	Impeller Pump
	DT 8	Inline 2-cylinder	12.8 (211)	2.13 x 1.81	Oil Injection	Suzuki P.E.I.	Manual w/ manual choke	Impeller Pump
	DT 9.9	Inline 2-cylinder	12.8 (211)	2.13 x 1.81	Oil Injection	Suzuki P.E.I.	Manual w/ manual choke	Impeller Pump
	DT 15	Inline 2-cylinder	17.3 (284)	2.32 x 2.05	Oil Injection	Digital I.C.	Manual/Electric w/ electric/manual choke	Impeller Pump
	DT 25	Inline 3-cylinder	33.1 (543)	2.44 x 2.36	Oil Injection	Digital I.C.	Manual/Electric w/ electric/manual choke	Impeller Pump / Thermostat Controlled
	DT 30	Inline 3-cylinder	33.1 (543)	2.44 x 2.36	Oil Injection	Digital I.C.	Manual/Electric w/ electric/manual choke	Impeller Pump / Thermostat Controlled
	DT 40	Inline 2-cylinder	42.5 (696)	3.11 x 2.80	Oil Injection	Suzuki P.E.I.	Electric/Manual w/ manual/electric choke	Impeller Pump / Thermostat Controlled
	DT 55	Inline 3-cylinder	54.4 (891)	2.87 x 2.80	Oil Injection	Suzuki P.E.I.	Electric w/ electric choke	Impeller Pump / Thermostat Controlled
	DT 65	Inline 3-cylinder	54.4 (891)	2.87 x 2.80	Oil Injection	Suzuki P.E.I.	Electric w/ electric choke	Impeller Pump / Thermostat Controlled
	DT 75	Inline 3-cylinder	73 (1197)	3.31 x 2.83	Oil Injection	Digital I.C.	Electric w/ electric choke	Impeller Pump / Thermostat Controlled
	DT 85	Inline 3-cylinder	73 (1197)	3.31 x 2.83	Oil Injection	Digital I.C.	Electric w/ electric choke	Impeller Pump / Thermostat Controlled
	DT 90	V-4 (70°)	86.6 (1419)	3.31 x 2.52	Oil Injection	Microlink	Electric w/ Suzuki Start System	Impeller Pump / Thermostat Controlled
	DT 100	V-4 (70°)	86.6 (1419)	3.31 x 2.52	Oil Injection	Microlink	Electric w/ Suzuki Start System	Impeller Pump / Thermostat Controlled
	DT 115	Inline 4-cylinder	108.2 (1773)	3.31 x 3.15	Oil Injection	Digital I.C.	Electric w/ Suzuki Start System	Impeller Pump / Thermostat Controlled
	DT 140	Inline 4-cylinder	108.2 (1773)	3.31 x 3.15	Oil Injection	Microlink	Electric w/ Suzuki Start System	Impeller Pump / Thermostat Controlled
	DT 150	V-6 (60°)	164.3 (2693)	3.31 x 3.19	Oil Injection	Microlink	Electric w/ Suzuki EFI	Impeller Pump / Thermostat Controlled
	DT 175	V-6 (60°)	164.3 (2693)	3.31 x 3.19	Oil Injection	Microlink	Electric w/ Suzuki EFI	Impeller Pump / Thermostat Controlled
	DT 225 EFI	V-6 (60°)	164.3 (2693)	3.31 x 3.19	Oil Injection	Microlink	Electric w/ Suzuki EFI	Impeller Pump / Thermostat Controlled
1991	DT 2	1-cylinder	3.05 (50)	1.61 x 1.49	100:1 Pre-Mix	Flywheel Magneto	Manual w/ manual choke	Impeller Pump
	DT 4	1-cylinder	5.5 (90)	1.97 x 1.81	100:1 Pre-Mix	Suzuki P.E.I.	Manual w/ manual choke	Impeller Pump
	DT 6	Inline 2-cylinder	10.1 (165)	1.97 x 1.65	100:1 Pre-Mix	Suzuki P.E.I.	Manual w/ manual choke	Impeller Pump
	DT 8	Inline 2-cylinder	12.8 (211)	2.13 x 1.81	Oil Injection	Suzuki P.E.I.	Manual w/ manual choke	Impeller Pump
	DT 9.9	Inline 2-cylinder	12.8 (211)	2.13 x 1.81	Oil Injection	Suzuki P.E.I.	Manual w/ manual choke	Impeller Pump
	DT 15	Inline 2-cylinder	17.3 (284)	2.32 x 2.05	Oil Injection	Digital I.C.	Manual/Electric w/ electric/manual choke	Impeller Pump
	DT 25	Inline 3-cylinder	33.1 (543)	2.44 x 2.36	Oil Injection	Digital I.C.	Manual/Electric w/ electric/manual choke	Impeller Pump
	DT 30	Inline 3-cylinder	33.1 (543)	2.44 x 2.36	Oil Injection	Digital I.C.	Manual/Electric w/ electric/manual choke	Impeller Pump
	DT 40	Inline 2-cylinder	42.5 (696)	3.11 x 2.80	Oil Injection	Suzuki P.E.I.	Electric/Manual w/ manual/electric choke	Impeller Pump
	DT 55	Inline 3-cylinder	54.4 (891)	2.87 x 2.80	Oil Injection	Suzuki P.E.I.	Electric w/ electric choke	Impeller Pump / Thermostat Controlled
	DT 65	Inline 3-cylinder	54.4 (891)	2.87 x 2.80	Oil Injection	Suzuki P.E.I.	Electric w/ electric choke	Impeller Pump / Thermostat Controlled
	DT 75	Inline 3-cylinder	73 (1197)	3.31 x 2.83	Oil Injection	Digital I.C.	Electric w/ electric choke	Impeller Pump / Thermostat Controlled
	DT 85	Inline 3-cylinder	73 (1197)	3.31 x 2.83	Oil Injection	Digital I.C.	Electric w/ electric choke	Impeller Pump / Thermostat Controlled
	DT 115	Inline 4-cylinder	108.2 (1773)	3.31 x 3.15	Oil Injection	Digital I.C.	Electric w/ Suzuki Start System	Impeller Pump / Thermostat Controlled
	DT 140	Inline 4-cylinder	108.2 (1773)	3.31 x 3.15	Oil Injection	Microlink	Electric w/ Suzuki Start System	Impeller Pump / Thermostat Controlled
	DT 150	V-6 (60°)	164.3 (2693)	3.31 x 3.19	Oil Injection	Microlink	Electric w/ Suzuki EFI	Impeller Pump / Thermostat Controlled
	DT 175	V-6 (60°)	164.3 (2693)	3.31 x 3.19	Oil Injection	Microlink	Electric w/ Suzuki EFI	Impeller Pump / Thermostat Controlled
	DT 200	V-6 (60°)	164.3 (2693)	3.31 x 3.19	Oil Injection	Microlink	Electric w/ Suzuki EFI	Impeller Pump / Thermostat Controlled
	DT 225 EFI	V-6 (60°)	164.3 (2693)	3.31 x 3.19	Oil Injection	Microlink	Electric w/ Suzuki EFI	Impeller Pump / Thermostat Controlled

MAINTENANCE 3-41

General Engine Specifications

Year	Model (Horsepower)	Engine Type	Displace cu.in. (cc)	Bore and Stroke	Oil Injection System	Ignition System	Starting System	Cooling System
1990	DT 2	1-cylinder	3.05 (50)	1.61 x 1.49	100:1 Pre-Mix	Flywheel Magneto	Manual w/ manual choke	Impeller Pump
	DT 4	1-cylinder	5.5 (90)	1.97 x 1.81	100:1 Pre-Mix	Suzuki P.E.I.	Manual w/ manual choke	Impeller Pump
	DT 6	Inline 2-cylinder	10.1 (165)	1.97 x 1.65	100:1 Pre-Mix	Suzuki P.E.I.	Manual w/ manual choke	Impeller Pump
	DT 8	Inline 2-cylinder	12.8 (211)	2.13 x 1.81	Oil Injection	Suzuki P.E.I.	Manual w/ manual choke	Impeller Pump
	DT 9.9	Inline 2-cylinder	12.8 (211)	2.13 x 1.81	Oil Injection	Suzuki P.E.I.	Manual w/ manual choke	Impeller Pump
	DT 15	Inline 2-cylinder	17.3 (284)	2.32 x 2.05	Oil Injection	Digital I.C.	Manual/Electric w/ electric/manual choke	Impeller Pump
	DT 25	Inline 3-cylinder	33.1 (543)	2.44 x 2.36	Oil Injection	Digital I.C.	Manual/Electric w/ electric/manual choke	Impeller Pump
	DT 30	Inline 3-cylinder	33.1 (543)	2.44 x 2.36	Oil Injection	Digital I.C.	Manual/Electric w/ manual/electric choke	Impeller Pump
	DT 40	Inline 2-cylinder	42.5 (696)	3.11 x 2.80	Oil Injection	Suzuki P.E.I.	Electric/Manual w/ manual/electric choke	Impeller Pump
	DT 55	Inline 3-cylinder	54.4 (891)	2.87 x 2.80	Oil Injection	Suzuki P.E.I.	Electric w/ electric choke	Impeller Pump / Thermostat Controlled
	DT 65	Inline 3-cylinder	54.4 (891)	2.87 x 2.80	Oil Injection	Suzuki P.E.I.	Electric w/ electric choke	Impeller Pump / Thermostat Controlled
	DT 75	Inline 3-cylinder	73 (1197)	3.31 x 2.83	Oil Injection	Digital I.C.	Electric w/ electric choke	Impeller Pump / Thermostat Controlled
	DT 85	Inline 3-cylinder	73 (1197)	3.31 x 2.83	Oil Injection	Digital I.C.	Electric w/ electric choke	Impeller Pump / Thermostat Controlled
	DT 90	V-4 (70°)	86.6 (1419)	3.31 x 2.52	Oil Injection	Microlink	Electric w/ Suzuki Start System	Impeller Pump / Thermostat Controlled
	DT 100	V-4 (70°)	86.6 (1419)	3.31 x 2.52	Oil Injection	Digital I.C.	Electric w/ Suzuki Start System	Impeller Pump / Thermostat Controlled
	DT 115	Inline 4-cylinder	108.2 (1773)	3.31 x 3.15	Oil Injection	Microlink	Electric w/ electric choke	Impeller Pump / Thermostat Controlled
	DT 140	Inline 4-cylinder	108.2 (1773)	3.31 x 3.15	Oil Injection	Microlink	Electric w/ Suzuki Start System	Impeller Pump / Thermostat Controlled
	DT 150	V-6 (60°)	164.3 (2693)	3.31 x 3.19	Oil Injection	Microlink	Electric w/ Suzuki EFI	Impeller Pump / Thermostat Controlled
	DT 175	V-6 (60°)	164.3 (2693)	3.31 x 3.19	Oil Injection	Microlink	Electric w/ Suzuki EFI	Impeller Pump / Thermostat Controlled
	DT 200	V-6 (60°)	164.3 (2693)	3.31 x 3.19	Oil Injection	Microlink	Electric w/ Suzuki EFI	Impeller Pump / Thermostat Controlled
	DT 225	V-6 (60°)	164.3 (2693)	3.31 x 3.19	Oil Injection	Microlink	Electric w/ Suzuki EFI	Impeller Pump / Thermostat Controlled
1989	DT 2	1-cylinder	3.05 (50)	1.61 x 1.49	100:1 Pre-Mix	Flywheel Magneto	Manual w/ manual choke	Impeller Pump
	DT 4	1-cylinder	5.5 (90)	1.97 x 1.81	100:1 Pre-Mix	Suzuki P.E.I.	Manual w/ manual choke	Impeller Pump
	DT 6	Inline 2-cylinder	10.1 (165)	1.97 x 1.65	100:1 Pre-Mix	Suzuki P.E.I.	Manual w/ manual choke	Impeller Pump
	DT 8	Inline 2-cylinder	12.8 (211)	2.13 x 1.81	Oil Injection	Suzuki P.E.I.	Manual w/ manual choke	Impeller Pump
	DT 9.9	Inline 2-cylinder	12.8 (211)	2.13 x 1.81	Oil Injection	Suzuki P.E.I.	Manual w/ manual choke	Impeller Pump
	DT 15	Inline 2-cylinder	17.3 (284)	2.32 x 2.05	Oil Injection	Digital I.C.	Manual/Electric w/ electric/manual choke	Impeller Pump
	DT 25	Inline 3-cylinder	33.1 (543)	2.44 x 2.36	Oil Injection	Digital I.C.	Manual/Electric w/ electric/manual choke	Impeller Pump
	DT 30	Inline 3-cylinder	33.1 (543)	2.44 x 2.36	Oil Injection	Digital I.C.	Manual/Electric w/ electric/manual choke	Impeller Pump
	DT 35	Inline 2-cylinder	42.5 (696)	2.44 x 2.36	Oil Injection	Suzuki P.E.I.	Electric w/ electric choke	Impeller Pump
	DT 40	Inline 2-cylinder	42.5 (696)	3.11 x 2.80	Oil Injection	Suzuki P.E.I.	Electric/Manual w/ manual/electric choke	Impeller Pump
	DT 55	Inline 3-cylinder	54.4 (891)	2.87 x 2.80	Oil Injection	Suzuki P.E.I.	Electric w/ electric choke	Impeller Pump / Thermostat Controlled
	DT 65	Inline 3-cylinder	54.4 (891)	2.87 x 2.80	Oil Injection	Suzuki P.E.I.	Electric w/ electric choke	Impeller Pump / Thermostat Controlled
	DT 75	Inline 3-cylinder	73 (1197)	3.31 x 2.83	Oil Injection	Digital I.C.	Electric w/ electric choke	Impeller Pump / Thermostat Controlled
	DT 85	Inline 3-cylinder	73 (1197)	3.31 x 2.83	Oil Injection	Digital I.C.	Electric w/ electric choke	Impeller Pump / Thermostat Controlled
	DT 90	Inline 3-cylinder	33.1 (543)	3.31 x 2.52	Oil Injection	Microlink	Electric w/ Suzuki Start System	Impeller Pump / Thermostat Controlled
	DT 100	V-4 (70°)	86.6 (1419)	3.31 x 2.52	Oil Injection	Microlink	Electric w/ Suzuki Start System	Impeller Pump / Thermostat Controlled
	DT 115	Inline 4-cylinder	108.2 (1773)	3.31 x 3.15	Oil Injection	Microlink	Electric w/ electric choke	Impeller Pump / Thermostat Controlled
	DT 140	Inline 4-cylinder	108.2 (1773)	3.31 x 3.15	Oil Injection	Microlink	Electric w/ electric choke	Impeller Pump / Thermostat Controlled
	DT 150	V-6 (60°)	164.3 (2693)	3.31 x 3.19	Oil Injection	Digital I.C.	Electric w/ electric choke	Impeller Pump / Thermostat Controlled
	DT 175	V-6 (60°)	164.3 (2693)	3.31 x 3.19	Oil Injection	Digital I.C.	Electric w/ Suzuki EFI	Impeller Pump / Thermostat Controlled
	DT 200	V-6 (60°)	164.3 (2693)	3.31 x 3.19	Oil Injection	Digital I.C.	Electric w/ Suzuki EFI	Impeller Pump / Thermostat Controlled
1988	DT 2	1-cylinder	3.1 (50)	1.61 x 1.5	100:1 Pre-Mix	Suzuki P.E.I.	Manual w/ manual choke	Impeller Pump
	DT 4	1-cylinder	5.5 (90)	1.97 x 1.81	100:1 Pre-Mix	Suzuki P.E.I.	Manual w/ manual choke	Impeller Pump
	DT 6	Inline 2-cylinder	10.1 (165)	1.97 x 1.65	100:1 Pre-Mix	Suzuki P.E.I.	Manual w/ manual choke	Impeller Pump
	DT 8	Inline 2-cylinder	12.8 (211)	2.13 x 1.81	Oil Injection	Suzuki P.E.I.	Manual w/ manual choke	Impeller Pump
	DT 9.9	Inline 2-cylinder	12.8 (211)	2.13 x 1.81	Oil Injection	Suzuki P.E.I.	Manual w/ manual choke	Impeller Pump
	DT 15	Inline 2-cylinder	17.3 (284)	2.32 x 2.05	Oil Injection	Digital I.C.	Manual/Electric w/ electric/manual choke	Impeller Pump
	DT 20	Inline 2-cylinder	27.1 (444)	2.64 x 2.48	Oil Injection	Suzuki P.E.I.	Manual/Electric w/ manual choke	Impeller Pump / Thermostat Controlled
	DT 25	Inline 3-cylinder	33.1 (543)	2.44 x 2.36	Oil Injection	Suzuki P.E.I.	Manual/Electric w/ electric/manual choke	Impeller Pump / Thermostat Controlled
	DT 30	Inline 3-cylinder	33.1 (543)	2.44 x 2.36	Oil Injection	Digital I.C.	Manual/Electric w/ electric/manual choke	Impeller Pump
	DT 35	Inline 2-cylinder	33.1 (543)	2.44 x 2.36	Oil Injection	Digital I.C.	Manual/Electric w/ manual/electric choke	Impeller Pump
	DT 40	Inline 2-cylinder	42.5 (696)	3.11 x 2.80	Oil Injection	Suzuki P.E.I.	Electric/Manual w/ manual/electric choke	Impeller Pump / Thermostat Controlled
	DT 55	Inline 3-cylinder	54.4 (891)	2.87 x 2.80	Oil Injection	Digital I.C.	Electric w/ electric choke	Impeller Pump / Thermostat Controlled

3-42 MAINTENANCE

General Engine Specifications

Year	Model (Horsepower)	Engine Type	Displace cu.in. (cc)	Bore and Stroke	Oil Injection System	Ignition System	Starting System	Cooling System
1988	DT 65	Inline 3-cylinder	54.4 (891)	2.87 x 2.80	Oil Injection	Digital I.C.	Electric w/ electric choke	Impeller Pump / Thermostat Controlled
	DT 75	Inline 3-cylinder	73 (1197)	3.31 x 2.83	Oil Injection	Digital I.C.	Electric w/ electric choke	Impeller Pump / Thermostat Controlled
	DT 85	Inline 3-cylinder	73 (1197)	3.31 x 2.83	Oil Injection	Digital I.C.	Electric w/ electric choke	Impeller Pump / Thermostat Controlled
	DT 115	Inline 4-cylinder	108.2 (1773)	3.31 x 3.15	Oil Injection	Digital I.C.	Electric w/ electric choke	Impeller Pump / Thermostat Controlled
	DT 140	Inline 4-cylinder	108.2 (1773)	3.31 x 3.15	Oil Injection	Digital I.C.	Electric w/ electric choke	Impeller Pump / Thermostat Controlled
	DT 150	V-6 (60°)	164.3 (2693)	3.31 x 3.19	Oil Injection	Digital I.C.	Electric w/ Suzuki Start System	Impeller Pump / Thermostat Controlled
	DT 175	V-6 (60°)	164.3 (2693)	3.31 x 3.19	Oil Injection	Digital I.C.	Electric w/ Suzuki Start System	Impeller Pump / Thermostat Controlled
	DT 200	V-6 (60°)	164.3 (2693)	3.31 x 3.19	Oil Injection	Digital I.C.	Electric w/ Suzuki Start System	Impeller Pump / Thermostat Controlled

MAINTENANCE 3-43

Serial Number Identification

Model	Year	Parts Designation	Serial No. Example	Sales Designation	Model Designation Example
DT2	1988	VJ	8XXXXX	J	DT 2 SJ
	1989	VK	9XXXXX	K	DT 2 SK
	1990	VL	011XXX	L	DT 2 LL
	1991	VM	131XXX	M	DT 2 SM
	1992	VN	231XXX	N	DT 2 SN
	1993	VP	351XXX	P	DT 2 SP
	1994	VR	461XXX	R	DT 2 SR
	1995	VS	581XXX	S	DT 2 SS
	1996	VT	651XXX	T	DT 2 ST
	1997	VV	751XXX	V	DT 2.2 SV
DT4	1988	VJ	8XXXXX	J	DT 4 LJ
	1989	VK	9XXXXX	K	DT 4 LK
	1990	VL	011XXX	L	DT 4 LL
	1991	VM	131XXX	M	DT 4 LM
	1992	VN	231 XXX	N	DT 4 LN
	1993	VP	351XXX	P	DT 4 LP
	1994	VR	461XXX	R	DT 4 LR
	1995	VS	581XXX	S	DT 4 LS
	1996	VT	651XXX	T	DT 4 LT
	1997	VV	751XXX	V	DT 4 LV
	1998	WW	861XXX	W	DT 4 LW
DT6	1988	VJ	8XXXXX	J	DT 6 SJ
	1989	VK	9XXXXX	K	DT 6 SK
	1990	VL	011XXX	L	DT 6 SL
	1991	VM	131XXX	M	DT 6 SM
	1992	VN	231XXX	N	DT 6 SN
	1993	VP	351XYX	P	DT 6 SP
	1994	VR	461XXX	R	DT 6 SR
	1995	VS	581XXX	S	DT 6 SS
	1996	VT	651XXX	T	DT 6 ST
	1997	VV	751XXX	V	DT 6 SV
	1998	WW	851XXX	W	DT 6 SW
	1999	XX	971XXX	X	DT 6 SX
DT8	1988	VJ	8XXXXX	J	DT 8 CSJ
	1989	VK	9XXXXX	K	DT 8 CLK
	1990	VL	011XXX	L	DT 8 CLL
	1991	VM	131XXX	M	DT 8 CSM
	1992	VN	231XXX	N	DT 8 MCSN
	1993	VP	351XXX	P	DT 8 MCSP
	1994	VR	461XXX	R	DT 8 MCSR
	1995	VS	581XXX	S	DT 8 MCSS
	1996	VT	651XXX	T	DT 8 MCST
	1997	VV	751XXX	V	DT 8 MCSV
DT9.9	1988	VJ	8XXXXX	J	DT 9.9 CESJ
	1989	VK	9XXXXX	K	DT 9.9 CELK
	1990	VL	011XXX	L	DT 9.9 CESL
	1991	VM	131XXX	M	DT 9.9 CESM

3-44 MAINTENANCE

Serial Number Identification

Model	Year	Parts Designation	Serial No. Example	Sales Designation	Model Designation Example
DT9.9	1992	VN	231XXX	N	DT 9.9 CESN
	1993	VP	351XXX	P	DT 9.9 CELP
	1994	VR	461XXX	R	DT 9.9 CELR
	1995	VS	581XXX	S	DT 9.9 CELS
	1996	VT	651XXX	T	DT 9.9 CELT
	1997	VV	751XXX	V	DT 9.9 CELV
DT15	1988	VJ	8XXXXX	J	DT 15 MLJ
	1989	VK	9XXXXX	K	DT 15 CESK
	1990	VL	011XXX	L	DT 15 CESL
	1991	VM	131XXX	M	DT 15 CESM
	1992	VN	231XXX	N	DT 15 CESN
	1993	VP	351XXX	P	DT 15 MCLP
	1994	VR	461XXX	R	DT 15 CESR
	1995	VS	581XXX	S	DT 15 CESS
	1996	VT	651XXX	T	DT 15 MCLT
	1997	VV	751XXX	V	DT 15 MCLV
DT20	1988	VJ	8XXXXX	J	DT 20 ESJ
DT25	1990	VL	011XXX	L	DT 25 CESL
	1991	VM	131XXX	M	DT 25 CESM
	1992	VN	231XXX	N	DT 25 CESN
	1993	VP	351XXX	P	DT 25 CELP
	1994	VR	461XXX	R	DT 25 CUR
	1995	VS	581XXX	S	DT 25 CELS
	1996	VT	651XXX	T	DT 25 CEST
	1997	VV	751XXX	V	DT 25 CESV
	1998	WW	861XXX	W	DT 25 CESW
	1999	XX	971XXX	X	DT 25 CESX
DT30	1988	VJ	8XXXXX	J	DT 30 MCLJ
	1989	VK	9XXXXX	K	DT 30 CESK
	1990	VL	011XXX	L	DT 30 CESL
	1991	VM	131XXX	M	DT 30 MCLM
	1992	VN	231XXX	N	DT 30 MCLN
	1993	VP	351XXX	P	DT 30 MCLP
	1994	VR	461XXX	R	DT 30 MCLR
	1995	VS	581XXX	S	DT 30 MCLS
	1996	VT	651XXX	T	DT 30 MCLT
	1997	VV	751XXX	V	DT 30 CRSV
DT35	1988	VJ	8XXXXX	J	DT 35 CRSJ
	1989	VK	9XXXXX	K	DT 35 CRSK
DT40	1989	VK	9XXXXX	K	DT 40 CELK
	1990	VL	011XXX	L	DT 40 CELL
	1991	VM	131XXX	M	DT 40 CRSM
	1992	VN	231XXX	N	DT 40 CRSN
	1993	VP	351XXX	P	DT 40 CRSP
	1994	VR	461XXX	R	DT 40 CRSR
	1995	VS	581XXX	S	DT 40 CRSS

MAINTENANCE 3-45

Serial Number Identification

Model	Year	Parts Designation	Serial No. Example	Sales Designation	Model Designation Example
DT90	1989	VK	9XXXXX	K	DT 90 TCLK
	1990	VL	011XXX	L	DT 90 TCLL
	1992	VN	231XXX	N	DT 90 TCLN
	1993	VP	351XXX	P	DT 90 TCLP
	1994	VR	461XXX	R	DT 90 TCLR
	1995	VS	581XXX	S	DT 90 TCLS
	1996	VT	651XXX	T	DT 90 TCLT
	1997	VV	751XXX	V	DT 90 TCLV
DT100	1989	VK	9XXXXX	K	DT 100 TCXK
	1990	VL	011XXX	L	DT100TCXL
	1992	VN	23IXXX	N	DT100TCLN
	1993	VP	351XXX	P	DT 100 TCLP
	1994	VR	461XXX	R	DT 100 TCLR
	1995	VS	581XXX	S	DT 100 TCLS
	1996	VT	651XXX	T	DT 100 TCLT
	1997	VV	751XXX	V	DT 100 TCLV
	1998	WW	861XXX	W	DT 100 TCLW
DT115	1988	VJ	8XXXXX	J	DT 115 TCXJ
	1989	VK	9XXXXX	K	DT 115 TCXK
	1990	VL	011XXX	L	DT 115 TCXL
	1991	VM	131 XXX	M	DT 115 TCXM
	1992	VN	231XXX	N	DT 115 TCXN
	1993	VP	351XXX	P	DT 115 TCXP
	1994	VR	461XXX	R	DT 115 TCXR
	1995	VS	581XXX	S	DT 115 TCXS
	1996	VT	651XXX	T	DT 115 TCXT
	1997	VV	751XXX	V	DT 115 TCXV
	1998	WW	861XXX	W	DT 115 STCLW
	1999	XX	971XXX	X	DT 115 STCLX
DT140	1988	VJ	8XXXXX	J	DT 140 TCLJ
	1989	VK	9XXXXX	K	DT140TCLK
	1990	VL	011XXX	L	DT 140 TCLL
	1991	VM	131XXX	M	DT 140 TCLM
	1992	VN	231XXX	N	DT 140 TCLN
	1993	VP	351XXX	P	DT 140 TCLP
	1994	VR	461XXX	R	DT 140 TCLR
	1995	VS	581XXX	S	DT 140 TCLS
	1996	VT	651XXX	T	DT 140 TCLT
	1997	VV	751XXX	V	DT 140 TCLV
	1998	WW	861XXX	W	DT 140 TCLW
	1999	XX	971XXX	X	DT140TCLX
DT150	1988	VJ	8XXXXX	J	DT 150 TCXJ
	1989	VK	9XXXXX	K	DT 150 TCXK
	1990	VL	011XXX	L	DT 150 TCXL
	1991	VM	131XXX	M	DT 150 TCXM
	1992	VN	231XXX	N	DT 150 TCXN
	1993	VP	351XXX	P	DT150TCXP

MAINTENANCE

Serial Number Identification

Model	Year	Parts Designation	Serial No. Example	Sales Designation	Model Designation Example
DT90	1989	VK	9XXXX	K	DT 90 TCLK
	1990	VL	011XXX	L	DT 90 TCLL
	1992	VN	231XXX	N	DT 90 TCLN
	1993	VP	351XXX	P	DT 90 TCLP
	1994	VR	461XXX	R	DT 90 TCLR
	1995	VS	581XXX	S	DT 90 TCLS
	1996	VT	651XXX	T	DT 90 TCLT
	1997	VV	751XXX	V	DT 90 TCLV
DT100	1989	VK	9XXXX	K	DT 100 TCXK
	1990	VL	011XXX	L	DT100TCXL
	1992	VN	23lXXX	N	DT100TCLN
	1993	VP	351XXX	P	DT 100 TCLP
	1994	VR	461XXX	R	DT 100 TCLR
	1995	VS	581XXX	S	DT 100 TCLS
	1996	VT	651XXX	T	DT 100 TCLT
	1997	VV	751XXX	V	DT 100 TCLV
	1998	WW	861XXX	W	DT 100 TCLW
DT115	1988	VJ	8XXXX	J	DT 115 TCXJ
	1989	VK	9XXXX	K	DT 115 TCXK
	1990	VL	011XXX	L	DT 115 TCXL
	1991	VM	131XXX	M	DT 115 TCXM
	1992	VN	231XXX	N	DT 115 TCXN
	1993	VP	351XXX	P	DT 115 TCXP
	1994	VR	461XXX	R	DT 115 TCXR
	1995	VS	581XXX	S	DT 115 TCXS
	1996	VT	651XXX	T	DT 115 TCXT
	1997	VV	751XXX	V	DT 115 TCXV
	1998	WW	861XXX	W	DT 115 STCLW
	1999	XX	971XXX	X	DT 115 STCLX
DT140	1988	VJ	8XXXX	J	DT 140 TCLJ
	1989	VK	9XXXX	K	DT140TCLK
	1990	VL	011XXX	L	DT140TCLX
	1991	VM	131XXX	M	DT 140 TCLM
	1992	VN	231XXX	N	DT 140 TCLN
	1993	VP	351XXX	P	DT 140 TCLP
	1994	VR	461XXX	R	DT 140 TCLR
	1995	VS	581XXX	S	DT 140 TCLS
	1996	VT	651XXX	T	DT 140 TCLT
	1997	VV	751XXX	V	DT 140 TCLV
	1998	WW	861XXX	W	DT 140 TCLW
	1999	XX	971XXX	X	DT140TCLX
DT150	1988	VJ	8XXXX	J	DT 150 TCXJ
	1989	VK	9XXXX	K	DT 150 TCXK
	1990	VL	011XXX	L	DT 150 TCXL
	1991	VM	131XXX	M	DT 150 TCXM
	1992	VN	231XXX	N	DT 150 TCXN
	1993	VP	351XXX	P	DT150TCXP

Serial Number Identification

Model	Year	Parts Designation	Serial No. Example	Sales Designation	Model Designation Example
DT150	1994	VR	461XXX	R	DT 150 TCXR
	1995	VS	581XXX	S	DT 150 TCXGS
	1996	VT	651XXX	T	DT 150 TCXGT
	1997	VV	751XXX	V	DT 150 TCXGV
	1998	WW	861XXX	W	DT 150 TCXGW
	1999	XX	971XXX	X	DT 150 TCXX
DT175	1988	VJ	8XXXX	J	DT 175 TCXJ
	1989	VK	9XXXX	K	DT 175 TCXK
	1990	VL	011XXX	L	DT 175 TCXL
	1991	VM	131XXX	M	DT 175 TCXM
	1992	VN	231XXX	N	DT175TCXN
DT200	1988	VJ	8XXXX	J	DT 200 TCXJ
	1989	VK	9XXXX	K	DT 200 TCXK
	1990	VL	011XXX	L	DT 200 TCXGL
	1991	VM	131XXX	M	DT 200 TCXGM
	1992	VN	231XXX	N	DT 200 TCXN
	1993	VP	351XXX	P	DT 200 TCXP
	1994	VR	461XXX	R	DT 200 TCXR
	1995	VS	581XXX	S	DT 200 TCXS
	1996	VT	651XXX	T	DT 200 TCXT
	1997	VV	751XXX	V	DT 200 TCXV
	1998	WW	861XXX	W	DT 200 TCXW
	1999	XX	971XXX	X	DT 200 TCXX
DT 225	1990	VL	011XXX	L	DT 225 TCXL
	1991	VM	131XXX	M	DT 225 TCXM
	1992	VN	231XXX	N	DT 225 TCXN
	1993	VP	351XXX	P	DT 225 TCXP
	1994	VR	461XXX	R	DT 225 TCXR
	1995	VS	581XXX	S	DT 225 TCXS
	1996	VT	651XXX	T	DT 225 TCXT
	1997	VV	751XXX	V	DT 225 TCXV
	1998	WW	861XXX	W	DT 225 TCXW
	1999	XX	971XXX	X	DT 225 TCXX

Note: Last letter of model designation indicates model year. All other letters are identified as follows:

C = Oil Injection
E = Electric Start
G = Counter Rotation
H = Tiller Control
L = 20" Shaft
M = Manual Start
N = Sail
PU = Jet Propulsion System
R = Remote Control
S = 15" Shaft
SS = Super Six
T = Tilt and Trim
X = 25" Shaft

FUEL AND COMBUSTION 4-2
FUEL 4-2
    RECOMMENDATIONS 4-2
    OCTANE RATING 4-2
    VAPOR PRESSURE AND
      ADDITIVES 4-2
    THE BOTTOM LINE WITH FUELS 4-2
    HIGH ALTITUDE OPERATION 4-2
    ALCOHOL-BLENDED FUELS 4-2
COMBUSTION 4-3
    ABNORMAL COMBUSTION 4-3
    FACTORS AFFECTING
      COMBUSTION 4-3
FUEL SYSTEM 4-3
CARBURETION 4-3
    GENERAL INFORMATION 4-3
    CARBURETOR CIRCUITS 4-4
    BASIC FUNCTIONS 4-5
    DUAL-THROAT CARBURETORS 4-6
    REMOVING FUEL FROM THE
      SYSTEM 4-6
FUEL PUMP 4-6
TROUBLESHOOTING 4-7
FUEL SYSTEM 4-7
    LOGICAL TROUBLESHOOTING 4-7
    COMMON PROBLEMS 4-8
FUEL PUMP 4-8
FUEL LINE 4-9
    COMMON PROBLEMS 4-10
COMBUSTION RELATED PISTON
  FAILURES 4-10
CARBURETOR SERVICE 4-11
DT2 AND DT2.2 4-11
    REMOVAL & INSTALLATION 4-11
    DISASSEMBLY 4-11
    CLEANING & INSPECTION 4-12
    ASSEMBLY 4-12
DT4 4-13
    REMOVAL & INSTALLATION 4-13
    DISASSEMBLY 4-13
    CLEANING & INSPECTION 4-13
    ASSEMBLY 4-14
DT6 AND 1988 DT8 4-14
    REMOVAL & INSTALLATION 4-14
    DISASSEMBLY 4-14
    CLEANING & INSPECTION 4-15
    ASSEMBLY 4-15
DT9.9, DT15 AND 1989-97 DT8 4-15
    REMOVAL & INSTALLATION 4-15
    DISASSEMBLY 4-16
    CLEANING & INSPECTION 4-16
    ASSEMBLY 4-17
DT20 TO DT85, DT115 AND DT140 4-17
    REMOVAL & INSTALLATION 4-19
    DISASSEMBLY 4-20
    CLEANING & INSPECTION 4-21
    ASSEMBLY 4-21
V4 & V6 POWERHEADS 4-22
    REMOVAL & INSTALLATION 4-23
    DISASSEMBLY 4-24
    FUEL LEVEL TEST 4-24
    CLEANING & INSPECTION 4-25
    ASSEMBLY 4-25

REED VALVE SERVICE 4-25
DT2, DT2.2 4-25
    REMOVAL & INSTALLATION 4-25
ALL OTHER MODELS 4-25
    REMOVAL & INSTALLATION 4-26
    INSPECTION & CLEANING 4-26
    REED & REED STOP
      REPLACEMENT 4-26
FUEL PUMP SERVICE 4-27
DIAPHRAGM TYPE FUEL PUMPS 4-27
    DESCRIPTION & OPERATION 4-27
    REMOVAL & INSTALLATION 4-27
    OVERHAUL 4-27
ELECTRONIC FUEL INJECTION 4-32
DESCRIPTION AND OPERATION 4-32
    FUEL INJECTION BASICS 4-32
    SUZUKI ELECTRONIC FUEL
      INJECTION 4-32
CYLINDER WALL TEMPERATURE
  SENSOR 4-36
    DESCRIPTION & OPERATION 4-36
    TESTING 4-38
    REMOVAL & INSTALLATION 4-39
AIR TEMPERATURE SENSOR 4-39
    DESCRIPTION & OPERATION 4-39
    TESTING 4-39
    REMOVAL & INSTALLATION 4-39
ATMOSPHERIC PRESSURE
  SENSOR 4-39
    DESCRIPTION & OPERATION 4-39
    TESTING 4-39
    REMOVAL & INSTALLATION 4-39
THROTTLE VALVE SENSOR 4-40
    DESCRIPTION & OPERATION 4-40
    TESTING 4-40
    ADJUSTMENT 4-42
    REMOVAL & INSTALLATION 4-45
GEAR COUNTER COIL (ENGINE SPEED
  SENSOR) 4-45
    DESCRIPTION & OPERATION 4-45
    TESTING 4-45
PULSER COIL 4-45
    DESCRIPTION & OPERATION 4-45
    TESTING 4-45
FUEL INJECTORS 4-45
    DESCRIPTION & OPERATION 4-45
    TESTING 4-45
FUEL PRESSURE REGULATOR 4-46
    TESTING 4-46
    REMOVAL & INSTALLATION 4-47
HIGH PRESSURE FUEL PUMP 4-47
    TESTING 4-48
ADDITIONAL INPUTS 4-48
SELF DIAGNOSTIC SYSTEM 4-48
    DESCRIPTION & OPERATION 4-48
    DIAGNOSIS PROCEDURE 4-49
    FAIL SAFE EMERGENCY
      BACKUP 4-49
SPECIFICATIONS CHART
    CIRCUIT/POSITION 4-7

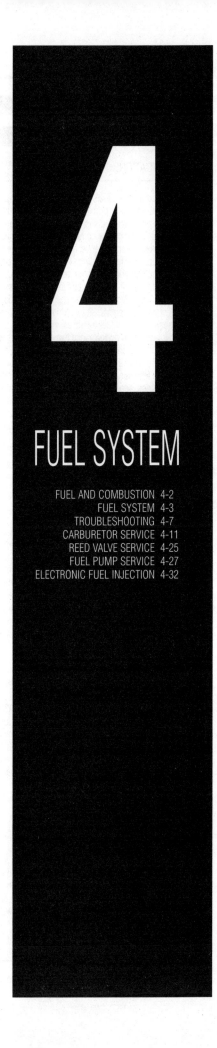

4
FUEL SYSTEM

FUEL AND COMBUSTION 4-2
FUEL SYSTEM 4-3
TROUBLESHOOTING 4-7
CARBURETOR SERVICE 4-11
REED VALVE SERVICE 4-25
FUEL PUMP SERVICE 4-27
ELECTRONIC FUEL INJECTION 4-32

4-2 FUEL SYSTEM

FUEL AND COMBUSTION

Fuel

RECOMMENDATIONS

Reformulated gasoline fuels are now found in many market areas. Current testing indicates no particular problems with using this fuel. Shelf life is shorter and, because of the oxygenates, a slight leaning out at idle may be experienced. This slightly lean condition can be compensated for by adjusting idle mixture screws.

Fuel recommendations have become more complex as the chemistry of modern gasoline changes. The major driving force behind the changes in gasoline chemistry is the search for additives to replace lead as an octane booster and lubricant. These new additives are governed by the types of emissions they produce in the combustion process. Also, the replacement additives do not always provide the same level of combustion stability, making a fuel's octane rating less meaningful.

In the search for new fuel additives, automobiles are used as the test medium. Not one high performance two cycle engine was tested in the process of determining the chemistry of today's gasoline.

In the 1960's and 1970's, leaded fuel was common. The lead served two functions. The lead served as an octane booster (combustion stabilizer) and, in four cycle engines, served as a valve seat lubricant. For two cycle engines, the primary benefit of lead was to serve as a combustion stabilizer. Lead served very well for this purpose, even in high heat applications.

Today, all lead has been removed from the gasoline process. This means that the benefit of lead as an octane booster has been eliminated. Several substitute octane boosters have been introduced in the place of lead. While many are adequate in an automobile, most do not perform nearly as well as lead did, even though the octane rating of the fuel is the same.

OCTANE RATING

A fuel's octane rating is a measurement of how stable the fuel is when heat is introduced. Octane rating is a major consideration when deciding whether a fuel is suitable for a particular application. For example, in an engine, we want the fuel to ignite when the spark plug fires and not before, even under high pressure and temperatures. Once the fuel is ignited, it must burn slowly and smoothly, even though heat and pressure are building up while the burn occurs. The unburned fuel should be ignited by the traveling flame front, not by some other source of ignition, such as carbon deposits or the heat from the expanding gasses. A fuel's octane rating is known as a measurement of the fuel's anti-knock properties (ability to burn without exploding).

Usually a fuel with a higher octane rating can be subjected to a more severe combustion environment before spontaneous or abnormal combustion occurs. To understand how two gasoline samples can be different, even though they have the same octane rating, we need to know how octane rating is determined.

The American Society of Testing and Materials (ASTM) has developed a universal method of determining the octane rating of a fuel sample. The octane rating you see on the pump at a gasoline station is known as the pump octane number. Look at the small print on the pump. The rating has a formula. The rating is determined by the R+M/2 method.

Therefore, the number you see on the pump is the average of two other octane ratings.

The Research Octane Reading is a measure of a fuel's anti-knock properties under a light load, or part throttle conditions. During this test, combustion heat is easily dissipated.

The Motor Octane Rating is a measure of a fuel's anti-knock properties under a heavy load, or full throttle conditions, when heat buildup is at maximum.

Because a two cycle engine has a power stroke every revolution, with heat buildup every revolution, it tends to respond more to the motor octane rating of the fuel than the research octane rating. Therefore, in an outboard motor, the motor octane rating of the fuel is the best indication of how it will perform, not the research octane. Unfortunately, the user has no way of knowing for sure the exact motor octane rating of the fuel.

VAPOR PRESSURE AND ADDITIVES

Two other factors besides octane rating affect how suitable the fuel is for a particular application.

Fuel vapor pressure is a measure of how easily a fuel sample evaporates. Many additives used in gasoline contain aromatics. Aromatics are light hydrocarbons distilled off the top of a crude oil sample. They are effective at increasing the research octane of a fuel sample, but can cause vapor lock on a very hot day. If you have an inconsistent running engine and you suspect vapor lock, use a piece of clear fuel line to look for bubbles, indicating that the fuel is vaporizing.

One negative side effect of aromatics is that they create additional combustion products such as carbon and varnish. If your engine requires high octane fuel to prevent detonation, de-carbon the engine more frequently with an internal engine cleaner to prevent ring sticking due to excessive varnish buildup.

Besides aromatics, two types of alcohol are used in fuel today as octane boosters, ethanol and methanol. Again, alcohol tends to raise the research octane of the fuel. This usually means they will have limited benefit in an outboard motor. Also, alcohol contains oxygen, which means that since it is replacing gasoline without oxygen content, alcohol fuel blends cause the fuel-air mixture to be leaner.

THE BOTTOM LINE WITH FUELS

If we could buy fuel of the correct octane rating, free of alcohol and aromatics, this would be our first choice.

Suzuki continues to recommend unleaded fuel. This is almost a redundant recommendation due to the near universal unavailability of any other type fuel.

According to the fuel recommendations that come with your outboard, there is no engine in the product line that requires more than 89 octane. Most Suzuki engines need only 86 octane or less. An 89 octane rating generally means middle grade unleaded. Premium unleaded is more stable under severe conditions, but also produces more combustion products. Therefore, when using premium unleaded, more frequent de-carboning is necessary.

Regardless of the fuel octane rating you choose, try to stay with a name brand fuel. You never know for sure what kinds of additives or how much is in off brand fuel.

HIGH ALTITUDE OPERATION

At elevated altitudes there is less oxygen in the atmosphere than at sea level. Less oxygen means lower combustion efficiency and less power output. Power output is reduced three percent for every thousand feet above sea level. At ten thousand feet, power is reduced 30 percent from that available at sea level.

Re-jetting for high altitude does not restore this lost power. Re-jetting simply corrects the air-fuel ratio for the reduced air density, and makes the most of the remaining available power. If you re-jet an engine, you are locked into the higher elevation. You cannot operate at sea level until you re-jet for sea level. Understand that going below the elevation jetted for your motor will damage the engine. As a general rule, jet for the lowest elevation anticipated. Spark plug insulator tip color is the best guide for high altitude jetting.

If you are in an area of known poor fuel quality, you may want to use fuel additives. Today's additives are mostly alcohol and aromatics, and their effectiveness may be limited. It is difficult to find additives without ethanol, methanol, or aromatics. If you use octane boosters frequent de-carboning may be necessary. If possible, the best policy is to use name brand pump fuel with no additional additives except Suzuki fuel conditioner and Ring-Free•.

ALCOHOL-BLENDED FUELS

The Environmental Protection Agency mandated a phase-out of the leaded fuels. Lead was used to boost the octane of fuel. By January of 1986, the maximum allowable amount of lead was 0.1 gm/gal, down from 1.1 gm/gal.

Gasoline suppliers, in general, feel that the 0.1 gm/gal limit is too low to make lead of any real use to improve octane. Therefore, alternate octane improvers are being used. There are multiple methods currently employed to improve octane but the most inexpensive additive seems to be alcohol.

There are, however, some special considerations due to the effects of alcohol in fuel. You should know about them and what steps to take when using alcohol-blended fuels commonly called gasohol.

Alcohol in fuel is either methanol (wood alcohol) or ethanol (grain alcohol). Either type can have serious effects when applied to outboard motor applications.

The leaching affect of alcohol will, in time, cause fuel lines and plastic components to become brittle to the point of cracking. Unless replaced, these cracked lines could leak fuel, increasing the potential for hazardous situations.

➡ **Suzuki fuel lines and plastic fuel system components have been specially formulated to resist alcohol leaching effects.**

When gasohol becomes contaminated with water, the water combines with the alcohol then settles to the bottom. This leaves the gasoline and the oil for models using premix, on a top layer. With alcohol-blended fuels, the amount of water necessary for this phase separation to occur is 0.5% by volume.

All fuels have chemical compounds added to reduce the tendency towards phase separation. If phase separation occurs, however, there is a possibility of a lean oil/fuel mixture with the potential for engine damage. With oil-injected outboards (Precision Blend models), phase separation will be less of a problem because the oil is injected separately rather than being premixed.

Combustion

A two cycle engine has a power stroke every revolution of the crankshaft. A four cycle engine has a power stroke every other revolution of the crankshaft. Therefore, the two cycle engine has twice as many power strokes for any given RPM. If the displacement of the two types of engines is identical, then the two cycle engine has to dissipate twice as much heat as the four cycle engine. In such a high heat environment, the fuel must be very stable to avoid detonation. If any parameters affecting combustion change suddenly (the engine runs lean for example), uncontrolled heat buildup occurs very rapidly in a two cycle engine.

ABNORMAL COMBUSTION

There are two types of abnormal combustion:
- Pre-ignition—Occurs when the air-fuel mixture is ignited by some other incandescent source other than the correctly timed spark from the spark plug.
- Detonation—Occurs when excessive heat and or pressure ignites the air/fuel mixture rather than the spark plug. The burn becomes explosive.

FACTORS AFFECTING COMBUSTION

The combustion process is affected by several interrelated factors. This means that when one factor is changed, the other factors also must be changed to maintain the same controlled burn and level of combustion stability.

Compression

Determines the level of heat buildup in the cylinder when the air-fuel mixture is compressed. As compression increases, so does the potential for heat buildup.

Ignition Timing

Determines when the gasses will start to expand in relation to the motion of the piston. If the gasses begin to expand too soon, such as they would during pre-ignition or in an overly advanced ignition timing, the motion of the piston opposes the expansion of the gasses, resulting in extremely high combustion chamber pressures and heat.

As ignition timing is retarded, the burn occurs later in relation to piston position. This means that the piston has less distance to travel under power to the bottom of the cylinder, resulting in less usable power.

Fuel Mixture

Determines how efficient the burn will be. A rich mixture burns slower than a lean one. If the mixture is too lean, it can't become explosive. The slower the burn, the cooler the combustion chamber, because pressure buildup is gradual.

Fuel Quality (Octane Rating)

Determines how much heat is necessary to ignite the mixture. Once the burn is in progress, heat is on the rise. The unburned poor quality fuel is ignited all at once by the rising heat instead of burning gradually as a flame front of the burn passing by. This action results in detonation (pinging).

Other Factors

In general, anything that can cause abnormal heat buildup can be enough to push an engine over the edge to abnormal combustion, if any of the four basic factors previously discussed are already near the danger point, for example, excessive carbon buildup raises the compression and retains heat as glowing embers.

FUEL SYSTEM

Carburetion

GENERAL INFORMATION

The carburetor is merely a metering device for mixing fuel and air in the proper proportions for efficient engine operation. At idle speed, an outboard engine requires a mixture of about 8 parts air to 1 part fuel. At high speed or under heavy duty service, the mixture may change to as much as 12 parts air to 1 part fuel.

Float Systems

♦ See Figure 1

A small chamber in the carburetor serves as a fuel reservoir. A float valve admits fuel into the reservoir to replace the fuel consumed by the engine. If the carburetor has more than one reservoir, the fuel level in each reservoir (chamber) is controlled by identical float systems.

Fuel level in each chamber is extremely critical and must be maintained accurately. Accuracy is obtained through proper adjustment of the floats. This adjustment will provide a balanced metering of fuel to each cylinder at all speeds.

Following the fuel through its course, from the fuel tank to the combustion chamber of the cylinder, will provide an appreciation of exactly what is taking place. In order to start the engine, the fuel must be moved from the tank to the carburetor by a squeeze bulb installed in the fuel line. This action is necessary because the fuel pump does not have sufficient pressure to draw fuel from the tank during cranking before the engine starts.

The fuel for some small horsepower units is gravity fed from a tank mounted at the rear of the powerhead. Even with the gravity feed method, a small fuel pump may be an integral part of the carburetor. After the engine starts, the fuel passes through the pump to the carburetor. All systems have some type of filter installed somewhere in the line between the tank and the carburetor. Many units have a filter as an integral part of the carburetor.

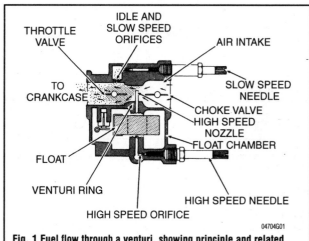

Fig. 1 Fuel flow through a venturi, showing principle and related parts controlling intake and outflow

4-4 FUEL SYSTEM

At the carburetor, the fuel passes through the inlet passage to the needle and seat, and then into the float chamber (reservoir). A float in the chamber rides up and down on the surface of the fuel. After fuel enters the chamber and the level rises to a predetermined point, a tang on the float closes the inlet needle and the flow entering the chamber is cut off. When fuel leaves the chamber as the engine operates, the fuel level drops and the float tang allows the inlet needle to move off its seat and fuel once again enters the chamber. In this manner, a constant reservoir of fuel is maintained in the chamber to satisfy the demands of the engine at all speeds.

A fuel chamber vent hole is located near the top of the carburetor body to permit atmospheric pressure to act against the fuel in each chamber. This pressure assures an adequate fuel supply to the various operating systems of the powerhead.

Air/Fuel Mixture

♦ See Figure 2

A suction effect is created each time the piston moves upward in the cylinder. This suction draws air through the throat of the carburetor. A restriction in the throat, called a venturi, controls air velocity and has the effect of reducing air pressure at this point.

The difference in air pressures at the throat and in the fuel chamber, causes the fuel to be pushed out of metering jets extending down into the fuel chamber. When the fuel leaves the jets, it mixes with the air passing through the venturi. This fuel/air mixture should then be in the proper proportion for burning in the cylinders for maximum engine performance.

In order to obtain the proper air/fuel mixture for all engine speeds, some models have high and low speed jets. These jets have adjustable needle valves which are used to compensate for changing atmospheric conditions. In almost all cases, the high-speed circuit has fixed high-speed jets and are not adjustable.

A throttle valve controls the flow of air/fuel mixture drawn into the combustion chambers. A cold powerhead requires a richer fuel mixture to start and during the brief period it is warming to normal operating temperature. A choke valve is placed ahead of the metering jets and venturi. As this valve begins to close, the volume of air intake is reduced, thus enriching the mixture entering the cylinders. When this choke valve is fully closed, a very rich fuel mixture is drawn into the cylinders.

The throat of the carburetor is usually referred to as the barrel. Carburetors with single, double, or four barrels have individual metering jets, needle valves, throttle and choke plates for each barrel. Single and two barrel carburetors are fed by a single float and chamber.

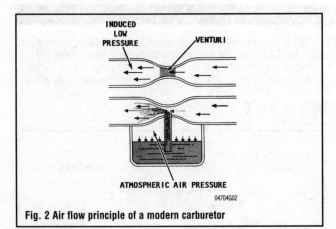
Fig. 2 Air flow principle of a modern carburetor

CARBURETOR CIRCUITS

The following section illustrates the circuit functions and locations of a typical marine carburetor.

Starting Circuit

♦ See Figure 3

The choke plate is closed, creating a partial vacuum in the venturi. As the piston rises, negative pressure in the crankcase draws the rich air-fuel mixture from the float bowl into the venturi and on into the engine.

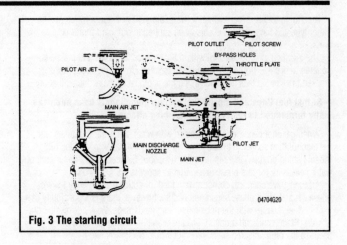
Fig. 3 The starting circuit

Low Speed Circuit

♦ See Figure 4

Zero–one-eighth throttle, when the pressure in the crankcase is lowered, the air-fuel mixture is discharged into the venturi through the pilot outlet because the throttle plate is closed. No other outlets are exposed to low venturi pressure. The fuel is metered by the pilot jet. The air is metered by the pilot air jet. The combined air-fuel mixture is regulated by the pilot air screw.

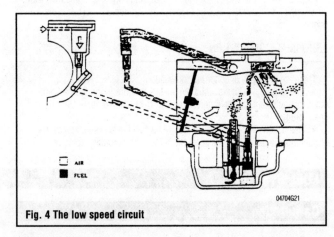
Fig. 4 The low speed circuit

Mid-Range Circuit

♦ See Figure 5

One-eighth–three-eighths throttle, as the throttle plate continues to open, the air-fuel mixture is discharged into the venturi through the bypass holes. As the throttle plate uncovers more bypass holes, increased fuel flow results because

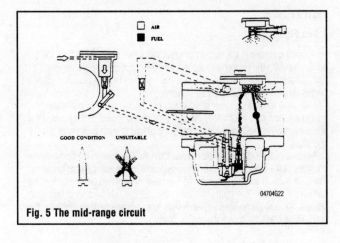
Fig. 5 The mid-range circuit

of the low pressure in the venturi. Depending on the model, there could be two, three or four bypass holes.

High Speed Circuit

♦ See Figure 6

Three-eighths–wide-open throttle, as the throttle plate moves toward wide open, we have maximum air flow and very low pressure. The fuel is metered through the main jet, and is drawn into the main discharge nozzle. Air is metered by the main air jet and enters the discharge nozzle, where it combines with fuel. The mixture atomizes, enters the venturi, and is drawn into the engine.

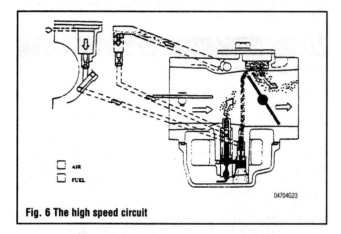

Fig. 6 The high speed circuit

BASIC FUNCTIONS

♦ See Figures 7, 8 and 9

The carburetor systems on in line engines require careful cleaning and adjustment if problems occur. These carburetors are complicated but not too complex to understand. All carburetors operate on the same principles.

Traditional carburetor theory often involves a number of laws and principles. To troubleshoot carburetors learn the basic principles, watch how the carburetor comes apart, trace the circuits, see what they do and make sure they are clean. These are the basic steps for troubleshooting and successful repair.

The diagram illustrates several carburetor basics. If you blow through the straw an atomized mixture (air and fuel droplets) comes out. When you blow through the straw a pressure drop is created in the straw column inserted in the liquid. In a carburetor this is mostly air and a little fuel. The actual ratio of air to fuel differs with engine conditions but is usually from 15 parts air to one part fuel at optimum cruise to as little as 7 parts air to one part fuel at full choke.

If the top of the container is covered and sealed around the straw what will happen? No flow. This is typical of a clogged carburetor bowl vent. If the base of the straw is clogged or restricted what will happen? No flow or low flow. This

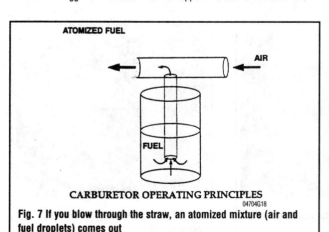

Fig. 7 If you blow through the straw, an atomized mixture (air and fuel droplets) comes out

Fig. 8 The V4 and V6 engines use a solenoid release valve and large volume chamber for fuel delivery into the intake system

represents a clogged main jet. If the liquid in the glass is lowered and you blow through the straw with the same force what will happen? Not as much fuel will flow. A lean condition occurs. If the fuel level is raised and you blow again at the same velocity what happens? The result is a richer mixture.

Suzuki carburetors control air flow semi-independently of RPM. This is done with a throttle plate. The throttle plate works in conjunction with other systems or circuits to deliver correct mixtures within certain RPM bands. The idle circuit pilot outlet controls from 0–1/8 throttle. The series of small holes in the carburetor throat called transition holes control the 1/8–3/8 throttle range. At wide open throttle the main jet handles most of the fuel metering chores, but the low and mid-range circuits continue to supply part of the fuel.

Enrichment is necessary to start a cold engine. Fuel and air mix does not want to vaporize in a cold engine. In order to get a little fuel to vaporize, a lot of fuel is dumped into the engine. On many older inline engines a choke plate is used for cold starts. This plate restricts air entering the engine and increases the fuel to air ratio.

The V4 and V6 engines use a solenoid release valve and large volume chamber for fuel delivery into the intake system behind No. 2 and No.3 carburetors (V6 models), and No.1 and No.2 (V4 models) to ensure easy starting under all conditions. Fuel for this system is delivered from the fuel pump (top pump on the V6) directly to the fuel starter valve assembly where it is controlled by a float and inlet needle valve. When the ignition key, in the ON position, is pushed in, the solenoid will open the solenoid release valve and fuel will flow to the two starter jet ports at the end of the carburetor bore. Turning the manual valve counterclockwise, to open, will allow fuel to flow in the event of an electrical problem. The manual valve must remain closed during normal engine operation.

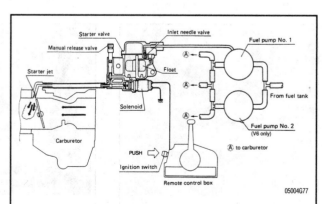

Fig. 9 Illustration of the V4 and V6 solenoid release valve starting system. The V4 engines have only one fuel pump

4-6 FUEL SYSTEM

DUAL-THROAT CARBURETORS

The carburetor systems on V4 and V6 engines require careful cleaning and adjustment if problems occur. These carburetors are not difficult to understand. All carburetors operate on the same principles. For best results, trace and analyze one circuit at a time.

Beginning in 1996, all Saltwater series 90 degree V engines have an additional jet in the carburetor. This pull over or enrichment jet improves mid-range response while maintaining fuel economy. Additional enrichment is necessary to start a cold engine. Fuel/air mixes to not want to vaporize in a cold engine. In order to get a little fuel to vaporize, a lot of fuel is dumped into the engine. On most V4 and V6 engines, a choke plate is used for cold starts. This plate restricts air entering the engine and increases the fuel/air ratio.

The enrichment system on the 90-degree 225 hp engines is controlled by a microprocessor. Temperature and throttle position are monitored and enrichment is automatic. A pair of injectors with different diameters are used to provide enrichment.

REMOVING FUEL FROM THE SYSTEM

♦ See Figures 10 and 11

For many years there has been the widespread belief that simply shutting off the fuel at the tank and then running the engine until it stops is the proper procedure before storing the engine for any length of time. Right? Wrong!

It is not possible to remove all of the fuel in the carburetor by operating the engine until it stops. Some fuel is trapped in the float chamber and other passages and in the line leading to the carburetor. The only guaranteed method of removing ALL of the fuel is to take the time to remove the carburetor, and drain the fuel.

If the engine is operated with the fuel supply shut off until it stops, the fuel and oil mixture inside the engine is removed, leaving bearings, pistons, rings, and other parts with little protective lubricant, during long periods of storage.

Proper procedure involves:
1. Shutting off the fuel supply at the tank.
2. Disconnecting the fuel line at the tank.
3. Operating the engine until it begins to run rough, then stopping the engine, which will leave some fuel/oil mixture inside.
4. Removing and draining the carburetor.

By disconnecting the fuel supply, all small passages are cleared of fuel even though some fuel is left in the carburetor. A light oil should be put in the combustion chamber as instructed in the owner's manual. On some model carburetors the high-speed jet plug can be removed to drain the fuel from the carburetor.

For short periods of storage, simply running the carburetor dry may help prevent severe gum and varnish from forming in the carburetor. This is especially true during hot weather.

Fuel Pump

♦ See Figures 12 thru 18

A fuel pump is a basic mechanical device that utilizes crankcase positive and negative pressures to pump fuel from the fuel tank to the carburetors.

This device contains a flexible diaphragm and two check valves (flappers or fingers) that control flow. As the piston goes up, crankcase pressure drops (negative pressure) and the inlet valve opens, pulling fuel from the tank. As the piston nears TDC, pressure in the pump area is neutral (atmospheric pressure). At this point both valves are closed. As the piston comes down, pressure goes up (positive pressure) and the fuel is pushed toward the carburetor bowl by the diaphragm through the now open outlet valve.

This is a reliable method to move fuel but can have several problems. Sometimes an engine backfire can rupture the diaphragm. The diaphragm and valves are moving parts subject to wear. The flexibility of the diaphragm material can go away, reducing or stopping flow. Rust or dirt can hang a valve open and reduce or stop fuel flow.

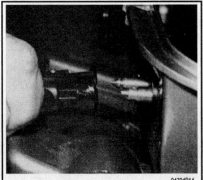
Fig. 10 Typical fuel line quick disconnect fitting

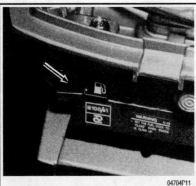
Fig. 11 Fuel shutoff knob on a 4 hp outboard

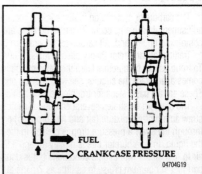
Fig. 12 A fuel pump is a basic mechanical device that utilizes crankcase positive and negative pressures to pump fuel

Fig. 13 The diaphragm is most subject to wear in a fuel pump

Fig. 14 Typical fuel pump mounting on the engine crankcase

Fig. 15 Make sure to inspect the fuel pump gasket . . .

FUEL SYSTEM 4-7

Fig. 16 . . . or O-ring for tears or damage

Fig. 17 When taking the pump apart, mark the pump sections for correct reassembly

Fig. 18 Exploded view of a diaphragm fuel pump—DT175

TROUBLESHOOTING

Fuel System

Troubleshooting fuel systems requires the same techniques used in other areas. A thorough, systematic approach to troubleshooting will pay big rewards. Build your troubleshooting checklist, with the most likely offenders at the top. Use your experience to adjust your list for local conditions. Everyone has been tempted to jump into the carburetor on a vague hunch. Pause a moment and review the facts when this urge occurs.

In order to accurately troubleshoot a carburetor or fuel system problem, you must first verify that the problem is fuel related. Many symptoms can have several different possible causes. Be sure to eliminate mechanical and electrical systems as the potential fault. Carburetion is the number one cause of most engine problems, but there are other possibilities.

One of the toughest tasks with a fuel system is the actual troubleshooting. Several tools are at your disposal for making this process very simple. A timing light works well for observing carburetor spray patterns. Look for the proper amount of fuel and for proper atomization in the two fuel outlet areas (main nozzle and bypass holes). The strobe effect of the lights helps you see in detail the fuel being drawn through the throat of the carburetor. On multiple carburetor engines, always attach the timing light to the cylinder you are observing so the strobe doesn't change the appearance of the patterns. If you need to compare two cylinders, change the timing light hookup each time you observe a different cylinder.

Pressure testing fuel pump output can determine whether the fuel spray is adequate and if the fuel pump diaphragms are functioning correctly. A pressure gauge placed between the fuel pumps and the carburetors will test the entire fuel delivery system. Normally a fuel system problem will show up at high speed where the fuel demand is the greatest. A common symptom of a fuel pump output problem is surging at wide open throttle, but normal operation at slower speeds. To check the fuel pump output, install the pressure gauge and accelerate the engine to wide open throttle. Observe the pressure gauge needle. It should always swing up to some value between 5–6 psi and remain steady. This reading would indicate a system that is functioning properly.

If the needle gradually swings down toward zero, fuel demand is greater than the fuel system can supply. This reading isolates the problem to the fuel delivery system (fuel tank or line). To confirm this, an auxiliary tank should be installed and the engine re-tested. Be aware that a bad anti-siphon valve on a built-in tank can create enough restriction to cause a lean condition and serious engine damage.

If the needle movement becomes erratic, suspect a ruptured diaphragm in the fuel pump.

A quick way to check for a ruptured fuel pump diaphragm is while the engine is at idle speed, to squeeze the primer bulb and hold steady firm pressure on it. If the diaphragm is ruptured, this will cause a rough running condition because of the extra fuel passing through the diaphragm into the crankcase. After performing this test you should check the spark plugs for cylinders that the fuel pump supplies. If the spark plugs are OK, but the fuel pumps are still suspected, you should remove the fuel pumps and completely disassemble them. Rebuild or replace the pumps as needed.

To check the boat's fuel system for a restriction, install a vacuum gauge in the line before the fuel pump. Run the engine under load at wide open throttle to get a reading. Vacuum should read no more than 4.5 in. Hg (15.2 kPa) for engines up to and including 200 hp, and should not exceed 6.0 in. Hg (20.2 kPa) for engines greater than 200 hp.

To check for air entering the fuel system, install a clear fuel hose between the fuel screen and fuel pump. If air is in the line, check all fittings back to the boat's fuel tank.

Spark plug tip appearance is a good indication of combustion efficiency. The tip should be a light tan. A white insulator or small beads on the insulator indicate too much heat. A dark or oil fouled insulator indicates incomplete combustion. To properly read spark plug tip appearance, run the engine at the RPM you are testing for about 15 second and then immediately turn the engine OFF without changing the throttle position.

Reading spark plug tip appearance is also the proper way to test jet verifications in high altitude.

The following chart explains the relationship between throttle position and carburetion circuits.

LOGICAL TROUBLESHOOTING

The following paragraphs provide an orderly sequence of tests to pinpoint problems in the fuel system.

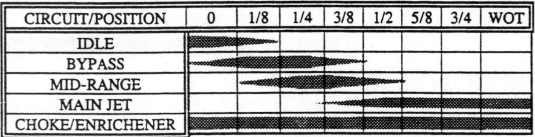

CIRCUIT/POSITION	0	1/8	1/4	3/8	1/2	5/8	3/4	WOT
IDLE	▬▬▬	▬						
BYPASS	▬▬▬	▬▬▬	▬▬▬					
MID-RANGE		▬▬▬	▬▬▬	▬▬▬				
MAIN JET				▬	▬▬▬	▬▬▬	▬▬▬	▬▬▬
CHOKE/ENRICHENER	▬▬▬							

4-8 FUEL SYSTEM

1. Gather as much information as you can.
2. Duplicate the condition. Take the boat out and verify the complaint.
3. If the problem cannot be duplicated, you cannot fix it. This could be a product operation problem.
4. Once the problem has been duplicated, you can begin troubleshooting. Give the entire unit a careful visual inspection. You can tell a lot about the engine from the care and condition of the entire rig. What's the condition of the propeller and the lower unit? Remove the hood and look for any visible signs of failure. Are there any signs of head gasket leakage. Is the engine paint discolored from high temperature or are there any holes or cracks in the engine block? Perform a compression and leak down test. While cranking the engine during the compression test, listen for any abnormal sounds. If the engine passes these simple tests we can assume that the mechanical condition of the engine is good. All other engine mechanical inspection would be too time consuming at this point.
5. Your next step is to isolate the fuel system into two sub-systems. Separate the fuel delivery components from the carburetors. To do this, substitute the boat's fuel supply with a known good supply. Use a 6 gallon portable tank and fuel line. Connect the portable fuel supply directly to the engine fuel pump, bypassing the boat fuel delivery system. Now test the engine. If the problem is no longer present, you know where to look. If the problem is still present, further troubleshooting is required.
6. When testing the engine, observe the throttle position when the problem occurs. This will help you pinpoint the circuit that is malfunctioning. Carburetor troubleshooting and repair is very demanding. You must pay close attention to the location, position and sometimes the numbering on each part removed. The ability to identify a circuit by the operating RPM it affects is important. Often your best troubleshooting tool is a can of cleaner. This can be used to trace those mystery circuits and find that last speck of dirt. Be careful and wear safety glasses when using this method.

COMMON PROBLEMS

Fuel Delivery

♦ See Figure 19

Many times fuel system troubles are caused by a plugged fuel filter, a defective fuel pump, or by a leak in the line from the fuel tank to the fuel pump. A defective choke may also cause problems. would you believe, a majority of starting troubles which are traced to the fuel system are the result of an empty fuel tank or aged sour fuel.

Fig. 19 An excellent way of protecting fuel hoses against contamination is an end cap filter

Sour Fuel

♦ See Figure 20

Under average conditions (temperate climates), fuel will begin to break down in about four months. A gummy substance forms in the bottom of the fuel tank and in other areas. The filter screen between the tank and the carburetor and small passages in the carburetor will become clogged. The gasoline will begin to give off an odor similar to rotten eggs. Such a condition can cause the owner much frustration, time in cleaning components, and the expense of replacement or overhaul parts for the carburetor.

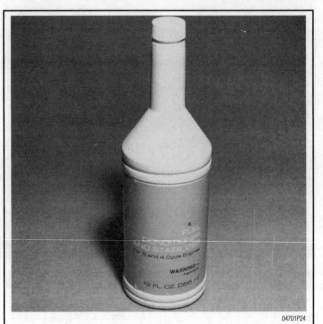
Fig. 20 The use of an approved fuel additive, such as this Suzuki Fuel Conditioner and Stabilizer, will prevent fuel from souring for up to twelve months

Even with the high price of fuel, removing gasoline that has been standing unused over a long period of time is still the easiest and least expensive preventative maintenance possible. In most cases, this old gas can be used without harmful effects in an automobile using regular gasoline.

The gasoline preservative additive Suzuki Fuel Conditioner and Stabilizer for 2 cycle engines, will keep the fuel fresh for up to twelve months. If this particular product is not available in your area, other similar additives are produced under various trade names.

Choke Problems

When the engine is hot, the fuel system can cause starting problems. After a hot engine is shut down, the temperature inside the fuel bowl may rise to 200 degrees F and cause the fuel to actually boil. All carburetors are vented to allow this pressure to escape to the atmosphere. However, some of the fuel may percolate over the high-speed nozzle.

If the choke should stick in the open position, the engine will be hard to start. If the choke should stick in the closed position, the engine will flood, making it very difficult to start.

In order for this raw fuel to vaporize enough to burn, considerable air must be added to lean out the mixture. Therefore, the only remedy is to remove the spark plugs, ground the leads, crank the powerhead through about ten revolutions, clean the plugs, reinstall the plugs, and start the engine.

If the needle valve and seat assembly is leaking, an excessive amount of fuel may enter the reed housing in the following manner. After the powerhead is shut down, the pressure left in the fuel line will force fuel past the leaking needle valve. This extra fuel will raise the level in the fuel bowl and cause fuel to overflow into the reed housing.

A continuous overflow of fuel into the reed housing may be due to a sticking inlet needle or to a defective float, which would cause an extra high level of fuel in the bowl and overflow into the reed housing.

Fuel Pump

♦ See Figure 21

Fuel pump testing is an excellent way to pinpoint air leaks, restricted fuel lines and fittings or other fuel supply related performance problems.

When a fuel starvation problem is suspected such as engine hesitation or engine stopping, perform the following fuel system test:
1. Connect the piece of clear fuel hose to a side barb of the "T" fitting O.

FUEL SYSTEM 4-9

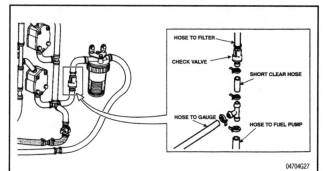

Fig. 21 Connecting a fuel pressure gauge inline in preparation for a fuel pump test

2. Connect one end of the long piece of fuel hose to the vacuum gauge and the other end to the center barb of the "T" fitting.

➡ Use a long enough piece of fuel hose so the vacuum gauge may be read at the helm.

3. Remove the existing fuel hose from the fuel tank side of the fuel pump, and connect the remaining barb of the "T" fitting to the fuel hose.
4. Connect the short piece of clear fuel hose to the fuel check valve leading from the fuel filter. If a check valve does not exist, connect the clear fuel hose directly to the fuel filter.
5. Check the vacuum gauge reading after running the engine long enough to stabilize at full power.

➡ The vacuum is to not exceed 4.5 in. Hg (15.2 kPa) for up to 200 hp engines. The vacuum is to not exceed 6.0 in. Hg (20.3 kPa) for engines greater than 200 hp.

6. An anti-siphon valve (required if the fuel system drops below the top of the fuel tank) will cause a 1.5 to 2.5 in. Hg (8.4 kPa) increase in vacuum.
7. If high vacuum is noted, move the T-fitting to the fuel filter outlet O and retest.
8. Continue to the fuel filter inlet and along the remaining fuel system until a large drop in vacuum locates the problem.
9. A good clean water separator fuel filter will increase vacuum about 0.5 in. Hg (1.7 kPa).
10. Small internal passages inside a fuel selector valve, fuel tank pickup, or fuel line fittings may cause excessive fuel restriction and high vacuum.
11. Unstable and slowly rising vacuum readings, especially with a full tank of fuel, usually indicates a restricted vent line.

➡ Bubbles in the clear fuel line section indicate an air leak, making for an inaccurate vacuum test. Check all fittings for tightened clamps and a tight fuel filter.

➡ Vacuum gauges are not calibrated and some may read as much as 2 in. Hg (6.8 kPa) lower than the actual vacuum. It is recommended to perform a fuel system test while no problems exist to determine vacuum gauge accuracy.

Fuel Line

♦ See Figures 22, 23 and 24

On most installations, the fuel line is provided with quick-disconnect fittings at the tank and at the engine. If there is reason to believe the problem is at the quick-disconnects, the hose ends should be replaced as an assembly. For a small additional expense, the entire fuel line can be replaced and thus eliminate this entire area as a problem source for many future seasons.

The primer squeeze bulb can be replaced in a short time. First, cut the hose line as close to the old bulb as possible. Slide a small clamp over the end of the fuel line from the tank. Next, install the small end of the check valve assembly into this side of the fuel line. The check valve always goes towards the fuel tank. Place a large clamp over the end of the check valve assembly. Use Primer Bulb Adhesive when the connections are made. Tighten the clamps. Repeat the procedure with the other side of the bulb assembly and the line leading to the engine.

Fig. 22 To test the fuel pickup in the fuel tank, operate the squeeze bulb and observe fuel flowing from the disconnected line at the fuel pump. Discharge fuel into an approved container.

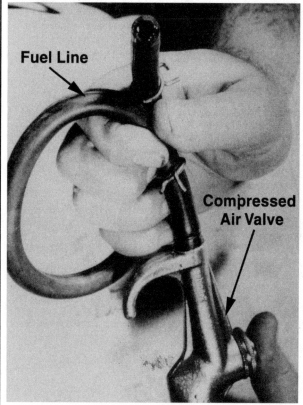

Fig. 23 Many times restrictions such as foreign material may be cleared from the fuel lines using compressed air. Ensure the open end of the hose is pointing in a clear direction to avoid personal injury

4-10 FUEL SYSTEM

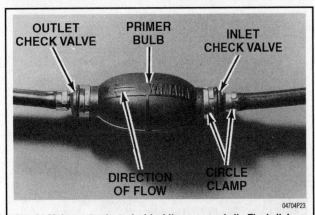

Fig. 24 Major parts of a typical fuel line squeeze bulb. The bulb is used to prime the fuel system until the powerhead is operating and the pump can deliver the required amount of fuel to run the engine

COMMON PROBLEMS

Rough Engine Idle

If an engine does not idle smoothly, the most reasonable approach to the problem is to perform a tune-up to eliminate such areas as:
- Defective points
- Faulty spark plugs
- Timing out of adjustment

Other problems that can prevent an engine from running smoothly include:
- An air leak in the intake manifold
- Uneven compression between the cylinders
- Sticky or broken reeds

Of course any problem in the carburetor affecting the air/fuel mixture will also prevent the engine from operating smoothly at idle speed. These problems usually include:
- Too high a fuel level in the bowl
- A heavy float
- Leaking needle valve and seat
- Defective automatic choke
- Improper adjustments for idle mixture or idle speed

Excessive Fuel Consumption

Excessive fuel consumption can be the result of any one of four conditions, or a combination of all.
- Inefficient engine operation.
- Faulty condition of the hull, including excessive marine growth.
- Poor boating habits of the operator.
- Leaking or out of tune carburetor.

If the fuel consumption suddenly increases over what could be considered normal, then the cause can probably be attributed to the engine or boat and not the operator.

Marine growth on the hull can have a very marked effect on boat performance. This is why sail boats always try to have a haul-out as close to race time as possible.

While you are checking the bottom, take note of the propeller condition. A bent blade or other damage will definitely cause poor boat performance.

If the hull and propeller are in good shape, then check the fuel system for possible leaks. Check the line between the fuel pump and the carburetor while the engine is running and the line between the fuel tank and the pump when the engine is not running. A leak between the tank and the pump many times will not appear when the engine is operating, because the suction created by the pump drawing fuel will not allow the fuel to leak. Once the engine is turned off and the suction no longer exists, fuel may begin to leak.

If a minor tune-up has been performed and the spark plugs, points, and timing are properly adjusted, then the problem most likely is in the carburetor and an overhaul is in order.

Check the needle valve and seat for leaking. Use extra care when making any adjustments affecting the fuel consumption, such as the float level or automatic choke.

Engine Surge

If the engine operates as if the load on the boat is being constantly increased and decreased, even though an attempt is being made to hold a constant engine speed, the problem can most likely be attributed to the fuel pump, or a restriction in the fuel line between the tank and the carburetor.

Combustion Related Piston Failures

▶ See Figure 25

When an engine has a piston failure due to abnormal combustion, fixing the mechanical portion of the engine is the easiest part. The hard part is determining what caused the problem, in order to prevent a repeat failure. Think back to the four basic areas that affect combustion to find the cause of the failure.

Since you probably removed the cylinder head. Inspect the failed piston, look for excessive deposit buildup that could raise compression, or retain heat in the combustion chamber. Statically check the wide open throttle timing. Be sure that the timing is not over advanced. It is a good idea to seal these adjustments with paint to detect tampering.

Look for a fuel restriction that could cause the engine to run lean. Don't forget to check the fuel pump, fuel tank and lines, especially if a built in tank is used. Be sure to check the anti-siphon valve on built in tanks.

If everything else looks good, the final possibility is poor quality fuel.

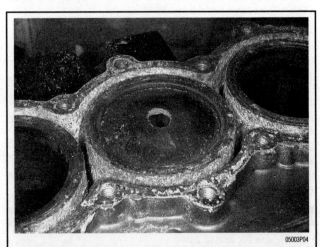

Fig. 25 This burned piston is typical of a combustion relate failure. The combustion chamber temperature got so hot that it melted the top of the piston (hole in the top of the piston)

FUEL SYSTEM 4-11

CARBURETOR SERVICE

DT2 and DT2.2

This carburetor is a single-barrel, float feed type with a manual choke. Fuel to the carburetor is gravity fed from a fuel tank mounted at the rear of the powerhead.

REMOVAL & INSTALLATION

♦ See accompanying illustrations

Good shop practice dictates a carburetor repair kit be purchased and new parts be installed any time the carburetor is disassembled.

Make an attempt to keep the work area organized and to cover parts after they have been cleaned. This practice will prevent foreign matter from entering passageways or adhering to critical parts.

1. Remove the port and starboard engine covers.

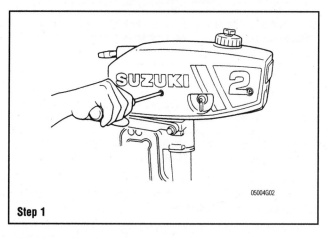
Step 1

2. With the fuel petcock lever in the OFF position (marked "S"), remove the fuel hose from the carburetor fitting and plug the hose to prevent fuel from leaking.
3. Loosen the choke knob set screw and remove the knob from the control panel.

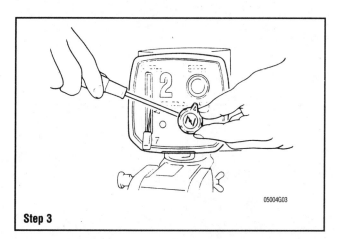
Step 3

4. Remove the screws holding the control panel to the carburetor and lift off the control panel and throttle link knob.
5. Loosen the carburetor clamp and remove the carburetor from the engine crankcase. Discard the O-ring.

To install:

6. Install the throttle lever post in the throttle valve with the anchor in the pocket at the bottom of the valve body.
7. Secure the carburetor in place by tightening the bolt and nut securely.
8. Install the control panel and the chock knob. Don't forget to tighten the set screw.

Slowly tighten the idle speed screw until it barely seats, then back it out the

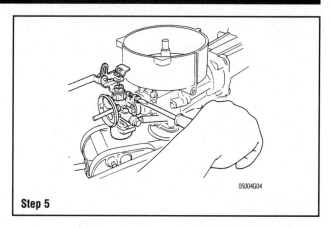
Step 5

same number of turns recorded during disassembly. If the number of turns was not recorded, back the screw out 1-¾ turns as a rough adjustment. Idle speed should be as specified in the "Tune-Up Specifications" chart.

9. Install the two halves of the cowling around the powerhead.
10. Secure the engine cover.

Mount the outboard unit in a test tank, or the boat in a body of water, or connect a flush attachment and hose to the lower unit. Start the engine and check the completed work. Allow the powerhead to warm to normal operating temperature. Adjust the idle speed to specification.

DISASSEMBLY

♦ See Figure 26

1. With the carburetor on clean working surface, remove the screws securing the fuel float bowl to the carburetor body. Discard the float bowl gasket.

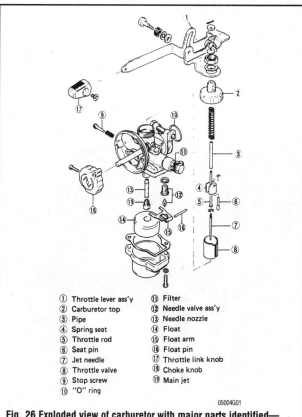

① Throttle lever ass'y
② Carburetor top
③ Pipe
④ Spring seat
⑤ Throttle rod
⑥ Seat pin
⑦ Jet needle
⑧ Throttle valve
⑨ Stop screw
⑩ "O" ring
⑪ Filter
⑫ Needle valve ass'y
⑬ Needle nozzle
⑭ Float
⑮ Float arm
⑯ Float pin
⑰ Throttle link knob
⑱ Choke knob
⑲ Main jet

Fig. 26 Exploded view of carburetor with major parts identified—DT2 and DT2.2 models

4-12 FUEL SYSTEM

2. Push the float pin free using a fine pointed awl and remove the float pin from the float arm and lift off the float and needle valve from the bowl.

3. Remove the main nozzle and the main jet assembly.

4. Remove the needle seat and gasket using a wide bladed screwdriver. Discard the gasket.

5. Loosen the throttle stop screw and unscrew the carburetor top with the throttle lever assembly as one unit.

6. Disconnect the throttle plunger from the needle valve and remove the valve, spring and retainer.

➡ It is not necessary to remove the E-clip from the jet needle, unless replacement is required or if the powerhead is to be operated at a significantly different elevation.

7. Turn in the throttle stop screw (counting the turns for reassembly) until it lightly seats. Turn out the screw and remove it and the attached spring.

8. Remove the fuel inlet fitting and fuel filter.

CLEANING & INSPECTION

♦ See Figures 27 and 28

❋❋ CAUTION

Never dip rubber parts, plastic parts, diaphragms, or pump plungers in carburetor cleaner. These parts should be cleaned only in solvent, and then blown dry with compressed air.

Place all metal parts in a screen-type tray and dip them in carburetor cleaner until they appear completely clean, then blow them dry with compressed air.

Blow out all passages in the castings with compressed air. Check all parts and passages to be sure they are not clogged or contain any deposits. Never use a piece of wire or any type of pointed instrument to clean drilled passages or calibrated holes in a carburetor.

Move the throttle shaft back and forth to check for wear. If the shaft appears to be too loose, replace the complete throttle body because individual replacement parts are not available.

Inspect the main body, airhorn, and venturi cluster gasket surfaces for cracks and burrs which might cause a leak. Check the float for deterioration. Check to be sure the float spring has not been stretched. If any part of the float is damaged, the unit must be replaced. Check the float arm needle contacting surface and replace the float if this surface has a groove worn in it.

Fig. 27 Metal parts from our disassembled 2 hp carburetor in a basket ready to be immersed in carburetor cleaner

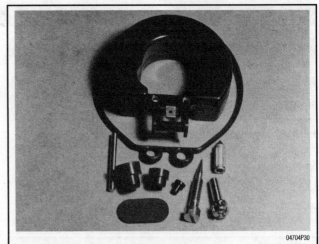

Fig. 28 A carburetor repair kit, like this one for our disassembled carburetor, are available at your local service dealer. They contain the necessary components to perform a carburetor overhaul

Inspect the tapered section of the idle adjusting needles and replace any that have developed a groove. As previously mentioned, most of the parts which should be replaced during a carburetor overhaul are included in overhaul kits available from your local marine dealer. One of these kits will contain a matched fuel inlet needle and seat. This combination should be replaced each time the carburetor is disassembled as a precaution against leakage.

ASSEMBLY

♦ See accompanying illustrations

1. Install a new carburetor O-ring into the carburetor body.
2. Apply an all-purpose lubricant to a new idle speed screw. Install the idle speed screw and spring.
3. Install the main jet into the main nozzle and tighten it just snug with a screwdriver.
4. Slide a new needle valve into the groove of the float arm.
5. Lower the float arm into position with the needle valve sliding into the needle valve seat. Now, push the float pin through the holes in the carburetor body and hinge using a small awl or similar tool.
6. Hold the carburetor body in a perfect upright position. Check the float hinge adjustment. The vertical distance between the float chamber mating face and the float should be 0.75–0.83 in. (19–21mm).

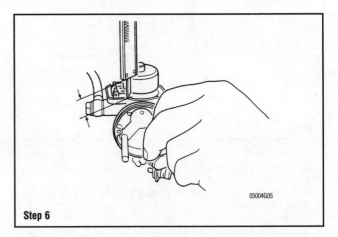

Step 6

7. Carefully, bend the hinge, if necessary, to achieve the required measurement.

➡ Make sure the gasket is removed when making the float height measurement.

FUEL SYSTEM   4-13

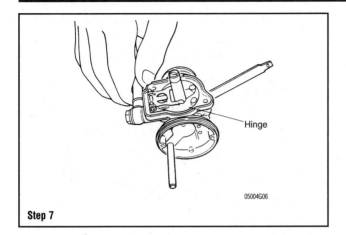

Step 7

8. Position a new float bowl gasket in place on the carburetor body. Install the float into the float bowl. Place the float bowl in position on the carburetor body, and then secure it with the two Phillips head screws.

9. If the E-clip on the jet needle is lowered, the carburetor will cause the powerhead to operate rich. Raising the E-clip will cause the powerhead to operate lean. Higher altitude raise E-clip to compensate for rarefied air. Standard E-clip setting is in the 3rd notch. Begin to assemble the throttle valve components by inserting the E-clip end of the jet needle into the throttle valve (the end with the recess for the throttle cable end). Next, place the needle retainer into the throttle valve over the E-clip and align the retainer slot with the slot in the throttle valve.

10. Reassemble the throttle valve assembly. Align the jet needle retainer should be positioned with the slot aligned.

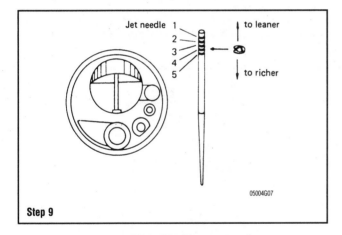

Step 9

DT4

♦ See Figure 29

This type of carburetor has been used in various configurations for many years. Most of the changes are in jetting calibration and control linkages.

The needle valve seat is not replaceable. If it is damaged or worn, the carburetor must be replaced as a complete unit.

REMOVAL & INSTALLATION

1. Remove the engine cover.
2. Pull the fuel hose off the carburetor. Plug the fuel hose to prevent leakage.
3. Remove the choke knob from the control panel.
4. Loosen the two nuts and remove the carburetor from the engine.

To install:

5. Position a new float bowl gasket in place on the carburetor body. Install the float into the float bowl. Place the float bowl in position on the carburetor body, and then secure it with the two Phillips head screws.
6. Secure the carburetor in place by tightening the bolt and nut securely.
7. Install the chock knob.

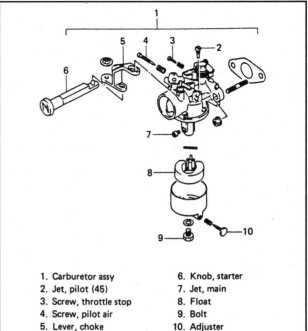

1. Carburetor assy
2. Jet, pilot (45)
3. Screw, throttle stop
4. Screw, pilot air
5. Lever, choke
6. Knob, starter
7. Jet, main
8. Float
9. Bolt
10. Adjuster

Fig. 29 Exploded view of the DT4 carburetor with major parts identified

Slowly tighten the idle speed screw until it barely seats, then back it out the same number of turns recorded during disassembly. If the number of turns was not recorded, back the screw out 1-½ turns as a rough adjustment. Idle speed should be as specified in the "Tune-Up Specifications" chart.

8. Secure the engine cover.

Mount the outboard unit in a test tank, or the boat in a body of water, or connect a flush attachment and hose to the lower unit. Start the engine and check the completed work. Allow the powerhead to warm to normal operating temperature. Adjust the idle speed to specification.

DISASSEMBLY

1. Remove the bolt and washer from the float bowl. Remove the float bowl and O-ring from the carburetor body. Discard the used O-ring.
2. Remove the float hinge pin and remove the float and pin assembly from the carburetor body.
3. Remove the inlet needle.

➡**Do not force removal of the inlet needle, on some models the inlet needle is permanently installed in the valve seat.**

4. Remove the main jet. Use jet removal tool or a wide blade screwdriver.
5. Remove the pilot (idle) jet.
6. Remove the air jet.
7. Turn in the pilot (idle) screw, counting the turns in (for reassembly later) until it lightly seats. Now, remove the pilot screw and spring.

CLEANING & INSPECTION

✲✲ CAUTION

Never dip rubber parts, plastic parts, diaphragms, or pump plungers in carburetor cleaner. These parts should be cleaned only in solvent, and then blown dry with compressed air.

Place all metal parts in a screen-type tray and dip them in carburetor cleaner until they appear completely clean, then blow them dry with compressed air.

Blow out all passages in the castings with compressed air. Check all parts and passages to be sure they are not clogged or contain any deposits. Never use a piece of wire or any type of pointed instrument to clean drilled passages or calibrated holes in a carburetor.

4-14 FUEL SYSTEM

Move the throttle shaft back and forth to check for wear. If the shaft appears to be too loose, replace the complete throttle body because individual replacement parts are not available.

Inspect the main body, air-horn, and venturi cluster gasket surfaces for cracks and burrs which might cause a leak. Check the float for deterioration. Check to be sure the float spring has not been stretched. If any part of the float is damaged, the unit must be replaced. Check the float arm needle contacting surface and replace the float if this surface has a groove worn in it.

Inspect the tapered section of the idle adjusting needles and replace any that have developed a groove. As previously mentioned, most of the parts which should be replaced during a carburetor overhaul are included in overhaul kits available from your local marine dealer. One of these kits will contain a matched fuel inlet needle and seat. This combination should be replaced each time the carburetor is disassembled as a precaution against leakage.

ASSEMBLY

1. Install a new carburetor O-ring into the carburetor body.
2. Apply an all-purpose lubricant to a new idle speed screw. Install the idle speed screw and spring.

➡ **The standard setting is: 1–1½ turns out.**

3. Install the main jet into the main nozzle and tighten it just snug with a screwdriver.
4. Slide a new needle valve into the groove of the float arm.
5. Lower the float arm into position with the needle valve sliding into the needle valve seat. Now, push the float pin through the holes in the carburetor body and hinge using a small awl or similar tool.
6. Hold the carburetor body in a perfect upright position. Check the float hinge adjustment. The vertical distance between the float chamber mating face and the float should be 0.47–0.55 in. (12–14mm). Carefully, bend the hinge, if necessary, to achieve the required measurement.

➡ **Make sure the gasket is removed when making the float height measurement.**

DT6 and 1988 DT8

♦ See Figure 30

REMOVAL & INSTALLATION

1. Remove the engine cover.

2. Pull the fuel hose off the carburetor. Plug the fuel hose to prevent leakage.
3. Remove the choke knob from the control panel.
4. Loosen the two nuts and remove the carburetor from the engine.

To install:
5. Secure the carburetor in place by tightening the bolt and nut securely.
6. Install the chock knob.

Slowly tighten the idle speed screw until it barely seats, then back it out the same number of turns recorded during disassembly. If the number of turns was not recorded, back the screw out 1-¾ turns as a rough adjustment. Idle speed should be as specified in the "Tune-Up Specifications" chart.

7. Secure the engine cover.

Mount the outboard unit in a test tank, or the boat in a body of water, or connect a flush attachment and hose to the lower unit. Start the engine and check the completed work. Allow the powerhead to warm to normal operating temperature. Adjust the idle speed to specification.

DISASSEMBLY

♦ See Figures 31 and 32

1. Remove the bolt and washer from the float bowl. Remove the float bowl and O-ring from the carburetor body. Discard the used O-ring.
2. Remove the float hinge pin and remove the float and pin assembly from the carburetor body.
3. Remove the inlet needle.

➡ **Do not force removal of the inlet needle, on some models the inlet needle is permanently installed in the valve seat.**

4. Remove the main jet. Use jet removal tool or a wide blade screwdriver.
5. Remove the pilot (idle) jet.
6. Remove the air jet.
7. Turn in the pilot (idle) screw, counting the turns in (for reassembly later) until it lightly seats. Now, remove the pilot screw and spring.

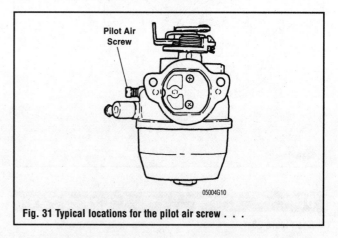

Fig. 31 Typical locations for the pilot air screw . . .

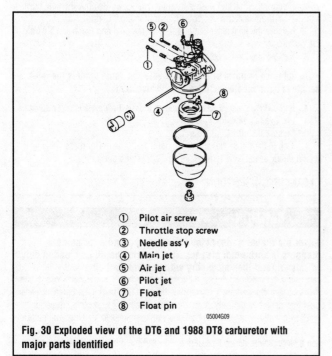

① Pilot air screw
② Throttle stop screw
③ Needle ass'y
④ Main jet
⑤ Air jet
⑥ Pilot jet
⑦ Float
⑧ Float pin

Fig. 30 Exploded view of the DT6 and 1988 DT8 carburetor with major parts identified

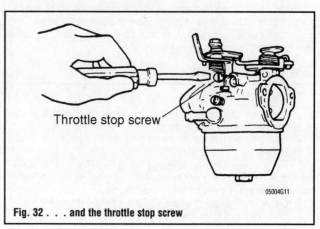

Fig. 32 . . . and the throttle stop screw

FUEL SYSTEM 4-15

CLEANING & INSPECTION

♦ See Figures 33 and 34

Never dip rubber parts, plastic parts, diaphragms, or pump plungers in carburetor cleaner. These parts should be cleaned only in solvent, and then blown dry with compressed air.

Place all metal parts in a screen-type tray and dip them in carburetor cleaner until they appear completely clean, then blow them dry with compressed air.

Blow out all passages in the castings with compressed air. Check all parts and passages to be sure they are not clogged or contain any deposits. Never use a piece of wire or any type of pointed instrument to clean drilled passages or calibrated holes in a carburetor.

Move the throttle shaft back and forth to check for wear. If the shaft appears to be too loose, replace the complete throttle body because individual replacement parts are not available.

Inspect the main body, airhorn, and venturi cluster gasket surfaces for cracks and burrs which might cause a leak. Check the float for deterioration. Check to be sure the float spring has not been stretched. If any part of the float is damaged, the unit must be replaced. Check the float arm needle contacting surface and replace the float if this surface has a groove worn in it.

Fig. 33 Good shop practice dictates a carburetor rebuild kit be purchased and new parts, especially gaskets and O-rings be installed any time the carburetor is disassembled. This photo includes parts in a repair kit for the 6 hp, 8 hp, 9.9 hp and 15 hp carburetor

Fig. 34 Remove all rubber and plastic parts before immersing metal parts of the 6 hp, 8 hp, 9.9 hp and 15 hp carburetor in cleaning solution

Inspect the tapered section of the idle adjusting needles and replace any that have developed a groove. As previously mentioned, most of the parts which should be replaced during a carburetor overhaul are included in overhaul kits available from your local marine dealer. One of these kits will contain a matched fuel inlet needle and seat. This combination should be replaced each time the carburetor is disassembled as a precaution against leakage.

ASSEMBLY

1. Install a new carburetor O-ring into the carburetor body.
2. Apply an all-purpose lubricant to a new idle speed screw. Install the idle speed screw and spring.

➡ The standard setting is:

- DT6: 1–1½ turns out.
- DT8: ¾–1–¼ turns out.

3. Install the main jet into the main nozzle and tighten it just snug with a screwdriver.
4. Slide a new needle valve into the groove of the float arm.
5. Lower the float arm into position with the needle valve sliding into the needle valve seat. Now, push the float pin through the holes in the carburetor body and hinge using a small awl or similar tool.
6. Hold the carburetor body in a perfect upright position. Check the float hinge adjustment. The vertical distance between the float chamber mating face and the float should be as follows:

- DT6 and DT8: 0.9–1.0 in. (22–26mm). Carefully, bend the hinge, if necessary, to achieve the required measurement.

➡ Make sure the gasket is removed when making the float height measurement.

7. Position a new float bowl gasket in place on the carburetor body. Install the float into the float bowl. Place the float bowl in position on the carburetor body, and then secure it with the two Phillips head screws.

DT9.9, DT15 and 1989-97 DT8

♦ See Figure 35

The fuel pump is constructed as an integral component of the carburetor. The pump is a diaphragm type, operating with the pressure pulses inside the engine crankcase. These pressure pulses are characteristic of a two-stroke type engine. The crankcase pressure becomes positive during the downward stroke of the piston and negative during the upward stroke. In response to these pressure pulses, the diaphragm will flex cyclically to pump the fuel from the fuel tank to the carburetor float bowls.

When the engine is started, positive and negative pressures are produced alternately in the crankcase, passing into the pump body which is mounted on the carburetor itself and actuating the diaphragm in the pump housing. The diaphragm pulsing action causes the fuel from the fuel filter to flow from the inlet into the pump. Then, the fuel inside the fuel pump is sent from the discharge outlet to the carburetor, which it passes through a valve hole and then into the float bowl.

If the engine speed is increased, the diaphragm cycles are increased proportionally, supplying the correct amount of fuel needed for that particular engine speed.

REMOVAL & INSTALLATION

1. Remove the engine cover.
2. Pull the fuel hose off the carburetor. Plug the fuel hose to prevent leakage.
3. Remove the choke knob from the control panel.
4. Loosen the two nuts and remove the carburetor from the engine.

To install:

5. Place a new carburetor mounting gasket on the powerhead studs. Install the carburetor and secure it in place with the washers and nuts. Tighten the nuts alternately and evenly.
6. Adjust the air screw setting. Baseline setting is: 1¼–1¾ turns out.
7. Install the choke knob onto the control panel.
8. Connect the fuel line onto the carburetor inlet fitting.
9. Install the engine cover.

4-16 FUEL SYSTEM

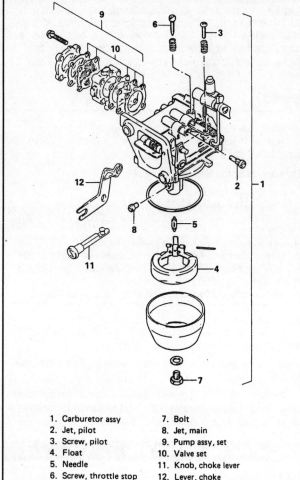

1. Carburetor assy
2. Jet, pilot
3. Screw, pilot
4. Float
5. Needle
6. Screw, throttle stop
7. Bolt
8. Jet, main
9. Pump assy, set
10. Valve set
11. Knob, choke lever
12. Lever, choke

Fig. 35 Exploded view of the DT9.9, DT15 and 1989-97 DT8 carburetor with major parts identified

5. Remove the main jet. Use jet removal tool or a wide blade screwdriver.
6. Remove the pilot (idle) jet with the proper tool.
7. Turn in the pilot (idle) screw, counting the turns in (for reassembly later) until it lightly seats. Now, remove the pilot screw and spring.
8. Clean and inspect all the parts.

CLEANING & INSPECTION

▶ See Figures 36 and 37

> **※※ CAUTION**
>
> **Never dip rubber parts, plastic parts, diaphragms, or pump plungers in carburetor cleaner. These parts should be cleaned only in solvent, and then blown dry with compressed air. Place all metal parts in a screen-type tray and dip them in carburetor cleaner until they appear completely clean, then blow them dry with compressed air.**

Blow out all passages in the castings with compressed air. Check all parts and passages to be sure they are not clogged or contain any deposits. Never use a piece of wire or any type of pointed instrument to clean drilled passages or calibrated holes in a carburetor.

Move the throttle shaft back and forth to check for wear. If the shaft appears to be too loose, replace the complete throttle body because individual replacement parts are not available.

Inspect the main body, airhorn, and venturi cluster gasket surfaces for cracks and burrs which might cause a leak. Check the float for deterioration. Check to be sure the float spring has not been stretched. If any part of the float is damaged, the unit must be replaced. Check the float arm needle contacting surface and replace the float if this surface has a groove worn in it.

Inspect all O-rings, seals and gaskets. All of these components become hard with age and tend to become brittle and deteriorate. This affects their ability to seal properly, so its always a good idea to replace them anytime the carburetor is apart for service.

Inspect the tapered section of the idle adjusting needles and replace any that have developed a groove.

Inspect the fuel pump diaphragm for tears. Any damage to the diaphragm means it must be replaced.

As previously mentioned, most of the parts which should be replaced during a carburetor overhaul are included in overhaul kits available from your local marine dealer. One of these kits will contain a matched fuel inlet needle and

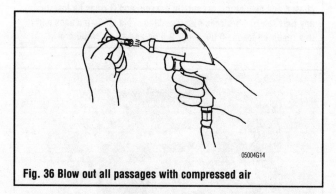

Fig. 36 Blow out all passages with compressed air

DISASSEMBLY

▶ See accompanying illustrations

1. Remove the fuel pump from the carburetor body by removing the screws.
2. Remove the bolt and washer from the float bowl. Remove the float bowl and O-ring from the carburetor body. Discard the used O-ring.
3. Remove the float hinge pin and remove the float and pin assembly from the carburetor body.
4. Remove the inlet needle.

➡The inlet needle valve seat is fixed to the carburetor body. If the seat is damaged, the carburetor body must be replaced.

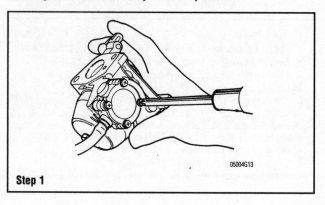

Step 1

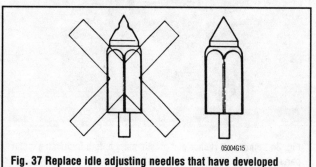

Fig. 37 Replace idle adjusting needles that have developed grooves

seat. This combination should be replaced each time the carburetor is disassembled as a precaution against leakage.

ASSEMBLY

♦ See accompanying illustrations

A float level inspection and adjustment are very important parts of an engine tune-up. The float level setting is correct when the "float height" is set to the proper specification.

1. Install the float and hinge pin into the carburetor body.
2. Invert the carburetor body, making sure that the float hinge pin doesn't fall out.
3. Keeping the float arm free, measure the distance between the carburetor body (with the gasket removed) to the bottom of the float. Correct float height is: 0.91–0.98 in. (23.0–25.0 mm).
4. Turn in the pilot (idle) screw, counting the turns in (for reassembly later) until it lightly seats.
5. Install the pilot (idle) jet with the proper tool.
6. Install the main jet. Use jet removal tool or a wide blade screwdriver.
7. Install the inlet needle.
8. Install the float bowl and new O-ring onto the carburetor body.
9. Install the fuel pump onto the carburetor body.

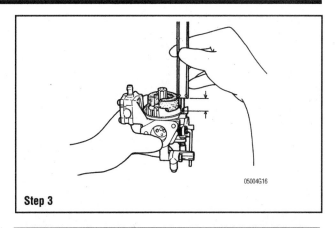

Step 3

DT20 to DT85, DT115 and DT140

♦ See Figures 38, 39 and 40

Different variations of these carburetor models have been used on Suzuki outboards, but they all operate in the same manner. The major differences in

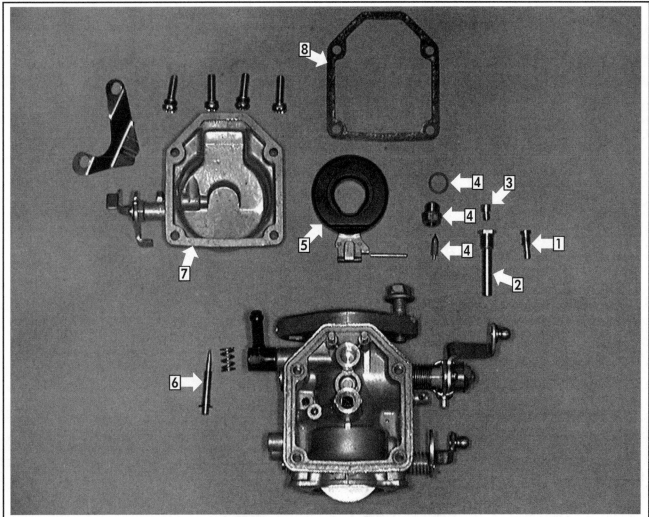

1. Pilot jet
2. Nozzle
3. Main jet
4. Needle valve assy.
5. Float
6. Pilot air screw
7. Float bowl
8. Float bowl gasket

Fig. 38 Exploded view of the single carburetor with all major parts identified

4-18 FUEL SYSTEM

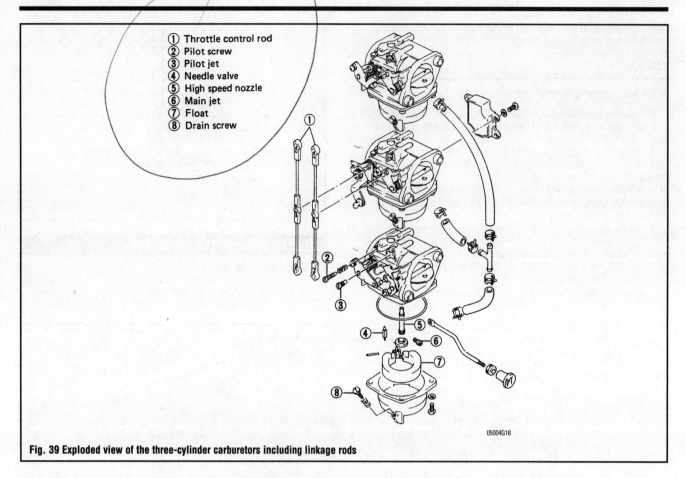

1. Throttle control rod
2. Pilot screw
3. Pilot jet
4. Needle valve
5. High speed nozzle
6. Main jet
7. Float
8. Drain screw

Fig. 39 Exploded view of the three-cylinder carburetors including linkage rods

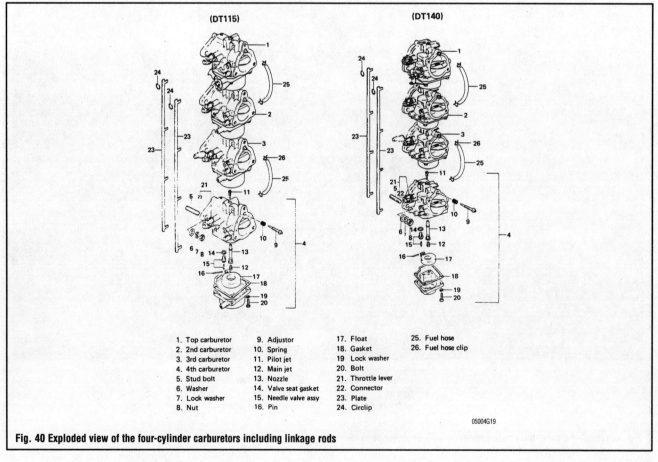

1. Top carburetor	9. Adjustor
2. 2nd carburetor	10. Spring
3. 3rd carburetor	11. Pilot jet
4. 4th carburetor	12. Main jet
5. Stud bolt	13. Nozzle
6. Washer	14. Valve seat gasket
7. Lock washer	15. Needle valve assy
8. Nut	16. Pin
17. Float	25. Fuel hose
18. Gasket	26. Fuel hose clip
19. Lock washer	
20. Bolt	
21. Throttle lever	
22. Connector	
23. Plate	
24. Circlip	

Fig. 40 Exploded view of the four-cylinder carburetors including linkage rods

FUEL SYSTEM 4-19

these carburetors are mostly confined to different calibrations and linkage setups. Removal procedures may vary slightly due to differences in linkage. These differences will be noted whenever they make a substantial change in a procedure.

REMOVAL & INSTALLATION

▶ See accompanying illustrations

1. Disconnect the negative battery cable.
2. Mark each individual carburetor on the float to aid in reinstallation.
3. Disconnect any linkage rods and control cables.
4. Disconnect any fuel or oil injection hoses.
5. Remove the air silencer cover from the carburetor assembly
6. Remove the silencer case from the carburetors.
7. After removing the carburetor nuts, lift the carburetors off the engine.

To install:

8. Identify each carburetor by the mark scribed on the float bowl during removal. Place the carburetors in-line on the work bench.
9. Install the throttle and choke linkage.
10. Place a new one piece gasket on the carburetor throats and align the silencer case to the carburetor assemblies.
11. Install the fuel hose from the fuel filter to the carburetors. Secure all connections with the clamps.
12. Install the silencer cover over the silencer case.
13. Install and tighten the screws securing the silencer cover in place.
14. Mount the outboard unit in a test tank, or put the boat in a body of water, or connect a flush attachment and hose to the lower unit. Connect a tachometer to the powerhead.

Step 3

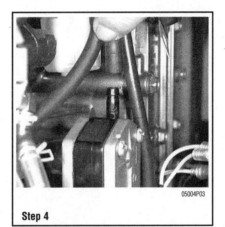
Step 4

Step 5

Step 6

Step 7

Step 9

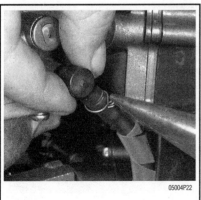
Step 11

Step 12

Step 13

4-20 FUEL SYSTEM

> ⚠️ **CAUTION**
>
> Never operate the engine at high speed with a flush device attached. The engine, operating at high speed with such a device attached, would runaway from lack of a load on the propeller, causing extensive damage.

Start the engine and check the completed work.

> ⚠️ **CAUTION**
>
> Water must circulate through the lower unit to the powerhead anytime the powerhead is operating to prevent damage to the water pump in the lower unit. Just five seconds without water will damage the water pump impeller.

Allow the powerhead to warm to normal operating temperature. Adjust the throttle stop screw until the powerhead idles at specification. Rotating the throttle stop screw clockwise increases powerhead speed, and rotating the screw counterclockwise decreases powerhead speed.

DISASSEMBLY

♦ See accompanying illustrations

The following procedures pick up the work after the carburetors have been removed from the powerhead. The procedures for each of the carburetors is identical. Any differences in float height measurement, pilot screw turns, jets, or any other adjustments will be clearly identified for the carburetor including location and model.

1. Remove the float bowl drain screw and spring and drain any remaining fuel into a suitable container.
2. Remove the screws securing the float bowl to the mixing chamber.
3. Remove and discard the O-ring or gasket from the float bowl.
4. Slide the float hinge pin to one side and remove the float and the pin from the carburetor body.
5. Grasp the float and gently remove it from the carburetor body.
6. Gently remove the needle valve from the needle seat.
7. Inspect the needle valve tip for damage.
8. Remove the needle seat and gasket assembly. Inspect the gasket for damage.

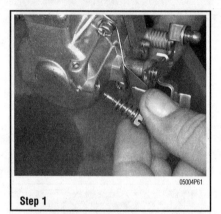

Step 1

Step 2

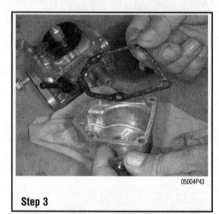

Step 3

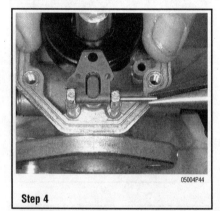

Step 4

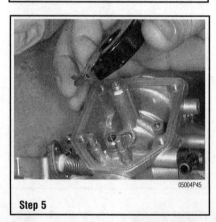

Step 5

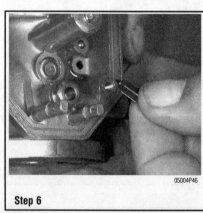

Step 6

Step 7

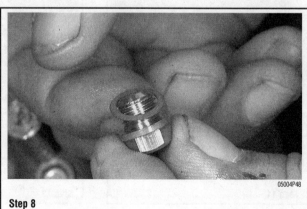

Step 8

FUEL SYSTEM 4-21

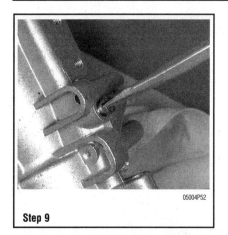
Step 9

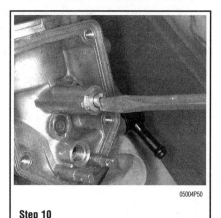
Step 10

Step 11

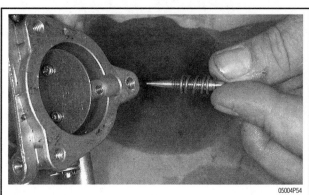
Step 15

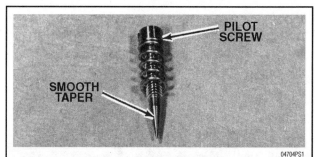

Fig. 41 Inspect the taper on the end of the pilot screw for ridges or signs of roughness. Good shop practice dictates a new pilot screw be installed each time the carburetor is overhauled

9. Remove the pilot jet.
10. Remove the main jet.
11. Inspect the main jet for damage or debris blocking the orifice. Look closely and you can read the jet number stamped on the jet itself.
12. Invert the mixing chamber and shake it, keeping a hand over the center turret. The main nozzle should fall free from the turret. If the nozzle refuses to fall out, gently reach in with a pick or similar instrument to raise the nozzle.
13. Obtain the correct size thin walled socket and remove the valve seat. Remove and discard the O-ring.
14. Remove the two Phillips head screws securing the top cover to the top of the mixing chamber. Lift off the cover.
15. Remove the pilot screw and spring from the carburetor. Count and record the number of turns in to a lightly seated position as a guide for installation. The specific number of turns will also be specified in the installation procedures.

CLEANING & INSPECTION

♦ See Figure 41

※※ CAUTION

Never dip rubber or plastic parts, in carburetor cleaner. These parts should be cleaned only in solvent, and then blown dry with compressed air.

Place all metal parts in a screen type tray and dip them in carburetor cleaner until they appear completely clean, then blow them dry with compressed air.

Blow out all passages in the castings with compressed air. Check all parts and passages to be sure they are not clogged or contain any deposits. Never use a piece of wire or any type of pointed instrument to clean drilled passages or calibrated holes in a carburetor.

Move the throttle and choke shafts back and forth to check for wear. If the shaft appears to be too loose, replace the complete mixing chamber because individual replacement parts are not available.

Inspect the mixing chamber, and fuel bowl gasket surfaces for cracks and burrs which might cause a leak. Check the float for deterioration. Check to be sure the needle valve spring has not been stretched. If any part of the float is damaged, the float must be replaced. Check the needle valve rubber tip contacting surface and replace the needle valve if this surface has a groove worn in it.

Inspect the tapered section of the pilot screw and replace the screw if it has developed a groove.

As previously mentioned, most of the parts which should be replaced during a carburetor overhaul are included in an overhaul kit available from your local marine dealer. One of these kits will contain a matched fuel inlet needle and seat. This combination should be replaced each time the carburetor is disassembled as a precaution against leakage.

ASSEMBLY

♦ See accompanying illustration

1. Slide a new spring over the pilot screw.
2. Install the pilot screw into the carburetor. Tighten the pilot screw until it barely seats. From this position, back out the screw the specified number of turns. Take notice, each year of manufacture has a different pilot screw setting. Furthermore, each carburetor has a different screw setting on certain models.
3. Slide a new O-ring over the shaft of the valve seat. Install and tighten the seat snugly, using a thin walled socket.
4. Insert the main nozzle into the aft hole on the center turret. Position the series of holes in the nozzle to face port and starboard when installed.
5. Install the main jet over the main nozzle. Tighten the jet until it seats snugly.
6. Install the pilot jet into the forward hole on the center turret. Tighten the jet until it seats snugly.
7. Check to be sure the wire clip is securely in position around the needle valve. Slide the clip over the tang on the float, and check to see if the needle valve can be moved freely.
8. Slide the hinge pin through the hole in the float.
9. Lower the float assembly over the center turret, guiding the needle valve into the needle seat and positioning the end of the hinge pin under its retaining screw. Tighten the screw securely.

4-22 FUEL SYSTEM

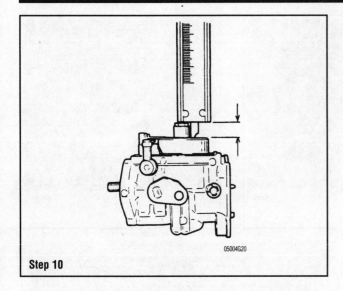

Step 10

10. Hold the mixing chamber in the inverted position, (as it has been held during the past few steps). Measure the distance between the carburetor body and the bottom of the float on all models with the exception of the 1989-99 DT25 and the 1988-97 DT30. These models are measured from the float bowl mating surface (gasket removed) to the bottom of the float. This distance should be as specified in the carburetor float height chart.

✷✷✷ CAUTION

Carefully bend the adjustment arm or tang when adjustment is necessary. DO NOT press down on the float. Downward pressure on the float will press the inlet needle into the valve seat and damage the needle tip.

V4 & V6 Powerheads

♦ See Figures 42 and 43

Suzuki V4 powerheads use two carburetors and V6 powerheads use three carburetors. Complete, detailed, illustrated, procedures to remove, service, and

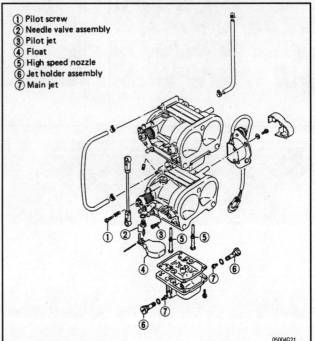

Fig. 42 Exploded view of the V4 carburetor assembly with the major components identified

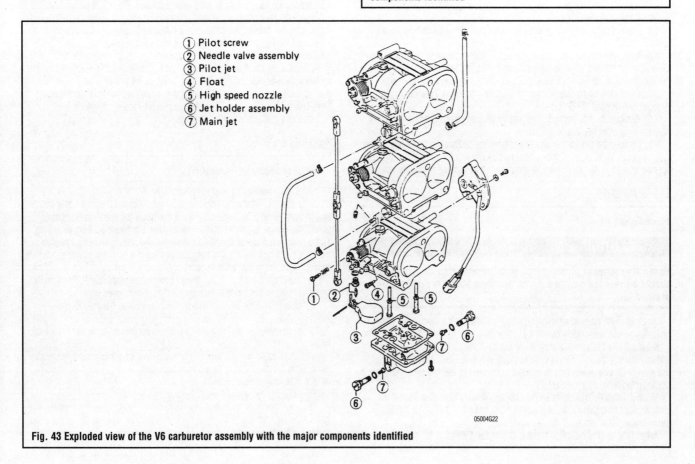

Fig. 43 Exploded view of the V6 carburetor assembly with the major components identified

FUEL SYSTEM   4-23

install the carburetors follow. Removal procedures may vary slightly due to differences in linkage.

As explained in the description of the carburetors, the float arrangement and adjustments differ due to a change in carburetor vendor. These differences are noted whenever they make a substantial change in a procedure.

REMOVAL & INSTALLATION

▶ **See accompanying illustrations**

1. Disconnect the battery negative cable.
2. Remove the engine cover.
3. Remove the fuel inlet hose from the fuel filter.
4. Disconnect the oil level switch lead wires
5. Disconnect the oil hoses.
6. Remove the oil tank.
7. Disconnect the throttle lever rod from the carburetors.
8. Remove the bolts and then remove the silencer cover.
9. Remove the throttle control lever and arm.
10. Snip the plastic hose retainers and gently pull off the fuel supply hoses from the carburetors.
11. After marking the position of the throttle sensor, remove the two securing screws and lift out the sensor from its plastic coupling. Slide the plastic coupling from the throttle shaft.

12. Identify each carburetor by inscribing or painting a 1, 2 and 3 (if applicable) on the float bowl cover to ensure each carburetor will be installed back into the same position from which it was removed.
13. Remove the mounting nuts, then remove the carburetors as an assembly. Each carburetor has a separate gasket which may either come away with the carburetor, or remain on the intake manifold. Remove and discard these gaskets.

Place the carburetor assembly on the workbench and remove each piece of linkage one at a time. Arrange the linkage on the workbench as it was installed on the carburetors, as an assist during assembling.

To install:

Install each marked carburetor (in the same location as disassembly) and gasket assembly on the engine and install the correct linkages.

14. Install the throttle sensor on the throttle shaft.
15. Install the fuel supply hoses to the carburetors and install the plastic hose retainers.
16. Innstall the throttle control lever and arm.
17. Connect the oil hoses.
18. Install the silencer cover and bolts.
19. Connect the throttle lever rod from the carburetors.
20. Connect the oil level switch lead wires and install the oil tank.
21. Install the fuel inlet hose from the fuel filter.
22. Install the engine cover.
23. Connect the battery negative cable.

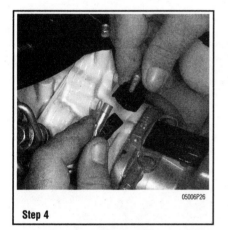
Step 4

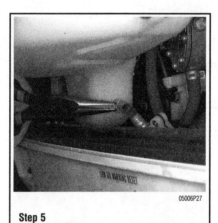
Step 5

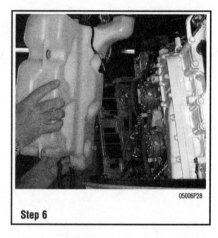
Step 6

Step 8

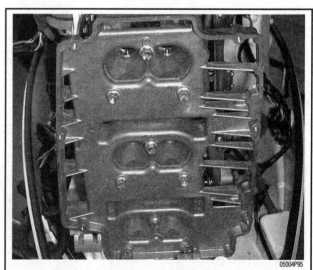
Step 13

4-24 FUEL SYSTEM

DISASSEMBLY

♦ See accompanying illustrations

1. Remove the float bowl attaching screws and remove the float bowl
2. Always discard the old gasket.
3. Slide the float hinge pin to one side and remove the pin assembly from the carburetor body.
4. Remove the float from the carburetor body.
5. Remove the inlet needle valve from the valve seat.
6. Remove the inlet needle valve seat using the appropriate tool.
7. Remove the pilot (idle) jet from the carburetor body:

➙ Before removing the pilot screw, turn it all the way in until it lightly seats. Record the number of turns needed to turn in the screw for reassembly.

8. Unscrew and remove the pilot (idle) screw and spring assembly.
9. Remove both high speed nozzles.
10. Remove the main jet holder from each side of the float bowl.
11. Remove the main jet from each main jet holder with the appropriate tool.
12. Remove and always discard the O-ring on each main jet holder.
13. Clean and inspect all the internal parts in the carburetor.

FUEL LEVEL TEST

For this test, you will need to acquire the following Suzuki special tools: Fuel Level Gauge Adaptor (p/n 09913–18711) and Fuel Level Gauge (p/n 09932–28211)

1. Remove the engine cover.
2. Connect a tachometer to the engine following the manufacturers instructions.
3. Remove the left-hand main jet holder from the carburetor float bowl.
4. Remove the main jet and O-ring from the main jet holder.
5. Install the main jet into the special tool adaptor.
6. Install the fuel level gauge into the and install the entire assembly into the carburetor float bowl.
7. Start the engine and let it idle. If necessary, readjust the idle to the speed specified in the "Tune-Up Specifications" chart.
8. The height of the fuel in the gauge should be as follows: 0.83–0.91 in. (21.0–23.0 mm).
9. If the fuel level is not to specification, adjust the fuel level by adjusting the float level tab.
10. Remove the main jet and O-ring from the adaptor.
11. Reinstall the main jet and O-ring onto the main jet holder and install them into the carburetor float bowl.
12. Repeat this procedure with the other carburetors.
13. Remove the tachometer.

Step 1

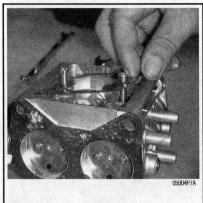

Step 2

Step 3

Step 4

Step 5

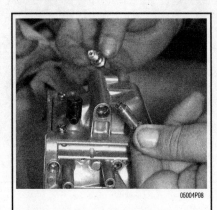

Step 10

Step 11

FUEL SYSTEM   4-25

CLEANING & INSPECTION

♦ See Figure 44

> ※ **CAUTION**
>
> **Never dip rubber or plastic parts in carburetor cleaner. These parts should be cleaned only in solvent, and then blown dry with compressed air.**

Place all metal parts in a screen type tray and dip them in carburetor cleaner until they appear completely clean, then blow them dry with compressed air.

Blow out all passages in the castings with compressed air. Check all parts and passages to be sure they are not clogged or contain any deposits. Never use a piece of wire or any type of pointed instrument to clean drilled passages or calibrated holes in a carburetor.

Move the throttle and choke shafts back and forth to check for wear. If the shaft appears to be too loose, replace the complete mixing chamber because individual replacement parts are not available.

Inspect the mixing chamber, and fuel bowl gasket surfaces for cracks and burrs which might cause a leak. Check the floats for deterioration. Check to be sure the needle valve loop has not been stretched. If any part of the float is damaged, the float must be replaced. Check the needle valve tip contacting surface and replace the needle valve if this surface has a groove worn in it.

Inspect the tapered section of the pilot screw and replace the screw if it has developed a groove.

As previously mentioned, most of the parts which should be replaced during a carburetor overhaul are included in an overhaul kit available from your local marine dealer. One of these kits will contain a matched fuel inlet needle and seat. This combination should be replaced each time the carburetor is disassembled as a precaution against leakage.

ASSEMBLY

♦ See accompanying illustration

1. Install the main jet in the main jet holder with a new O-ring.
2. Install the main jet holders into the float bowl.
3. Install the high speed nozzles.
4. Install the pilot screw and spring. Turn in the screw until it lightly seats, then back it out the correct number of turns.
5. Install the pilot (idle) jet.
6. Install the fuel inlet needle valve and valve seat into the carburetor body.
7. Install the float.
8. Invert the carburetor and lower the float until the adjusting tab on the float just touches the needle. Hold the float in the is position and measure the height between the carburetor body (with gasket removed) and the bottom of the float.
9. If the float level is not within specification, adjust the level by bending the adjusting tab as needed.

Step 8

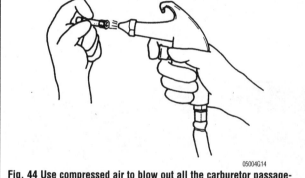

Fig. 44 Use compressed air to blow out all the carburetor passageways

REED VALVE SERVICE

The reed valves operate in response to changes in crankcase pressure. Located between the intake manifold and the crankcase, the reed valves admit the air-fuel mixture into the crankcase and during the scavenging stroke, act as a one-way valve to prevent the mixture from flowing back into the intake manifold. The travel of the reed itself is limited by the reed stop. By this action, the scavenging action is improved and the engine will produce greater power.

On all Suzuki models, except the DT2 and DT2.2, the reed valves are located between the intake manifold and the crankcase. On the DT2 models the reed valves are an integral part of the front crankcase half of the engine.

DT2, DT2.2

REMOVAL & INSTALLATION

On these models, the reed valves are located in the front half of the engine crankcase. In order to access the reed valves on this model, the engine must be removed and the crankcase separated. See "Powerhead Overhaul".

All Other Models

♦ See Figures 45 and 46

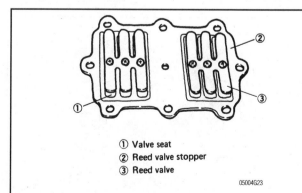

① Valve seat
② Reed valve stopper
③ Reed valve

Fig. 45 Typical in-line reed valve assembly

4-26 FUEL SYSTEM

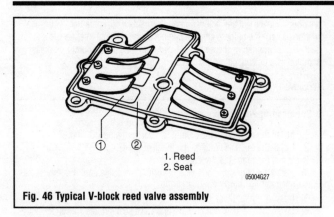

1. Reed
2. Seat

Fig. 46 Typical V-block reed valve assembly

REMOVAL & INSTALLATION

♦ See accompanying illustrations

1. Remove the carburetor assembly.
2. Disconnect any hose attached to the intake manifold.

➥On some models, the reed valve assembly is secured by separate fasteners. On models with this design, the intake manifold must first be removed in order to remove the reed valve and gasket assembly.

3. Remove the intake manifold fasteners holding the intake manifold to the crankcase cover.
4. Remove the intake manifolds.
5. Remove the gaskets and reed valve assemblies from the crankcase. Make sure to discard all used gaskets.
6. Clean all the mounting surfaces of any sealant or gasket residue.

To install:

7. Make sure during installation that the reeds are installed in the correct direction.
8. Install new gaskets and tighten the reed valve assembly to the crankcase.
9. Reconnect any hoses to the intake manifold.
10. Install the carburetor assembly.

INSPECTION & CLEANING

♦ See Figures 47 and 48

1. Check the reeds for sign of cracking, wear or any other damage. Replace the reeds if any damage is found.
2. Check the reeds to see if they lie flat on the valve seat with no preload on them.
3. To check the flatness of the reed, gently push each reed out from the seat. Constant resistance should be felt while pushing the reed.
4. Check the clearance between the reed and the seat with a feeler gauge. If the clearance is greater than 0.008 in. (0.20 mm), you will need to replace the reed set.
5. Measure the distance between the reed stop and the valve seat. If the measurement is not within specifications, check the valve seat for warpage and replace it as required. If the seat is okay, replace the reed stop assembly.

REED & REED STOP REPLACEMENT

♦ See Figure 49

1. Remove the screws holding the reed stop and reeds to the valve seat.
2. Remove the reed stop and reeds.
3. Place the new reed on the valve seat and check the assembly for flatness.
4. Center the reed over the valve seat openings.
5. Before installing the reed stop screws, apply a thread locker.
6. Check the reed tension and range of motion.

Step 3

Step 4

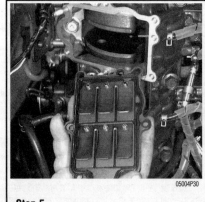

Step 5

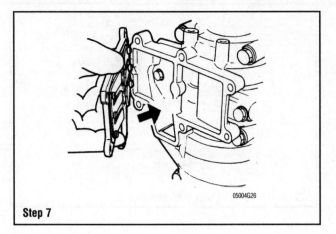

Step 7

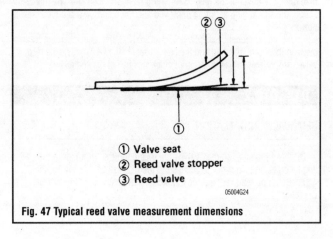

① Valve seat
② Reed valve stopper
③ Reed valve

Fig. 47 Typical reed valve measurement dimensions

FUEL SYSTEM 4-27

Fig. 48 Check the clearance between the reed and the seat with a feeler gauge

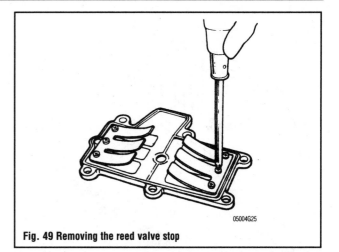

Fig. 49 Removing the reed valve stop

FUEL PUMP SERVICE

Diaphragm Type Fuel Pumps

DESCRIPTION & OPERATION

The smaller capacity powerheads do not have a fuel pump of any type. Fuel is provided to the carburetor by gravity flow from the fuel tank atop the powerhead.

The rest of the carbureted models use a diaphragm-type fuel pump. These pumps are operated by crankcase pressures. These pressure pulses are created by the movement of the piston in the crankcase and are directed to the pump by a passageway in the crankcase wall.

The piston moving upwards creates a low-pressure on the diaphragm in the pump body. This low pressure opens a check valve in the pump body, allowing fuel to be drawn from the fuel line to the pump. At the same time, the low pressure in the crankcase draws the air-fuel mixture into the crankcase from the intake manifold and carburetors.

Downward motion of the piston creates a high-pressure on the diaphragm. This high-pressure closes the inlet check valve in the pump body and open the outlet check valve, forcing the fuel in the fuel lines into the carburetor float bowl and moving the fuel-air mixture in the crankcase into the combustion chamber where ignition occurs.

This type of fuel pump is not powerful enough to draw fuel from the fuel tank during cranking. In that case, it is necessary to introduce fuel into the carburetor with the priming bulb mounted on the fuel line.

Suzuki uses both powerhead mounted, remote pumps and integral carburetor/fuel pumps. Both designs are very simple and reliable. The most common failure is to the diaphragm, and dirty fuel can block the check valve, causing it to fail. Most fuel systems are equipped with a separate fuel filter and some are equipped with a fuel/water separating filter assembly.

REMOVAL & INSTALLATION

Remote Mounted

1. Loosen the hose clamps and slide the fuel lines off the pump body. Plug the hoses to prevent fuel leaks.
2. Remove the screws holding the pump to the powerhead and pull the fuel pump off the power head.
3. If so equipped, remove the fuel pump insulator from the powerhead.
4. Throughly clean the pump mounting surface of any gasket or sealer residue. Do not scratch or gouge the sealing surface of the powerhead.

To install:
5. If needed, install the fuel pump insulator and new O-ring.
6. Install the pump onto the powerhead with a new gasket and O-ring in the pump body.

Carburetor Mounted

1. Remove the fuel hose at the pump inlet cover and plug the line.
2. Remove the retaining screws that hold the pump to the carburetor.
3. Remove the pump as an assembly.

To install:
4. Reassemble the pump components in the reverse order of disassembly, using new gaskets.
5. Mount the pump assembly back on the side of the carburetor. Do not overtighten the mounting screws.
6. Unplug and reinstall the fuel hoses.

OVERHAUL

Remote Mounted

▶ See Figures 50 thru 61

There are several variations in the pumps used on the different models. All operate in the same manner, but there are differences in the internal

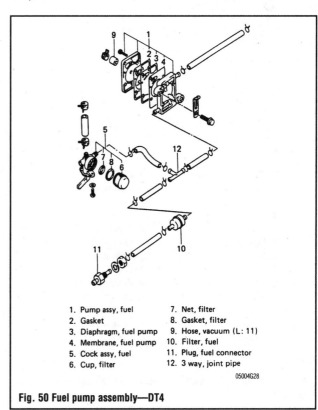

1. Pump assy, fuel
2. Gasket
3. Diaphragm, fuel pump
4. Membrane, fuel pump
5. Cock assy, fuel
6. Cup, filter
7. Net, filter
8. Gasket, filter
9. Hose, vacuum (L: 11)
10. Filter, fuel
11. Plug, fuel connector
12. 3 way, joint pipe

Fig. 50 Fuel pump assembly—DT4

4-28 FUEL SYSTEM

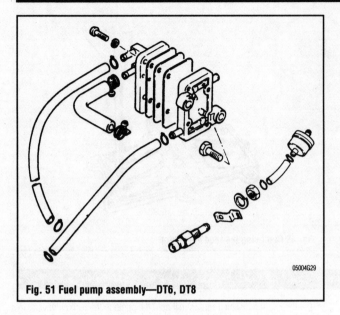

Fig. 51 Fuel pump assembly—DT6, DT8

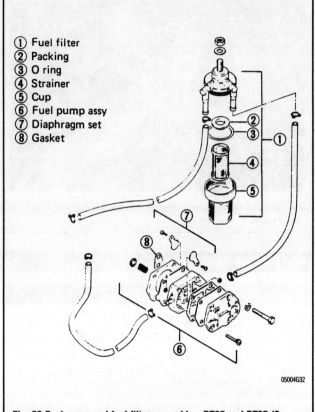

① Fuel filter
② Packing
③ O ring
④ Strainer
⑤ Cup
⑥ Fuel pump assy
⑦ Diaphragm set
⑧ Gasket

Fig. 53 Fuel pump and fuel filter assembly—DT25 and DT35 (3-cylinder)

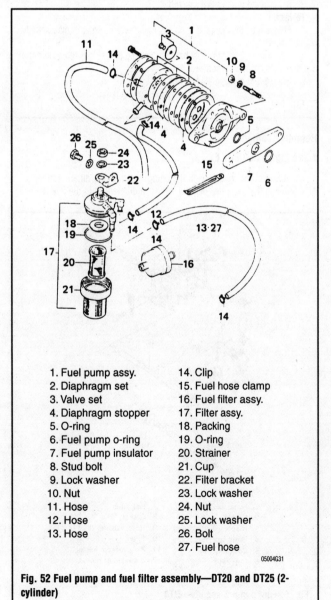

1. Fuel pump assy.
2. Diaphragm set
3. Valve set
4. Diaphragm stopper
5. O-ring
6. Fuel pump o-ring
7. Fuel pump insulator
8. Stud bolt
9. Lock washer
10. Nut
11. Hose
12. Hose
13. Hose
14. Clip
15. Fuel hose clamp
16. Fuel filter assy.
17. Filter assy.
18. Packing
19. O-ring
20. Strainer
21. Cup
22. Filter bracket
23. Lock washer
24. Nut
25. Lock washer
26. Bolt
27. Fuel hose

Fig. 52 Fuel pump and fuel filter assembly—DT20 and DT25 (2-cylinder)

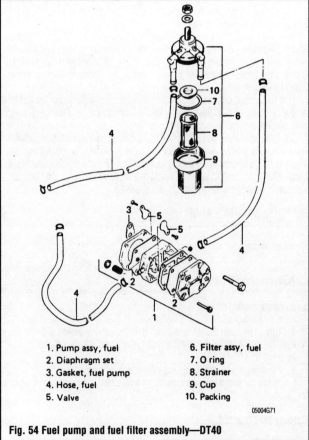

1. Pump assy, fuel
2. Diaphragm set
3. Gasket, fuel pump
4. Hose, fuel
5. Valve
6. Filter assy, fuel
7. O ring
8. Strainer
9. Cup
10. Packing

Fig. 54 Fuel pump and fuel filter assembly—DT40

FUEL SYSTEM 4-29

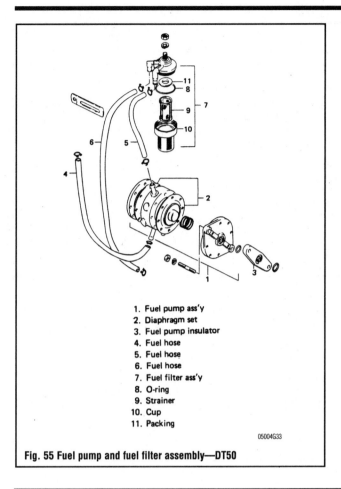

1. Fuel pump ass'y
2. Diaphragm set
3. Fuel pump insulator
4. Fuel hose
5. Fuel hose
6. Fuel hose
7. Fuel filter ass'y
8. O-ring
9. Strainer
10. Cup
11. Packing

Fig. 55 Fuel pump and fuel filter assembly—DT50

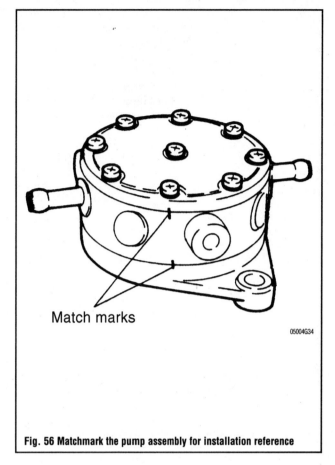

Fig. 56 Matchmark the pump assembly for installation reference

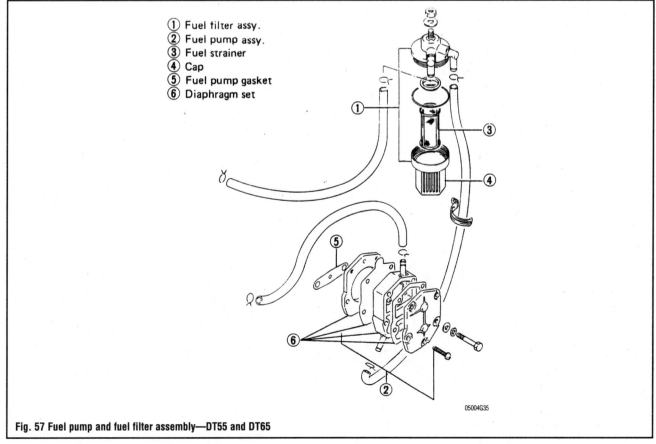

① Fuel filter assy.
② Fuel pump assy.
③ Fuel strainer
④ Cap
⑤ Fuel pump gasket
⑥ Diaphragm set

Fig. 57 Fuel pump and fuel filter assembly—DT55 and DT65

4-30 FUEL SYSTEM

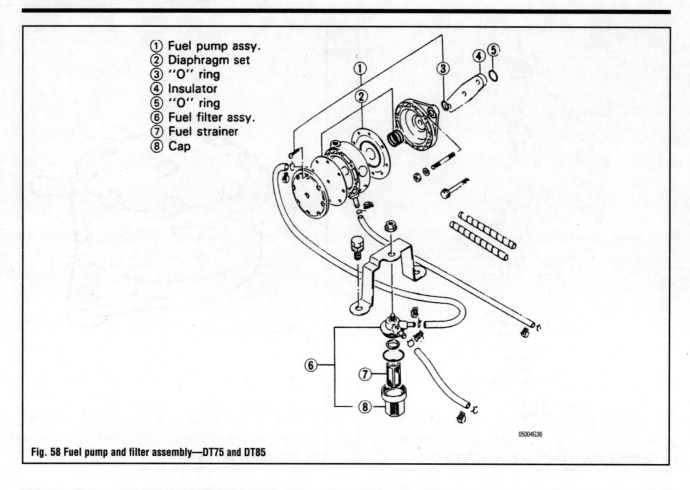

Fig. 58 Fuel pump and filter assembly—DT75 and DT85

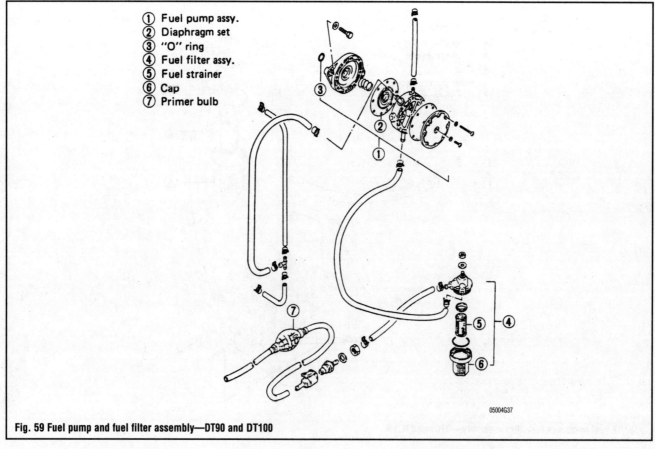

Fig. 59 Fuel pump and fuel filter assembly—DT90 and DT100

FUEL SYSTEM   4-31

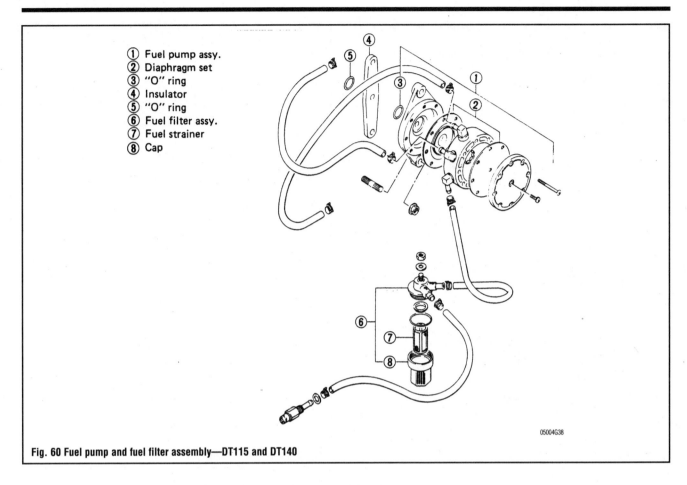

Fig. 60 Fuel pump and fuel filter assembly—DT115 and DT140

① Fuel pump assy.
② Diaphragm set
③ "O" ring
④ Insulator
⑤ "O" ring
⑥ Fuel filter assy.
⑦ Fuel strainer
⑧ Cap

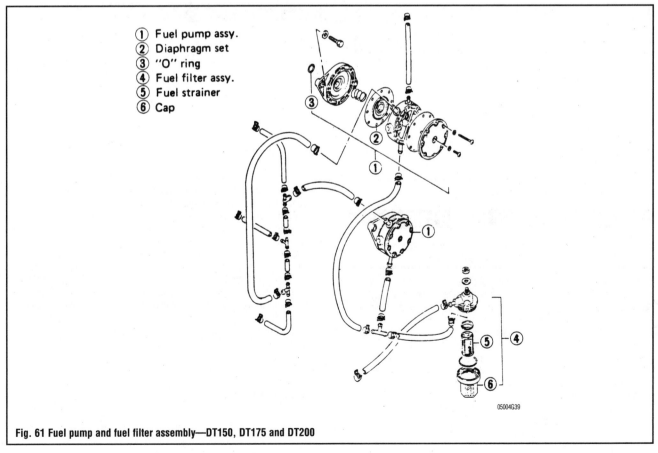

Fig. 61 Fuel pump and fuel filter assembly—DT150, DT175 and DT200

① Fuel pump assy.
② Diaphragm set
③ "O" ring
④ Fuel filter assy.
⑤ Fuel strainer
⑥ Cap

4-32 FUEL SYSTEM

parts. Make sure to note the location of any of these parts during disassembly.

Mark the pump assembly prior to disassembly. This will ensure that the pump will go back together in the order that it was taken apart.

Refer to the exploded view drawings in this section during overhaul.

Carburetor Mounted

♦ See Figures 62 and 63

Mark the pump assembly prior to disassembly. This will ensure that the pump will go back together in the order that it was taken apart.

Refer to the exploded view drawing in this section during overhaul.

Separate the pump cover, diaphragm, valve assembly and pump body. Discard the used gaskets.

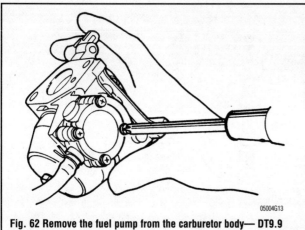

Fig. 62 Remove the fuel pump from the carburetor body— DT9.9 and DT15

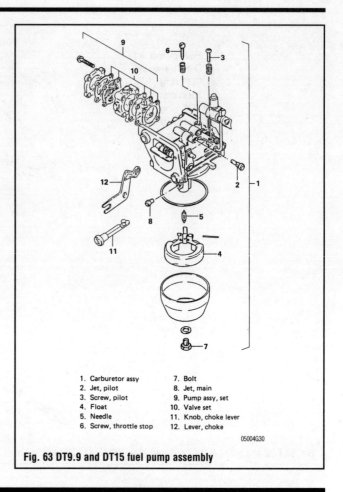

1. Carburetor assy
2. Jet, pilot
3. Screw, pilot
4. Float
5. Needle
6. Screw, throttle stop
7. Bolt
8. Jet, main
9. Pump assy, set
10. Valve set
11. Knob, choke lever
12. Lever, choke

Fig. 63 DT9.9 and DT15 fuel pump assembly

ELECTRONIC FUEL INJECTION

Description and Operation

The fuel injection control system on the DT115, DT140, DT150, DT200 and DT225 models, detects the current condition of the engine through various sensors mounted on the engine. These sensors send information to the micro computer as electrical signals. Based on these electrical signals, the computer determines the optimal amount of fuel needed for the engines current condition and calculates the injection duration time. The fuel injectors receive this duration signal and inject the fuel into the intake manifold.

FUEL INJECTION BASICS

♦ See Figure 64

※※ CAUTION

Precautions must be taken not to cause damage to expensive electronic components during these service procedures.

Fuel injection is not a new invention. Even as early as the 1950s, various automobile manufacturers experimented with mechanical-type injection systems. There was even a vacuum tube equipped control unit offered for one system! This might have been the first "electronic fuel injection system." Early problems with fuel injection revolved around the control components. The electronics were not very smart or reliable. These systems have steadily improved since. Today's fuel injection technology, responding to the need for better economy and emission control, has become amazingly reliable and efficient. Computerized engine management, the brain of fuel injection, continues to get more reliable and more precise.

Components needed for a basic computer-controlled system are as follows:
• A computer-controlled engine manager, which is the Electronic Control Unit (or ECU), with a set of internal maps to follow.
• A set of input devices to inform the ECU of engine performance parameters.
• A set of output devices. Each device is controlled by the ECU. These devices modify fuel delivery and timing. Changes to fuel and timing are based on input information matched to the map programs.

This list gets a little more complicated when you start to look at specific components. Some fuel injection systems may have twenty or more input devices. On many systems, output control can extend beyond fuel and timing. The Suzuki Fuel Injection System provides more than just the basic functions, but is still straight forward in its layout. There are twelve input devices and six output controls. The diagram on the following page shows the input and output devices with their functions.

There are several fuel injection delivery methods. Throttle body injection is relatively inexpensive and was used widely in early automotive systems. This is usually a low pressure system running at 15 PSI or less. Often an engine with a single carburetor was selected for throttle body injection. The carburetor was recast to hold a single injector, and the original manifold was retained. Throttle body injection is not as precise or efficient as port injection.

Multi-port fuel injection is defined as one or more electrically activated solenoid injectors for each cylinder. Multi-port injection generally operates at higher pressures than throttle body systems. The Suzuki system operates at 35.5 PSI.

Port injectors can be triggered two ways. One system uses simultaneous injection. All injectors are triggered at once. The fuel "hangs around" until the pressure drop in the cylinder pulls the fuel into the combustion chamber.

The second type is more precise and follows the firing order of the engine. Each cylinder gets a squirt of fuel precisely when needed.

SUZUKI ELECTRONIC FUEL INJECTION

♦ See Figure 65

Suzuki uses the multi-point, sequential method of fuel injection. The whole system can be divided into three areas: air intake system, fuel system and control system.

FUEL SYSTEM 4-33

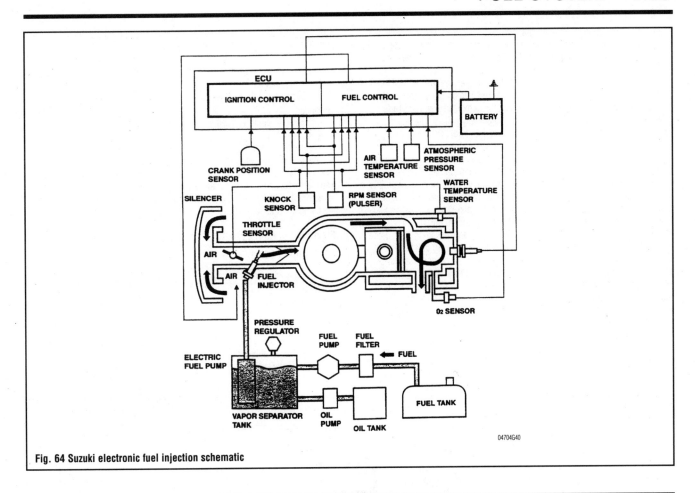

Fig. 64 Suzuki electronic fuel injection schematic

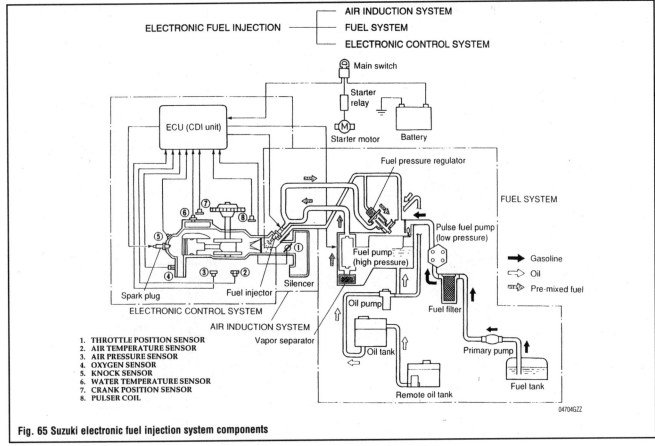

Fig. 65 Suzuki electronic fuel injection system components

4-34 FUEL SYSTEM

Air flows through the throttle body and surge tank to the intake manifold, where the intake air is mixed with a combined fuel/oil mixture that has been delivered through the fuel injector. The throttle sensor assesses the throttle opening angle and sends the corresponding signal to the control unit.

Fuel is drawn through the primer bulb to the low pressure fuel pump. Upon reaching the vapor separator, the gasoline is mixed with oil (supplied directly from the oil pump to the vapor separator) and the combined fuel/oil mixture goes through the high-pressure fuel pump and delivery gallery. The pressure regulator maintains the fuel pressure in the feed line, from the high pressure fuel pump to the fuel injector. This pressure, maintained at a constant level, is higher than the manifold pressure (injector nozzle ambient pressure). When the higher fuel pressure in the feed line exceeds the pressure in the manifold by more than approximately 36.6 psi (2.55 kg-cm), the valve in the pressure regulator will open, thereby returning fuel to the vapor separator through the fuel return hose. The fuel/oil mixture is injected into the intake manifold by the fuel injector when the signal is supplied from the engine control unit (ECU).

The electronic fuel injection control system is primarily based on the signals supplied from both the engine speed sensor (gear counter coil) and the throttle valve sensor (TPS). To compensate for other conditions, a cylinder wall temperature sensor, air temperature sensor and atmospheric pressure sensor are also used. Based on these sensor signals, the engine control unit (ECU) determines the injection time duration and sends the injection signal to the injector. The ECU's sequential multi-point programming provides individual control of each injector operating duration and timing.

An additional feature of the Suzuki system is return-to-port capability or fail safe mode. If there is a major sensor failure that prevents the ECU from processing, the engine is run on a minimum performance map. If either the counter coil or pulser coil fail, the ECU will not provide an injection signal without a reference signal from these coils. Under this condition, the engine can be cranked but it will not start without fuel injection.

The only input that will shut the engine down completely is battery voltage. If the battery is disconnected or the battery voltage falls below 9 volts, the fuel pump quits pumping so the engine stops.

The Suzuki fuel injection system is divided into three component groups. These are:
- Electronic control system
- Fuel system
- Air intake system

The air induction and fuel systems are the delivery agents for the fuel/air mixture. The fuel delivery system is controlled by the electronic control system. The electronic control system can be further subdivided into input sensors (the informants), outputs (the workers that make adjustments to fuel and timing), and the computer (the boss or decision maker).

Electronic Control System

▶ See Figure 66

The electronic control system has the largest number of individual components. Each component is connected to the control unit. The control unit functions in several modes

FAIL SAFE MODE

When any of the sensors of the ECU fails, the self diagnostic system will indicate the failure location by means of the indicator lamp flashing out the code for the failure. In the event of such a failure, another system, called the back-up sys-

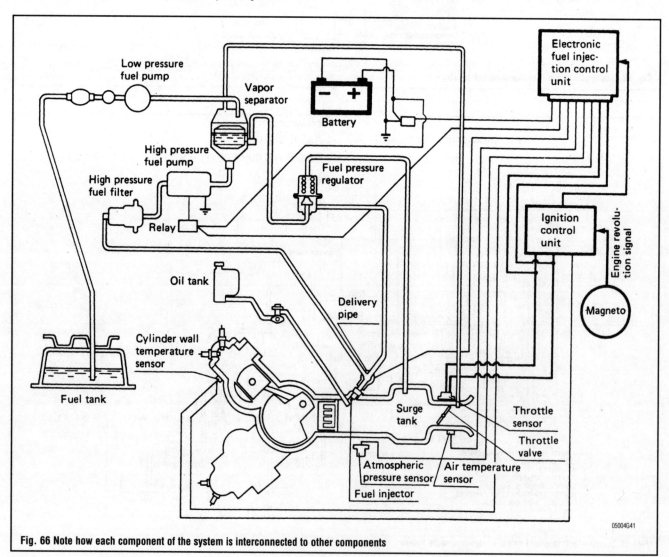

Fig. 66 Note how each component of the system is interconnected to other components

FUEL SYSTEM 4-35

tem, will come into operation. This system provides alternative signals to compensate for the ECU or sensor failure so that the engine will not stop running and continue to operate. Because of this provision, an emergency, return-to-dock operation will be possible, but the engine will run in a much reduced state of tune.

FUEL PUMP CONTROL

♦ See Figure 67

To supply the optimal amount of fuel, the ECM controls the duty cycle signal for the fuel pump drive, which repeats the ON/OFF cycle at a specified rate (1,000 times a second). Based on engine speed, the ECM determines the optimal duty (repeating the "ON" time rate within a cycle) and sends the signal to the fuel pump.

INJECTION MODE

♦ See Figure 68

During starting conditions, fuel is sequentially injected into each cylinder while cranking.

After the powerhead is started, fuel control falls under one of three modes.

1. Warm-up mode. For the first 3 minutes upon start-up, the ECU controls the fuel injection duration as the "Warm-up (Enrichment) mode". This mode is

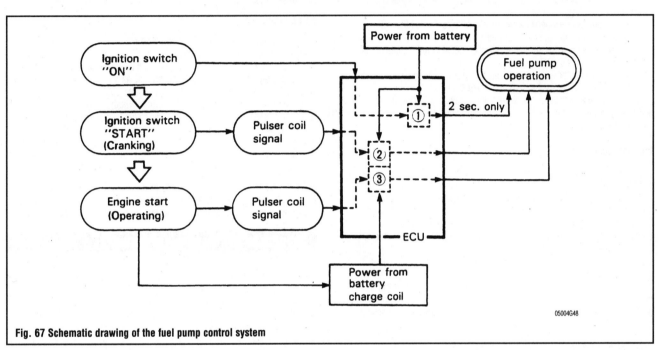

Fig. 67 Schematic drawing of the fuel pump control system

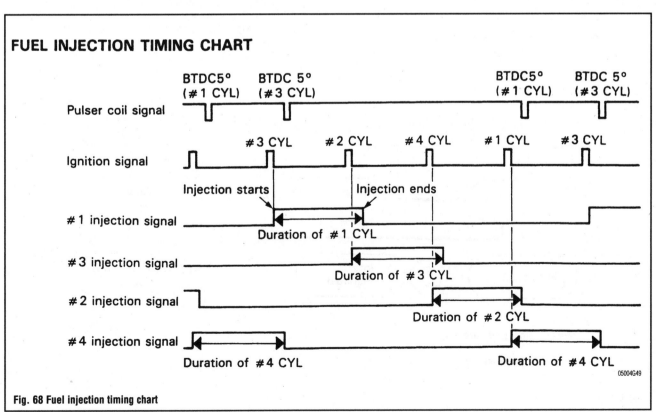

Fig. 68 Fuel injection timing chart

4-36 FUEL SYSTEM

based on the map control in relation to engine speed, throttle valve opening angle and cylinder wall temperature.

2. Normal operating mode. Three minutes after starting, the ECU changes to the "Normal operating mode". This mode is based on the map control in relation to engine speed, throttle valve opening angle, cylinder wall temperature, air temperature and atmospheric pressure.

3. Fail safe mode. Each sensor has an assigned default value programmed into the ECU. In the event of a sensor failure, the monitor gauge will flash a code indicating the failure and the engine will continue to operate, but with reduced performance. Injection duration during sensor failure automatically defaults to the following control methods:

- Throttle valve sensor failure: Injection duration will be automatically set according to engine speed
- Cylinder wall temperature sensor failure: injection duration will be automatically set if the sensor senses 30°C
- Air temperature sensor failure: Injection duration will be automatically set if the sensor senses 20°C
- Atmospheric pressure sensor failure: Injection duration will be automatically set if the sensor senses 763 mm Hg (101.7 kPa)

If either the gear counter coil or the pulser coil fails, the ECU will not provide and injection signal without a reference from these coils. Under this condition, the engine can be cranked but it will not start due to no fuel being injected.

Fuel Delivery System

Fuel injectors require clean, water-free, pressurized gasoline. Fuel is supplied through the primer bulb, fuel filter and low pressure fuel pump to the fuel vapor separator. From the fuel vapor separator, the fuel goes through the high pressure fuel pump and on to the fuel rail. The pressure regulator maintains the fuel pressure in the feed line, from the high pressure fuel pump to the fuel injector. This pressure, maintained at a constant level, is higher than the barometric pressure. When the higher fuel pressure in the feed line exceeds the barometric pressure by more than approximately 36.3 psi (255 kPa), the valve in the fuel pressure regulator will open, allowing fuel to return to the fuel vapor separator through the fuel return hose. The fuel is injected into the intake manifold by the fuel injector when the signal is supplied by the ECU. Oil mixing is accomplished with a standard oil pump. This pump injects oil into the vapor separator tank.

FUEL SUPPLY AND VAPOR SEPARATOR TANK

♦ See Figures 69, 70 and 71

The vapor separator incorporates a float system which maintains a constant fuel level inside the body of the unit. When the fuel level decreases, more fuel will flow into the vapor separator from the low pressure fuel pump. The function of this unit is to separate the vapor from the fuel delivered by the low pressure fuel pump, or fuel that has returned from the pressure regulator. This vapor is "bled" to the throttle body through the hose on the top of the vapor separator.

OIL INJECTION

The separator also functions as the mixing chamber for oil and fuel. Oil is injected into the separator tank by the engine-driven oil pump. The amount of oil delivered to the separator tank is dependent upon engine RPM and throttle opening. The supply side of the oil injection system is similar to previous models.

Air Intake System

♦ See Figure 72

The throttle body assembly including the throttle valve sensor, is a precision-made part which detects the throttle valve opening angle and supplies a signal to the ECU. Therefore, when replacing any of the components within the throttle body assembly, special care should be taken so as not to alter the precise clearances and operating efficiency of the throttle body.

The air which ahs passed through the throttle body, flows into the surge tank/intake manifold and is then distributed to the reed valves. The throttle body adjusts the air intake amount using the throttle valve which is connected to the throttle lever and Throttle Position (TP) sensor. The TP sensor is installed on the top of the throttle body and informs the ECM of the current throttle valve opening angle and is used as a basic signal for the engine control system.

→Do not try to adjust or remove any of the component parts (sensor base plate, throttle valve, throttle linkage lever, etc.). The TP sensor must be installed and adjusted by following the proper procedure.

Cylinder Wall Temperature Sensor

DESCRIPTION & OPERATION

The cylinder wall temperature sensor (a thermistor type) is located on the cylinder bank port side. The thermistor's ability to change resistance with temperature is used to measure engine temperature. Voltage flow through the sensor, it is modified (increased/decreased) by resistance, returns to the ECU as a voltage signal measurement of the engine temperature. The ECU map then compensates by effecting a temperature based change in fuel injection duration. For example, a cold engine requires a richer fuel mixture for starting than a fully warmed up engine. This temperature related resistance change also serves as the signal for compensating ignition timing and overheat detection.

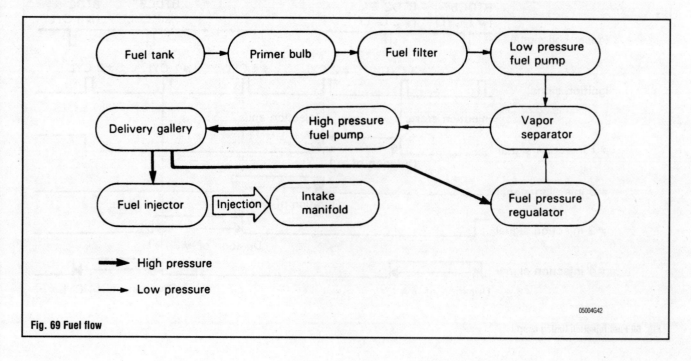

Fig. 69 Fuel flow

FUEL SYSTEM 4-37

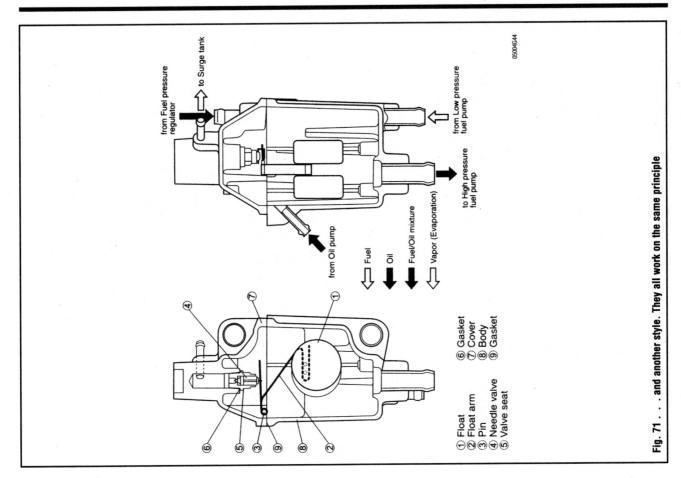

Fig. 71 ... and another style. They all work on the same principle

① Float ⑥ Gasket
② Float arm ⑦ Cover
③ Pin ⑧ Body
④ Needle valve ⑨ Gasket
⑤ Valve seat

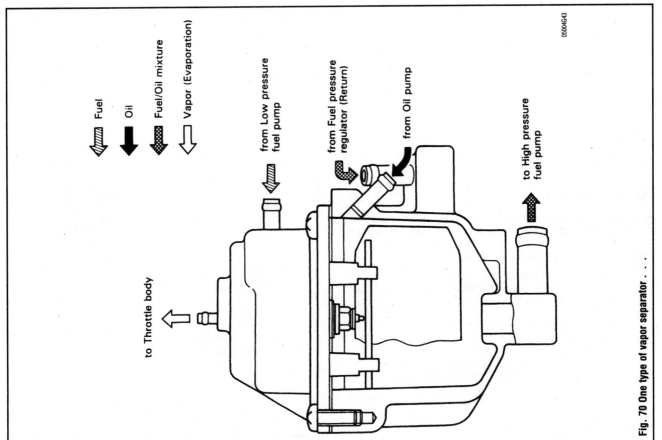

Fig. 70 One type of vapor separator ...

4-38 FUEL SYSTEM

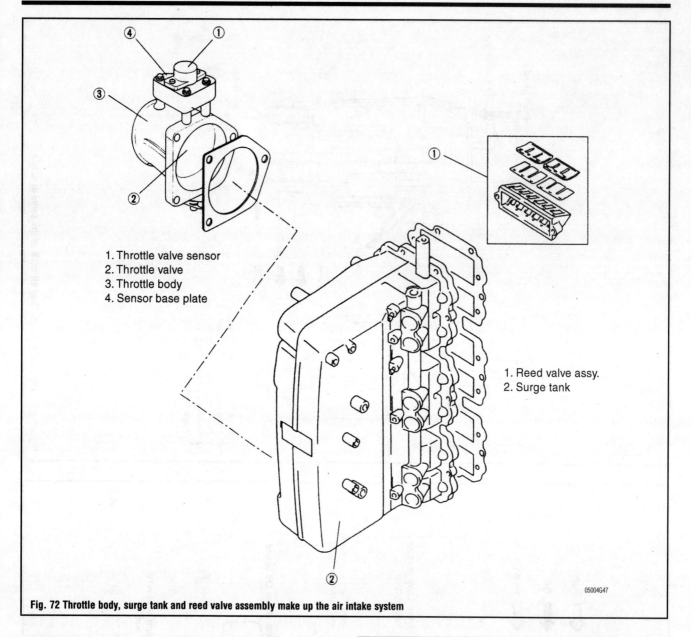

Fig. 72 Throttle body, surge tank and reed valve assembly make up the air intake system

1. Throttle valve sensor
2. Throttle valve
3. Throttle body
4. Sensor base plate

1. Reed valve assy.
2. Surge tank

TESTING

♦ See Figures 73 and 74

➡ The sensor inspection must be performed with the battery connected. If an out of range condition exists in the sensor signal sent back to the ECU, the self-diagnosis system will be indicated by the "CHECK ENGINE" monitor gauge lamp flashing the code.

1. Connect the multimeter probes to each of the sensor leads.
2. Immerse the sensor's tip in water and gradually heat the water while monitoring the changes in sensor resistance. Check if resistance matches specifications:
   - 32°F (0°C): 5.3–6.6 kilo ohms
   - 77°F (25°C): 1.8–2.3 kilo ohms
   - 135°F (75°C): 0.33–0.45 kilo ohms
3. On the DT225, check for continuity through the engine harness between the ECU 2-pin connector Lg/W terminal and the sensor Lg/W lead as shown.
4. Disconnect the harness to the ECU 2-pin connector and sensor Lg./W lead.
5. Connect the tester as shown.
6. Check for an indication of continuity (0 Resistance)

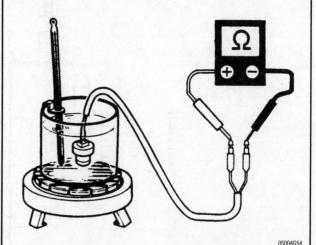

Fig. 73 Immerse the sensor's tip in water and gradually heat the water while monitoring the changes in sensor resistance

FUEL SYSTEM 4-39

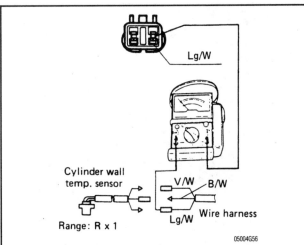

Fig. 74 On the DT225, check for continuity through the engine harness between the ECU 2-pin connector Lg/W terminal and the sensor Lg/W lead as shown

REMOVAL & INSTALLATION

1. Disconnect the battery negative cable.
2. Remove the engine cover.
3. Locate the sensor on the engine and remove it using the appropriate tools.

To install:

4. Replace the gasket or O-ring as equipped.
5. Connect the sensor lead connector securely.
6. Install the sensor on the engine.
7. Replace the engine cover.
8. Connect the battery negative cable.

Air Temperature Sensor

DESCRIPTION & OPERATION

The air temperature sensor, also a thermistor, is located above the throttle body. Voltage flow through the sensor, is modified (increased/decreased) by resistance, returns to the ECU as a voltage signal measurement of the air temperature flowing into the throttle body. As air intake temperature changes, air density also changes, which affects the air/fuel mixture ratio. The ECU map then compensates for these air density changes to the mixture by adjusting the fuel injector duration time.

TESTING

1. Connect the multimeter probes to each of the sensor leads.
2. Immerse the sensor's tip in water and gradually heat the water while monitoring the changes in sensor resistance. Check if resistance matches specifications:
   - 32°F (0°C): 5.3–6.6 kilo ohms
   - 77°F (25°C): 1.8–2.3 kilo ohms
   - 135°F (75°C): 0.33–0.45 kilo ohms

REMOVAL & INSTALLATION

1. Disconnect the battery negative cable.
2. Remove the engine cover.
3. Locate the sensor on the engine and remove it using the appropriate tools.

To install:

4. Replace the gasket or O-ring as equipped.
5. Connect the sensor lead connector securely.
6. Install the sensor on the engine.
7. Replace the engine cover.
8. Connect the battery negative cable.

Atmospheric Pressure Sensor

DESCRIPTION & OPERATION

The atmospheric pressure sensor is located above the throttle body and measures the ambient air pressure in which the engine is operating. As pressure affects the sensors ability to conduct voltage, the sensor voltage signal returning to the ECU is used as a measure of current air pressure. As altitude changes, atmospheric pressure changes, this affects the air portion of the fuel/air mixture. The ECU map then compensates for these changes by effecting a pressure based change in fuel injection duration. For example, an engine operating at sea level with more air pressure requires a richer fuel mixture than an engine operating in lower air pressure at 5,000 feet altitude.

TESTING

DT115 and DT140

The following special tool is required to perform the sensor output test:
- 6-pin connector test cord (09930–89251)

1. With the multimeter set on DC volts, connect the tester positive (red) to the gray lead on the test cord and the tester negative (black) to ground.
2. Turn the ignition on and check the voltage. Output should measure approximately 3.64 volts at 14.6 psi (101.3 kPa).
3. If the sensor does not meet this specification, replace the sensor.

➡ **Make sure the wiring harness assembly between the ECU and sensor has continuity.**

DT150, DT150 and 1988–93 DT200

1. With the multimeter set on DC volts, connect the negative (black) to the sensor black wire and the tester positive (red) to the sensor white wire.
2. Turn the ignition on and check the voltage. Output should measure approximately 3.64 volts at 14.6 psi (101.3 kPa).
3. If the sensor does not meet this specification, replace the sensor.

DT225 and 1994–99 DT200

The following special tool is required to perform the sensor voltage test:
- 3-pin connector test cord (3-branch harness) (09930-89220)

1. Turn the ignition switch off.
2. Connect the 3-pin connector test cord between the atmospheric pressure sensor and the main wiring harness.
3. Connect the multimeter as follows:
   - Tester positive (+): atmospheric pressure sensor white lead
   - Tester negative (–): atmospheric pressure sensor green lead (black/white lead in wiring harness)
4. Turn the ignition on and check the voltage. Output should measure approximately 3.8 volts at 14.6 psi (101.3 kPa).
5. If the sensor does not meet this specification, replace the sensor.

REMOVAL & INSTALLATION

1. Remove the engine cover.
2. Locate the sensor and remove the wiring cover.
3. Disconnect the appropriate wiring harness connectors and install the test cord assembly if needed.

To install:

4. Remove any test cords from the wiring harness and securely reconnect the wiring harness connectors.
5. Use wire ties to bind the wiring harness in place.
6. Replace the wiring covers on the engine.
7. Replace the engine cover.

4-40 FUEL SYSTEM

Throttle Valve Sensor

DESCRIPTION & OPERATION

♦ See Figure 75

The throttle valve sensor (TVS) is located on the top of the throttle body and is connected to the upper end of the throttle valve shaft. As the throttle valve shaft turns, the sensor resistance changes. the resistance change of the voltage returning to the ECU is a measure of the throttle valve opening angle. The throttle valve opening angle signal serves as one of the signals for determining ignition timing and fuel injection duration time (fuel injection amount) in the ECU.

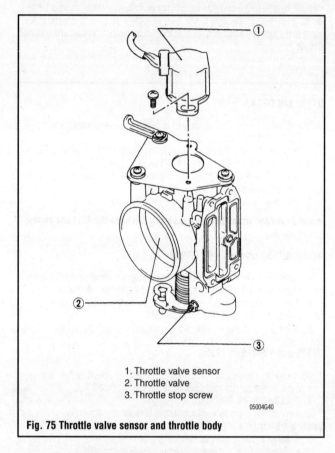

1. Throttle valve sensor
2. Throttle valve
3. Throttle stop screw

Fig. 75 Throttle valve sensor and throttle body

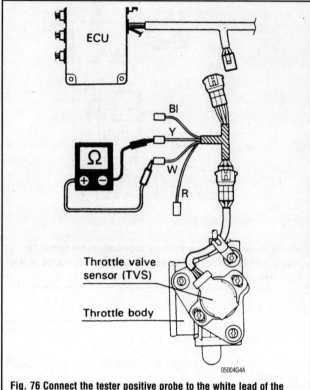

Fig. 76 Connect the tester positive probe to the white lead of the test cord and the negative lead to the yellow lead

TESTING

➡The throttle valve sensor (TVS) inspection/adjustment except for the "TVS RESISTANCE CHECK" must be performed with the battery connected. If an out-of-range condition exists in the TVS signal returned to the ECU, the self-diagnosis system will be indicated by the monitor gauge "CHECK ENGINE" lamp flash code. However, this system is unable to detect an incorrect TVS adjustment.

Resistance Check

DT115 AND DT140

♦ See Figure 76

The following special tools must be obtained to perform the sensor check:
- 4-pin connector test cord (09930–89240)
- Multimeter

1. Connect the tester positive probe to the white lead of the test cord and the negative lead to the yellow lead. Set the tester to the ohms k scale.

➡Do not connect the ECU side connector of the test cord.

2. Slowly move the throttle lever to the open position, then check to see if the resistance changes linearly within the specifications according to the TVS shaft turning angle:
- TVS resistance: 0–6 kilo ohms (linear change)

➡The above resistance range specification is a minimum /maximum reference only for an uninstalled TVS. When the TVS is installed on the throttle body, the actual resistance may be shown as a narrower range than the above specification. If the resistance changes suddenly at any point, the TVS must be replaced.

DT150 AND DT200

♦ See Figure 77

The following special tool must be obtained to perform the sensor check:
- 12-pin connector test cord (09930–89940)
1. Set the multimeter on the ohms scale.
2. Connect the negative tester lead to the light green/black test cord lead and the positive tester lead to the orange/yellow test cord lead.

➡Do not connect the ECM side connector of the test cord.

3. Slowly move the throttle lever to open, and check if resistance changes linearly within specification, according to the TVS shaft turning angle:
- TVS resistance: 0–6 kilo ohms (linear change)

➡The above resistance range specification is a minimum /maximum reference only for an uninstalled TVS. When the TVS is installed on the throttle body, the actual resistance may be shown as a narrower range than the above specification. If the resistance changes suddenly at any point, the TVS must be replaced.

Voltage Check

DT225 AND DT200

♦ See Figures 78 and 79

The following special tool must be obtained to perform the sensor check:
- 3-pin connector test cord (09930–89230)
1. Set the multimeter on the DC voltage scale.

FUEL SYSTEM 4-41

➡ While performing these tests, the throttle valve must be in the fully closed position. Make sure the throttle stop screw is not holding the throttle valve open.

2. Install the test cord between the TVS and the harness connector. Using the multimeter, perform the following procedures with the ignition switch in the ON position and the throttle stop screw turned counter-clockwise until it no longer touches the throttle lever.

3. Connect the negative tester lead to the black test cord lead and the positive tester lead to brown/yellow test cord lead. This voltage is referred to as V1 and must be between 3.64–4.39 volts.

➡ The V1 voltage is not adjustable and is supplied to the TVS through the ICU (Ignition Control Unit). If the V1 voltage is not within the above specifications, the ICU must be replaced. Under normal conditions the diagnostic indicator lamp will show a TVS failure mode if the V1 voltage is below specification.

4. Voltage V2 is a check voltage and is determined by multiplying the above V1 voltage by 0.125. Example: 4.05 (V1) x 0.125 = 0.506 (V2).

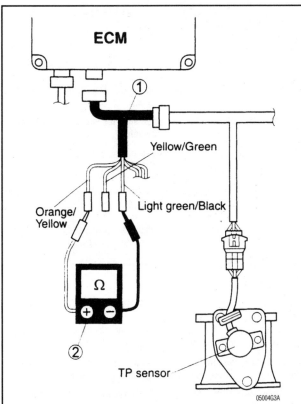

Fig. 77 Connect the negative tester lead to the light green/black test cord lead and the positive tester lead to the orange/yellow test cord lead

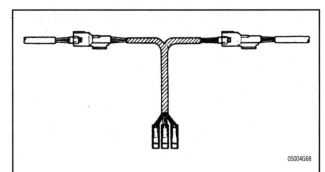

Fig. 78 Install the test cord between the TVS and the harness connector

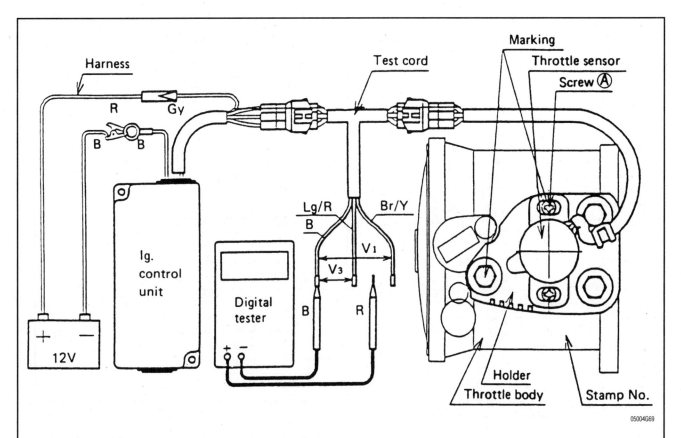

Fig. 79 Connect the negative tester lead to the black test cord lead and the positive tester lead to brown/yellow test cord lead

4-42 FUEL SYSTEM

5. Connect the multimeter positive lead (red) to the light green/red test cord lead and the tester negative lead to the test cord black lead. This voltage is referred to as V3. The difference of the voltage, V2 minus V3 must be within +0.00—0.01 volts.

➥ If the only difference between the V2 and V3 measurements is not specified, it will be necessary to perform the following adjustment procedure.

ADJUSTMENT

➥ Proper idling and acceleration are dependent on the correct V2 voltage. Because of the sensitivity of V2, even the slightest amount of movement, fastening screws too tightly or securing screws unevenly may alter this reading. Therefore, following installation or adjustment of the TVS, recheck V2 to ensure accuracy.

DT115 AND DT140

♦ See Figure 80

The following special tools must be obtained to perform the sensor check:
- 4-pin connector test cord (09930–89240)

1. Loosen the locknut and unscrew the idle adjusting screw until it is fully backed out and not touching the throttle lever.

➥ Manually flock the throttle valve open and closed 2–3 times by hand, allowing the spring tension to snap the valve fully closed.

2. Set the multimeter on the DC volts setting and connect the 4-pin test cord as follows: tester positive probe (red) to the test cord white lead and the tester negative probe (black) to the test cord yellow lead.
3. Turn the ignition switch to the ON position, then check the voltage (V2) to see if it matches that specified for the TVS power supply (V1). Refer to the TVS voltage specifications.
4. If the voltage (V2) not correct, loosen the TVS set screws and gently rotate the TVS until the voltage indicated matches that specified in the chart. Then retighten the set screws using a thread locking compound.

DT150 AND DT200

The following special tools must be obtained to perform the sensor check:
- 12-pin connector test cord (09930–89940)

1. Loosen the locknut and unscrew the idle adjusting screw until it is fully backed out and not touching the throttle lever.

➥ Manually flock the throttle valve open and closed 2–3 times by hand, allowing the spring tension to snap the valve fully closed.

2. Set the multimeter on the DC volts setting and connect the 12-pin test cord as follows: tester positive probe (red) to the test cord orange/yellow lead and the tester negative probe (black) to the test cord light green/black lead.
3. Turn the ignition switch to the ON position, then check the voltage (V2) to see if it matches that specified for the TVS power supply (V1).
4. If the voltage (V2) not correct, loosen the TVS set screws and gently rotate the TVS until the voltage indicated matches that specified in the chart. Then retighten the set screws using a thread locking compound.

DT200 AND DT225

♦ See Figures 81, 82, 83 and 84

The following special tools must be obtained to perform the sensor check:
- 3-pin connector test cord (09930–89230)

1. Loosen the locknut and unscrew the idle adjusting screw until it is fully backed out and not touching the throttle lever.

➥ Manually flock the throttle valve open and closed 2–3 times by hand, allowing the spring tension to snap the valve fully closed.

2. Set the multimeter on the DC volts setting and connect the 3-pin test cord as follows: tester positive probe (red) to the test cord brown/yellow lead and the tester negative probe (black) to the test cord black lead.

(Unit: volt)

V_1 (Bl–Y)	$V_2 \pm 0.01$ (W–Y)	Vt (W–Y)	V_1 (Bl–Y)	$V_2 \pm 0.01$ (W–Y)	Vt (W–Y)
4.75	0.70		5.01	0.74	
4.76	0.70		5.02	0.74	
4.77	0.70		5.03	0.74	1.19
4.78	0.71		5.04	0.74	
4.79	0.71		5.05	0.75	
4.80	0.71		5.06	0.75	
4.81	0.71		5.07	0.75	
4.82	0.71		5.08	0.75	
4.83	0.71	1.15	5.09	0.75	
4.84	0.71		5.10	0.75	
4.85	0.72		5.11	0.75	
4.86	0.72		5.12	0.76	
4.87	0.72		5.13	0.76	
4.88	0.72		5.14	0.76	
4.89	0.72		5.15	0.76	
4.90	0.72		5.16	0.76	
4.91	0.72		5.17	0.76	1.23
4.92	0.73		5.18	0.76	
4.93	0.73		5.19	0.77	
4.94	0.73		5.20	0.77	
4.95	0.73		5.21	0.77	
4.96	0.73	1.19	5.22	0.77	
4.97	0.73		5.23	0.77	
4.98	0.74		5.24	0.77	
4.99	0.74		5.25	0.77	
5.00	0.74				

Fig. 80 TVS voltage specifications—DT115 AND DT140

FUEL SYSTEM 4-43

$$\text{DT225TCL}: V_3 + (V_1 \times 0.021) = V_4$$

Example: 0.506 (V_3)
+ 0.085 (4.05 × 0.021) ······ (from example above)
0.591 (V_4)

$$\text{DT225TCUL}: V_3 + (V_1 \times 0.010) = V_4$$

Example: 0.506 (V_3)
+ 0.040 (4.05 × 0.010) ······ (from example above)
0.546 (V_4)

Fig. 81 TVS adjustment formula—DT200 and DT225

V_1: TVS input voltage
V_2: TVS output voltage at F.C.T ($V_1 \times 0.125$)
V max: TVS output voltage at 4° throttle valve angle ($V_1 \times 0.1583$)

V_1	V_2	V max	V_1	V_2	V max	V_1	V_2	V max
3.64	0.455	0.576	3.90	0.488	0.617	4.20	0.525	0.665
.65	0.456	0.578	.91	0.489	0.619	.21	0.526	0.666
.66	0.458	0.579	.92	0.490	0.621	.22	0.528	0.668
.67	0.459	0.581	.93	0.491	0.622	.23	0.529	0.670
.68	0.460	0.583	.94	0.493	0.624	.24	0.530	0.671
.69	0.461	0.584	.95	0.494	0.625	.25	0.531	0.673
3.70	0.463	0.586	.96	0.495	0.627	.26	0.532	0.674
.71	0.464	0.587	.97	0.496	0.628	.27	0.534	0.676
.72	0.465	0.589	.98	0.498	0.630	.28	0.535	0.678
.73	0.466	0.590	.99	0.499	0.632	.29	0.536	0.679
.74	0.468	0.592	4.00	0.500	0.633	4.30	0.538	0.681
.75	0.469	0.594	.01	0.501	0.635	.31	0.539	0.682
.76	0.470	0.595	.02	0.503	0.636	.32	0.540	0.684
.77	0.471	0.597	.03	0.504	0.638	.33	0.541	0.685
.78	0.473	0.598	.04	0.505	0.640	.34	0.543	0.687
.79	0.474	0.600	.05	0.506	0.641	.35	0.544	0.689
3.80	0.475	0.602	.06	0.508	0.643	.36	0.545	0.690
.81	0.476	0.603	.07	0.509	0.644	.37	0.546	0.692
.82	0.478	0.605	.08	0.510	0.646	.38	0.548	0.693
.83	0.479	0.606	.09	0.511	0.647	.39	0.549	0.695
.84	0.480	0.608	4.10	0.513	0.649			
.85	0.481	0.609	.11	0.514	0.651			
.86	0.483	0.611	.12	0.515	0.652			
.87	0.484	0.613	.13	0.516	0.654			
.88	0.485	0.614	.14	0.518	0.655			
.89	0.486	0.616	.15	0.519	0.657			
			.16	0.520	0.659			
			.17	0.521	0.660			
			.18	0.522	0.662			
			.19	0.524	0.663			

Fig. 82 TVS voltage specifications—1988–93 DT200

4-44 FUEL SYSTEM

(Unit: volt)

V1 (Y/G – Lg/B)	V2 ±0.01 (O/Y – Lg/B)	Vt ±0.01 (O/Y – Lg/B)	V1 (Y/G – Lg/B)	V2 ±0.01 (O/Y – Lg/B)	Vt ±0.01 (O/Y – Lg/B)
4.80	0.576	0.806	5.00	0.600	0.840
4.81	0.577	0.808	5.01	0.601	0.842
4.82	0.578	0.810	5.02	0.602	0.843
4.83	0.580	0.811	5.03	0.604	0.845
4.84	0.581	0.813	5.04	0.605	0.847
4.85	0.582	0.815	5.05	0.606	0.848
4.86	0.583	0.816	5.06	0.607	0.850
4.87	0584	0.818	5.07	0.608	0.852
4.88	0.586	0.820	5.08	0.610	0.853
4.89	0.587	0.822	5.09	0.611	0.855
4.90	0.588	0.823	5.10	0.612	0.857
4.91	0.589	0.825	5.11	0.613	0.858
4.92	0.590	0.827	5.12	0.614	0.860
4.93	0.592	0.828	5.13	0.616	0.862
4.94	0.593	0.830	5.14	0.617	0.864
4.95	0.594	0.832	5.15	0.618	0.865
4.96	0.595	0.833	5.16	0.619	0.867
4.97	0.596	0.835	5.17	0.620	0.869
4.98	0.598	0.837	5.18	0.622	0.870
4.99	0.599	0.838	5.19	0.623	0.872

Fig. 83 TVS voltage specifications—DT225 and 1994-99 DT200

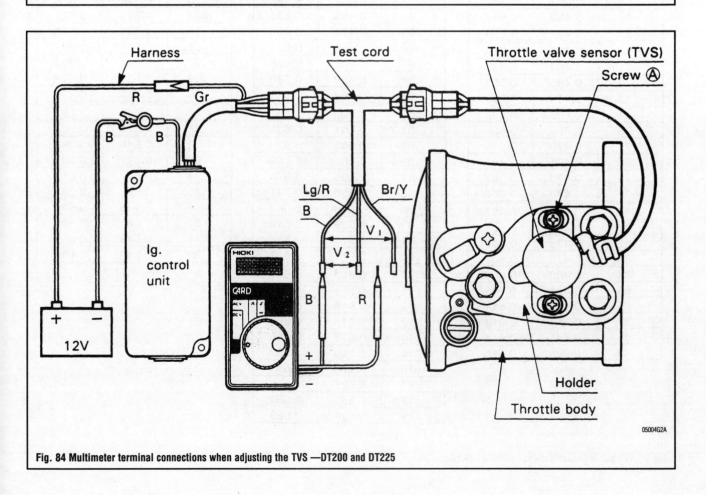

Fig. 84 Multimeter terminal connections when adjusting the TVS —DT200 and DT225

FUEL SYSTEM   4-45

3. Turn the ignition switch to the ON position, then check the voltage (V2) to see if it matches that specified for the TVS power supply (V1). Refer to the TVS voltage specifications.
   - Since V1 voltage is not adjustable, a reading of 3.64–4.39 volts can be considered correct. If the voltage is not between these numbers, it may indicate either a fault in the ICU or a short circuit of the TVS (although that is a relatively uncommon occurrence). Both components therefore must be replaced and checked sequentially.
4. If the voltage (V2) not correct, loosen the TVS set screws and gently rotate the TVS until the voltage indicated matches that specified in the chart. Then retighten the set screws using a thread locking compound.
5. With the test cord still attached, reattach the emergency switch lock plate and start the engine, letting it warm up thoroughly.
6. When the engine is at operating temperature, adjust the throttle stop screw. Idle speed should be as specified in the "Tune-Up Specifications" chart.
7. Securely lock the throttle stop screw with the nut.
   The idle speed has been correctly adjusted, check the TVS output voltage. The reading must be lower than an upper limit referred to as V max. V max is a designated value at throttle opening of 4° and when adjusting the in-gear idle speed, the throttle opening should be 4°at most.

➥Value V max corresponds directly to (and is listed alongside) inter-related V1 and V2. Having checked V1 and V2 according to the TVS voltage chart. V max is the third relevant throttle valve factor. If V2 and V max are accurate, and within specification respectively, and if in-gear idle speed is as specified in the "Tune-Up Specifications" chart, the TVS is adjusted correctly.

8. If the output voltage exceeding V max is required to obtain the specified idle speed, then this abnormally wide throttle valve opening may indicate mechanical, fuel delivery or other electrical system problems. Such circumstances should be investigated immediately.

➥If the above situation occurs when TVS input voltage (V1) and output voltage at fully closed throttle (V2) are known to be correct, then the ICU and TVS are operating correctly and should not be checked.

REMOVAL & INSTALLATION

1. Remove the engine cover.
2. Disconnect the wiring to the TVS mounted on the throttle body.
3. Remove the TVS from the throttle body.

To install:

➥If the TVS has been moved even the slightest amount or taken off the throttle body, you must perform the TVS adjustment procedure.

4. Install the TVS onto the throttle body.
5. Perform the TVS adjustment procedure covered in detail above.
6. Make sure to use thread locking compound on the screws.
7. Connect the TVS wiring, making sure the connections are fastened tightly.
8. Install the engine cover.

Gear Counter Coil (Engine Speed Sensor)

DESCRIPTION & OPERATION

The gear counter coil is located on top of the engine at the rear of the flywheel. Voltage pulses induced in the coil by the passing flywheel ring gear teeth are signals used by the ECU to determine engine speed. This signal also serves as one of the basic signals used in determining ignition timing and fuel injector duration time (amount of fuel injected) in the ECU.

TESTING

Gear counter coil testing is covered in detail in "Ignition and Electrical".

Pulser Coil

DESCRIPTION & OPERATION

The pulser coil is located under the flywheel rotor. Voltage pulses induced in the coil by the passing reluctor bar which is attached to the flywheel rotor are signals used by the ECU to determine crankshaft angle. This is the base information from which the ECU computes the ignition spark signal in the correctly sequenced firing order. Fuel injection timing is then set by this ignition signal.

TESTING

Pulser coil testing is covered in detail in the in "Ignition and Electrical".

Fuel Injectors

DESCRIPTION & OPERATION

♦ See Figure 85

The fuel injector is an electromagnet fuel injection valve operated by the injection signal supplied by the fuel injection control unit. The coil used in the injector is a high pressure resistant type. The fuel injection control unit determines the optimal fuel injection time duration on the basis of the signal input from the various sensors mounted on the engine.

When the injection signal is sent to the fuel injector, it energizes the coil and pulls up the needle valve, thereby opening the valve and injecting fuel. Because fuel pressure (pressure differential between fuel line and manifold) is kept constant, the amount of fuel injected is determined by the duration time of the open.

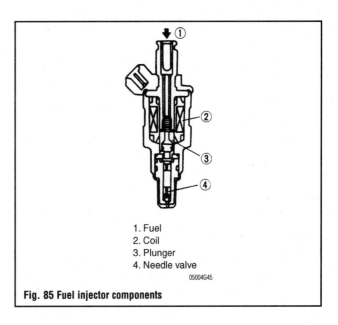

1. Fuel
2. Coil
3. Plunger
4. Needle valve

Fig. 85 Fuel injector components

TESTING

Operational Sound

DT115, DT140, DT150 AND DT200

➥The fuel injector inspection must be performed with the battery terminals connected.

1. Using a mechanics stethoscope or a long bladed screwdriver, touch the injector connector.

4-46 FUEL SYSTEM

2. Crank the engine with the spark plugs removed and listen for the sound of the injector clicking.
3. If there is no sound, then the injector is not operating and it will need to be replaced.

Obtain the following special tool:
- Injector test cord "A" (09930-99420)

➡It is not necessary to remove the injectors from the engine to perform this test.

4. Disconnect the injector wire and connect the test cord.
5. Touch the stethoscope or screwdriver to the injector connector.
6. Momentarily touch the red wire of the test cord to the starter motor relay left terminal (which is connected to the battery positive terminal) and listen for the sound of the injector clicking.
7. If there is no sound, then the injector is not operating and it will need to be replaced.

DT225 (MULTIPLE INJECTORS)

1. Obtain the following special tool to perform the test:
   - 6-pin test harness (09930-89251).
2. Disconnect the 6-pin ECU connector containing black/yellow leads and install the test harness.
3. Turn ignition switch to the ON position and touch one black/yellow lead to ground. Listen for the sound of the injector clicking from 3 injectors each time the black/yellow lead is touched to ground. This means the injector is operating properly. Repeat the test for the other black/yellow lead.

➡3 injectors are controlled by each black/yellow ECU lead.

➡Another way to tell if the injector is to place your finger on the injector itself and feel for the clicking motion inside the injector.

DT225 (INDIVIDUAL INJECTORS)

If injector operation is not heard or felt, you will need to obtain the following special tool to perform the check:
- 2-pin connector test cord (09930-89260)

1. Disconnect the fuel injector's 2-pin connector and connect the 2-pin connector test cord to the injector. Connect the black test lead first to the battery negative terminal. Next, touch the positive test lead to the battery positive terminal and then remove. Repeat this on and off contact action several times and check for the sound of the injector operating. If the clicks are heard as this on and off contact action is made, the injector is operating normally. If the clicks are not heard, there is a failure in the injector and it will need to be replaced.

✱✱ WARNING

Never attempt to disconnect or remove the fuel hose during this test, or fuel under high pressure will spray out, causing an extremely hazardous condition.

Power Supply

DT115, DT140, DT150 AND DT200 (NO LOAD)

1. Obtain the following special tool:
   - Injector test cord "B" (09930-99430)
2. Connect the test cord as shown in the illustration to the injector wiring harness.
3. Connect the test meter positive probe to the test cord red lead and the negative lead to ground.
4. Turn the ignition to the ON position and check for voltage. The injector voltage should read approximately .11 volts.

DT115, DT140, DT150 AND DT200TC (PEAK VOLT CHECK)

1. Obtain the following special tools:
   - Injector test cord "B" (09930-99430)
   - Peak reading voltage meter
2. Set the peak volt meter to NEG 50.
3. Connect the test cord positive probe (+) to the test cord black lead and the negative test probe to the starter relay left terminal (connected to the battery positive terminal).
4. Crank the engine with the spark plugs removed and check the indicated voltage. The injector operating signal should indicate approximately 8–10 volts.

DT225 (NO LOAD)

Obtain the following special tool:
- 2-Pin connector test cord "B" (09930-89270)

1. Turn the ignition switch to the OFF position.
2. With the injectors installed on the engine, remove all six of the 2-pin connectors. Connect the 2-pin connector test cord between the injector being checked and the wiring harness as illustrated.

➡When inserting the 2-pin connector test cord connector, make sure the same color lead are connected together.

3. Connect a peak voltage meter (set on Positive 50V) as follows:
   - Tester positive lead to the gray lead of the 2-pin connector test cord
   - Tester negative lead to an engine ground
4. If the tester indication is approximately 12 volts with the ignition switch in the ON position, the condition of the power system is normal. If the voltage is not as specified, check the following:
   - Loose or discolored connectors
   - Open circuits of the gray lead
   - Disconnected connector or lead, failure of the main relay etc.
5. Check the other five injectors in the same manner.

DT225 (PEAK VOLT CHECK)

1. Peform the first and second procedures of the previous power supply check (no load).
2. Connect the peak voltage meter (set on POSITIVE 50V) as follows:
   - Tester positive lead to the black/yellow lead of the 2-pin test connector cord.
   - Tester negative lead to an engine ground.
3. If the tester indication is approximately 12 volts with the ignition switch in the ON position, the operating condition is normal.
4. Check for voltage with the engine being cranked. If more than approximately 20 volts is indicated during engine cranking, the injector control signal is normal. If the voltage being indicated is less than approximately 20 volts, check for connector looseness or the disconnection of the black/yellow lead.
5. Check the other five injectors in the same manner. If less than 20 volts is indicated at all six injectors, perform an ECU power supply and ground checks.

Injector Resistance

1. Disconnect the injector connector.
2. Using a multimeter, measure the resistance between the injector terminals. Resistance should measure approximately 12.2–15.1 ohms at 68°(20°C).
3. If the measured resistance does not meet specifications, replace the injector.
4. Reconnect the injector connector.

Fuel Pressure Regulator

♦ See Figure 86

The fuel pressure regulator is used in the system for the purpose of maintaining the fuel pressure relative to the manifold at a constant level. The regulator diaphragm chamber is connected with the surge tank to keep the pressure balanced, while the fuel pressure is adjusted by the regulator to be constantly higher than the surge tank pressure by approximately 36.3 psi (2.55 kg-cm), the diaphragm is pushed up, allowing fuel to flow through the return pipe to the vapor separator. When fuel returns to the vapor separator, the excess fuel pressure is relieved, thus keeping the pressure adjusted to a constant level.

TESTING

1. Obtain the following special tools:
   - Hand vacuum/air pump
   - Air Pressure gauge Attachment (09940-44130)
2. Remove the fuel pressure regulator from the engine.
3. Connect the attachment, gauge and pump to the fuel pressure regulator as shown in the illustration.

FUEL SYSTEM 4-47

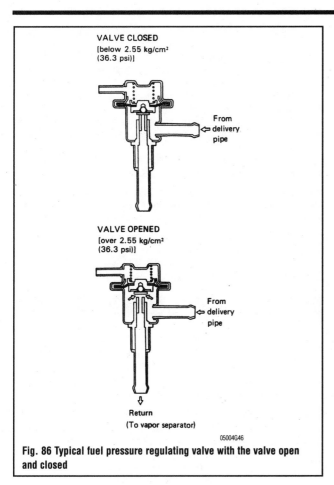

Fig. 86 Typical fuel pressure regulating valve with the valve open and closed

4. Using the pump, pump air into the regulator until the air is released through the outlet.
5. As air is released from the regulator, check the gauge to see if the pressure meets specification:
   - 34.1–38.4 psi (2.4–2.7 kg/cm2)

REMOVAL & INSTALLATION

1. Remove the engine cover.
2. Turn the ignition switch to the OFF position and then disconnect the battery cables from the battery terminals.
3. Loosen the fasteners securing the oil tank to the engine and move the oil tank to the outside.
4. Place a rag over the banjo bolt fitting on the fuel line and then slowly loosen the bolt relieving the fuel pressure in the fuel line.
5. Wipe up any spilled fuel immediately.
6. Pinch the fuel line with your fingers to make sure all the pressure has been relieved.

To install:

7. Reinstall all the hoses on the regulator.
8. Tighten the banjo bolt to 29 ft. lbs. (40 Nm)
9. Install the oil tank back in its place on the side of the engine and tighten the fasteners.
10. Reconnect the battery cables to the battery terminals.
11. Install the engine cover.

High Pressure Fuel Pump

♦ See Figure 87

This an "In-line" type pump which has all the pump mechanisms located within the fuel line.

The pump unit consists of the housing, rotor and rollers. As the rotor rotates, the centrifugal force pushes the rollers outward so that they revolve along the inside housing wall. During rotation, the space formed by these three parts:

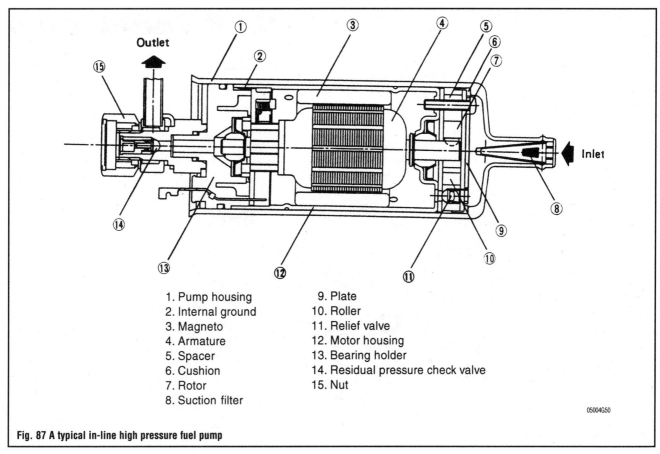

1. Pump housing
2. Internal ground
3. Magneto
4. Armature
5. Spacer
6. Cushion
7. Rotor
8. Suction filter
9. Plate
10. Roller
11. Relief valve
12. Motor housing
13. Bearing holder
14. Residual pressure check valve
15. Nut

Fig. 87 A typical in-line high pressure fuel pump

FUEL SYSTEM

rotor, rollers and housing will alternate, first larger, then smaller. This space variation is utilized for fuel suction and feed pressure.

Residual pressure check valve A check valve is provided to keep residual pressure in the fuel line after the engine has been shut off.

Relief valve A relief valve is provided to prevent overpressure in the high-pressure side of the fuel line.

Suction filter To prevent fuel tank or sediment from entering the high-pressure fuel pump, there is a fine mesh suction filter at the pump inlet.

TESTING

Fuel Pressure

1. Obtain the following special tools:
   - Fuel pressure gauge
   - 3-way joint and hose
   - Fuel pressure hose
2. After following the procedures for relieving the fuel pressure in the fuel lines, connect the joint & hose and gauge as shown in the illustration.
3. Use a 50:1 fuel/oil mixture for pressure testing.
4. Turn the ignition switch to the ON and OFF positions several times to activate the high pressure fuel pump and fill the fuel hoses with fuel.
5. Inspect all the fuel line connections for leaks and correct any problems before proceeding with the tests.
6. Check if the fuel pressure indicated on the gauge is within specification at both idle speed and cranking speed.

➡ **To prevent damage to the starter motor, do not crank the engine for more than 20 seconds at a time.**

7. Fuel pressure should be approximately 36.3 psi (2.55 kg/cm2).
8. If the fuel pressure is not within specification, recheck all the fuel line connections for leaks. If there are no leaks, replace the fuel pump assembly.

Residual Fuel Pressure

1. After performing the fuel pressure check, shut off the engine and wait 5 minutes.
2. Check the fuel pressure indicated on the fuel gauge. Pressure should be 28.4 psi (2.0 kg/cm2) or more.

Power Supply

➡ **The fuel pump inspection must be performed with the battery connected to the terminals.**

1. Turn the ignition switch to the ON position and listen for the sound of the fuel pump operating. It should sound for approximately 2 seconds only.

The following special tool must be obtained to perform the fuel pump power supply check:
   - 2-pin connector test cord (09930–89210)

2. Connect the multimeter positive probe to the pink test cord lead and the multimeter negative probe to ground.
3. With the tester set on DC volts, fuel pump voltage should measure approximately 12 volts (battery voltage).

To perform the fuel pump 2 second operating signal check, connect the multimeter positive probe to the pink test cord lead and the tester negative probe to the black test cord lead.

4. Turn the ignition switch to the ON position and measure the voltage. Voltage should measure approximately 12 volts (battery voltage) for 2 seconds only.

➡ **Excepting the first 2 second after the ignition switch is turned ON, the tester must indicate 0 voltage. If not, the pump has failed and needs to be replaced.**

5. To perform the fuel pump operating signal check, crank the engine with the spark plugs removed and check the voltage.
6. The fuel pump operating signal should be approximately 9–10 volts during cranking.

Additional Inputs

Additional inputs include:
- Neutral switch, which informs the ECU of the shifter position
- The key switch, which tells the computer when to begin the program
- The thermoswitch (overheat) for overheat information
- Oil level sensor information
- The shift cut switch
- The lanyard switch
- The gray over-rev loop lead

Although not strictly inputs, battery power and ground are essential for computer operation. This computer also monitors battery voltage. As battery voltage drops, the injectors would open more slowly, decreasing fuel delivery volume. Therefore, when the computer detects low battery voltage, it makes adjustments to injector "time on" to keep fuel delivery volume correct.

Self Diagnostic System

DESCRIPTION & OPERATION

▶ See Figures 88, 89, 90, 91 and 92

The ECU is provided with a self diagnostic function. When a failure occurs in the system and no input signal is sent from the sensors, the ECU stores this information and activates the "Check Engine" LED on the monitor gauge. The

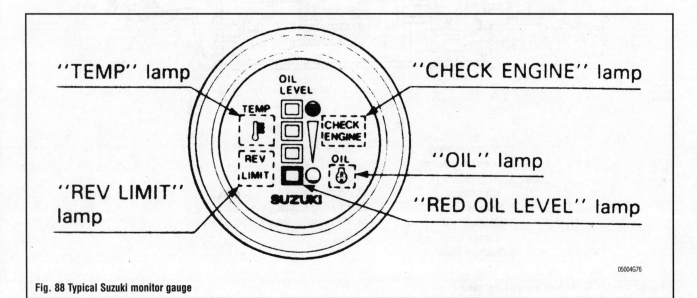

Fig. 88 Typical Suzuki monitor gauge

FUEL SYSTEM 4-49

location of the problem can be determined by the flashing light sequence of the monitor LED as shown in the illustration.

DIAGNOSIS PROCEDURE

1. Turn on the ignition switch.
2. Check the "check engine" LED flashing sequence and count the number of flashes. From the observed flashes, determine which one of the following codes it corresponds to.

FAIL SAFE EMERGENCY BACKUP

When any of the ECU system fails, the self-diagnostic system will indicate the failure location by means of the indicator lamp's flashing code. In the event of such a failure condition, another system, called a back-up system, will come into operation. This system provides alternative signals to compensate for the ECU or sensor failure so that the engine will not stop operating but will be continuously operable. Because of this provision, emergency, return-to-port operation will be possible, but in a much reduced performance mode.

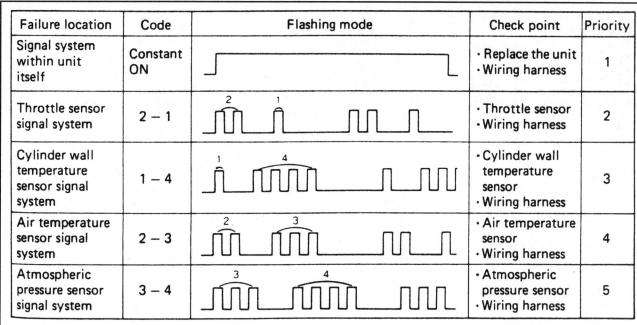

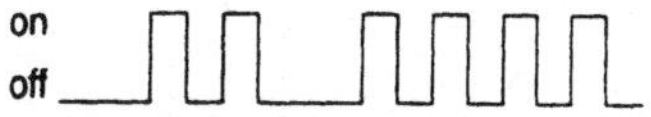

Fig. 89 Self-diagnostic system chart —DT225

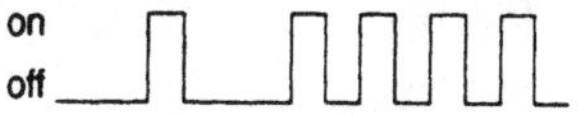

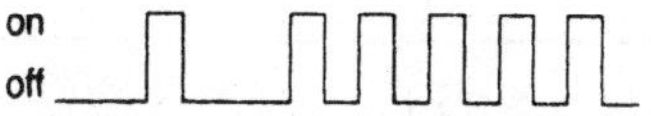

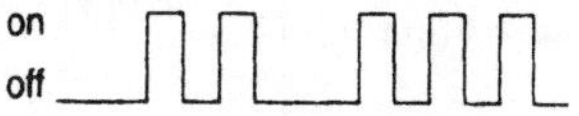

Fig. 90 Self-diagnostic system chart —DT150 AND 1988–93 DT200

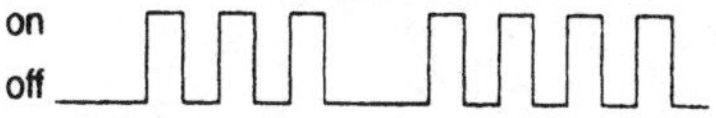

Failure location	Code	Flashing mode	Check point	Priority
Signal system within unit itself	Constant ON		· Replace the unit · Wiring harness	1
Throttle sensor signal system	2 – 1	2 1	· Throttle sensor · Wiring harness	2
Cylinder wall temperature sensor signal system	1 – 4	1 4	· Cylinder wall temperature sensor · Wiring harness	3
Air temperature sensor signal system	2 – 3	2 3	· Air temperature sensor · Wiring harness	4
Atmospheric pressure sensor signal system	3 – 4	3 4	· Atmospheric pressure sensor · Wiring harness	5

05004G74

Fig. 91 Self-diagnostic system chart—1994–99 DT200

Malfunction	Range	Code	"CHECK ENGINE" lamp flashing pattern	*Priority
Battary	**More than 9.9 volts (DT115STC)/ 10.2 volts (DT140TC)	CON.	Constant "ON" ON OFF	1
Throttle valve sensor	0.6 to 4.7 volts	2–1	1 cycle ON OFF	2
Cylinder wall temperature sensor	0.1 to 4.7 volts	1–4	1 cycle ON OFF	3
Air temperature sensor	0.5 to 4.7 volts	2–3	1 cycle ON OFF	4
Atmospheric pressure sensor	2.2 to 4.4 volts	3–4	1 cycle ON OFF	5

05004G75

Fig. 92 Self-diagnostic system chart—DT115 and DT140

UNDERSTANDING AND
TROUBLESHOOTING ELECTRICAL
SYSTEMS 5-2
BASIC ELECTRICAL THEORY 5-2
   HOW ELECTRICITY WORKS: THE
      WATER ANALOGY 5-2
   OHM'S LAW 5-2
ELECTRICAL COMPONENTS 5-2
   POWER SOURCE 5-2
   GROUND 5-3
   PROTECTIVE DEVICES 5-3
   SWITCHES & RELAYS 5-3
   LOAD 5-3
   WIRING & HARNESSES 5-4
   CONNECTORS 5-4
TEST EQUIPMENT 5-4
   JUMPER WIRES 5-4
   TEST LIGHTS 5-5
   MULTIMETERS 5-5
TROUBLESHOOTING THE ELECTRICAL
   SYSTEM 5-6
TESTING 5-6
   VOLTAGE 5-6
   VOLTAGE DROP 5-6
   RESISTANCE 5-6
   OPEN CIRCUITS 5-7
   SHORT CIRCUITS 5-7
WIRE AND CONNECTOR REPAIR 5-7
ELECTRICAL SYSTEM
   PRECAUTIONS 5-7
**BREAKER POINTS IGNITION
(MAGNETO IGNITION) 5-7**
SYSTEM TESTING 5-8
BREAKER POINTS 5-8
   POINT GAP ADJUSTMENT 5-8
   TESTING 5-9
   REMOVAL & INSTALLATION 5-9
CONDENSER 5-10
   DESCRIPTION & OPERATION 5-10
   TESTING 5-10
   REMOVAL & INSTALLATION 5-10
IGNITION COIL 5-10
   DESCRIPTION & OPERATION 5-10
   TESTING 5-11
   REMOVAL & INSTALLATION 5-11
**CAPACITOR DISCHARGE IGNITION
(CDI) SYSTEM 5-11**
DESCRIPTION AND OPERATION 5-11
   SINGLE-CYLINDER IGNITION 5-11
   SUZUKI PEI IGNITION 5-12
SYSTEM TESTING 5-14
   PROCEDURE 5-14
PULSAR/CHARGING/GEAR COUNTER
   COILS 5-14
   DESCRIPTION & OPERATION 5-14
   TESTING 5-15
   REMOVAL & INSTALLATION 5-20
IGNITION COILS 5-25
   DESCRIPTION & OPERATION 5-25
   TESTING 5-25
   REMOVAL & INSTALLATION 5-26
CDI UNIT 5-28

DESCRIPTION & OPERATION 5-28
   TESTING 5-28
   REMOVAL & INSTALLATION 5-34
RECTIFIER 5-37
   DESCRIPTION & OPERATION 5-37
   TESTING 5-37
   REMOVAL & INSTALLATION 5-38
REGULATOR 5-38
   DESCRIPTION & OPERATION 5-38
   TESTING 5-38
   REMOVAL & INSTALLATION 5-38
ELECTRONIC IGNITION 5-38
DESCRIPTION AND OPERATION 5-38
WARNING SYSTEMS 5-38
   OVER REVOLUTION 5-38
   OIL LEVEL 5-38
   OIL FLOW 5-38
   OVERHEAT 5-38
   BATTERY VOLTAGE 5-39
CHARGING CIRCUIT 5-39
DESCRIPTION AND OPERATION 5-39
   SINGLE PHASE CHARGING
      SYSTEM 5-39
   THREE-PHASE CHARGING
      SYSTEM 5-39
   PRECAUTIONS 5-39
TROUBLESHOOTING THE CHARGING
   SYSTEM 5-40
   OVERCHARGING 5-40
   UNDERCHARGING 5-40
ALTERNATOR (STATOR) 5-40
   TESTING 5-40
BATTERY 5-41
   MARINE BATTERIES 5-41
   BATTERY CONSTRUCTION 5-41
   BATTERY RATINGS 5-42
   BATTERY LOCATION 5-42
   BATTERY SERVICE 5-42
   BATTERY TERMINALS 5-44
   SAFETY PRECAUTIONS 5-44
   BATTERY CHARGERS 5-44
   BATTERY CABLES 5-45
   BATTERY STORAGE 5-45
STARTING CIRCUIT 5-45
DESCRIPTION AND OPERATION 5-45
TROUBLESHOOTING THE STARTING
   SYSTEM 5-46
STARTER MOTOR 5-46
   DESCRIPTION & OPERATION 5-46
   TESTING 5-48
   REMOVAL & INSTALLATION 5-48
   OVERHAUL 5-49
STARTER MOTOR RELAY SWITCH 5-50
   DESCRIPTION & OPERATION 5-50
   TESTING 5-51
   REMOVAL & INSTALLATION 5-51
**IGNITION AND ELECTRICAL WIRING
DIAGRAMS 5-52**

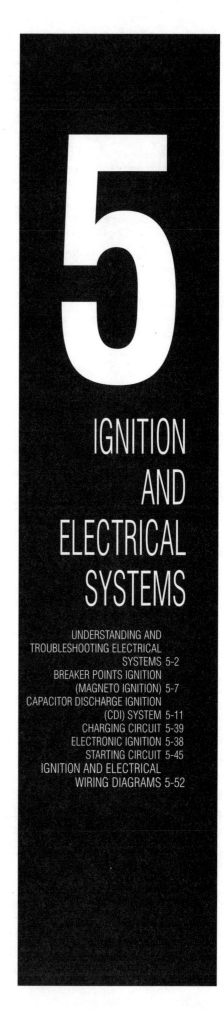

5

IGNITION
AND
ELECTRICAL
SYSTEMS

UNDERSTANDING AND
TROUBLESHOOTING ELECTRICAL
SYSTEMS 5-2
BREAKER POINTS IGNITION
(MAGNETO IGNITION) 5-7
CAPACITOR DISCHARGE IGNITION
(CDI) SYSTEM 5-11
CHARGING CIRCUIT 5-39
ELECTRONIC IGNITION 5-38
STARTING CIRCUIT 5-45
IGNITION AND ELECTRICAL
WIRING DIAGRAMS 5-52

5-2 IGNITION AND ELECTRICAL SYSTEMS

UNDERSTANDING AND TROUBLESHOOTING ELECTRICAL SYSTEMS

Basic Electrical Theory

♦ See Figure 1

For any 12 volt, negative ground, electrical system to operate, the electricity must travel in a complete circuit. This simply means that current (power) from the positive terminal (+) of the battery must eventually return to the negative terminal (-) of the battery. Along the way, this current will travel through wires, fuses, switches and components. If, for any reason, the flow of current through the circuit is interrupted, the component fed by that circuit will cease to function properly.

Perhaps the easiest way to visualize a circuit is to think of connecting a light bulb (with two wires attached to it) to the battery—one wire attached to the negative (-) terminal of the battery and the other wire to the positive (+) terminal. With the two wires touching the battery terminals, the circuit would be complete and the light bulb would illuminate. Electricity would follow a path from the battery to the bulb and back to the battery. It's easy to see that with longer wires on our light bulb, it could be mounted anywhere. Further, one wire could be fitted with a switch so that the light could be turned on and off.

The normal marine circuit differs from this simple example in two ways. First, instead of having a return wire from each bulb to the battery, the current travels through a single ground wire which handles all the grounds for a specific circuit. Secondly, most marine circuits contain multiple components which receive power from a single circuit. This lessens the amount of wire needed to power components.

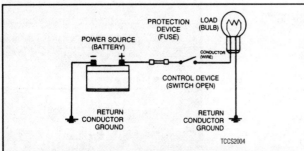

Fig. 1 This example illustrates a simple circuit. When the switch is closed, power from the positive (+) battery terminal flows through the fuse and the switch, and then to the light bulb. The light illuminates and the circuit is completed through the ground wire back to the negative (-) battery terminal.

HOW ELECTRICITY WORKS: THE WATER ANALOGY

Electricity is the flow of electrons—the sub-atomic particles that constitute the outer shell of an atom. Electrons spin in an orbit around the center core of an atom. The center core is comprised of protons (positive charge) and neutrons (neutral charge). Electrons have a negative charge and balance out the positive charge of the protons. When an outside force causes the number of electrons to unbalance the charge of the protons, the electrons will split off the atom and look for another atom to balance out. If this imbalance is kept up, electrons will continue to move and an electrical flow will exist.

Many people have been taught electrical theory using an analogy with water. In a comparison with water flowing through a pipe, the electrons would be the water and the wire is the pipe.

The flow of electricity can be measured much like the flow of water through a pipe. The unit of measurement used is amperes, frequently abbreviated as amps (a). You can compare amperage to the volume of water flowing through a pipe. When connected to a circuit, an ammeter will measure the actual amount of current flowing through the circuit. When relatively few electrons flow through a circuit, the amperage is low. When many electrons flow, the amperage is high.

Water pressure is measured in units such as pounds per square inch (psi); The electrical pressure is measured in units called volts (v). When a voltmeter is connected to a circuit, it is measuring the electrical pressure.

The actual flow of electricity depends not only on voltage and amperage, but also on the resistance of the circuit. The higher the resistance, the higher the force necessary to push the current through the circuit. The standard unit for measuring resistance is an ohm (Ω). Resistance in a circuit varies depending on the amount and type of components used in the circuit. The main factors which determine resistance are:

• Material—some materials have more resistance than others. Those with high resistance are said to be insulators. Rubber materials (or rubber-like plastics) are some of the most common insulators used, as they have a very high resistance to electricity. Very low resistance materials are said to be conductors. Copper wire is among the best conductors. Silver is actually a superior conductor to copper and is used in some relay contacts, but its high cost prohibits its use as common wiring. Most marine wiring is made of copper.

• Size—the larger the wire size being used, the less resistance the wire will have. This is why components which use large amounts of electricity usually have large wires supplying current to them.

• Length—for a given thickness of wire, the longer the wire, the greater the resistance. The shorter the wire, the less the resistance. When determining the proper wire for a circuit, both size and length must be considered to design a circuit that can handle the current needs of the component.

• Temperature—with many materials, the higher the temperature, the greater the resistance (positive temperature coefficient). Some materials exhibit the opposite trait of lower resistance with higher temperatures (negative temperature coefficient). These principles are used in many of the sensors on the engine.

OHM'S LAW

There is a direct relationship between current, voltage and resistance. The relationship between current, voltage and resistance can be summed up by a statement known as Ohm's law.

• Voltage (E) is equal to amperage (I) times resistance (R): $E = I \times R$
• Other forms of the formula are $R = E/I$ and $I = E/R$

In each of these formulas, E is the voltage in volts, I is the current in amps and R is the resistance in ohms. The basic point to remember is that as the resistance of a circuit goes up, the amount of current that flows in the circuit will go down, if voltage remains the same.

The amount of work that the electricity can perform is expressed as power. The unit of power is the watt (w). The relationship between power, voltage and current is expressed as:

• Power (W) is equal to amperage (I) times voltage (E): $W = I \times E$

This is only true for direct current (DC) circuits; The alternating current formula is a tad different, but since the electrical circuits in most vessels are DC type, we need not get into AC circuit theory.

Electrical Components

POWER SOURCE

Power is supplied to the vessel by two devices: The battery and the alternator. The battery supplies electrical power during starting or during periods when the current demand of the vessel's electrical system exceeds the output capacity of the alternator. The alternator supplies electrical current when the engine is running. The alternator does not just supply the current needs of the vessel, but it recharges the battery.

The Battery

In most modern vessels, the battery is a lead/acid electrochemical device consisting of six 2 volt subsections (cells) connected in series, so that the unit is capable of producing approximately 12 volts of electrical pressure. Each subsection consists of a series of positive and negative plates held a short distance apart in a solution of sulfuric acid and water.

The two types of plates are of dissimilar metals. This sets up a chemical reaction, and it is this reaction which produces current flow from the battery when its positive and negative terminals are connected to an electrical load. The power removed from the battery is replaced by the alternator, restoring the battery to its original chemical state.

IGNITION AND ELECTRICAL SYSTEMS

Charging System

When the imbedded magnets in the flywheel rotate past the charging coils or A/C lighting coil(s), it creates alternating current. This current is sent to the rectifier or combination rectifier/voltage regulator, where it is then converted to D/C and then supplied to the battery or electrical accessories through a fused link.

A malfunction in the battery charging system will result in the battery being undercharged.

GROUND

Two types of grounds are used in marine electric circuits. Direct ground components are grounded to the electrically conductive metal through their mounting points. All other components use some sort of ground wire which leads back to the battery. The electrical current runs through the ground wire and returns to the battery through the ground (-) cable; if you look, you'll see that the battery ground cable connects between the battery and a heavy gauge ground wire.

➥It should be noted that a good percentage of electrical problems can be traced to bad grounds.

PROTECTIVE DEVICES

♦ See Figure 2

It is possible for large surges of current to pass through the electrical system of your vessel. If this surge of current were to reach the load in the circuit, the surge could burn it out or severely damage it. It can also overload the wiring, causing the harness to get hot and melt the insulation. To prevent this, fuses, circuit breakers and/or fusible links are connected into the supply wires of the electrical system. These items are nothing more than a built-in weak spot in the system. When an abnormal amount of current flows through the system, these protective devices work as follows to protect the circuit:

- Fuse—when an excessive electrical current passes through a fuse, the fuse "blows" (the conductor melts) and opens the circuit, preventing the passage of current.
- Circuit Breaker—a circuit breaker is basically a self-repairing fuse. It will open the circuit in the same fashion as a fuse, but when the surge subsides, the circuit breaker can be reset and does not need replacement.
- Fusible Link—a fusible link (fuse link or main link) is a short length of special, high temperature insulated wire that acts as a fuse. When an excessive electrical current passes through a fusible link, the thin gauge wire inside the link melts, creating an intentional open to protect the circuit. To repair the circuit, the link must be replaced. Some newer type fusible links are housed in plug-in modules, which are simply replaced like a fuse, while older type fusible links must be cut and spliced if they melt. Since this link is very early in the electrical path, it's the first place to look if nothing on the vessel works, yet the battery seems to be charged and is properly connected.

✱✱ CAUTION

Always replace fuses, circuit breakers and fusible links with identically rated components. Under no circumstances should a component of higher or lower amperage rating be substituted.

SWITCHES & RELAYS

♦ See Figure 3

Switches are used in electrical circuits to control the passage of current. The most common use is to open and close circuits between the battery and the various electric devices in the system. Switches are rated according to the amount of amperage they can handle. If a sufficient amperage rated switch is not used in a circuit, the switch could overload and cause damage.

Some electrical components which require a large amount of current to operate use a special switch called a relay. Since these circuits carry a large amount of current, the thickness of the wire in the circuit is also greater. If this large wire were connected from the load to the control switch, the switch would have to carry the high amperage load and the space needed for wiring in the vessel would be twice as big to accommodate the increased size of the wiring harness. To prevent these problems, a relay is used.

Relays are composed of a coil and a set of contacts. When the coil has a current passed though it, a magnetic field is formed and this field causes the contacts to move together, completing the circuit. Most relays are normally open, preventing current from passing through the circuit, but they can take any electrical form depending on the job they are intended to do. Relays can be considered "remote control switches." They allow a smaller current to operate devices that require higher amperages. When a small current operates the coil, a larger current is allowed to pass by the contacts. Some common circuits which may use relays are horns, lights, starter, electric fuel pumps and other high draw circuits.

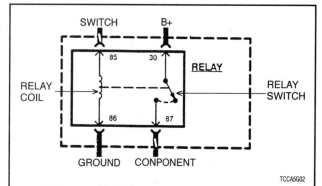

Fig. 3 Relays are composed of a coil and a switch. These two components are linked together so that when one operates, the other operates at the same time. The large wires in the circuit are connected from the battery to one side of the relay switch (B+) and from the opposite side of the relay switch to the load (component). Smaller wires are connected from the relay coil to the control switch for the circuit and from the opposite side of the relay coil to ground

LOAD

Every electrical circuit must include a "load" (something to use the electricity coming from the source). Without this load, the battery would attempt to deliver its entire power supply from one pole to another. This is called a "short circuit". All this electricity would take a short cut to ground and cause a great amount of damage to other components in the circuit by developing a tremendous amount of heat. This condition could develop sufficient heat to melt the insulation on all the surrounding wires and reduce a multiple wire cable to a lump of plastic and copper.

Fig. 2 Fuses protect the vessel's electrical system from abnormally high amounts of current flow

5-4 IGNITION AND ELECTRICAL SYSTEMS

WIRING & HARNESSES

The average vessel contains miles of wiring, with hundreds of individual connections. To protect the many wires from damage and to keep them from becoming a confusing tangle, they are organized into bundles, enclosed in plastic or taped together and called wiring harnesses. Different harnesses serve different parts of the vessel. Individual wires are color coded to help trace them through a harness where sections are hidden from view.

Marine wiring or circuit conductors can be either single strand wire, multi-strand wire or printed circuitry. Single strand wire has a solid metal core and is usually used inside such components as alternators, motors, relays and other devices. Multi-strand wire has a core made of many small strands of wire twisted together into a single conductor. Most of the wiring in a marine electrical system is made up of multi-strand wire, either as a single conductor or grouped together in a harness. All wiring is color coded on the insulator, either as a solid color or as a colored wire with an identification stripe. A printed circuit is a thin film of copper or other conductor that is printed on an insulator backing. Occasionally, a printed circuit is sandwiched between two sheets of plastic for more protection and flexibility. A complete printed circuit, consisting of conductors, insulating material and connectors is called a printed circuit board. Printed circuitry is used in place of individual wires or harnesses in places where space is limited, such as behind instrument panels.

Since marine electrical systems are very sensitive to changes in resistance, the selection of properly sized wires is critical when systems are repaired. A loose or corroded connection or a replacement wire that is too small for the circuit will add extra resistance and an additional voltage drop to the circuit.

The wire gauge number is an expression of the cross-section area of the conductor. Vessels from countries that use the metric system will typically describe the wire size as its cross-sectional area in square millimeters. In this method, the larger the wire, the greater the number. Another common system for expressing wire size is the American Wire Gauge (AWG) system. As gauge number increases, area decreases and the wire becomes smaller. An 18 gauge wire is smaller than a 4 gauge wire. A wire with a higher gauge number will carry less current than a wire with a lower gauge number. Gauge wire size refers to the size of the strands of the conductor, not the size of the complete wire with insulator. It is possible, therefore, to have two wires of the same gauge with different diameters because one may have thicker insulation than the other.

It is essential to understand how a circuit works before trying to figure out why it doesn't. An electrical schematic shows the electrical current paths when a circuit is operating properly. Schematics break the entire electrical system down into individual circuits. In a schematic, usually no attempt is made to represent wiring and components as they physically appear on the vessel; switches and other components are shown as simply as possible. Face views of harness connectors show the cavity or terminal locations in all multi-pin connectors to help locate test points.

CONNECTORS

♦ See Figures 4, 5 and 6

Three types of connectors are commonly used in marine applications—weatherproof, molded and hard shell.

- Weatherproof—these connectors are most commonly used where the connector is exposed to the elements. Terminals are protected against moisture and dirt by sealing rings which provide a weather tight seal. All repairs require the use of a special terminal and the tool required to service it. Unlike standard blade type terminals, these weatherproof terminals cannot be straightened once they are bent. Make certain that the connectors are properly seated and all of the sealing rings are in place when connecting leads.
- Molded—these connectors require complete replacement of the connector if found to be defective. This means splicing a new connector assembly into the harness. All splices should be soldered to insure proper contact. Use care when probing the connections or replacing terminals in them, as it is possible to create a short circuit between opposite terminals. If this happens to the wrong terminal pair, it is possible to damage certain components. Always use jumper wires between connectors for circuit checking and NEVER probe through weatherproof seals.
- Hard Shell—unlike molded connectors, the terminal contacts in hard-shell connectors can be replaced. Replacement usually involves the use of a special terminal removal tool that depresses the locking tangs (barbs) on the connector terminal and allows the connector to be removed from the rear of the shell. The connector shell should be replaced if it shows any evidence of burning, melting, cracks, or breaks. Replace individual terminals that are burnt, corroded, distorted or loose.

Test Equipment

Pinpointing the exact cause of trouble in an electrical circuit is most times accomplished by the use of special test equipment. The following sections describe different types of commonly used test equipment and briefly explain how to use them in diagnosis. In addition to the information covered below, the tool manufacturer's instruction manual (provided with most tools) should be read and clearly understood before attempting any test procedures.

JUMPER WIRES

♦ See Figure 7

✳✳✳ CAUTION

Never use jumper wires made from a thinner gauge wire than the circuit being tested. If the jumper wire is of too small a gauge, it may overheat and possibly melt. Never use jumpers to bypass high resistance loads in a circuit. Bypassing resistances, in effect, creates a short circuit. This may, in turn, cause damage and fire. Jumper wires should only be used to bypass lengths of wire or to simulate switches.

Jumper wires are simple, yet extremely valuable, pieces of test equipment. They are basically test wires which are used to bypass sections of a circuit. Although jumper wires can be purchased, they are usually fabricated from lengths of standard marine wire and whatever type of connector (alligator clip, spade connector or pin connector) that is required for the particular application being tested. In cramped, hard-to-reach areas, it is advisable to have insulated boots over the jumper wire terminals in order to prevent accidental grounding. It is also advisable to include a standard marine fuse in any jumper wire. This is commonly referred to as a "fused jumper". By inserting an in-line fuse holder

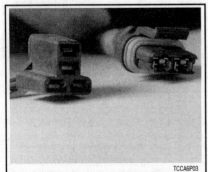

Fig. 4 Hard shell (left) and weatherproof (right) connectors have replaceable terminals

Fig. 5 Weatherproof connectors are most commonly used in the engine compartment or where the connector is exposed to the elements

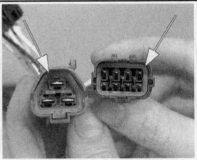

Fig. 6 The seals on weatherproof connectors must be kept in good condition to prevent the terminals from corroding

IGNITION AND ELECTRICAL SYSTEMS 5-5

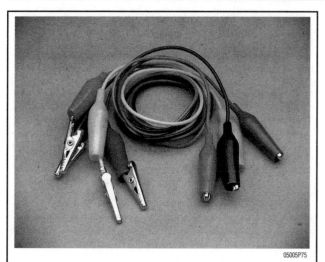

Fig. 7 Jumper wires are simple, yet extremely valuable, pieces of test equipment

between a set of test leads, a fused jumper wire can be used for bypassing open circuits. Use a 5 amp fuse to provide protection against voltage spikes.

Jumper wires are used primarily to locate open electrical circuits, on either the ground (-) side of the circuit or on the power (+) side. If an electrical component fails to operate, connect the jumper wire between the component and a good ground. If the component operates only with the jumper installed, the ground circuit is open. If the ground circuit is good, but the component does not operate, the circuit between the power feed and component may be open. By moving the jumper wire successively back from the component toward the power source, you can isolate the area of the circuit where the open is located. When the component stops functioning, or the power is cut off, the open is in the segment of wire between the jumper and the point previously tested.

You can sometimes connect the jumper wire directly from the battery to the "hot" terminal of the component, but first make sure the component uses 12 volts in operation. Some electrical components, such as fuel injectors or sensors, are designed to operate on about 4 to 5 volts, and running 12 volts directly to these components will cause damage.

TEST LIGHTS

♦ See Figure 8

The test light is used to check circuits and components while electrical current is flowing through them. It is used for voltage and ground tests. To use a 12 volt test light, connect the ground clip to a good ground and probe wherever necessary with the pick. The test light will illuminate when voltage is detected. This does not necessarily mean that 12 volts (or any particular amount of voltage) is present; it only means that some voltage is present. It is advisable before using the test light to touch its ground clip and probe across the battery posts or terminals to make sure the light is operating properly.

✼✼ WARNING

Do not use a test light to probe electronic ignition, spark plug or coil wires. Never use a pick-type test light to probe wiring on electronically controlled systems unless specifically instructed to do so. Any wire insulation that is pierced by the test light probe should be taped and sealed with silicone after testing.

Like the jumper wire, the 12 volt test light is used to isolate opens in circuits. But, whereas the jumper wire is used to bypass the open to operate the load, the 12 volt test light is used to locate the presence of voltage in a circuit. If the test light illuminates, there is power up to that point in the circuit; if the test light does not illuminate, there is an open circuit (no power). Move the test light in successive steps back toward the power source until the light in the handle illuminates. The open is between the probe and a point which was previously probed.

The self-powered test light is similar in design to the 12 volt test light, but contains a 1.5 volt penlight battery in the handle. It is most often used in place of a multimeter to check for open or short circuits when power is isolated from the circuit (continuity test).

The battery in a self-powered test light does not provide much current. A weak battery may not provide enough power to illuminate the test light even when a complete circuit is made (especially if there is high resistance in the circuit). Always make sure that the test battery is strong. To check the battery, briefly touch the ground clip to the probe; if the light glows brightly, the battery is strong enough for testing.

➡ **A self-powered test light should not be used on any electronically controlled system or component. The small amount of electricity transmitted by the test light is enough to damage many electronic marine components.**

MULTIMETERS

♦ See Figure 9

Multimeters are an extremely useful tool for troubleshooting electrical problems. They can be purchased in either analog or digital form and have a price range to suit any budget. A multimeter is a voltmeter, ammeter and ohmmeter (along with other features) combined into one instrument. It is often used when testing solid state circuits because of its high input impedance (usually 10

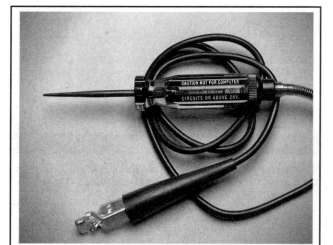

Fig. 8 A 12 volt test light is used to detect the presence of voltage in a circuit

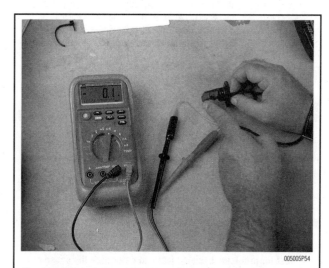

Fig. 9 Multimeters are essential for diagnosing faulty wires, switches and other electrical components

5-6 IGNITION AND ELECTRICAL SYSTEMS

megohms or more). A brief description of the multimeter main test functions follows:

• **Voltmeter**—the voltmeter is used to measure voltage at any point in a circuit, or to measure the voltage drop across any part of a circuit. Voltmeters usually have various scales and a selector switch to allow the reading of different voltage ranges. The voltmeter has a positive and a negative lead. To avoid damage to the meter, always connect the negative lead to the negative (-) side of the circuit (to ground or nearest the ground side of the circuit) and connect the positive lead to the positive (+) side of the circuit (to the power source or the nearest power source). Note that the negative voltmeter lead will always be black and that the positive voltmeter will always be some color other than black (usually red).

• **Ohmmeter**—the ohmmeter is designed to read resistance (measured in ohms) in a circuit or component. Most ohmmeters will have a selector switch which permits the measurement of different ranges of resistance (usually the selector switch allows the multiplication of the meter reading by 10, 100, 1,000 and 10,000). Some ohmmeters are "auto-ranging" which means the meter itself will determine which scale to use. Since the meters are powered by an internal battery, the ohmmeter can be used like a self-powered test light. When the ohmmeter is connected, current from the ohmmeter flows through the circuit or component being tested. Since the ohmmeter's internal resistance and voltage are known values, the amount of current flow through the meter depends on the resistance of the circuit or component being tested. The ohmmeter can also be used to perform a continuity test for suspected open circuits. In using the meter for making continuity checks, do not be concerned with the actual resistance readings. Zero resistance, or any ohm reading, indicates continuity in the circuit. Infinite resistance indicates an opening in the circuit. A high resistance reading where there should be none indicates a problem in the circuit. Checks for short circuits are made in the same manner as checks for open circuits, except that the circuit must be isolated from both power and normal ground. Infinite resistance indicates no continuity, while zero resistance indicates a dead short.

※※ WARNING

Never use an ohmmeter to check the resistance of a component or wire while there is voltage applied to the circuit.

• **Ammeter**—an ammeter measures the amount of current flowing through a circuit in units called amperes or amps. At normal operating voltage, most circuits have a characteristic amount of amperes, called "current draw" which can be measured using an ammeter. By referring to a specified current draw rating, then measuring the amperes and comparing the two values, one can determine what is happening within the circuit to aid in diagnosis. An open circuit, for example, will not allow any current to flow, so the ammeter reading will be zero. A damaged component or circuit will have an increased current draw, so the reading will be high. The ammeter is always connected in series with the circuit being tested. All of the current that normally flows through the circuit must also flow through the ammeter; if there is any other path for the current to follow, the ammeter reading will not be accurate. The ammeter itself has very little resistance to current flow and, therefore, will not affect the circuit, but it will measure current draw only when the circuit is closed and electricity is flowing. Excessive current draw can blow fuses and drain the battery, while a reduced current draw can cause motors to run slowly, lights to dim and other components to not operate properly.

Troubleshooting the Electrical System

When diagnosing a specific problem, organized troubleshooting is a must. The complexity of a modern marine vessel demands that you approach any problem in a logical, organized manner. There are certain troubleshooting techniques, however, which are standard:

• **Establish when the problem occurs**. Does the problem appear only under certain conditions? Were there any noises, odors or other unusual symptoms? Isolate the problem area. To do this, make some simple tests and observations, then eliminate the systems that are working properly. Check for obvious problems, such as broken wires and loose or dirty connections. Always check the obvious before assuming something complicated is the cause.

• **Test for problems systematically to determine the cause once the problem area is isolated**. Are all the components functioning properly? Is there power going to electrical switches and motors. Performing careful, systematic checks will often turn up most causes on the first inspection, without wasting time checking components that have little or no relationship to the problem.

• **Test all repairs after the work is done to make sure that the problem is fixed**. Some causes can be traced to more than one component, so a careful verification of repair work is important in order to pick up additional malfunctions that may cause a problem to reappear or a different problem to arise. A blown fuse, for example, is a simple problem that may require more than another fuse to repair. If you don't look for a problem that caused a fuse to blow, a shorted wire (for example) may go undetected.

Experience has illustrated that most problems tend to be the result of a fairly simple and obvious cause, such as loose or corroded connectors, bad grounds or damaged wire insulation which causes a short. This makes careful visual inspection of components during testing essential to quick and accurate troubleshooting.

Testing

VOLTAGE

This test determines voltage available from the battery and should be the first step in any electrical troubleshooting procedure after visual inspection. Many electrical problems, especially on electronically controlled systems, can be caused by a low state of charge in the battery. Excessive corrosion at the battery cable terminals can cause poor contact that will prevent proper charging and full battery current flow.

1. Set the voltmeter selector switch to the 20V position.
2. Connect the multimeter negative lead to the battery's negative (-) post or terminal and the positive lead to the battery's positive (+) post or terminal.
3. Turn the ignition switch **ON** to provide a load.
4. A well charged battery should register over 12 volts. If the meter reads below 11.5 volts, the battery power may be insufficient to operate the electrical system properly.

VOLTAGE DROP

When current flows through a load, the voltage beyond the load drops. This voltage drop is due to the resistance created by the load and also by small resistances created by corrosion at the connectors and damaged insulation on the wires. The maximum allowable voltage drop under load is critical, especially if there is more than one load in the circuit, since all voltage drops are cumulative.

1. Set the voltmeter selector switch to the 20 volt position.
2. Connect the multimeter negative lead to a good ground.
3. Operate the circuit and check the voltage prior to the first component (load).
4. There should be little or no voltage drop in the circuit prior to the first component. If a voltage drop exists, the wire or connectors in the circuit are suspect.
5. While operating the first component in the circuit, probe the ground side of the component with the positive meter lead and observe the voltage readings. A small voltage drop should be noticed. This voltage drop is caused by the resistance of the component.
6. Repeat the test for each component (load) down the circuit.
7. If a large voltage drop is noticed, the preceding component, wire or connector is suspect.

RESISTANCE

※※ WARNING

Never use an ohmmeter with power applied to the circuit. The ohmmeter is designed to operate on its own power supply. The normal 12 volt electrical system voltage could damage the meter!

1. Isolate the circuit from the vessel's power source.
2. Ensure that the ignition key is **OFF** when disconnecting any components or the battery.
3. Where necessary, also isolate at least one side of the circuit to be checked, in order to avoid reading parallel resistances. Parallel circuit resistances will always give a lower reading than the actual resistance of either of the branches.

IGNITION AND ELECTRICAL SYSTEMS

4. Connect the meter leads to both sides of the circuit (wire or component) and read the actual measured ohms on the meter scale. Make sure the selector switch is set to the proper ohm scale for the circuit being tested, to avoid misreading the ohmmeter test value.

OPEN CIRCUITS

♦ See Figure 10

This test already assumes the existence of an open in the circuit and it is used to help locate the open portion.

1. Isolate the circuit from power and ground.
2. Connect the self-powered test light or ohmmeter ground clip to the ground side of the circuit and probe sections of the circuit sequentially.
3. If the light is out or there is infinite resistance, the open is between the probe and the circuit ground.
4. If the light is on or the meter shows continuity, the open is between the probe and the end of the circuit toward the power source.

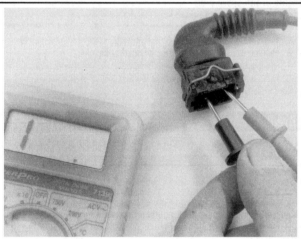

Fig. 10 The infinite reading on this multimeter (1 .) indicates that the circuit is open

SHORT CIRCUITS

➞Never use a self-powered test light to perform checks for opens or shorts when power is applied to the circuit under test. The test light can be damaged by outside power.

1. Isolate the circuit from power and ground.
2. Connect the self-powered test light or ohmmeter ground clip to a good ground and probe any easy-to-reach point in the circuit.

3. If the light comes on or there is continuity, there is a short somewhere in the circuit.
4. To isolate the short, probe a test point at either end of the isolated circuit (the light should be on or the meter should indicate continuity).
5. Leave the test light probe engaged and sequentially open connectors or switches, remove parts, etc. until the light goes out or continuity is broken.
6. When the light goes out, the short is between the last two circuit components which were opened.

Wire and Connector Repair

Almost anyone can replace damaged wires, as long as the proper tools and parts are available. Wire and terminals are available to fit almost any need. Even the specialized weatherproof, molded and hard shell connectors are now available from aftermarket suppliers.

Be sure the ends of all the wires are fitted with the proper terminal hardware and connectors. Wrapping a wire around a stud is never a permanent solution and will only cause trouble later. Replace wires one at a time to avoid confusion. Always route wires in the same manner of the manufacturer.

When replacing connections, make absolutely certain that the connectors are certified for marine use. Automotive wire connectors may not meet United States Coast Guard (USCG) specifications.

➞If connector repair is necessary, only attempt it if you have the proper tools. Weatherproof and hard shell connectors require special tools to release the pins inside the connector. Attempting to repair these connectors with conventional hand tools will damage them.

Electrical System Precautions

- Wear safety glasses when working on or near the battery.
- Don't wear a watch with a metal band when servicing the battery or starter. Serious burns can result if the band completes the circuit between the positive battery terminal and ground.
- Be absolutely sure of the polarity of a booster battery before making connections. Connect the cables positive-to-positive, and negative-to-negative. Connect positive cables first, and then make the last connection to ground on the body of the booster vessel so that arcing cannot ignite hydrogen gas that may have accumulated near the battery. Even momentary connection of a booster battery with the polarity reversed will damage alternator diodes.
- Disconnect both vessel battery cables before attempting to charge a battery.
- Never ground the alternator or generator output or battery terminal. Be cautious when using metal tools around a battery to avoid creating a short circuit between the terminals.
- When installing a battery, make sure that the positive and negative cables are not reversed.
- Always disconnect the battery (negative cable first) when charging.
- Never smoke or expose an open flame around the battery . Hydrogen gas accumulates near the battery and is highly explosive.

BREAKER POINTS IGNITION (MAGNETO IGNITION)

♦ See Figures 11, 12 and 13

➞All Suzuki outboard engines use a pointless electronic ignition system with the exception of the pre-1990 DT2 engines which use a breaker point type magneto.

This ignition system uses a mechanically switched, collapsing field to induce spark at the plug. A magnet moving by a coil produces current in the primary coil winding. The current in the primary winding creates a magnetic field. When the points are closed the current goes to ground. As the breaker points open the primary magnetic field collapses across the secondary coil winding. This induces (transforms) a high voltage potential in the secondary coil winding. This high voltage current travels to the spark plug and jumps the gap.

The breaker point ignition system contains a condenser that works like a sponge in the circuit. Current that is flowing through the primary circuit tries to keep going. When the breaker point switch opens the current will arc over the widening gap. The condenser is wired in parallel with the points. The condenser absorbs some of the current flow as the points open. This reduces arc over and extends the life of the points.

The breaker point ignition consists of the rotor assembly, contact point assembly, ignition coil, condenser spark plug, spark plug cap and the engine stop switch.

As the pole pieces of the magnet pass over the heels of the coil, a magnetic field is built up about the coil, causing a current to flow through the primary winding.

At the proper time, the breaker points are separated by action of a cam designed into the collar of the crankshaft and the primary circuit is broken. When the circuit is broken, the flow of primary current stops and causes the magnetic field about the coil to break down instantly. At this precise moment, an electrical current of extremely high voltage is induced in the fine secondary windings of the coil. This high voltage is conducted to the spark plug where it jumps the gap between the points of the plug to ignite the compressed charge of air-fuel mixture in the cylinder.

5-8 IGNITION AND ELECTRICAL SYSTEMS

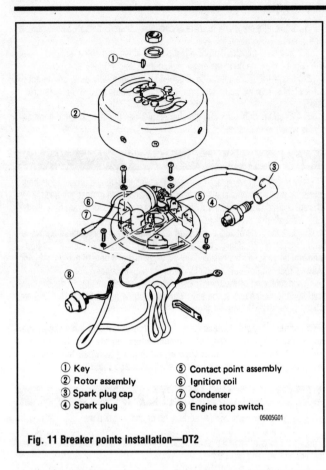

① Key
② Rotor assembly
③ Spark plug cap
④ Spark plug
⑤ Contact point assembly
⑥ Ignition coil
⑦ Condenser
⑧ Engine stop switch

Fig. 11 Breaker points installation—DT2

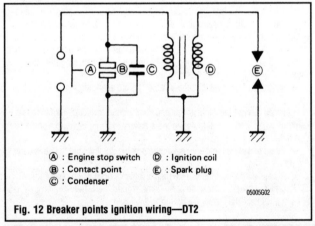

Ⓐ : Engine stop switch
Ⓑ : Contact point
Ⓒ : Condenser
Ⓓ : Ignition coil
Ⓔ : Spark plug

Fig. 12 Breaker points ignition wiring—DT2

The breaker points must be aligned accurately to provide the best contact surface. This is the only way to assure maximum contact area between the point surfaces; accurate setting of the point gap; proper synchronization; and satisfactory point life. If the points are not aligned properly, the result will be premature wear or pitting. This type of damage may change the cam angle, although the actual distance will remain the same.

Magnetos installed on outboard engines will usually operate over extremely long periods of time without requiring adjustment or repair. However, if ignition system problems are encountered, and the usual corrective actions such as replacement of spark plugs does not correct the problem, the magneto output should be checked to determine if the unit is functioning properly.

System Testing

Perform a spark test if you suspect the ignition system of not working properly.

※※ WARNING

When checking the spark, make sure there is no fuel on either the engine or the spark plug. Also keep your hands away from high voltage electrical components.

1. Remove the spark plug and ground the plug electrode to the engine.
2. Pull the recoil starter and check for spark at the plug.

If there is a good spark at the plug, the ignition system should be performing properly. If there is no spark, precede to the next step in Troubleshooting the ignition system problem.

Breaker Points

POINT GAP ADJUSTMENT

▶ See Figures 14 and 15

→Before checking the ignition timing, be sure that the contact point faces are in good condition. Smoothen and make parallel the two faces by grinding with an oil stone as much as necessary and then clean the points by wiping them with cloth dampened with a suitable solvent. Then apply a small amount of grease to the breaker shaft.

1. Remove the engine cover to access the engine.
2. Remove the hand rewind starter assembly.

→There are slots in the flywheel rotor in which to insert the feeler gauge and measure the points gap without removing the flywheel itself.

3. Rotate the stator base until it is at the wide open throttle position.
4. Turn the crankshaft using a wrench on the flywheel nut **clockwise** until the breaker point rubbing block touches the high point on the cam. At this point the points will be wide open.
5. Measure the point gap with a feeler gauge. There should be a slight drag on the feeler gauge if the gap is correct. The point gap should measure: 0.012–0.016 in. (0.3–0.4 mm). If the gap is out of specification, adjustment will be necessary.

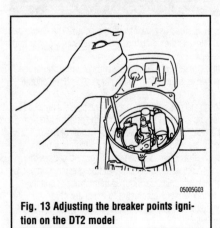

Fig. 13 Adjusting the breaker points ignition on the DT2 model

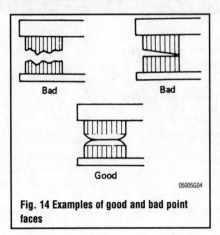

Fig. 14 Examples of good and bad point faces

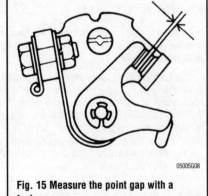

Fig. 15 Measure the point gap with a feeler gauge

IGNITION AND ELECTRICAL SYSTEMS 5-9

TESTING

▶ See accompanying illustrations

1. Remove the flywheel.
2. Remove the spark plug.
3. With the contact points set right, now check the ignition timing by using the timing gauge (09931–00112). Remove the spark plug and install this gauge in the spark plug hole as illustrated. Bring the piston up to TDC and set the indicating hand of the gauge to read zero millimeters.
4. Obtain a timing digital multimeter, also know as a buzz box: (09900–27003).
5. Disconnect the breaker point leads from the stator base.
6. Connect the positive (red) digital multimeter lead to the black lead of the contact breaker and the negative digital multimeter lead to an engine ground.
7. Gently turn the flywheel clockwise (with the digital multimeter switch turned ON) until the digital multimeter starts buzzing.

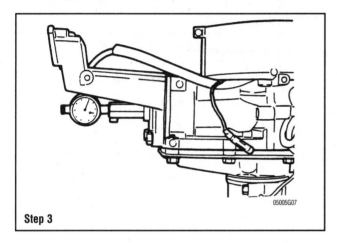
Step 3

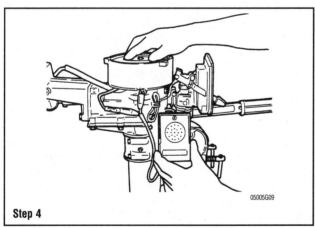
Step 4

REMOVAL & INSTALLATION

▶ See accompanying illustrations

1. Remove the engine cover.
2. Remove the fuel tank assembly.
3. Remove the recoil starter assembly.
4. Remove the starter cup and magneto insulator.
5. With a flywheel holder or a commonly available strap wrench, hold the flywheel and loosen the retaining nut.
6. With the flywheel rotor remover (09930-30713) remove the flywheel. Make sure to keep track of the flywheel key when removing the flywheel.
7. Disconnect the plug cap and the two lead wires (that are coming from the stator) and remove the stator.
8. A typical points set with major components identified.
9. Point faces must be in good condition and aligned correctly.

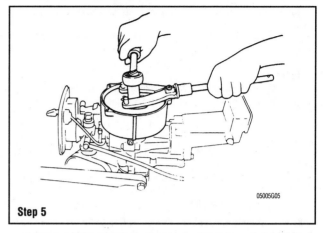
Step 5

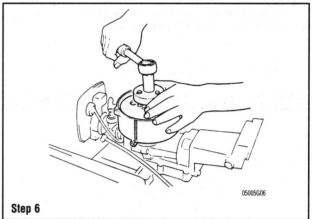
Step 6

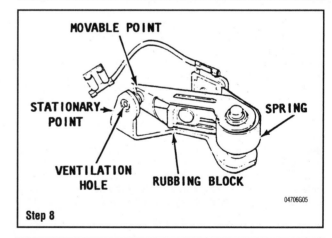
Step 8

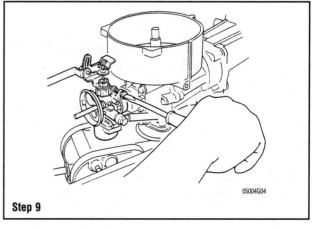
Step 9

5-10 IGNITION AND ELECTRICAL SYSTEMS

10. Remove the stator base screw which holds the breaker point assembly to the base.
11. Disconnect the coil and condenser leads at the breaker point. Now remove the breaker point assembly.
12. Remove the condenser from the stator base by removing the screw.

To install:

13. Install the replacement breaker point set on the stator base. Make sure the pivot point on the bottom of the point set engages the hole in the stator base. Now, install, but don't tighten the retaining screw.
14. Install the condenser on the stator base and now tighten the retaining screw securely.
15. Reconnect the stator leads. Inspect the connectors and clean off any corrosion before connecting.
16. Check the lubrication felt for dryness. If it dry, add a couple of drops of 30w engine oil.
17. Reconnect the spark plug lead.

➡ **Before installing the flywheel, thoroughly inspect the crankshaft and flywheel tapers. These surfaces must be absolutely clean and free of oil, grease and dirt. Use solvent and a lint free cloth to clean the surfaces and then blow dry with compressed air.**

18. Install the flywheel key, starter cup and flywheel and flywheel bolt. Tighten the bolt to 30–36 ft. lbs. (40–50 Nm.)
19. Install the fuel and engine cover.

Condenser

DESCRIPTION & OPERATION

♦ See Figure 16

In simple terms, a condenser is composed of two sheets of tin or aluminum foil laid one on top of the other, but separated by a sheet of insulating material such as waxed paper, etc. The sheets are rolled into a cylinder to conserve space and then inserted into a metal case for protection and to permit easy assembling.

The purpose of the condenser is to prevent excessive arcing across the points and to extend their useful life. When the flow of primary current is brought to a sudden stop by the opening of the points, the magnetic field in the primary windings collapses instantly, and is not allowed to fade away, which would happen if the points were allowed to arc.

The condenser stores the electricity that would have arced across the points and discharges that electricity when the points close again. This discharge is in the opposite direction to the original flow, and tends to smooth out the current. The more quickly the primary field collapses, the higher the voltage produced in the secondary windings and delivered to the spark plugs. In this way, the condenser (in the primary circuit), affects the voltage (in the secondary circuit) at the spark plugs.

Modern condensers seldom cause problems, therefore, it is not necessary to install a new one each time the points are replaced. However, if the points show evidence of arcing, the condenser may be at fault and should be replaced. A faulty condenser may not be detected without the use of special test equipment. Testing will reveal any defects in the condenser, but will not predict the useful life left in the unit.

The modest cost of a new condenser justifies its purchase and installation to eliminate this item as a source of trouble.

TESTING

1. Remove the flywheel.
2. Disconnect the condenser lead from the breaker points assembly.
3. Connect on test lead to the condenser lead. Connect the other test lead to the stator base.
4. Set the digital multimeter controls according to the manufacturers instructions. Check the condenser for resistance, leakage and capacity.
5. Compare the results in the previous step with the ignition digital multimeter. Replace the condenser if it does not pass any one of the three tests.

REMOVAL & INSTALLATION

1. Remove the engine cover.
2. Remove the fuel tank assembly.
3. Remove the recoil starter assembly.
4. Remove the starter cup and magneto insulator.
5. With a flywheel holder or a commonly available strap wrench, hold the flywheel and loosen the retaining nut.
6. With the flywheel rotor remover (09930-30713) remove the flywheel. Make sure to keep track of the flywheel key when removing the flywheel.
7. Disconnect the plug cap and the two lead wires (that are coming from the stator) and remove the stator.
8. Remove the stator base screw which holds the breaker point assembly to the base.
9. Disconnect the coil and condenser leads at the breaker point. Now remove the breaker point assembly.
10. Remove the condenser from the stator base by removing the screw.

To install:

11. Install the replacement breaker point set on the stator base. Make sure the pivot point on the bottom of the point set engages the hole in the stator base. Now, install, but don't tighten the retaining screw.
12. Install the condenser on the stator base and now tighten the retaining screw securely.
13. Reconnect the stator leads. Inspect the connectors and clean off any corrosion before connecting.
14. Check the lubrication felt for dryness. If it dry, add a couple of drops of 30w engine oil.
15. Reconnect the spark plug lead.

➡ **Before installing the flywheel, thoroughly inspect the crankshaft and flywheel tapers. These surfaces must be absolutely clean and free of oil, grease and dirt. Use solvent and a lint free cloth to clean the surfaces and then blow dry with compressed air.**

16. Install the flywheel key, starter cup and flywheel and flywheel bolt. Tighten the bolt to 30–36 ft. lbs. (40–50 Nm.)
17. Install the fuel and engine cover.

Ignition Coil

DESCRIPTION & OPERATION

The coil is the heart of the ignition system. Essentially, it is nothing more than a transformer which takes the relatively low voltage (12 volts) available from the primary coil and increases it to a point where it will fire the spark plug as much as 20,000 volts.

Once the voltage is discharged from the ignition coil the secondary circuit begins and only stretches from the ignition coil to the spark plugs via extremely large high tension leads. At the spark plug end, the voltage arcs in the form of a spark, across from the center electrode to the outer electrode, and then to ground via the spark plug threads. This completes the ignition circuit.

Fig. 16 This sketch illustrates how waxed paper, aluminum foil and insulation are rolled in the manufacture of a typical condenser

IGNITION AND ELECTRICAL SYSTEMS

TESTING

1. Remove the flywheel.
2. For coil primary resistance:
   a. Disconnect the black primary ignition coil lead at the connector.
   b. Disconnect the secondary coil lead (spark plug wire) at the spark plug.
   c. Make sure the ohmmeter is on the low-ohm scale. Connect the meter between the primary coil lead and an engine ground.
   d. Check the resistance reading in the "Ignition Coil Resistance" chart.
3. For coil secondary resistance:
   a. Make sure the ohmmeter is on the high-ohms scale.
   b. Connect the meter between the secondary coil lead and an engine ground.
   c. Check the resistance reading in the "Ignition Coil Resistance" chart.
4. Replace the ignition coil if the either the primary or secondary resistance does not meet specifications.

REMOVAL & INSTALLATION

1. Remove the engine cover.
2. Remove the fuel tank assembly.
3. Remove the recoil starter assembly.
4. Remove the starter cup and magneto insulator.
5. With a flywheel holder or a commonly available strap wrench, hold the flywheel and loosen the retaining nut.
6. With the flywheel rotor remover (09930-30713) remove the flywheel. Make sure to keep track of the flywheel key when removing the flywheel.
7. Disconnect the plug cap and the two lead wires (that are coming from the stator) and remove the stator.
8. Remove the stator base screw which holds the breaker point assembly to the base.
9. Disconnect the coil and condenser leads at the breaker point. Now remove the breaker point assembly.
10. Remove the condenser from the stator base by removing the screw.

To install:

11. Install the replacement breaker point set on the stator base. Make sure the pivot point on the bottom of the point set engages the hole in the stator base. Now, install, but don't tighten the retaining screw.
12. Install the condenser on the stator base and now tighten the retaining screw securely.
13. Reconnect the stator leads. Inspect the connectors and clean off any corrosion before connecting.
14. Check the lubrication felt for dryness. If it dry, add a couple of drops of 30w engine oil.
15. Reconnect the spark plug lead.

➡Before installing the flywheel, thoroughly inspect the crankshaft and flywheel tapers. These surfaces must be absolutely clean and free of oil, grease and dirt. Use solvent and a lint free cloth to clean the surfaces and then blow dry with compressed air.

16. Install the flywheel key, starter cup and flywheel and flywheel bolt. Tighten the bolt to 30–36 ft. lbs. (40–50 Nm.)
17. Install the fuel and engine cover.
18. Remove the spark plug caps
19. With an ohmmeter, measure the resistance between the spark plug wires. Resistance should measure 6.4K–9.6K ohms.
20. Measure the resistance between the primary terminal and the coil mounting lug (for ground).
21. Resistance should measure 0.46–0.66 ohms.

CAPACITOR DISCHARGE IGNITION (CDI) SYSTEM

Description and Operation

SINGLE-CYLINDER IGNITION

♦ See Figure 17

In its simplest form, a CDI ignition is composed of the following elements:
- Magneto
- Pulser coil
- Charge, or source coil
- Igniter (CDI) box
- Ignition coil
- Spark plug

Other components such as main switches, stop switches, or computer systems may be included, though, these items are not necessary for basic CDI operation.

To understand basic CDI operation, it is important to understand the basic theory of induction. Induction theory states that if we move a magnet (magnetic field) past a coil of wire(or the coil by the magnet), AC current will be generated in the coil.

The amount of current produced depends on several factors:
- How fast the magnet moves past the coil
- The size of the magnet(strength)
- How close the magnet is to the coil
- Number of turns of wire and the size of the windings

When the flywheel rotates, the electrical power generated at the exciter coil is rectified by the diode and charged into the ignition condenser. The thyristor is off at this time.

When the magnet on the crankshaft passes the pulser coil, the electric pulser coil signal is emitted by the magnetic force. This signal passes the gate circuit, turns on the thyristor, and discharges the electric charge from the condenser. When the discharged current flows through the ignition coil primary circuit, high voltage is generated in the secondary circuit and the spark plug sparks.

The spark advance is handled by electronic advance spark system, which advances the ignition timing when the gate circuit turns on the thyristor according to the engine speed to obtain high speed power.

The current produced in the charge coil goes to the CDI box. On the way in, it is converted to DC current by a diode. This DC current is stored in the capacitor located inside the box. As the charge coil produces current, the capacitor stores it.

At a specific time in the magneto's revolution, the magnets go past the pulser coil. The pulser coil is smaller than the charge coil so it has less current output. The current from the pulser also goes into the CDI box. This current signals the CDI box when to fire the capacitor (the pulser may be called a trigger coil for obvious reasons). The current from the capacitor flows out to the ignition coil and spark plug. The pulser acts much like the points in older ignitions systems.

When the pulser signal reaches the CDI box, all the electricity stored in the capacitor is released at once. This current flows through the ignition coil's primary windings.

The ignition coil is a step-up transformer. It turns the relatively low voltage entering the primary windings into high voltage at the secondary windings. This occurs due to a phenomena known as induction.

The high voltage generated in the secondary windings leaves the ignition coil and goes to the spark plug. The spark in turn ignites the air-fuel charge in the combustion chamber.

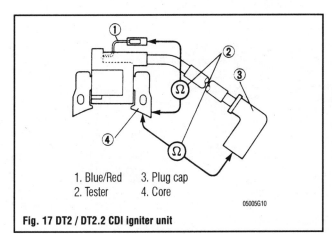

1. Blue/Red
2. Tester
3. Plug cap
4. Core

Fig. 17 DT2 / DT2.2 CDI igniter unit

5-12 IGNITION AND ELECTRICAL SYSTEMS

Once the complete cycle has occurred, the spinning magneto immediately starts the process over again.

Main switches, engine stop switches, and the like are usually connected on the wire in between the CDI box and the ignition coil. When the main switch or stop switch is turned to the OFF position, the switch is closed. This closed switch short-circuits the charge coil current to ground rather than sending it through the CDI box. With no charge coil current through the CDI box, there is no spark and the engine stops or, if the engine is not running, no spark is produced.

SUZUKI PEI IGNITION

The Suzuki PEI (Pointless Electronic Ignition) is a magneto CDI type system.

DT4

♦ See Figure 18

The DT4 models have an ignition which uses 1 ignition coil, a single primary coil, one pulser coil, an optional lighting coil. The CDI unit itself electronically advances the ignition timing when firing the spark plugs. The secondary ignition coils and CDI unit are separate components.

DT6

The DT6 models use a single condenser charge coil, lighting coil and single pulser coil which in turn supply current to a combined CDI unit and single secondary coil with two spark plug wires. The stator base moves according to the throttle opening to obtain the proper ignition timing. For this reason, metal (brass) is cast in the spigot joint of the oil seal housing and stator base.

Parts of the stator base include a coil which charges a capacitor of the CDI unit, a pulser coil which sends a signal to the CDI unit at ignition timing, and a lighting coil which generates a lighting output of 12V and 80W.

DT9.9 and DT15

♦ See Figure 19

The DT9.9 and DT15 are equipped with a single ignition coil for two cylinders in their ignition systems. On these models the CDI unit electronically advances the ignition timing when the spark plugs are firing. The CDI unit and secondary ignition coils are separate components.

DT25 and DT30

The DT25 and DT30 use the I.C. ignition system. A built in I.C. control unit monitors the degree of throttle opening and the engine rpm, then determines ideal spark timing. This not only improves acceleration, but by maintaining optimum carburetion and ignition synchronization, the engines run smoother and respond quicker to changes in throttle operation.

The ignition system is comprised of three pulser coils, a battery charge coil, condenser charging coil and a gear counter coil. These coils are connected to the CDI unit and along with the throttle position sensor and engine temperature sensor determines the ignition sequence.

➡ In the 91-on models, a single sensor called a cooling water & engine temperature sensor was used. This sensor served two purposes: one is to detect the engine temperature with which to select either of the two different starting ignition timing duration's at 5°BTDC (3 seconds for a warm engine and 15 seconds for a cold engine) so as to improve engine

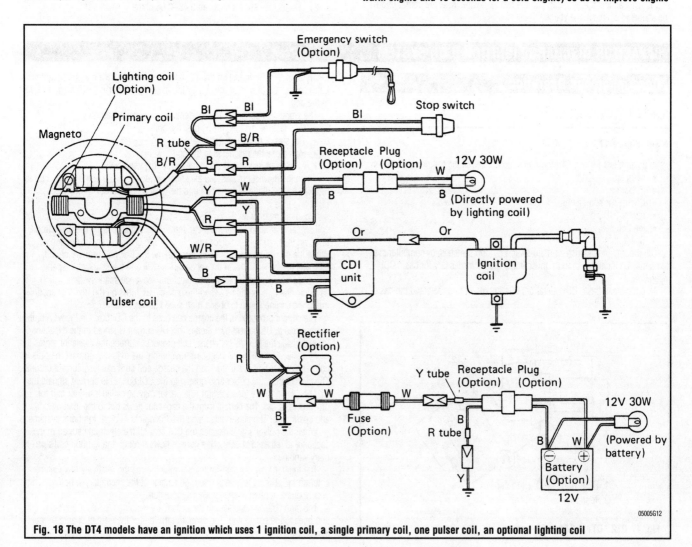

Fig. 18 The DT4 models have an ignition which uses 1 ignition coil, a single primary coil, one pulser coil, an optional lighting coil

IGNITION AND ELECTRICAL SYSTEMS   5-13

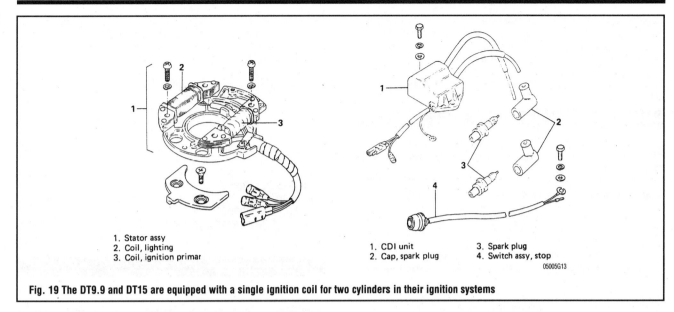

Fig. 19 The DT9.9 and DT15 are equipped with a single ignition coil for two cylinders in their ignition systems

starting. The sensor's other function is to determine if the cooling water level reaches the powerhead. From the 92-on models, this sensor has been eliminated. To replace this sensor, two separate sensors, the engine starting ignition timing sensor and overheat sensor, have been installed.

DT35 and DT40

♦ See Figure 20

The DT35 and DT40 models use a 2-cylinder, simultaneous ignition CDI system. The system is made up of the magneto, CDI unit. A condenser charge coil, timing coil and a lighting coil are mounted on the stator. The ignition coil is contained in the CDI unit. The ignition timing characteristics are the result of the advance angle of the sliding stator. An electronic advance system employing the I.C. ignition has been adopted in the advance angle of the magneto to assure highly precise timing characteristics. The CDI unit includes the over-rev limiter and oil warning circuit.

DT55, DT65, DT75 and DT85

The ignition systems on these models are equipped with the Digital IC Ignition system. This ignition system automatically alters the ignition timing electronically based on throttle valve position and engine speed. This system is comprised of 3 pulser coils, battery charging coil, condenser charge coil. There is also a gear counter coil to control ignition timing at all rpm and an ignition timing sensor that provides an automatic 5°BTDC advance for engine starting (15 seconds when cold, 3 seconds when warm) and a throttle valve position sensor.

DT90, DT100

These models are equipped with the Micro-Link ignition system. This system incorporates a microcomputer to improve engine performance by maximizing combustion control. This system uses information from sensors and switches located a various positions on the engine to monitor throttle valve opening, engine rpm, shift lever position and operator selected idle speed. The microcomputer constantly evaluates this information and instantly provides the optimal spark timing for the current engine running conditions.

The Micro-Link system also monitors the caution system sensors for oil level, oil flow, water flow and engine over-rev. If any of these sensors indicate a malfunction, the microcomputer will activate a warning buzzer and/or monitor gauge indicator lamp and then operate the engine under reduced power.

The magneto consists of the following components. Each coil functions as follows:

• The condenser charging coil charges the condenser for the spark primary power source in the CDI unit.

• The pulser coils, being positioned at intervals of 70° and 110°, input a reference pulse, for the corresponding cylinder, causing the condenser to be discharged into the microcomputer inside the CDI unit.

• The counter coil causes a reference pulse to decide the ignition timing to be input into the microcomputer.

• The battery charging coils, No.1 and No.2, perform power generating necessary to charge the battery.

In addition, the microcomputer gets it's 12 volt operating power from the battery.

Carbureted DT115 and DT140

The ignition systems on these models is the Suzuki Digital IC Ignition system. This ignition system automatically alters the ignition timing electronically

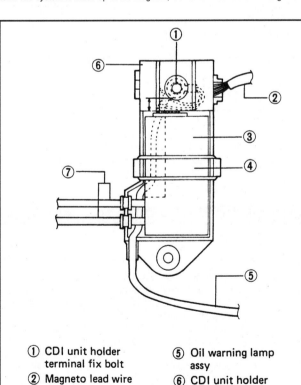

Fig. 20 The DT35 and DT40 models use a 2-cylinder, simultaneous ignition CDI system

5-14 IGNITION AND ELECTRICAL SYSTEMS

based on throttle valve position and engine speed. This system is comprised of 4-pulser coils, battery charging coil, condenser charge coil. There is also a gear counter coil to control ignition timing at all rpm, an ignition timing sensor that provides an automatic 7°BTDC advance for engine starting. It will remain at 7°BTDC for 12–15 seconds, at which time the ignition timing will return to the idle speed circuit and whatever position the "Idle Speed Adjustment Switch" is set at.

> ✲✲✲ **CAUTION**
>
> Due to the higher idle speed created by this automatic starting device, DO NOT shift the engine until the engine rpm has returned to idle speed.

Fuel Injected DT115 and DT140

The fuel injected models use the Suzuki Micro-Link ignition system. This system uses sensors to monitor specific engine operating conditions and supplies signals to an Engine Control Unit (ECU) for ignition and warning operation. The primary ignition sensors are the throttle valve sensor (TVS) for determining throttle valve opening angle and the engine speed sensor (gear count coil) that determines engine speed.

Based on these signals, the ECU determines the ignition timing necessary for the engine's current requirements and delivers voltage to the ignition coils thus producing ignition spark.

Two compensation sensors, cylinder wall temperature and air temperature, supply signals used by the ECU to compensate ignition timing based on temperature related conditions. A pulser coil supplies crankshaft angle signals which the ECU uses when determining a trigger signal for the ignition.

DURING START

At engine start, the ignition timing is set at 5°BTDC Sequential operation while cranking (below 440 rpm).

AFTER START

After engine start, keeping the throttle lever at the idle position will allow the ECU "Warm-up mode" map to control the ignition timing in relation to cylinder wall temperature. In this mode, ignition timing stays advanced above normal until the cylinder wall temperature reaches 113° (45°C) the timer, which was set according to cylinder wall temperature, expires.

NORMAL OPERATING MODE

When the "Warm-up mode" ends, the ECU changes to the "Normal operating mode". This mode is based on the map control in relation to engine speed, throttle valve opening angle, cylinder wall temperature and air temperature.

FAIL SAFE MODE

Each sensor has an assigned default value programmed into the ECU. In the event of a sensor failure, the monitor gauge flash code will indicate the failure and the engine will continue to operate, but with much reduced performance, ignition timing during a sensor failure will automatically default to the following method of control:

- Throttle valve sensor failure: Ignition timing will be automatically set according to the engine speed.
- Cylinder wall temperature sensor failure: Ignition timing will be automatically set as if the sensor senses 86°F (30°C).
- Air temperature sensor failure: Ignition timing will be automatically set as if the sensor senses 68°F (20°C).
- If either gear counter coil or pulser coil fails, the ECU will not provide and injection signal without a reference from these coils. Under this condition, the engine can be cranked, but it will not start due to no fuel injection pressure.

DT150, DT175, DT200 and DT225

These models are equipped with the Micro-Link ignition system. This system incorporates a microcomputer to improve engine performance by maximizing combustion control. This system uses information from sensors and switches located a various positions on the engine to monitor throttle valve opening, engine rpm, shift lever position and operator selected idle speed. The microcomputer constantly evaluates this information and instantly provides the optimal spark timing for the current engine running conditions.

The Micro-Link system also monitors the caution system sensors for oil level, oil flow, water flow and engine over-rev. If any of these sensors indicate a malfunction, the microcomputer will activate a warning buzzer and/or monitor gauge indicator lamp and then operate the engine under reduced power.

The magneto consists of the following components. Each coil functions as follows:

- The condenser charging coil charges the condenser for the spark primary power source in the CDI unit.
- The pulser coils, positioned 120°apart. When the pulser coil voltage enters the delay circuit, the CDI unit begins to count the voltage signals from the gear counter coil and will release the signal from the delay circuit when the flywheel indicates the proper piston position according to the idle speed adjustment switch setting.
- The counter coil causes a reference pulse to decide the ignition timing to be input into the microcomputer.
- The battery charging coils, No.1 and No.2, perform power generating necessary to charge the battery.

In addition, the microcomputer gets it's 12 volt operating power from the battery.

System Testing

PROCEDURE

Perform a visual inspection of the wiring connections and grounds. Determine if the problem affects all or just certain cylinders and perform a spark check using a spark gap tool ("spark tester") and then check ignition timing with a timing light.

If the problem affects all the cylinders, check the capacitor charge coil output, engine stop switch, CDI unit output and pulse coil output.

If the problem affects individual cylinders only, check the components whose failure would affect that particular cylinder such as the pulse coil and ignition coil performance.

If the problem is timing related, check the mechanical part of the system, such as the pulse coil or CDI box itself and then check the electronic timing advance components of the system, the throttle position sensor (if applicable), the pulse coil(s) and the CDI module.

CDI troubleshooting can be performed with a peak reading voltmeter. This will check the CDI voltage to the ignition coils.

- If CDI voltage is good, isolate individual ignition coils or spark plugs and check output voltage.
- If the CDI voltage is bad, check all CDI input voltages.
- Check the pulse coil output to the CDI unit.
- Check the capacitor charge coil output to the CDI unit.
- Check the pulser coil output to the CDI unit.

If all the input voltages are normal, the problem has now been isolated to the CDI unit itself. If any input voltage is abnormal, check the appropriate coil for winding resistance and insulation breakdown. If the problem is timing related, check all the timing inputs to the CDI unit, such as the throttle position sensor. If the timing inputs are good, the problem is isolated to the CDI unit.

Pulsar/Charging/Gear Counter Coils

DESCRIPTION & OPERATION

The second circuit used in CDI systems is the pulsar circuit. The pulsar circuit has its own flywheel magnet, a pulsar coil, a diode, and a thyristor. A thyristor is a solid state electronic switching device which permits voltage to flow only after it is triggered by another voltage source.

At the point in time when the ignition timing marks align, an alternating current is induced in the pulsar coil, in the same manner as previously described for the charge coil. This current is then passed to a second diode located in the CDI unit where it becomes DC current and flows on to the thyristor. This voltage triggers the thyristor to permit the voltage stored in the capacitor to be discharged. The capacitor voltage passes through the thyristor and on to the primary windings of the ignition coil.

In this manner, a spark at the plug may be accurately timed by the timing marks on the flywheel relative to the magnets in the flywheel and to provide as many as 100 sparks per second for a powerhead operating at 6000 rpm.

A system of battery charging is standard on all electric start model engines.

IGNITION AND ELECTRICAL SYSTEMS

The battery charging system is made up of charge coil(s) or an A/C lighting coil on some models. The permanent magnets located on the flywheel edge, a rectifier (voltage regulator/rectifier on some models), a battery and the associated wiring and fuses.

Rotation of the flywheel magnets past the charge coil or A/C lighting coils will create alternating current. This current is then sent to the rectifier or voltage regulator/rectifier, where it is then converted to direct current and then supplied to the battery or electrical accessories through a fuse.

The capacitor charge coil supplies electricity to the CDI unit. As the flywheel rotates past the capacitor charge coil, alternating current is produced. This voltage is supplied to the CDI unit and passes through a diode where it is rectified to direct current. This D/C voltage is then stored in the capacitor.

The gear counter coil (if equipped) is the same construction and operation as the pulser coil, but it uses each ring gear tooth as a pulser bar to generate a signal voltage to send to the CDI unit.

TESTING

DT4, DT6 and DT8

♦ See accompanying illustrations

1. Remove the engine cover.
2. Disconnect the wires at the connectors between the stator plate and the CDI unit.
3. To test the pulser coil, connect the digital multimeter between the white/red and black wires. Resistance should read 12.0–18.0 ohms for the DT4, and 20.5–25.1 ohms for the DT6 and DT8.
4. To test the lighting coil, connect the digital multimeter between the yellow and red wires. Resistance should read between 0.5–1.5 ohms for the DT4 and 0.37–0.45 ohms for the DT6 and DT8.
5. If the resistance reading is not within specification, replace the faulty coil.

DT9.9 and DT15

♦ See accompanying illustrations

1. The two pulser coils are located at 180° angles to each other and the condenser and battery charge coils are also located opposite of each other under the flywheel.
2. Remove the engine cover.
3. On models equipped with electric starters, disconnect the negative battery cable.
4. Disconnect the connector and black ground wire.
5. On the No.1 pulser coil, connect the digital multimeter between the black/red wire terminal in the connector and the black ground wire. And the No.2 coil between the white/green wire terminal in the connector and the black ground wire. Resistance should be 260–380 ohms.
6. On the condenser charge coil, measure the resistance between the black/red and green wires. Resistance should be between 170–250 ohms.
7. If the resistance reading is not within specification, replace the faulty coil.
8. On the battery charge coil, measure the resistance between the yellow/red and yellow wires, red and yellow wires (DT15MC) or the red and yellow wires (DT15CE). Resistance should measure between 0.2 and 0.5 ohms.
9. If the resistance reading is not within specification, replace the faulty coil.

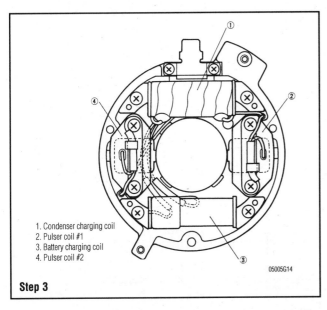

1. Condenser charging coil
2. Pulser coil #1
3. Battery charging coil
4. Pulser coil #2

Step 3

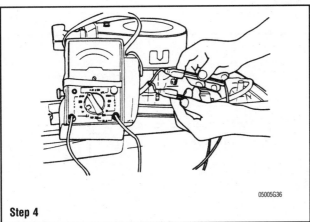

Step 4

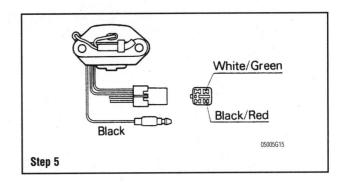

Step 5

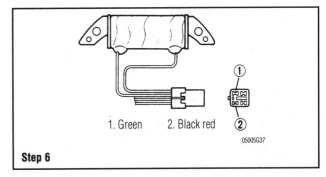

1. Green   2. Black red

Step 6

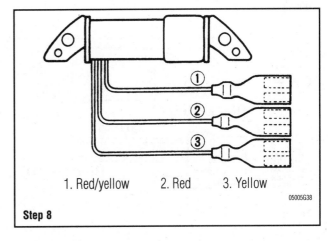

1. Red/yellow   2. Red   3. Yellow

Step 8

5-16 IGNITION AND ELECTRICAL SYSTEMS

DT25 and DT30

▶ See accompanying illustrations

1. The three pulser coils are located at 120° angles to each other underneath the flywheel along with the condenser charge coil, battery charge coil and the gear counting coil.
2. Remove the engine cover.
3. On models equipped with electric starters, disconnect the negative battery cable.
4. Disconnect the 6-pin connector and set the digital multimeter on the ohms scale.
5. For the No.1 pulser coil, connect the digital multimeter between the red/black wire terminal and ground. For the No.2 pulser coil, connect the digital multimeter between the white/black wire terminal and ground and for the No.3 pulser coil, connect the digital multimeter between the red/white wire terminal and ground. Resistance should measure 170–250 ohms on the multimeter.
6. For the condenser charge coil, measure the resistance between the black/green and green wires. Resistance should measure between 170–250 ohms.
7. For the battery charge coil, measure the resistance between the yellow/red and yellow wires and red and yellow wires (DT30MC) and the red and yellow wires (DT30CR). Resistance should measure between 0.2–0.6 ohms.
8. For the gear counter coil, measure the resistance between the orange/green and black wires. Resistance should measure between 160–240 ohms.
9. Replace the faulty coil if the resistance reading is not within specification.

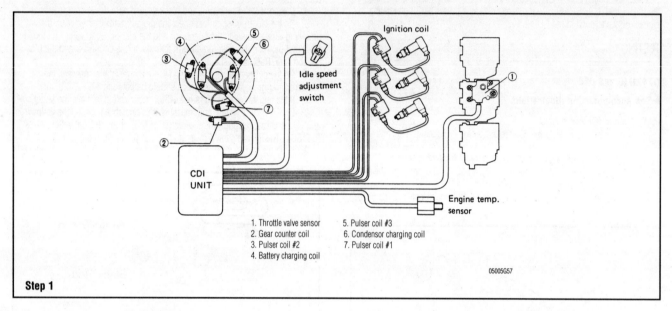

1. Throttle valve sensor
2. Gear counter coil
3. Pulser coil #2
4. Battery charging coil
5. Pulser coil #3
6. Condensor charging coil
7. Pulser coil #1

Step 1

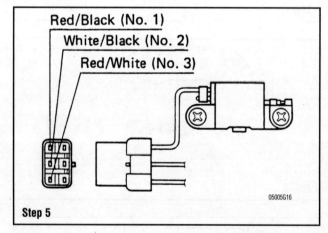

Step 5

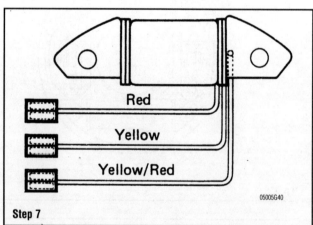

Step 7

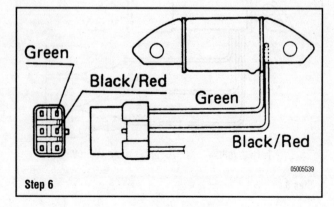

Step 6

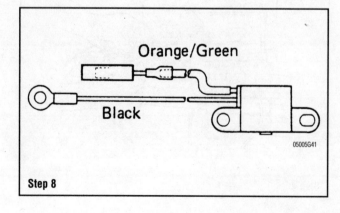

Step 8

IGNITION AND ELECTRICAL SYSTEMS 5-17

DT35 and DT40

1. The pulser (timing coil) and lighting (battery) charge coil and condenser charge coil are all located underneath the flywheel on the stator base.
2. Remove the engine cover.
3. On models equipped with electric starters, disconnect the negative battery cable.
4. For the pulser coil, measure the resistance between the white/red wire lead and the black ground. Resistance should read 175–210 ohms.
5. For the battery charge coil, measure the resistance between the red and yellow wires. Resistance should measure between 0.20–210 ohms.
6. For the condenser charge coil, measure the resistance between the green and black wires. Resistance should measure between 230–280 ohms.
7. Replace the faulty coil if the resistance reading is not within specification.

DT55 and DT65

▶ See accompanying illustrations

1. The three pulser coils, condenser charge coil and battery charge coil are all located underneath the flywheel.
2. Remove the engine cover.
3. On models equipped with electric starters, disconnect the negative battery cable.
4. Disconnect each coil at the bullet connector.
5. For the pulser coils, connect the digital multimeter to the pulser leads as follows:
  - No.1: Connect the tester between the red/black wire and the black ground wire
  - No.2: Connect the tester between the white/black wire and black ground wire
  - No.3: Connect the tester between the red/white wire and the black ground wire
6. Resistance should read between 170–250 ohms.
7. Connect the tester between each pulser coil colored wire lead and the metal mounting stay. There should be no continuity on the meter.
8. On the condenser charge coil, measure the resistance between the green and black/red wires. The resistance should read between 190–270 ohms.
9. On the battery charge coil, measure the resistance between the red and yellow wires. Resistance should read 0.2–0.6 ohms.
10. On the gear counter coil, measure the resistance between the orange/green and black/green wires. Resistance should measure between 170–250 ohms.
11. If the resistance reading is not within specification, replace the faulty coil.

Step 1

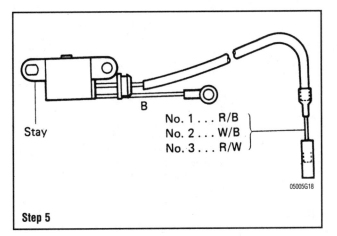

Step 5

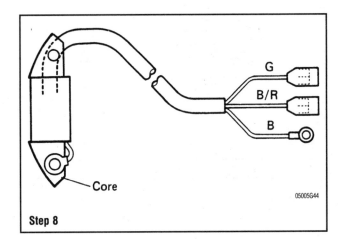

Step 8

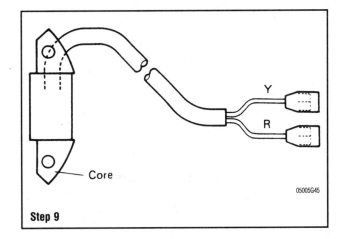

Step 9

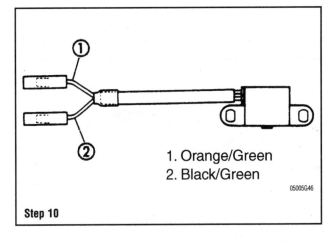

1. Orange/Green
2. Black/Green

Step 10

5-18 IGNITION AND ELECTRICAL SYSTEMS

DT75 and DT85

▶ See accompanying illustrations

1. The three pulser coils are located at 120° angles to each other underneath the flywheel.
2. Remove the engine cover.
3. Disconnect the negative battery cable.
4. Disconnect the pulser coil wires which are located inside the electrical junction box.
5. Connect the digital multimeter to the pulser coil leads as follows:
  • No.1: Connect the tester red lead between the red/black wire and the black ground wire
  • No.2: Connect the tester red between the white/black wire and black ground wire
  • No.3: Connect the tester red between the red/white wire and the black ground wire
6. Resistance for the pulser coils should read between 160–240 ohms.
7. On the gear counting coil, measure the resistance between the orange/green and black/green wires. The resistance should read between 160–240 ohms.
8. On the condenser charging coil, measure the resistance between the green and black (red tube) wires. Resistance should read 170–250 ohms.
9. On the battery charging coil, measure the resistance between the red and yellow wires. Resistance should read 0.2–0.6 ohms.
10. If the resistance reading is not within specification, replace the faulty coil.

DT90 and DT100

▶ See accompanying illustrations

1. The four pulser coils, battery charging coil, condenser charging coil and gear counting coil are all located underneath the flywheel, mounted on the stator base.
2. Remove the engine cover.
3. Disconnect the negative battery cable.
4. Disconnect the pulser coil wires which are located inside the electrical junction box.
5. Connect the digital multimeter to the pulser coil leads as follows:
  • No.1: Connect the tester red lead between the red/green wire and ground
  • No.2: Connect the tester red lead between the white/black wire and ground

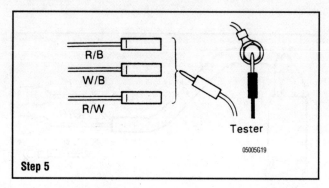
Step 5

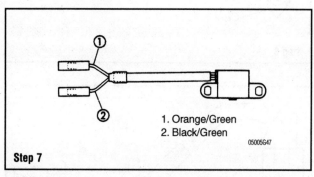
Step 7

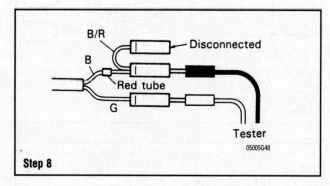
Step 8

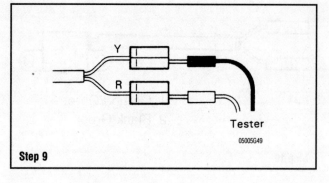
Step 9

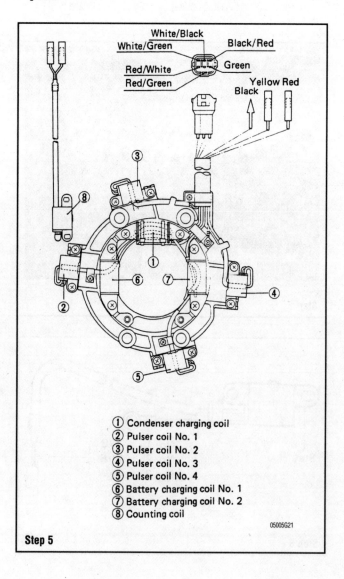

① Condenser charging coil
② Pulser coil No. 1
③ Pulser coil No. 2
④ Pulser coil No. 3
⑤ Pulser coil No. 4
⑥ Battery charging coil No. 1
⑦ Battery charging coil No. 2
⑧ Counting coil

Step 5

IGNITION AND ELECTRICAL SYSTEMS   5-19

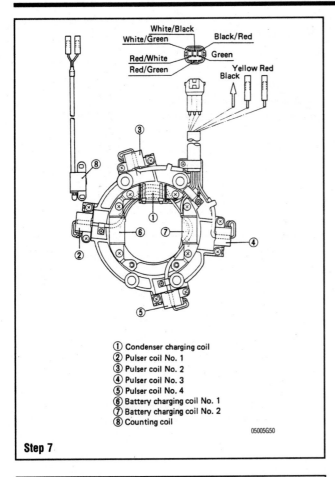

① Condenser charging coil
② Pulser coil No. 1
③ Pulser coil No. 2
④ Pulser coil No. 3
⑤ Pulser coil No. 4
⑥ Battery charging coil No. 1
⑦ Battery charging coil No. 2
⑧ Counting coil

Step 7

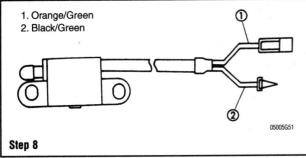

1. Orange/Green
2. Black/Green

Step 8

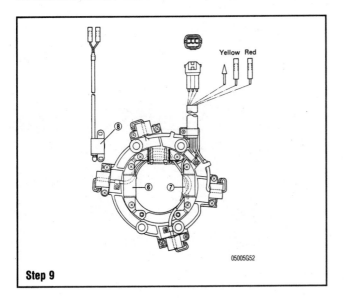

Step 9

• No.3: Connect the tester red lead between the red/white wire and ground
• No.4: Connect the tester red lead between the white/green wire and ground

6. Resistance should read between 160–230 ohms.
7. On the condenser charging coil, measure the resistance between the black/red and green wires. Resistance should read 180–270 ohms.
8. On the gear counting coil, measure the resistance between the orange/green and black/green wires. Resistance should measure between 160–230 ohms.
9. On the battery charging coil, measure the resistance on both coils between the red and yellow wires. Resistance should measure between 0.4–0.06 ohms.
10. If the resistance reading is not within specification, replace the faulty coil.

DT115 and DT140

♦ See accompanying illustrations

1. The four pulser coils, battery charging coil, condenser charging coil and gear counting coil are located, underneath the flywheel, mounted on the stator base,.
2. Remove the engine cover.
3. Disconnect the negative battery cable.
4. Disconnect the pulser coil wires which are located inside the electrical junction box.
5. Connect the digital multimeter to the pulser coil leads as follows:
   • No.1: Connect the tester red lead between the red/green wire and the black ground wire
   • No.2: Connect the tester red lead between the white/black wire and black ground wire
   • No.3: Connect the tester red lead between the red/white wire and the black ground wire
   • No.4: Connect the tester red lead between the white/green wire and the black ground wire. Resistance should read between 170–250 ohms.

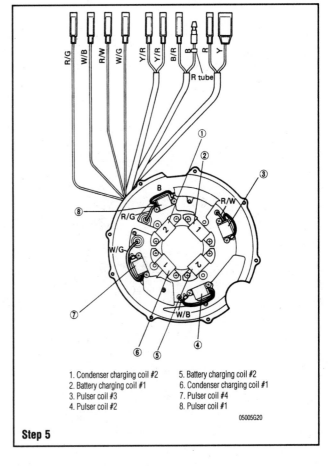

1. Condenser charging coil #2
2. Battery charging coil #1
3. Pulser coil #3
4. Pulser coil #2
5. Battery charging coil #2
6. Condenser charging coil #1
7. Pulser coil #4
8. Pulser coil #1

Step 5

5-20 IGNITION AND ELECTRICAL SYSTEMS

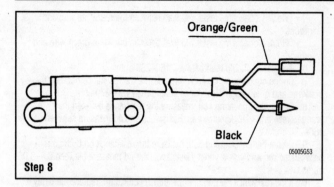

Step 8

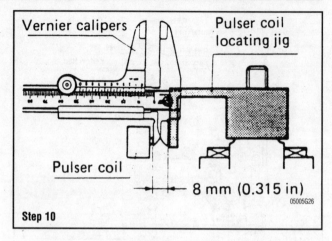

Step 10

6. On the battery charging coils, measure the resistance on No.1 between the yellow/red and yellow/red wires and on No.2 between the red and yellow wires. Resistance should measure between 0.1–0.3 ohms.

7. On the condenser charging coils, measure the resistance on No.1 between the black/red and green wires and on No.2 between black/red and brown wires. Resistance should measure between 170–250 ohms.

8. On the gear counting coil, measure the resistance between the orange/green and black wires. The resistance should measure between 170–250 ohms.

9. If the resistance reading is not within specification, replace the faulty coil.

DT150, DT175, DT200 and DT225

♦ See accompanying illustrations

1. The three pulser coils, condenser coil and battery charging coil are all mounted on the stator base underneath the flywheel assembly.
2. Remove the engine cover
3. Disconnect the negative battery cable
4. Disconnect the pulser coil wires which are located inside the electrical junction box.
5. Connect the digital multimeter to the pulser coil leads as follows:

• No.1: Connect the tester red lead between the red/black wire and the black to ground
• No.2: Connect the tester red lead between the white/black wire and the black to ground
• No.3: Connect the tester red lead between the red/white wire and the black to ground. Resistance should read between 160–240 ohms.

6. On the gear counter coil, measure the resistance between the coil wires. Resistance should read 160–240 ohms (220–340 ohms for the "W" models '98 150,175, 200).

7. On the condenser charge coil, measure the resistance between the charge coil wires for the low speed (325–575 ohms) and high speed (50–100 ohms).

8. On the battery charge coil, measure the resistance between each of the coil wires. Resistance should read between 0.1–0.4 ohms.

9. If the resistance reading is not within specification, replace the faulty coil.

10. If the coil is removed or disturbed, it must be reassembled using the pulser coil locating jig (09931–88711) to be sure that the coil-to-flywheel pulser bar air gap of 0.32 in. (8 mm) is maintained.

REMOVAL & INSTALLATION

DT4, DT6 and DT8

♦ See accompanying illustrations

1. Remove the engine cover from the engine.
2. Remove the built-in fuel tank (if equipped).
3. After removing the bolts, remove the recoil starter assembly.
4. Remove the starter cup. If the screws are hard to loosen, use an impact drive to remove them.
5. Use the flywheel holder (09930–40113) to remove the flywheel nut.

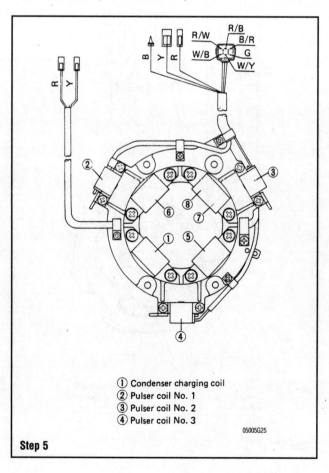

① Condenser charging coil
② Pulser coil No. 1
③ Pulser coil No. 2
④ Pulser coil No. 3

Step 5

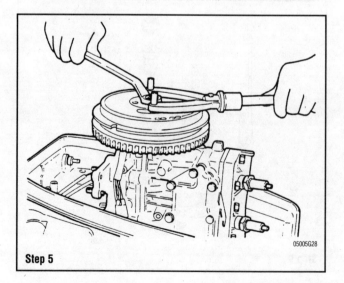

Step 5

IGNITION AND ELECTRICAL SYSTEMS 5-21

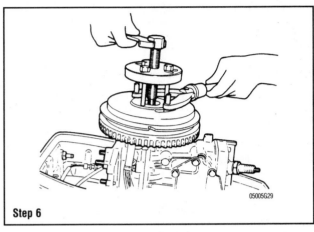
Step 6

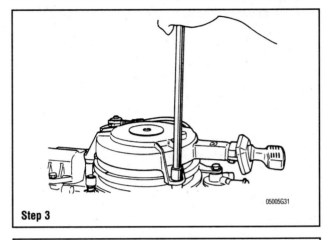
Step 3

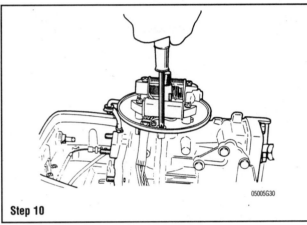
Step 10

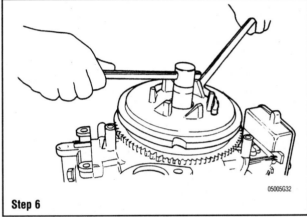
Step 6

6. Use the flywheel holder and flywheel rotor remover (09930–30713) to remove the flywheel.
7. Make sure to remove the flywheel key from the crankshaft.
8. Disconnect the stator wire leads.
9. Remove the screws and remove the ignition coil and pulser coil.

To install:
10. Install the pulser coil and ignition coil onto the engine.
11. Reconnect all the wires in their proper order.
12. Install the flywheel key into the keyway on the crankshaft. Make sure the key is seated correctly into the keyway.
13. Install the flywheel onto the crankshaft.
14. Using the flywheel holder, install the flywheel nut and tighten to 32.5 ft. lbs. (45 Nm).
15. Install the starter cup onto the flywheel and tighten the screws.
16. Install the recoil starter.
17. Install the fuel tank.
18. Install the engine cover.

DT9.9 and DT15

♦ **See accompanying illustrations**

1. Remove the engine cover from the engine.
2. Remove the two nuts and disconnect the battery(s) cables and the neutral switch wire (if equipped).
3. Remove the recoil starter assembly (if equipped).
4. Disconnect the wire lead extending from the stator assembly to the rectifier assembly.
5. Remove the two bolts and remove the starter motor from the engine.
6. Using the flywheel holder (09930–49310), remove the flywheel nut.
7. Using the flywheel holder and the flywheel remover plate (09930–30713), remove the flywheel.
8. If any difficulty is experienced in removing the flywheel, tap the head of the bolt with a hammer. This will usually help in the removal operation.
9. Remove the key from the crankshaft keyway.

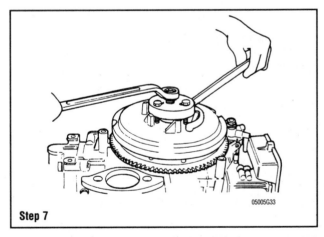
Step 7

Step 8

5-22 IGNITION AND ELECTRICAL SYSTEMS

10. After the flywheel is removed, remove the stator base retaining screws and lift of the stator assembly.
11. Remove the coils from the from the stator base.

To install:

12. Install the coils onto the stator plate.
13. Install the stator base on the engine.

➡ **Make sure that the stator's alignment mark is in line with the stopper's alignment mark.**

14. Install the key securely into the crankshaft keyway.
15. Install the flywheel onto the crankshaft.
16. Using the flywheel holder, tighten the flywheel nut to 58–65 ft. lbs. (80–90 Nm)
17. Install the starter motor back onto the engine and securely tighten the bolts (if equipped).
18. Reconnect the starter/battery cables and reconnect the neutral safety switch lead wire.
19. Install the recoil starter assembly (if equipped)
20. Connect the stator wires making sure all connections are free from corrosion and are securely fastened.
21. Install the engine cover.

DT25 and DT30

◆ See accompanying illustrations

1. Disconnect the negative battery cable lead to prevent accidental engine start.
2. Remove the engine cover.
3. Disconnect the wire leads in the electrical junction box, leading from the stator assembly.
4. Remove the recoil starter assembly.
5. Using a screwdriver to hold the flywheel, remove the starter pulley bolts and lift of the starter pulley.
6. Using a flywheel holder (09930–48720), remove the flywheel nut.
7. Using a flywheel holder and a flywheel remover (09930–39411), remove the flywheel from the engine.
8. Remove the four screws and lift of the stator assembly and wiring harness.
9. Remove the coils from the stator base.

To install:

10. Install the coils onto the stator base and use a small amount of thread locking agent on the screw threads .
11. Install the stator base onto the engine and using a thread locking agent, tighten the screws securely.
12. Throughly clean the mating surface of the flywheel and crankshaft taper with cleaning solvent. Install the key onto the crankshaft securely.
13. Using the flywheel holder, tighten the flywheel nut to 94–108 ft. lbs. (130–150 Nm).
14. Measure the air gap between the coil and flywheel with a feeler gauge. Clearance should be 0.03 in. (0.75 mm). Adjust the coil as necessary to obtain the correct air gap clearance.
15. Connect the stator wire leads to their proper connections.

16. Install the starter pulley onto the flywheel.
17. Install the recoil starter assembly.
18. Connect the battery negative battery cable.
19. Install the engine cover.

DT35 and DT40

1. Disconnect the negative battery cable lead to prevent accidental engine start.
2. Remove the engine cover.
3. Disconnect the wire leads in the electrical junction box and CDI/control unit holder, leading from the stator assembly.
4. Remove the recoil starter assembly.
5. Using a flywheel holder (09930–39520), remove the flywheel nut.
6. Using a flywheel holder and a flywheel remover (09930–39410), remove the flywheel and key.
7. Remove the stator screws and lift off the stator assembly. If necessary, use a impact driver to loosen the screws.
8. Remove the coils from the stator base.

To install:

9. Install the coil onto the stator base and secure the screws with a small amount of thread locking compound.
10. Install the stator assembly onto the engine and secure the screws with a small amount of thread locking compound.
11. Throughly clean the mating surface of the flywheel and crankshaft taper with cleaning solvent. Install the key onto the crankshaft securely.
12. Install the flywheel onto thew crankshaft.
13. Using the flywheel holder, tighten the flywheel nut to 144.7–151.9 ft. lbs. (200–210 Nm).
14. Connect the stator wire leads to their proper connections.
15. Install the starter pulley onto the flywheel.
16. Install the recoil starter assembly.
17. Connect the battery negative battery cable.
18. Install the engine cover.

DT55, DT65, DT75 and DT85

◆ See accompanying illustrations

1. The magneto assembly includes three pulser coils, condensing charge coil, battery charging coil.
2. Remove the engine cover.
3. Remove the electrical junction box cover and disconnect the stator leads.
4. Remove the flywheel cover.
5. Using a flywheel holder (09930–39520) and flywheel & propeller shaft housing remover (09930–39410), remove the flywheel nut.
6. Using the special tools, remove the flywheel from the crankshaft.

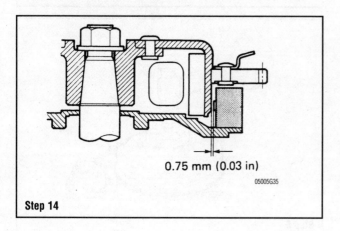

Step 14

Step 6

IGNITION AND ELECTRICAL SYSTEMS 5-23

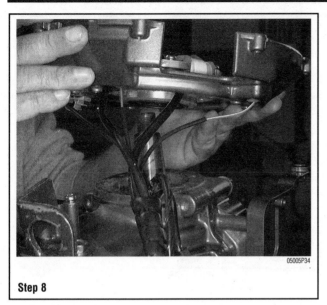

Step 8

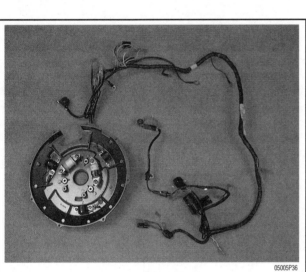

Step 9

Step 11

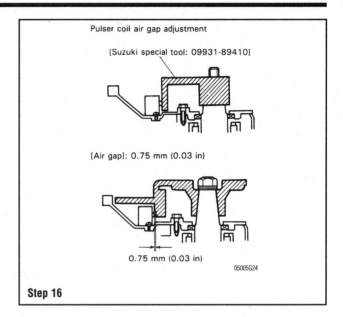

Step 16

7. Make sure to remove the flywheel key before removing the magneto case to prevent tearing the seal.
8. Remove the magneto case from the engine.
9. Remove the coil(s) from the magneto case

To install:
10. Install the coil(s) in the magneto case and secure the screws with a small amount of thread locking compound.
11. Inspect the magneto case seal for damage or tears.
12. Install the magneto case on the engine and tighten the bolts using thread locking compound.
13. Throughly clean the mating surface of the flywheel and crankshaft taper with cleaning solvent. Install the key onto the crankshaft securely.
14. Install the flywheel onto the crankshaft.
15. Using the special tools, tighten the flywheel nut to 144.5–152 ft. lbs. (200–210 Nm).
16. If the pulser coil is removed or disturbed, it must be reassembled using the pulser coil locating jig (09931–89410) to be sure that the coil-to-flywheel pulser bar air gap of 0.03 in. (0.75 mm) is maintained.
17. Connect the stator wire leads to their proper connections.
18. Connect the battery negative battery cable.
19. Install the engine cover.

DT90, DT100, DT115 and DT140

♦ **See accompanying illustrations**

1. Remove the engine cover.
2. Remove the electrical junction box cover and disconnect the stator leads.
3. Remove the flywheel cover.
4. Using a flywheel holder (09930–48720), remove the flywheel nut.
5. Using a flywheel holder, flywheel remover (09930–39411) and flywheel bolts (09930–39420), remove the flywheel.
6. Remove the flywheel from the crankshaft.
7. Remove the stator assembly and gear counter coil from the engine.
8. Remove the coils from the stator assembly and wiring harness.

To install:
9. Install the coils onto the stator assembly and secure the screws with thread locking compound.
10. Install the stator assembly onto the engine and secure the bolts with thread locking compound.
11. Throughly clean the mating surface of the flywheel and crankshaft taper with cleaning solvent. Install the key onto the crankshaft securely.
12. Install the flywheel onto the crankshaft.
13. Using the flywheel holder, tighten the flywheel nut to 181–188 ft. lbs. (250–260 Nm).

5-24 IGNITION AND ELECTRICAL SYSTEMS

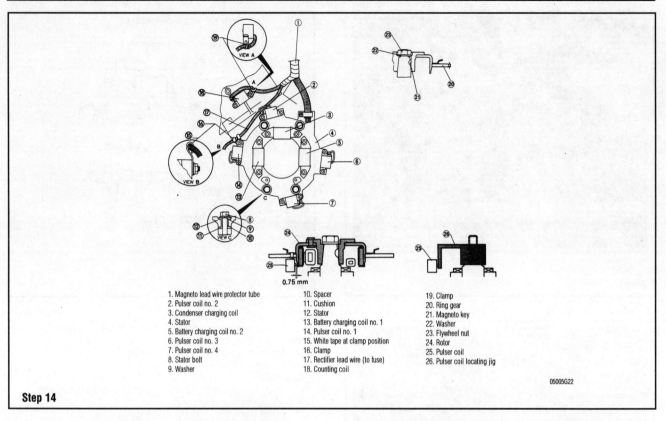

1. Magneto lead wire protector tube
2. Pulser coil no. 2
3. Condenser charging coil
4. Stator
5. Battery charging coil no. 2
6. Pulser coil no. 3
7. Pulser coil no. 4
8. Stator bolt
9. Washer
10. Spacer
11. Cushion
12. Stator
13. Battery charging coil no. 1
14. Pulser coil no. 1
15. White tape at clamp position
16. Clamp
17. Rectifier lead wire (to fuse)
18. Counting coil
19. Clamp
20. Ring gear
21. Magneto key
22. Washer
23. Flywheel nut
24. Rotor
25. Pulser coil
26. Pulser coil locating jig

Step 14

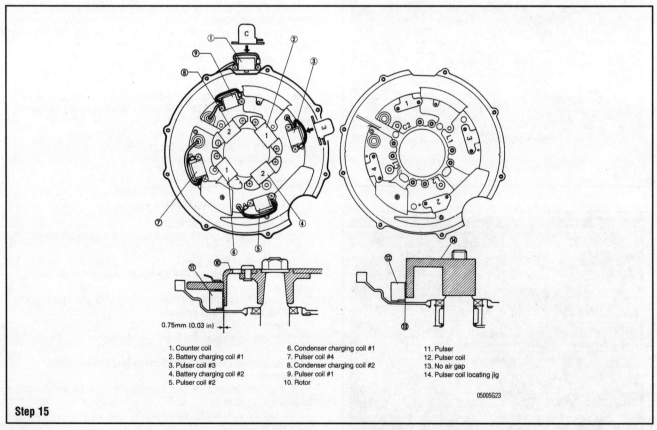

1. Counter coil
2. Battery charging coil #1
3. Pulser coil #3
4. Battery charging coil #2
5. Pulser coil #2
6. Condenser charging coil #1
7. Pulser coil #4
8. Condenser charging coil #2
9. Pulser coil #1
10. Rotor
11. Pulser
12. Pulser coil
13. No air gap
14. Pulser coil locating jig

Step 15

14. If the pulser coil is removed or disturbed, it must be reassembled using the pulser coil locating jig (09931–88710) to be sure that the coil-to-flywheel pulser bar air gap of 0.03 in. (0.75 mm) is maintained.

15. Make sure to use the proper jig, as the one for the DT115/140 (09931–89421) is different from other models
16. Connect the stator wire leads to their proper connections.
17. Connect the battery negative battery cable.
18. Install the engine cover.

IGNITION AND ELECTRICAL SYSTEMS

DT150, DT175, DT200 and DT225

▶ See accompanying illustrations

1. Remove the engine cover.
2. Remove the electrical junction box cover and disconnect the stator leads.
3. Remove the flywheel cover.
4. Using a flywheel holder (09930–48720), remove the flywheel nut.
5. Using a flywheel holder, flywheel remover (09930–39411) and flywheel bolts (09930–39420), remove the flywheel.
6. Remove the flywheel from the crankshaft.
7. Remove the coils from the stator assembly and wiring harness.

To install:

8. Install the coils onto the stator assembly and secure the screws with thread locking compound.
9. Fit the pulser coil locating jig onto the crankshaft as illustrated.
10. Locate the pulser coil 0.32 in. (8 mm) from the end of the pulser coil locating jig using vernier calipers. With the pulser coil held in this position, secure the attaching screws using a small amount of thread locking compound. Install the other two pulser coils in this manner.
11. Throughly clean the mating surface of the flywheel and crankshaft taper with cleaning solvent. Install the key onto the crankshaft securely.
12. Install the flywheel onto the crankshaft.
13. Using the flywheel holder, tighten the flywheel nut to 181–188 ft. lbs. (250–260 Nm).
14. Connect the stator wire leads to their proper connections.
15. Connect the battery negative battery cable.
16. Install the engine cover.

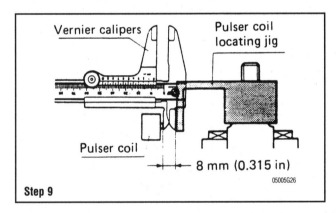
Step 9

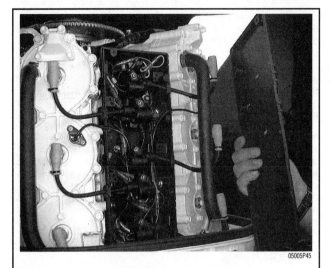
Fig. 21 The coil is the heart of the ignition system

Fig. 22 Once the voltage is discharged from the ignition coil the secondary circuit begins and only stretches from the ignition coil to the spark plugs via extremely large high tension leads

Ignition Coils

DESCRIPTION & OPERATION

▶ See Figures 21 and 22

The coil is the heart of the ignition system. Essentially, it is nothing more than a transformer which takes the relatively low voltage (12 volts) available from the primary coil and increases it to a point where it will fire the spark plug as much as 20,000 volts.

Once the voltage is discharged from the ignition coil the secondary circuit begins and only stretches from the ignition coil to the spark plugs via extremely large high tension leads. At the spark plug end, the voltage arcs in the form of a spark, across from the center electrode to the outer electrode, and then to ground via the spark plug threads. This completes the ignition circuit.

TESTING

▶ See Figure 23

1. Although the best test for an ignition coil is on a dynamic ignition coil digital multimeter, resistance checks can also be done.
2. There are two circuits in an ignition coil, the primary winding circuit and the secondary winding circuit. Both need to be checked.

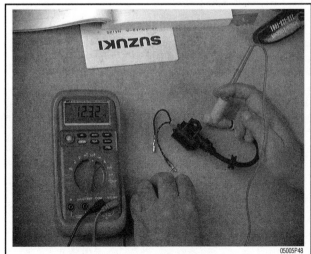
Fig. 23 The digital multimeter connection procedure for a continuity check will depend on how the coil is constructed

5-26 IGNITION AND ELECTRICAL SYSTEMS

3. The digital multimeter connection procedure for a continuity check will depend on how the coil is constructed. Generally, the primary circuit is the small gauge wire or wires, while the secondary circuit contains the high tension or plug lead.

4. Some ignition coils have the primary and/or secondary circuits grounded on one end. On these type coils, only the continuity check is done. On ignition coils that are not grounded on one end, the short-to-ground test must also be done. Regardless of the coil type, compare the resistance with the "Ignition Coil Resistance" chart.

➡ **When checking the secondary side, make sure to read the procedure for your particular engine. Some models require the spark plug caps be removed and other require them to remain on. In some cases the cap is bad, not the coil. Bad resistor caps can be the cause of high-speed misfire. Unscrew the cap and check the resistance (5 kilo-ohms).**

5. The other method used to test ignition coils is with a Dynamic Ignition Coil Digital multimeter. Since the output side of the ignition coil has very high voltage, a regular voltmeter can not be used. While resistance reading can be valuable, the best tool for checking dynamic coil performance is a dynamic ignition coil digital multimeter.

6. Connect the coil to the digital multimeter according to the manufacturer's instructions.
7. Set the spark gap according to the specifications.
8. Operate the coil for about 5 minutes.
9. If the spark jumps the gap with the correct spark color, the coil is probably good.

REMOVAL & INSTALLATION

DT4

1. Remove engine cover.
2. Disconnect the ignition coil leads and spark plug wires.
3. Remove the retaining bolts and remove the ignition coil.

To install:

4. The ignition coil core should always be sanded with emery cloth before installing on the engine. Also sand the area on the cylinder block. This ensures a good ground.
5. Install the retaining bolts and tighten snugly.
6. Connect the ignition coil lead wires. Make sure the connections are tight and corrosion free.
7. Install the spark plug wires on the spark plugs.
8. Install the engine cover.

DT6 and DT8

▶ See accompanying illustrations

1. Remove the engine cover.
2. Disconnect the spark plug wires.

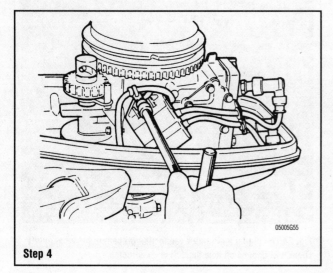

Step 4

3. Disconnect the CDI unit/ignition coil leads.
4. Loosen the bolts and remove the CDI unit/ignition coil.

➡ **The ignition coil and CDI unit are integrated into one assembly.**

To install:

5. Install the CDI unit/ignition coil onto the engine.
6. Connect the CDI unit/ignition coil leads. Make sure all connections are tight and free of corrosion.
7. Install the spark plug wires.
8. Install the engine cover.

DT9.9 and DT15

▶ See accompanying illustrations

1. Remove the engine cover.
2. Disconnect the spark plug wires.
3. Disconnect the two pole connector and the lead wire extending from the CDI unit/ignition coil and remove from the engine.

➡ **The ignition coil and CDI unit are integrated into one assembly.**

To install:

4. Install the CDI unit/ignition coil on the engine and connect the two-pole connector.
5. Connect the CDI unit/ignition coil lead wire. Make sure all connections are tight and free of corrosion.
6. Connect the spark plug wires.
7. Install the engine cover.

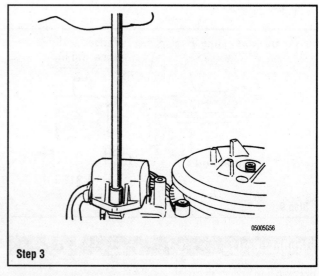

Step 3

DT25 and DT30

1. Disconnect the battery negative cable.
2. Remove the engine cover.
3. Disconnect the spark plug wires.
4. Open the electrical parts holder and disconnect the ignition coil leads at the CDI unit.
5. Remove the ignition coil retaining bolts and remove the coils from the engine.

To install:

6. Install the ignition coil on the engine and tighten the retaining bolts.
7. Connect the ignition coil leads at the CDI unit. Make sure all connections are tight and free of corrosion.
8. Replace the electrical parts holder cover. Make sure the cover is on properly and no wires or sealing ring is being pinched by the cover.
9. Connect the spark plug wires.
10. Install the engine cover.

DT35 and DT40

1. Remove the engine cover.
2. Disconnect the spark plug wires.

IGNITION AND ELECTRICAL SYSTEMS   5-27

3. Disconnect the CDI unit/ignition coil leads.

➡ The ignition coil and CDI unit are integrated into one assembly.

4. Remove the elastic band and lift the CDI unit/ignition coil from the electric parts holder.

To install:

5. Set the CDI unit/ignition coil assembly into the electric parts holder and secure it with the elastic band.
6. Connect the CDI unit/ignition coil leads. Make sure all connections are tight and free of corrosion.
7. Connect the spark plug wires.
8. Install the engine cover.

DT55 and DT65

♦ See accompanying illustrations

1. Remove the engine cover.
2. Disconnect the spark plug wires.
3. Open the electric parts holder and disconnect the ignition coil leads from the CDI unit.
4. Unbolt the ignition coils and remove them from the electric parts holder.

To install:

5. Install the ignition coils in the electric parts holder. Make sure the retaining bolts are secure.
6. Connect the ignition coil leads to the CDI unit. Make sure all the connections are tight and free of corrosion.
7. Install the cover on the electric parts holder.
8. Install the spark plug wires.
9. Install the engine cover.

DT75 and DT85

♦ See accompanying illustrations

1. Remove the engine cover.
2. Disconnect the spark plug wires.
3. Open the electric parts holder and disconnect the ignition coil leads from the CDI unit.
4. Unbolt the ignition coils and remove them from the electric parts holder.

To install:

5. Install the ignition coils in the electric parts holder. Make sure the retaining bolts are secure.
6. Connect the ignition coil leads to the CDI unit. Make sure all the connections are tight and free of corrosion.
7. Install the cover on the electric parts holder.
8. Install the spark plug wires.
9. Install the engine cover.

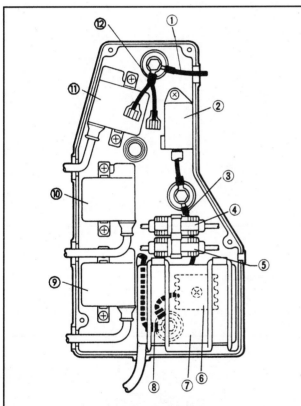

1. Magneto ground lead wire
2. Low oil warning unit
3. CDI unit ground lead wire
4. Fuse case assy. (battery)
5. Fuse case ass. (rectifier)
6. Rectifier assy.
7. CDI unit
8. Wiring harness ground lead wire
9. Ignition coil no. 3
10. Ignition coil no. 2
11. Ignition coil no. 1
12. Ignition coil ground lead wire

Step 3

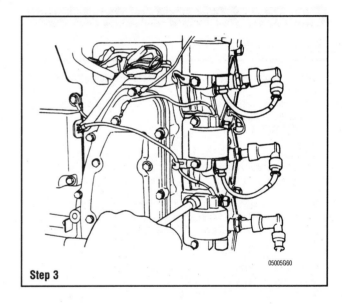

Step 3

DT90 and DT100

1. Remove the engine cover.
2. Disconnect the spark plug wires.
3. Open the electric parts holder on the front of the engine and disconnect the ignition coil leads from the CDI unit.
4. Unbolt the ignition coils and remove them from the electric parts holder.

To install:

5. Install the ignition coils in the electric parts holder. Make sure the retaining bolts are secure.
6. Connect the ignition coil leads to the CDI unit. Make sure all the connections are tight and free of corrosion.
7. Install the cover on the electric parts holder.
8. Install the spark plug wires.
9. Install the engine cover.

DT115 and DT140

1. Remove the engine cover.
2. Disconnect the spark plug wires.
3. Open the electric parts holder on the side of the engine and disconnect the ignition coil leads from the wiring harness leading to the CDI unit.

5-28 IGNITION AND ELECTRICAL SYSTEMS

4. Unbolt the ignition coils and remove them from the electric parts holder.

To install:

5. Install the ignition coils in the electric parts holder. Make sure all ground connections are in place and the retaining bolts are tight.
6. Connect the ignition coil leads to the CDI unit. Make sure all the connections are tight and free of corrosion.
7. Install the cover on the electric parts holder.
8. Install the spark plug wires.
9. Install the engine cover.

DT150, DT175, DT200 and DT225

1. Remove the engine cover.
2. Disconnect the spark plug wires.
3. Open the electric parts holder on the front of the engine and disconnect the ignition coil leads from the wiring harness leading to the CDI unit.
4. Unbolt the ignition coils and remove them from the electric parts holder.

To install:

5. Install the ignition coils in the electric parts holder. Make sure all ground connections are in place and the retaining bolts are tight.
6. Connect the ignition coil leads to the CDI unit. Make sure all the connections are tight and free of corrosion.
7. Install the cover on the electric parts holder.
8. Install the spark plug wires.
9. Install the engine cover.

CDI Unit

DESCRIPTION & OPERATION

In its simplest form, a CDI ignition is composed of the following elements:
- Magneto
- Pulser coil
- Charge, or source coil
- Igniter (CDI) box
- Ignition coil
- Spark plug

Other components such as main switches, stop switches, or computer systems may be included, though, these items are not necessary for basic CDI operation.

To understand basic CDI operation, it is important to understand the basic theory of induction. Induction theory states that if we move a magnet (magnetic field) past a coil of wire(or the coil by the magnet), AC current will be generated in the coil.

The amount of current produced depends on several factors:
- How fast the magnet moves past the coil
- The size of the magnet(strength)
- How close the magnet is to the coil
- Number of turns of wire and the size of the windings

The current produced in the charge coil goes to the CDI box. On the way in, it is converted to DC current by a diode. This DC current is stored in the capacitor located inside the box. As the charge coil produces current, the capacitor stores it.

At a specific time in the magneto's revolution, the magnets go past the pulser coil. The pulser coil is smaller than the charge coil so it has less current output. The current from the pulser also goes into the CDI box. This current signals the CDI box when to fire the capacitor (the pulser may be called a trigger coil for obvious reasons). The current from the capacitor flows out to the ignition coil and spark plug. The pulser acts much like the points in older ignitions systems.

When the pulser signal reaches the CDI box, all the electricity stored in the capacitor is released at once. This current flows through the ignition coil's primary windings.

The ignition coil is a step-up transformer. It turns the relatively low voltage entering the primary windings into high voltage at the secondary windings. This occurs due to a phenomena known as induction.

The high voltage generated in the secondary windings leaves the ignition coil and goes to the spark plug. The spark in turn ignites the air-fuel charge in the combustion chamber.

Once the complete cycle has occurred, the spinning magneto immediately starts the process over again.

Main switches, engine stop switches, and the like are usually connected on the wire in between the CDI box and the ignition coil. When the main switch or stop switch is turned to the OFF position, the switch is closed. This closed switch short-circuits the charge coil current to ground rather than sending it through the CDI box. With no charge coil current through the CDI box, there is no spark and the engine stops or, if the engine is not running, no spark is produced.

TESTING

The unit may remain installed on the powerhead, or it may be removed for testing. In either case, the testing procedures are identical.

Measure the continuity between the CDI unit terminals. If the any of the readings are not within specifications, the CDI unit must be replaced.

DT2 and DT2.2

▶ **See accompanying illustrations**

The DT2 model uses a combined CDI unit/ignition coil.

1. Using a multimeter, measure the resistance in the primary (core–blue/red wire:0.8–1.2 ohms) and the secondary (core–plug cap: 7–10 kilo-ohms)

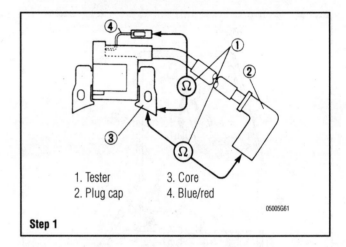

1. Tester
2. Plug cap
3. Core
4. Blue/red

Step 1

DT4

▶ **See accompanying illustrations**

1. Obtain a Suzuki CDI tester (09930–99810) and test cord (09930–40113).
2. Before using the tester, be sure to select the proper voltage range on the

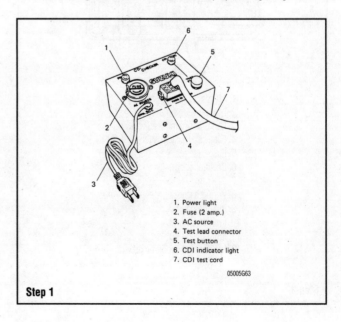

1. Power light
2. Fuse (2 amp.)
3. AC source
4. Test lead connector
5. Test button
6. CDI indicator light
7. CDI test cord

Step 1

IGNITION AND ELECTRICAL SYSTEMS 5-29

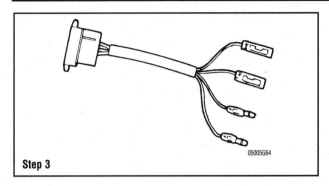

Step 3

voltage selector (100v, 117v, 220v and 240v). If the selector socket is not set at the proper voltage range, remove the fuse and pull out the voltage selector and reinsert into the unit so that the proper voltage scale is visible in the cutaway.
3. Disconnect the CDI unit leads and install the test leads in their place.
4. Connect the CDI unit to test lead to the tester.
5. Plug the tester into a power outlet.
6. Push the "Test" button.
   • Both the power light and the CDI indicator light should come on. This indicates that the CDI unit is functioning correctly.
   • If the power light is ON and the CDI indicator light is OFF, the CDI unit is not functioning correctly and needs to be replaced.
   • If both lights are OFF, check the fuse (replace if blown) and the A/C power source.

DT6 and DT8

♦ See accompanying illustrations

1. Obtain a Suzuki CDI tester (09930–99830) and test cord (09930–89812).
2. Make sure the correct voltage range is chosen on the tester "1".

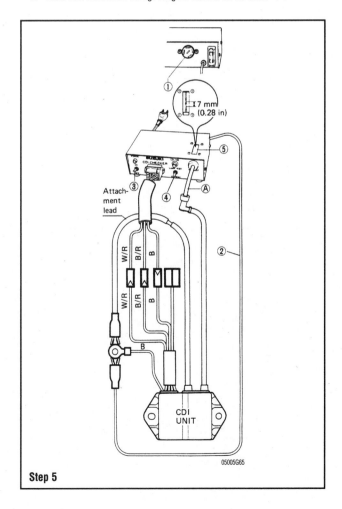

Step 5

3. Connect the CDI test cord to the tester.
4. Connect the A/C cord to a power outlet.
5. Connect the CDI tester, CDI test cord and CDI unit as illustrated.

➡ When checking the ignition coil assembly with a plug cap left detached, insert the spark plug wire directly into the tester without using the attachment "A".

✲✲ WARNING

Failure to connect this ground lead "2" wire may cause an electric shock upon touching the tester.

Turn the main switch "3" of the tester to the ON position. Turning the switch "4" to the "CDI TEST" side, check if a spark occurs across the needles in the indicator window "5". If the spark occurs, it means that the CDI unit is functioning normally. If there is no spark indicated, the CDI unit is not operating normally and will need to be replaced.

DT9.9, DT15, DT25 and DT30

♦ See Figure 24

1. Obtain a CDI tester (09930–99830) and CDI test cord (DT9.9/15:09930–88910; DT25/30: 09930–89630).
   • ON: Meter reading should be 100k ohms or less
   • OFF: Meter reading should be more than 100k ohms
   • CON: The meter pointer swings slightly and immediately return to the original position. Watch the meter carefully. If you have failed to see the pointer moving, wait for about 10 minutes and repeat the test.

		⊖ lead of tester				
		Green	Black	Blue/Red	H_1	H_2
⊕ lead of tester	Green		ON	OFF	OFF	OFF
	Black	CON		OFF	OFF	OFF
	Blue/Red	ON	ON		OFF	OFF
	H_1	OFF	OFF	OFF		2500Ω
	H_2	OFF	OFF	OFF	2500Ω	

Fig. 24 Use the CDI unit test chart to determine the condition of the CDI unit

CHECK BY INDICATOR LAMP

♦ See Figure 25

1. Before using the tester first be sure to select the proper voltage range on the voltage selector "1". (100V, 117V, 220V and 240V)
2. Connect the CDI test cord to the tester.
3. Connect the A/C power cord to a power outlet.
4. Connect the CDI test cord to the CDI unit lead wire as follows: (check each cylinder individually)
   • DT9.9/15: No.1 cylinder: orange, Connectors, black; No.2 cylinder: gray, Connectors, black
   • DT25/30: No1 cylinder: orange, connectors; No.2 cylinder: blue, connectors; No.3 cylinder: gray, connectors
5. Turn the main switch "2" of the tester to the ON position.
6. Turn the test switch "3" to the "LAMP TEST" side. Make sure the CDI indicator lamp "4" comes ON.
7. Turning the switch "3" to the "CDI TEST" side, check if the lamp "4" comes on. If the lamp is ON, the CDI unit is good. If the lamp is OFF, the CDI unit is bad and needs to be replaced.

5-30 IGNITION AND ELECTRICAL SYSTEMS

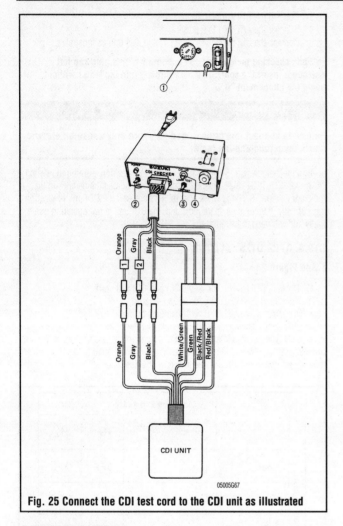

Fig. 25 Connect the CDI test cord to the CDI unit as illustrated

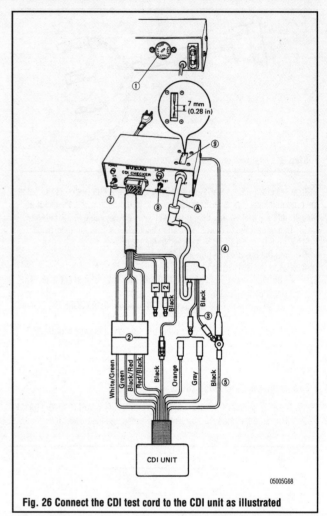

Fig. 26 Connect the CDI test cord to the CDI unit as illustrated

CHECKING BY SPARK

♦ See Figure 26

1. Before using the tester first be sure to select the proper voltage range on the voltage selector "1". (100V, 117V, 220V and 240V)
2. Connect the CDI test cord to the tester.
3. Connect the A/C power cord to a power outlet.
4. Connect the connectors "2" of the CDI test cord to the CDI unit lead wires as illustrated.
5. Inserting attachment "A" into the tester, connect each ignition coil assembly as illustrated.

➟When checking the ignition coil assembly with a plug cap left detached, insert the spark plug wire of the ignition coil assembly directly into the tester without using the attachment "A".

6. Connect the black lead wire "3" of the ignition coil assembly to the ground lead wire "4" of the tester to the black lead wire "5" of the CDI unit.

※※ WARNING

Failure to connect this ground lead "4" wire may cause an electric shock upon touching the tester.

7. Connect the lead wire "6" of the ignition coil assembly to the same color lead wire of the CDI unit.
   • DT9.9/15: No.1 orange; No.2 gray
   • DT25/30: No.1 orange; No.2 blue; No.3 gray
8. Turning the main switch "7" of the tester to the ON position. Turning the switch "8" to the "CDI TEST" position, check if a spark occurs across the needles in the indicator window "9". If the spark occurs, the CDI unit is functioning correctly. If there is no spark, the CDI unit is not functioning and will need to be replaced.

DT35 and DT40

USING THE SUZUKI POCKET TESTER

♦ See Figure 27

1. Use the CDI unit test chart to determine the condition of the CDI unit.
   • ON: Meter reading should be 100k ohms or less.
   • OFF: Meter reading should be more than 100k ohms.
   • CON: Meter swings once and returns immediately.

		\- lead of tester					
		Green	White/red	Black	Blue/red	Blue	Pink
+ lead of tester	Green		ON	ON	OFF	OFF	ON
	White/red	ON		ON	OFF	OFF	ON
	Black	ON	ON		OFF	OFF	ON
	Blue/red	ON	ON	ON		OFF	ON
	Blue	ON	ON	ON	OFF		ON
	Pink	ON	ON	ON	OFF	OFF	

Fig. 27 Use the CDI unit test chart to determine the condition of the CDI unit

IGNITION AND ELECTRICAL SYSTEMS

➡ Set the meter to the R x 100 range. After checking CON, allow some time before checking the reading.

USING THE SUZUKI CDI TESTER (09930–99810)

♦ See Figure 28

1. Before using the tester, make sure to set the tester to the proper voltage range (100V, 117V, 220V and 240V). If the selector is not set at the proper voltage range, remove the fuse and pull out the voltage selector and reinsert into unit so that the proper voltage scale is visible in the cutaway.
2. Connect the CDI unit to the test cord and the test cord to the CDI tester.
3. Connect the A/C cord to a power source.
4. Push the "TEST" button.
   • Both the power light and CDI indicator light should come on. This indicates that the CDI unit is functioning properly.
   • If the power light is ON and the CDI indicator light is OFF, the CDI unit is not working properly and need to be replaced.
   • If both the lights on the CDI tester are out, check to see if the tester unit fuse is blown, and if so find the cause and then replace the fuse. Also check the A/C power source going to the tester.

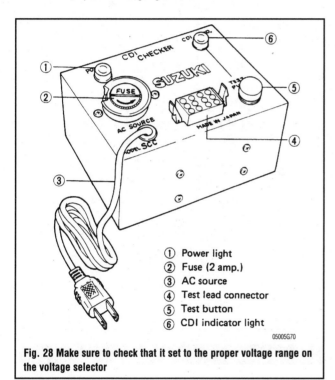

① Power light
② Fuse (2 amp.)
③ AC source
④ Test lead connector
⑤ Test button
⑥ CDI indicator light

Fig. 28 Make sure to check that it set to the proper voltage range on the voltage selector

DT55 and DT65

♦ See Figure 29

1. Obtain a CDI tester (09930–99830) and test cord (09930–89480)

CHECK BY INDICATOR LAMP

1. Before using the tester, make sure the tester is set on the proper voltage range (100V, 117V, 220V AND 240V).
2. Connect the CDI test cord tot he tester.
3. Connect the A/C power cord to an electrical outlet.
4. Turn ON the main switch "2" on the tester.
5. Turn the test switch "3" to the "LAMP TEST" position. Make sure the CDI indicator lamp "4" comes on then.
6. Turn the switch "3" to the "CDI TEST" position, then check if the lamp "4" comes on.
   • ON: the CDI unit is functioning properly
   • OFF: the CDI unit is not functioning properly and will need to be replaced.

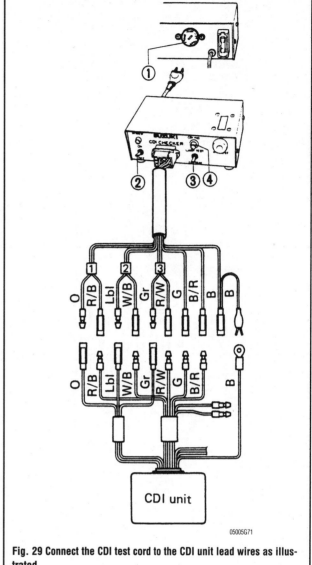

Fig. 29 Connect the CDI test cord to the CDI unit lead wires as illustrated

CHECKING BY SPARK

♦ See Figure 30

1. Before using the tester, make sure the tester is set on the proper voltage range (100V, 117V, 220V and 240V).
2. Connect the CDI test cord tot he tester.
3. Connect the A/C power cord to an electrical outlet.
4. Connect the CDI test cord to the CDI unit lead wires as illustrated.
5. Connect the ground lead wire "2" of the tester to the black lead wire "3" of the CDI unit.

✱✱ CAUTION

Failure to connect this ground lead wire "2" may cause an electric shock upon touching the tester.

6. Inserting attachment "A" into the checker, connect each ignition coil assembly as illustrated in the illustration.

➡ When checking the ignition coil assembly with a plug cap left detached, insert the spark plug wire directly into the checker without using the attachment "A".

7. Turning ON the main power switch of the tester. If a spark occurs

5-32 IGNITION AND ELECTRICAL SYSTEMS

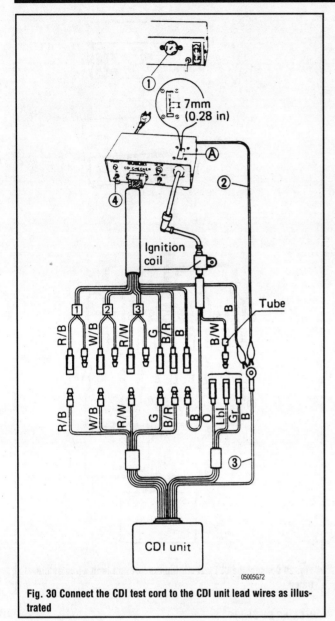

Fig. 30 Connect the CDI test cord to the CDI unit lead wires as illustrated

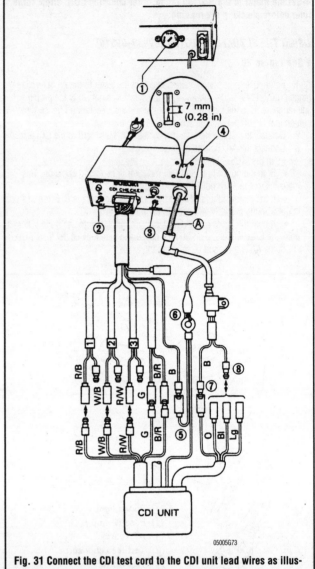

Fig. 31 Connect the CDI test cord to the CDI unit lead wires as illustrated

across the needles in the indicator window "5" the CDI unit is operating normally.

8. If there is no spark in the window, the CDI unit is not operating properly and will need to be replaced.

DT75 and DT85

♦ See Figure 31

Testing the CDI unit with the CDI tester.
1. Obtain a CDI tester (09930–99830) and test cord (09930–89521).
2. Before using the tester, make sure the tester is set on the proper voltage range (100V, 117V, 220V or 240V).
3. Connect the CDI test cord to the tester.
4. Connect the A/C power cord to an electrical outlet.

※※ WARNING

Make sure the tester main switch "2" is OFF before connecting the power supply or test leads.

5. Connect the CDI test cord to the CDI unit lead wires as illustrated (check each cylinder separately).

※※ WARNING

Failure to connect the ground wires may cause an electric shock upon touching the tester.

➡When checking the ignition coil with the spark plug cap removed, insert the high tension cord of the ignition coil directly into the checker without using the attachment "A".

6. Individually connect the cylinder circuits to be checked.
7. Turn the tester main switch "2" ON.
8. Turn the tester switch "3" to the "CDI TEST" position. Check for spark in the "spark indicator" window. A continuous spark will mean that the circuit being tested is performing normally.
9. If there is no spark indicated, check the tester fuse and the power supply. If these are operational, then there is a failure in the CDI unit and it will need to be replaced.

DT90 and DT100

Test the CDI peak voltage output using a Stevens CD-77 peak voltage tester.
1. Remove all spark plugs to eliminate variables at cranking speed.
2. Use a 12-volt, 70AH fully charged battery.

IGNITION AND ELECTRICAL SYSTEMS   5-33

3. Crank the engine using the electric starter for no more than 20 seconds at a time.

4. CDI peak voltage should measure 102 volts or over at 300 rpm, and 121 volts at 500 rpm.

DT115 and DT140

The CDI unit can be checked using either a Stevens CD-77 peak voltage tester or a Suzuki CDI tester (09930–99830) and CDI test cord (09930–99410).

CHECK BY INDICATOR LAMP

♦ See Figure 32

1. Before using the CDI tester, make sure the tester is set at the proper voltage for your location.
2. Connect the CDI test cord to the tester.
3. Plug the tester into a power outlet.
4. Connect the CDI test cord to the CDI unit lead wires as illustrated.
5. Turn ON the main power switch "2".
6. Turn the test switch "2" of the tester to the "LAMP TEST" position. Make sure the CDI indicator lamp "4" comes ON then.
7. Turn the switch "3" to the "CDI TEST" position, now check to see if the lamp "4" comes ON.
8. If the lamp is on, the CDI unit is functioning correctly. If the lamp does not come on, check the tester fuse and power supply. If both are working, then the CDI unit is not working and will need to be replaced.

CHECK BY SPARK

♦ See Figure 33

1. Connect the CDI unit lead, CDI test cord and ignition coil lead as illustrated.
2. Connect the ground lead wire "2" of the tester to the black lead wire "3" of the CDI unit.

✱✱✱ CAUTION

Failure to connect this ground lead "2" wire may cause an electric shock upon touching the tester.

3. Inserting the attachment "A" into the tester, connect each ignition coil assembly as illustrated in the illustration.

➡ When checking the ignition coil assembly with a plug cap left detached, insert the high tension cord of the ignition coil assembly directly into the tester without using the attachment "A".

4. Turn the main power switch "4" of the checker to the ON position and check for a spark between the needles in the spark indicator window "5".
5. If a spark occurs, the CDI unit is functioning properly. If there is no spark, the unit is not working and will need to be replaced.

To check voltage on the EFI equipped models, check the ECU using a Stevens Model CD-77 peak voltage tester and test cord.

6. Remove all the spark plugs to eliminate variables at cranking speed.
7. Install the 6-pin connector between the CD-77 and the ECU.

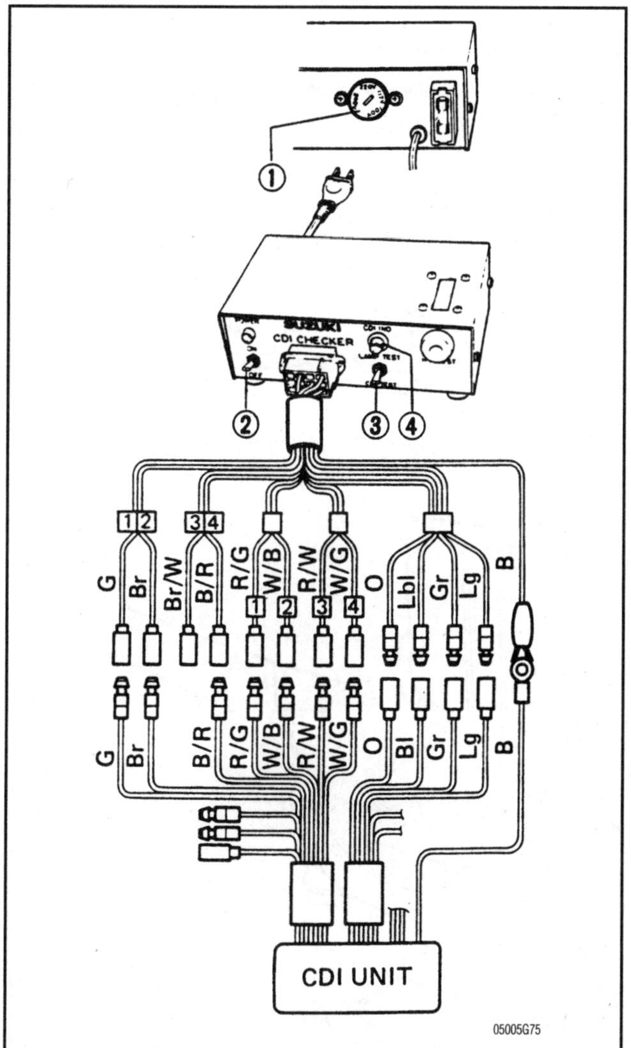

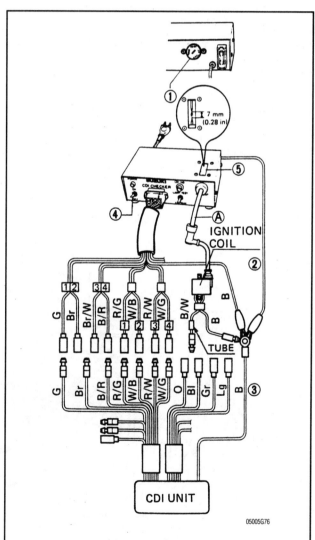

Fig. 32 Connect the CDI test cord to the CDI unit lead wires as illustrated

Fig. 33 Connect the CDI test cord to the CDI unit lead wires as illustrated

5-34 IGNITION AND ELECTRICAL SYSTEMS

8. Connect each of the ECU leads to the positive tester probe and ground the negative test probe.
9. Crank the engine using the electric starter. DO NOT crank the engine for more than 20 seconds at a time.
10. Peak voltage measured at the CD-77 should be 104 volts or over. Make each measurement at least three times and if the voltage is too low, the ECU will need to be replaced.

DT150, DT175, DT200 and DT225

Check the ECU/CDI using a Stevens Model CD-77 peak voltage tester.
1. Remove all the spark plugs to eliminate variables at cranking speed.
2. Connect each of the ECU leads to the positive tester probe and ground the negative test probe.
3. Crank the engine using the electric starter. DO NOT crank the engine for more than 20 seconds at a time.
4. Peak voltage measured at the CD-77 should be 120 volts or over. Make each measurement at least three times and if the voltage is too low, the ECU will need to be replaced.

REMOVAL & INSTALLATION

DT2 and DT2.2

▶ See Figure 34

1. Remove the engine cover.
2. With the fuel petcock in the closed position (marked "S"), remove the fuel hose from the petcock.
3. Unbolt the fuel tank and remove it from the engine.
4. Disconnect the spark plug wire.
5. Disconnect the CDI unit lead.
6. Unbolt the CDI unit and remove.

To install:
7. When installing the CDI/ignition coil unit, measure the clearance between the flywheel magneto and the ignition unit.
   • Clearance should measure 0.016 in. (0.4 mm).

DT4

1. Remove the engine cover.
2. Disconnect the spark plug lead.
3. Disconnect the CDI leads and remove the retaining bolt.
4. Remove the CDI unit.

To install:
5. Install the CDI unit. Make sure that the ground wire is connected and the retaining bolt is tightened.
6. Connect the CDI unit leads making sure they are tight and corrosion free.
7. Connect the spark plug wire.
8. Install the engine cover.

DT6 and DT8

1. The CDI unit and ignition coil are integrated into one unit. Removal and installation are covered in the ignition coil section.

DT9.9 and DT15

1. The CDI unit and ignition coil are integrated into one unit. Removal and installation are covered in the ignition coil section.

DT25 and DT30

1. The removal and installation of the CDI unit is covered in the ignition coil section.

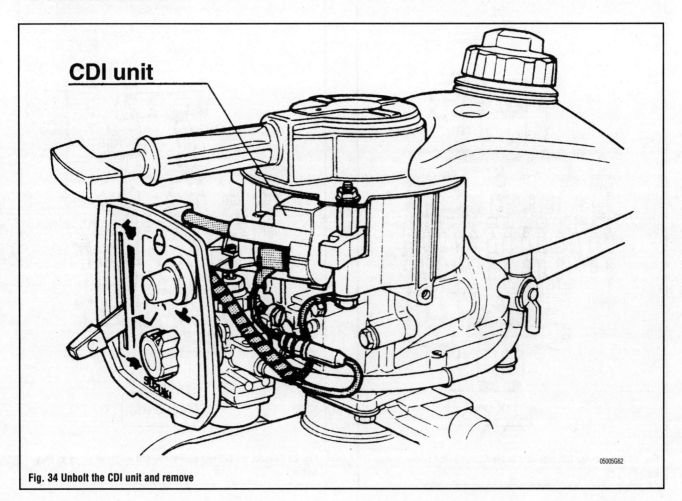

Fig. 34 Unbolt the CDI unit and remove

IGNITION AND ELECTRICAL SYSTEMS 5-35

DT35 and DT40

1. The CDI unit and ignition coils are integrated into one unit. The removal and installation of the CDI unit is covered in the ignition coil section.

DT55 and DT65

♦ See accompanying illustrations

1. Remove the engine cover.
2. Disconnect the spark plug wires.
3. Open the electric parts holder.
4. Disconnect the CDI unit leads.
5. Remove the elastic bands and remove the CDI unit from the electric parts holder.

To install:

6. Install the CDI unit in the electric parts holder. Make sure the elastic bands are secure.
7. Connect the CDI unit leads. Make sure all the connections are tight and free of corrosion.
8. Install the cover on the electric parts holder.
9. Install the spark plug wires.
10. Install the engine cover.

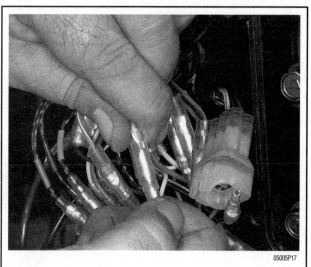
Step 4

Step 2

Step 5

DT75 and DT85

1. Remove the engine cover.
2. Disconnect the spark plug wires.
3. Open the electric parts holder and disconnect the CDI unit from the ignition coil leads.
4. Release the elastic bands or unscrew the bolts holding the CDI unit in place and remove the CDI unit from the electrical parts holder

To install:

5. Install the ignition coils in the electric parts holder. Make sure the retaining bolts are secure.
6. Connect all the wire leads to the CDI unit. Make sure all the connections are tight and free of corrosion.
7. Install the cover on the electric parts holder.
8. Install the spark plug wires.
9. Install the engine cover.

DT90 and DT100

♦ See Figure 35

1. Disconnect the negative battery cable.
2. Remove the engine cover.

Step 3

5-36 IGNITION AND ELECTRICAL SYSTEMS

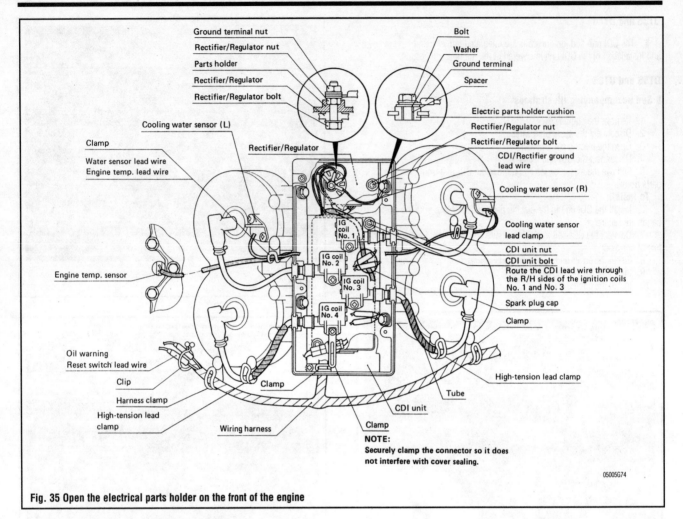

Fig. 35 Open the electrical parts holder on the front of the engine

3. Open the electrical parts holder on the front of the engine. The CDI unit is mounted behind the ignition coils.
4. Loosen the ignition coil mounting bolts and move them out of the way.
5. Disconnect the CDI wire leads.
6. Unbolt the CDI unit and remove it from the electric parts holder.

To install:

7. Install the CDI unit into the electrical parts holder. Make sure all connections are tight and free of corrosion. Also make sure all ground connections are made tight.
8. Install the ignition coils.
9. Replace the electric parts holder cover.
10. Install the engine cover.
11. Connect the negative battery cable.

DT115 and DT140

♦ See Figures 36 and 37

1. Disconnect the negative battery cable.
2. Remove the engine cover.
3. Open the electrical parts holder on the side of the engine to access the CDI unit.
4. The same procedure is used to access the ECU on the fuel injected models.
5. Disconnect the ECU/CDI wire leads and grounds.
6. Loosen the retaining screws and remove the unit from the electrical parts holder.

To install:

7. Install the ECU/CDI in the electrical parts holder. Make sure all ground points are tight, clean and free of corrosion. This is especially important on the EFI equipped engines.
8. Reconnect the ECU/CDI leads. Again, make sure all connections are tight and free of corrosion.

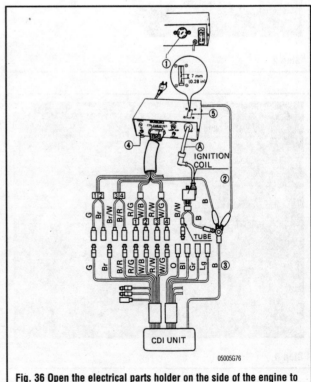

Fig. 36 Open the electrical parts holder on the side of the engine to access the CDI unit

IGNITION AND ELECTRICAL SYSTEMS 5-37

9. Use wire ties to bind together the wiring harnesses.
10. Install the electrical parts holder cover.
11. Install the engine cover.
12. Connect the battery negative cable.

DT150, DT175, DT200 and DT225

1. Disconnect the negative battery cable.
2. Remove the engine cover.
3. The CDI unit is mounted behind the ignition coils on the front of the engine.
4. Disconnect the ignition coil leads and remove the entire ignition coil assembly as a unit.
5. Disconnect the leads and remove the CDI unit from the back of the ignition coil assembly.

To install:

6. Install the CDI unit onto the back of the ignition coil assembly.
7. Install the ignition coil assembly onto the engine and connect the CDI unit and ignition coil wire leads.
8. Install the ignition cover.
9. Install the engine cover.
10. Attach the negative battery cable.

Rectifier

DESCRIPTION & OPERATION

The rectifier consists of a series of diodes or one-way electrical valves. It rectifies or corrects the alternating current (AC) produced within the windings to charge the direct current (DC) battery.

TESTING

1. The unit may remain installed on the powerhead, or it may be removed for testing. In either case, the testing procedures are identical.
2. Disconnect the red, white, yellow and black rectifier leads.

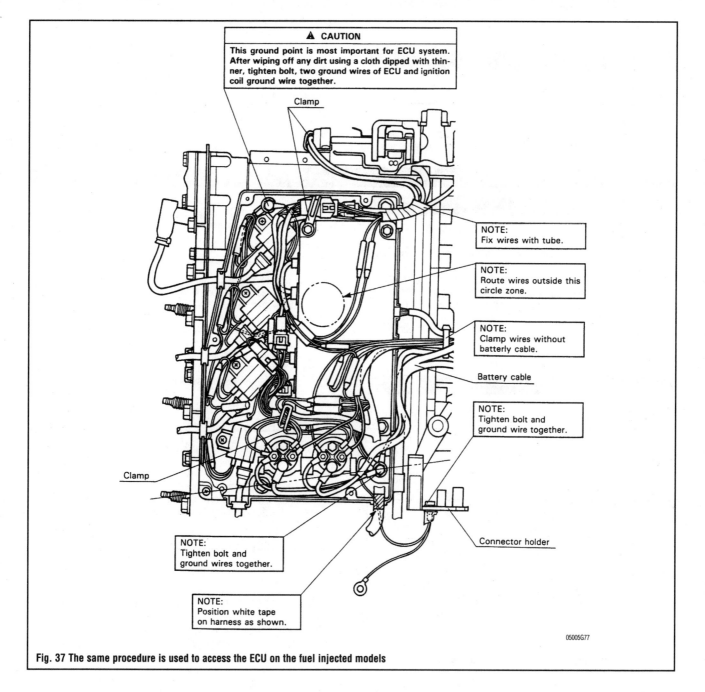

Fig. 37 The same procedure is used to access the ECU on the fuel injected models

IGNITION AND ELECTRICAL SYSTEMS

3. Connect a test light between the yellow and white leads and then reverse the test light leads. The light should light up when tested in one direction and not the other.
4. Test the red and white, black and red, and yellow and black rectifier leads.
5. If the test light does not light as described in any one of the diode directions, the rectifier must be replaced.

REMOVAL & INSTALLATION

1. Disconnect the negative battery cable.
2. Disconnect the rectifier or voltage regulator/rectifier connectors.
3. Remove the fastener attaching the unit to the engine or electrical component and remove the unit.
4. If so equipped, remove the ground lead and disconnect it from the powerhead.

To install:
5. Install the rectifier or voltage regulator/rectifier onto the powerhead or electrical component.
6. Connect all leads and make sure all connections are tight and free of corrosion.
7. Connect the negative battery cable.

Regulator

DESCRIPTION & OPERATION

The voltage regulator controls the alternators field voltage by grounding one end of the field windings very rapidly. The frequency varies according to current demand the more the field is grounded, the more voltage and current the alternator produces. Voltage is maintained at about 13.5–15 volts. During high engine speeds and low current demands, the regulator will adjust the voltage of the alternator field to lower the alternator output voltage. Conversely, when the engine is idling and the current demands may be high, the regulator will increase field voltage, increasing the output of the alternator.

TESTING

1. Make sure that the battery is fully charged.
2. Remove the engine cover.
3. Connect a tachometer according to the manufacturers instructions.
4. Start the engine and let it warm up to normal operating temperature.
5. Connect a voltmeter across the battery terminals.
6. Slowly increase engine speed to approximately 5,000 rpm. Note the voltmeter reading, if it is not 14–15 volts, the voltage regulator must be replaced.

REMOVAL & INSTALLATION

1. Disconnect the negative battery cable.
2. Disconnect the voltage regulator connections.
3. Remove fasteners holding the voltage regulator to the powerhead or electrical component and remove the regulator.
4. If equipped, remove the ground wire and fastener.

To install:
5. Install the regulator onto the powerhead or electrical component and tighten the fastener(s).
6. If equipped, reattach the ground wire and fastener.
7. Connect the regulator leads, making sure that they are all tight and free of corrosion.
8. Connect the negative battery cable

ELECTRONIC IGNITION

Description And Operation

The fuel injected engines use an electronic ignition system which, when combined with the fuel injection system, becomes an integrated electronic engine management system that improves fuel consumption, performance and exhaust emissions.

The ignition timing control system controls the ignition timing by first determining the basic ignition timing based on an engine speed signal sent by the pulser coil and the intake manifold vacuum signal sent by the MAP sensor.

The ignition timing control system, with inputs from the Throttle Position Sensor (TPS), the Engine Temperature Compensation Sensor (ECT), and the Suction Air Temperature Sensor (IAT), compensates the base timing for optimum ignition timing.

Warning Systems

OVER REVOLUTION

▶ See Figure 38

Once engine speed reaches 6,100 rpm and remains over 5,900 rpm for 10-seconds, engine speed will automatically be reduced to 3,000 rpm. While the caution system is activated, the "REV LIMIT" lamp will light up when the engine speed is above 2750 rpm. The system can be reset by reducing engine speed below 2,500 rpm for one-second.

OIL LEVEL

If the oil level switch is turned on for 10-seconds, engine speed will automatically be reduced to 3,000 rpm. While the caution system is activated, the buzzer will sound and the red "OIL LEVEL" lamp will light up. The "REV LIMIT" lamp will also light when the engine speed is above 2,750 rpm. This system can be reset to normal operation when the oil level switch is turned off (oil tank is refilled) and engine speed is reduced below 2,500 rpm.

Only engine speed limit control of this caution system can be cancelled for emergency use by pushing the reset switch. This status can be reset by stopping the engine or by resetting the system to normal operation as described above.

OIL FLOW

If the oil flow sensor is kept on for 2-seconds, engine speed will automatically be reduced to 3,000 rpm. While the caution system is activated, the buzzer will sound and the "OIL" lamp will light up. The "REV LIMIT" lamp will also light up when the engine speed is above 2,750 rpm.

The system can be reset to normal operation when the oil flow sensor is turned off and engine speed is reduced below 2,500 rpm.

OVERHEAT

Engine speed is reduced to 2,000 rpm when the engine temperature reaches 149°F (65°C) or the engine temperature rising rate exceeds the level preset in the system. While the caution system is activated, the buzzer will sound and the "TEMP" lamp will light up. The "REV LIMIT" lamp will also light up when the engine speed is above 1,750 rpm.

The system can be reset when the engine temperature drops lower than 140°F (60°C) and engine speed is reduced below 1,500 rpm.

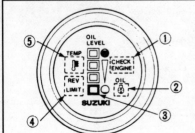

1. "Check engine" lamp
2. "Oil" lamp
3. "Red oil level" lamp
4. "Rev limit" lamp
5. "Temp" lamp

Fig. 38 Typical Suzuki monitor gauge with warning lamps

IGNITION AND ELECTRICAL SYSTEMS

BATTERY VOLTAGE

Engine speed is reduced to either 2,000 or 3,000 rpm when the system detects low battery voltage which is preset by engine speed. (Example: 8 volts @ 2,500 rpm, 9.8 volts @ 5,500 rpm).

CHARGING CIRCUIT

Description and Operation

The voltage regulator controls the alternator's field voltage by grounding one end of the field windings very rapidly. The frequency varies according to current demand. The more the field is grounded, the more voltage and current the alternator produces. Voltage is maintained at about 13.5–15 volts. During high engine speeds and low current demands, the regulator will adjust the voltage of the alternator field to lower the alternator output voltage.

SINGLE PHASE CHARGING SYSTEM

The single-phase charging system found on inline engines provides basic battery maintenance. Single-phase, full wave systems like these are found on a variety of products. Many outboard engines, water vehicles, motorcycles, golf cars and snowmobiles use similar systems.

This charging system produces electricity by moving a magnet past a fixed coil. Alternating current is produced by this method. Since a battery cannot be charged by AC (alternating current), the AC current produced by the lighting coil is rectified or changed into DC (direct current) to charge the battery.

To control the charging rate an additional device called a regulator is used. When the battery voltage reaches approximately 14.6 volts the regulator sends the excess current to ground. This prevents the battery from overcharging and boiling away the electrolyte.

The charging system consists of the following components:
- A flywheel containing magnets
- The lighting coil or alternator coil
- The battery, fuse assembly and wiring
- A regulator/rectifier

The lighting coil is usually a bright exposed copper wire with a lacquer-type coating. Lighting coils are built in with the ignition charge coils on some models. If the charge coil or lighting coil fails the whole stator assembly must be replaced.

The flywheel contains the magnets. The number of magnets determines the number of poles. Each magnet has two poles, so a 4-pole system has two magnets. You need to know the number of poles in order to set the tachometer correctly.

Servicing charging systems is not difficult if you follow a few basic rules. Always start by verifying the problem. If the complaint is that the battery will not stay charged do not automatically assume that the charging system is at fault. Something as simple as an accessory that draws current with the key off will convince anyone they have a bad charging system. Another culprit is the battery. Remember to clean and service your battery regularly. Battery abuse is the number one charging system problem.

The regulator/rectifier is the brains of the charging system. This assembly controls current flow in the charging system. If battery voltage is below 14.6 volts the regulator sends the available current from the rectifier to the battery. If the battery is fully charged (about 15 volts) the regulator diverts most of the current from the rectifier back to the lighting coil through ground.

Do not expect the regulator/rectifier to send current to a fully charged battery. You may find that you must pull down the battery voltage below 12.5 volts to test charging system output. Running the power trim and tilt will reduce the battery voltage. Even a pair of 12 volt sealed beam lamps hooked to the battery will reduce the battery voltage quickly.

In the charging system the regulator/rectifier is the most complex item to troubleshoot. You can avoid Troubleshooting the regulator/rectifier by checking around it. Check the battery and charge or replace it as needed. Check the AC voltage output of the lighting coil. If AC voltage is low check the charge coil for proper resistance and insulation to ground. If these check OK measure the resistance of the Black wire from the rectifier/regulator to ground and for proper voltage output on the Red lead coming from the rectifier/regulator going to the battery. If all the above check within specification replace the rectifier/regulator and verify the repair by performing a charge rate test. This same check around method is used on other components like the CDI unit.

While the caution system is activated, the buzzer will sound and the "CHECK ENGINE" lamp will light. The "REV LIMIT" lamp will also light up when the engine speed is above either 1,750 or 2,750 rpm. The system can be reset by not using any electrical equipment which consumes high electric energy such as the power tilt and trim system, hydraulic jack plate, etc. and engine speed is reduced to idle for one second.

THREE-PHASE CHARGING SYSTEM

Three-phase systems have two more coils in the stator and one more wire than single-phase charging systems. They create higher amperage output than single-phase in nearly the same space.

➡ **If you do not have a solid grasp of single-phase charging systems, please read the description and operation for single-phase systems before continuing.**

AC current is generated identically in both three-phase and single-phase systems. These charging systems produce AC (alternating current) by moving magnets past a fixed set of coils. Since a battery cannot be charged by AC, the AC produced by the lighting coils is rectified or changed into DC (direct current). The rate at which the battery receives this rectified current is controlled by the regulator.

The two additional lighting coils found in a three-phase charging system add complexity to circuit tracing and troubleshooting. Some systems also incorporate a battery isolator. These additional components can make these systems intimidating.

When attempting to troubleshoot these systems, apply a divide and conquer method to demystify this system. Once you have separated the components and circuitry into digestible blocks the system will be much easier to understand.

The charging system consists of the following components:
- A flywheel containing magnets
- The stator assembly, consisting of three individual lighting coils tied together in a "Y" configuration
- The battery, fuse assemblies and wiring
- A battery isolator and wiring, if so equipped

Servicing this system requires a consistent approach using a reliable checklist. If you are not systematic you may forget to check a critical component.

PRECAUTIONS

To prevent damage to the on-board computer, alternator and regulator, the following precautionary measures must be taken when working with the electrical system:

- Wear safety glasses when working on or near the battery.
- Don't wear a watch with a metal band when servicing the battery. Serious burns can result if the band completes the circuit between the positive battery terminal and ground.
- Be absolutely sure of the polarity of a booster battery before making connections. Connect the cables positive-to-positive, and negative-to-negative. Connect positive cables first, and then make the last connection to ground on the body of the booster vehicle so that arcing cannot ignite hydrogen gas that may have accumulated near the battery. Even momentary connection of a booster battery with the polarity reversed will damage alternator diodes.
- Never ground the alternator or generator output or battery terminal. Be cautious when using metal tools around a battery to avoid creating a short circuit between the terminals.
- Never ground the field circuit between the alternator and regulator.
- Never run an alternator or generator without load unless the field circuit is disconnected.
- Never attempt to polarize an alternator.
- When installing a battery, make sure that the positive and negative cables are not reversed.
- When jump-starting the boat, be sure that like terminals are connected. This also applies to using a battery charger. Reversed polarity will burn out the alternator and regulator in a matter of seconds.
- Never operate the alternator with the battery disconnected or on an otherwise uncontrolled open circuit.
- Do not short across or ground any alternator or regulator terminals.
- Do not try to polarize the alternator.

5-40 IGNITION AND ELECTRICAL SYSTEMS

- Do not apply full battery voltage to the field (brown) connector.
- Always disconnect the battery ground cable before disconnecting the alternator lead.
- Always disconnect the battery (negative cable first) when charging it.
- Never subject the alternator to excessive heat or dampness. If you are steam cleaning the engine, cover the alternator.
- Never use arc-welding equipment on the car with the alternator connected.

Troubleshooting the Charging System

The charging system should be inspected if:
- A Diagnostic Trouble Code (DTC) is set relating to the charging system
- The charging system warning light is illuminated
- The voltmeter on the instrument panel indicates improper charging (either high or low) voltage
- The battery is overcharged (electrolyte level is low and/or boiling out)
- The battery is undercharged (insufficient power to crank the starter)

The starting point for all charging system problems begins with the inspection of the battery, related wiring and the alternator drive belt (if equipped). The battery must be in good condition and fully charged before system testing. If a Diagnostic Trouble Code (DTC) is set, diagnose and repair the cause of the trouble code first.

If equipped, the charging system warning light will illuminate if the charging voltage is either too high or too low. The warning light should light when the key is turned to the **ON** position as a bulb check. When voltage is produced due to the engine starting, the light should go out. A good sign of voltage that is too high are lights that burn out and/or burn very brightly. Over-charging can also cause damage to the battery and electronic circuits.

A thorough, systematic approach to troubleshooting will pay big rewards. Build your troubleshooting check list with the most likely offenders at the top. Do not be tempted to throw parts at a problem without systematically Troubleshooting the system first.

Do a visual check of the battery, wiring and fuses. Are there any new additions to the wiring? An excellent clue might be, "Everything was working OK until I added that live well pump." With a comment like this you would know where to check first.

The regulator/rectifier assembly is the brains of the charging system. The regulator controls current flow in the charging system. If battery voltage is below about 14.6 volts the regulator sends the available current to the battery. If the battery is fully charged (about 14.5 to 15 volts) the regulator diverts the current/amps to ground.

Do not expect the regulator to send current to a fully charged battery. Check the battery for a possible draw with the key off. This draw may be the cumulative effect of several radio and/or clock memories. If these accessories are wired to the cranking battery then a complaint of charging system failure may really be excessive draw. Draw over about 25 milliamps should arouse your suspicions. The fuel management gauge memory and speedometer clock draw about 10 milliamps each. Remember that a milliamp is 1/1000 of an amp. Check battery condition thoroughly because it is the #1 culprit in charging system failures.

Do not forget to check through the fuses. It can be embarrassing to overlook a blown fuse.

You must pull the battery voltage down below 12.5 volts to test charging system output. Running the power trim and tilt will reduce the battery voltage. A load bank or even a pair of 12-volt sealed-beam headlamps hooked to the battery can also be used to reduce the battery voltage.

Once the battery's good condition is verified and it has been reduced to below 12.5 volts you can test further.

Install an ammeter to check actual amperage output. Several tool manufacturers produce a shunt adapter that will attach to your multi-meter and allow you to read the amp output. Verify that the system is delivering sufficient amperage. Too much amperage and a battery that goes dry very quickly indicates that the rectifier/regulator should be replaced.

If the system does not put out enough amperage, then test the lighting coil. Isolate the coil and test for correct resistance and short to ground.

During these test procedures the regulator/rectifier has not been bench checked. Usually it is advisable to avoid troubleshooting the regulator/rectifier directly. The procedures listed so far have focused on checking around the rectifier/regulator. If you verify that all other systems stator are good then what is left in the system to cause the verified problem? The process of elimination has declared the rectifier/regulator bad.

This check around method is also useful on other components that can not be checked directly or involve time-consuming test procedures. This is the same method suggested for checking the capacitor discharge ignition box.

OVERCHARGING

There is really only one cause for this type of failure, the regulator is not working. It isn't controlling charging output to the battery. Since there is no repair of this part, replace it.

UNDERCHARGING

If there is an undercharge condition after running the DC amperage check at the fuse assembly, then disconnect the stator coupling from the harness and perform AC voltage checks between the three stator leads. Check between two stator leads at a time. There are three volt checks done to cover all possible combinations.

At idle, there is typically 14+ volts on each test. It can be higher if the idle is higher. All three readings should be equal, within a volt or two. Stator shorts to ground can be checked by doing a voltage test between one stator lead and ground, engine running. There should be roughly half the normal stator voltage check reading.

If the readings are all within specification, the stator is working correctly. If any or all readings are below normal, turn the engine **OFF** and check the stator windings using an ohmmeter. An isolated continuity check and a short to ground check should be done. If the stator is bad, replace it since it can't be repaired.

➡ **The charging system is an integral part of the ignition system. For information on service procedures, please refer to the "Ignition" section of this manual.**

Alternator (Stator)

TESTING

➡ **Before testing, make sure all connections and mounting bolts are clean and tight. Many charging system problems are related to loose and corroded terminals or bad grounds. Don't overlook the engine ground connection to the body, or the tension of the alternator drive belt.**

Voltage Drop Test

1. Make sure the battery is in good condition and fully charged.
2. Perform a voltage drop test of the positive side of the circuit as follows:
   a. Start the engine and allow it to reach normal operating temperature.
   b. Turn the headlamps, heater blower motor and interior lights on.
   c. Bring the engine to about 2,500 rpm and hold it there.
   d. Connect the negative (-) voltmeter lead directly to the battery positive (+) terminal.
   e. Touch the positive voltmeter lead directly to the alternator B+ output stud, not the nut. The meter should read no higher than about 0.5 volts. If it does, then there is higher than normal resistance between the positive side of the battery and the B+ output at the alternator.
   f. Move the positive (+) meter lead to the nut and compare the voltage reading with the previous measurement. If the voltage reading drops substantially, then there is resistance between the stud and the nut.

➡ **The theory is to keep moving closer to the battery terminal one connection at a time in order to find the area of high resistance (bad connection).**

3. Perform a voltage drop test of the negative side of the circuit as follows:
   a. Start the engine and allow it to reach normal operating temperature.
   b. Turn the headlamps, heater blower motor and interior lights on.
   c. Bring the engine to about 2,500 rpm and hold it there.
   d. Connect the negative (-) voltmeter lead directly to the negative battery terminal.
   e. Touch the positive (+) voltmeter lead directly to the alternator case or ground connection. The meter should read no higher than about 0.3 volts. If it does, then there is higher than normal resistance between the battery ground terminal and the alternator ground.

IGNITION AND ELECTRICAL SYSTEMS

f. Move the positive (+) meter lead to the alternator mounting bracket, if the voltage reading drops substantially then you know that there is a bad electrical connection between the alternator and the mounting bracket.

➡ The theory is to keep moving closer to the battery terminal one connection at a time in order to find the area of high resistance (bad connection).

Current Output Test

♦ See Figure 39

1. Perform a current output test as follows:

➡ The current output test requires the use of a volt/amp digital multimeter with battery load control and an inductive amperage pick-up. Follow the manufacturer's instructions on the use of the equipment.

    a. Start the engine and allow it to reach normal operating temperature.
    b. Turn **off** all electrical accessories.
    c. Connect the digital multimeter to the battery terminals and cable according to the instructions.
    d. Bring the engine to about 2,500 rpm and hold it there.
    e. Apply a load to the charging system with the rheostat on the digital multimeter. Do not let the voltage drop below 12 volts.
    f. The alternator should deliver to within 10% of the rated output. If the amperage is not within 10% and all other components test good, replace the alternator.

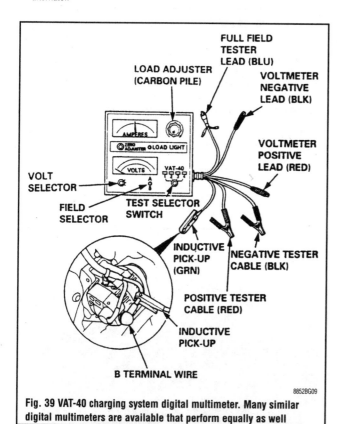

Fig. 39 VAT-40 charging system digital multimeter. Many similar digital multimeters are available that perform equally as well

Battery

The battery is one of the most important parts of the electrical system. In addition to providing electrical power to start the engine, it also provides power for operation of the running lights, radio, and electrical accessories.

Because of its job and the consequences (failure to perform in an emergency), the best advice is to purchase a well-known brand, with an extended warranty period, from a reputable dealer.

The usual warranty covers a pro-rated replacement policy, which means the purchaser would be entitled to a consideration for the time left on the warranty period if the battery should prove defective before its time.

Do not consider a battery of less than 70- amp/hour or 100-minute reserve capacity. If in doubt as to how large the boat requires, make a liberal estimate and then purchase the one with the next higher amp rating. Outboards equipped with an onboard computer, should be equipped with a battery of at least 100 to 105 amp/hour capacity.

MARINE BATTERIES

♦ See Figure 40

Because marine batteries are required to perform under much more rigorous conditions than automotive batteries, they are constructed differently than those used in automobiles or trucks. Therefore, a marine battery should always be the No. 1 unit for the boat and other types of batteries used only in an emergency.

Marine batteries have a much heavier exterior case to withstand the violent pounding and shocks imposed on it as the boat moves through rough water and in extremely tight turns. The plates are thicker and each plate is securely anchored within the battery case to ensure extended life. The caps are spill proof to prevent acid from spilling into the bilges when the boat heels to one side in a tight turn, or is moving through rough water. Because of these features, the marine battery will recover from a low charge condition and give satisfactory service over a much longer period of time than any type intended for automotive use.

✳✳ WARNING

Never use a Maintenance-free battery with an outboard engine that is not voltage regulated. The charging system will continues to charge as long as the engine is running and it is possible that the electrolyte could boil out if periodic checks of the cell electrolyte level are not done.

Fig. 40 A fully charged battery, filled to the proper level with electrolyte, is the heart of the ignition and electrical systems. Engine cranking and efficient performance of electrical items depend on a full rated battery

BATTERY CONSTRUCTION

♦ See Figure 41

A battery consists of a number of positive and negative plates immersed in a solution of diluted sulfuric acid. The plates contain dissimilar active materials and are kept apart by separators. The plates are grouped into elements. Plate

5-42 IGNITION AND ELECTRICAL SYSTEMS

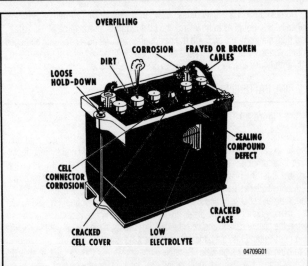

Fig. 41 A visual inspection of the battery should be made each time the boat is used. Such a quick check may reveal a potential problem in its early stages. A dead battery in a busy waterway or far from assistance could have serious consequences

straps on top of each element connect all of the positive plates and all of the negative plates into groups.

The battery is divided into cells holding a number of the elements apart from the others. The entire arrangement is contained within a hard plastic case. The top is a one-piece cover and contains the filler caps for each cell. The terminal posts protrude through the top where the battery connections for the boat are made. Each of the cells is connected to its neighbor in a positive-to-negative manner with a heavy strap called the cell connector.

BATTERY RATINGS

♦ See Figure 42

Three different methods are used to measure and indicate battery electrical capacity:
- Amp/hour rating
- Cold cranking performance
- Reserve capacity

The amp/hour rating of a battery refers to the battery's ability to provide a set amount of amps for a given amount of time under test conditions at a constant temperature. Therefore, if the battery is capable of supplying 4 amps of current for 20 consecutive hours, the battery is rated as an 80 amp/hour battery. The amp/hour rating is useful for some service operations, such as slow charging or battery testing.

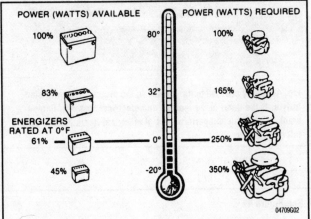

Fig. 42 Comparison of battery efficiency and engine demands at various temperatures

Cold cranking performance is measured by cooling a fully charged battery to 0°F (-17°C) and then testing it for 30 seconds to determine the maximum current flow. In this manner the cold cranking amp rating is the number of amps available to be drawn from the battery before the voltage drops below 7.2 volts.

The illustration depicts the amount of power in watts available from a battery at different temperatures and the amount of power in watts required of the engine at the same temperature. It becomes quite obvious—the colder the climate, the more necessary for the battery to be fully charged.

Reserve capacity of a battery is considered the length of time, in minutes, at 80°F (27°C), a 25 amp current can be maintained before the voltage drops below 10.5 volts. This test is intended to provide an approximation of how long the engine, including electrical accessories, could operate satisfactorily if the stator assembly or lighting coil did not produce sufficient current. A typical rating is 100 minutes.

If possible, the new battery should have a power rating equal to or higher than the unit it is replacing.

BATTERY LOCATION

Every battery installed in a boat must be secured in a well protected, ventilated area. If the battery area lacks adequate ventilation, hydrogen gas, which is given off during charging, is very explosive. This is especially true if the gas is concentrated and confined.

BATTERY SERVICE

♦ See Figures 43, 44 and 45

Batteries require periodic servicing and a definite maintenance program will ensure extended life. If the battery should test satisfactorily but still fails to perform properly, one of four problems could be the cause.

1. An accessory might have accidentally been left on overnight or for a long period during the day. Such an oversight would result in a discharged battery.

2. Using more electrical power than the stator assembly or lighting coil can replace would result in an undercharged condition.

3. A defect in the charging system. A faulty stator assembly or lighting coil, defective rectifier, or high resistance somewhere in the system could cause the battery to become undercharged.

4. Failure to maintain the battery in good order. This might include a low level of electrolyte in the cells, loose or dirty cable connections at the battery terminals or possibly an excessively dirty battery top.

Fig. 43 Explosive hydrogen gas is normally released from the cells under a wide range of circumstances. This battery exploded when the gas ignited from someone smoking in the area when the caps were removed. Such an explosion could also be caused by a spark from the battery terminals

IGNITION AND ELECTRICAL SYSTEMS 5-43

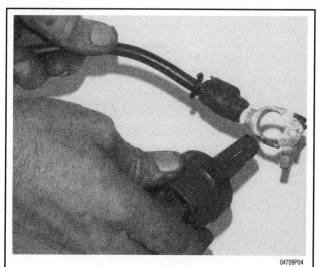
Fig. 44 A two part battery cable cleaning tool will do an excellent job of cleaning the inside of the cable connectors

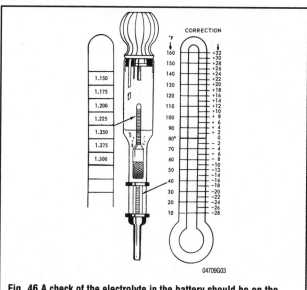
Fig. 46 A check of the electrolyte in the battery should be on the maintenance schedule for any boat

Fig. 45 The second part of the battery cable cleaning tool contains a brush for cleaning the battery terminals

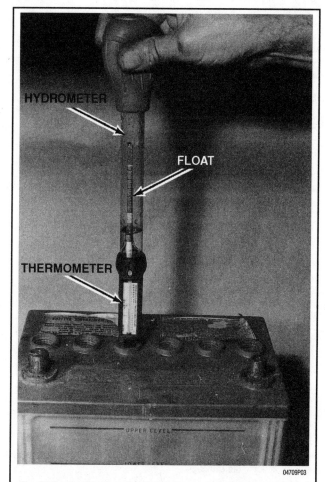
Fig. 47 Testing the electrolyte's specific gravity using a temperature corrected hydrometer

Electrolyte Level

▶ See Figures 46 and 47

The most common procedure for checking the electrolyte level in a battery is to remove the cell caps and visually observe the level in the cells. The bottom of each cell has a split vent which will cause the surface of the electrolyte to appear distorted when it makes contact. When the distortion first appears at the bottom of the split vent, the electrolyte level is correct.

During hot weather and periods of heavy use, the electrolyte level should be checked more often than during normal operation. Add distilled water to bring the level of electrolyte in each cell to the proper level. Take care not to overfill, because adding an excessive amount of water will cause loss of electrolyte and any loss will result in poor performance, short battery life, and will contribute quickly to corrosion.

➡ **Never add electrolyte from another battery. Use only distilled water.**

Battery Testing

A hydrometer is a device to measure the percentage of sulfuric acid in the battery electrolyte in terms of specific gravity. When the condition of the battery drops from fully charged to discharged, the acid leaves the solution and enters the plates, causing the specific gravity of the electrolyte to drop.

It may not be common knowledge, but hydrometer floats are calibrated for use at 80°F (27°C). If the hydrometer is used at any other temperature, hotter or colder, a correction factor must be applied.

➡ **Remember, a liquid will expand if it is heated and will contract if cooled. Such expansion and contraction will cause a definite change in the specific gravity of the liquid, in this case the electrolyte.**

A quality hydrometer will have a thermometer/temperature correction table in the lower portion, as illustrated in the accompanying illustration. By knowing the air temperature around the battery and from the table, a correction factor may be applied to the specific gravity reading of the hydrometer float. In this manner, an accurate determination may be made as to the condition of the battery.

When using a hydrometer, pay careful attention to the following points:
1. Never attempt to take a reading immediately after adding water to the battery. Allow at least ¼ hour of charging at a high rate to thoroughly mix the electrolyte with the new water. This time will also allow for the necessary gases to be created.
2. Always be sure the hydrometer is clean inside and out as a precaution against contaminating the electrolyte.
3. If a thermometer is an integral part of the hydrometer, draw liquid into it several times to ensure the correct temperature before taking a reading.
4. Be sure to hold the hydrometer vertically and suck up liquid only until the float is free and floating.
5. Always hold the hydrometer at eye level and take the reading at the surface of the liquid with the float free and floating.
6. Disregard the slight curvature appearing where the liquid rises against the float stem. This phenomenon is due to surface tension.
7. Do not drop any of the battery fluid on the boat or on your clothing, because it is extremely caustic. Use water and baking soda to neutralize any battery liquid that does accidentally drop.
8. After drawing electrolyte from the battery cell until the float is barely free, note the level of the liquid inside the hydrometer. If the level is within the Green band range for all cells, the condition of the battery is satisfactory. If the level is within the white band for all cells, the battery is in fair condition.
9. If the level is within the Green or white band for all cells except one, which registers in the red, the cell is shorted internally. No amount of charging will bring the battery back to satisfactory condition.
10. If the level in all cells is about the same, even if it falls in the Red band, the battery may be recharged and returned to service. If the level fails to rise above the Red band after charging, the only solution is to replace the battery.

Battery Cleaning

Dirt and corrosion should be cleaned from the battery as soon as it is discovered. Any accumulation of acid film or dirt will permit current to flow between the terminals. Such a current flow will drain the battery over a period of time.

Clean the exterior of the battery with a solution of diluted ammonia or a paste made from baking soda to neutralize any acid which may be present. Flush the cleaning solution off with clean water.

➡ **Take care to prevent any of the neutralizing solution from entering the cells, by keeping the caps tight.**

A poor contact at the terminals will add resistance to the charging circuit. This resistance will cause the voltage regulator to register a fully charged battery, and thus cut down on the stator assembly or lighting coil output adding to the low battery charge problem.

Scrape the battery posts clean with a suitable tool or with a stiff wire brush. Clean the inside of the cable clamps to be sure they do not cause any resistance in the circuit.

BATTERY TERMINALS

At least once a season, the battery terminals and cable clamps should be cleaned. Loosen the clamps and remove the cables, negative cable first. On batteries with top mounted posts, the use of a puller specially made for this purpose is recommended. These are inexpensive and available in most parts stores.

Clean the cable clamps and the battery terminal with a wire brush, until all corrosion, grease, etc., is removed and the metal is shiny. It is especially important to clean the inside of the clamp thoroughly (a wire brush is useful here), since a small deposit of foreign material or oxidation there will prevent a sound electrical connection and inhibit either starting or charging. It is also a good idea to apply some dielectric grease to the terminal, as this will aid in the prevention of corrosion.

After the clamps and terminals are clean, reinstall the cables, negative cable last, do not hammer the clamps onto battery posts. Tighten the clamps securely, but do not distort them. Give the clamps and terminals a thin external coating of grease after installation, to retard corrosion.

Check the cables at the same time that the terminals are cleaned. If the insulation is cracked or broken, or if its end is frayed, that cable should be replaced with a new one of the same length and gauge.

SAFETY PRECAUTIONS

Always follow these safety precautions when charging or handling a battery:
- Wear eye protection when working around batteries. Batteries contain corrosive acid and produce explosive gas a byproduct of their operation. Acid on the skin should be neutralized with a solution of baking soda and water made into a paste. In case acid contacts the eyes, flush with clear water and seek medical attention immediately.
- Avoid flame or sparks that could ignite the hydrogen gas produced by the battery and cause an explosion. Connection and disconnection of cables to battery terminals is one of the most common causes of sparks.
- Always turn a battery charger **OFF**, before connecting or disconnecting the leads. When connecting the leads, connect the positive lead first, then the negative lead, to avoid sparks.
- When lifting a battery, use a battery carrier or lift at opposite corners of the base.
- Ensure there is good ventilation in a room where the battery is being charged.
- Do not attempt to charge or load-test a maintenance-free battery when the charge indicator dot is indicating insufficient electrolyte.
- Disconnect the negative battery cable if the battery is to remain in the boat during the charging process.
- Be sure the ignition switch is **OFF** before connecting or turning the charger **ON**. Sudden power surges can destroy electronic components.
- Use proper adapters to connect charger leads to batteries with non-conventional terminals.

BATTERY CHARGERS

Before using any battery charger, consult the manufacturer's instructions for its use. Battery chargers are electrical devices that change Alternating Current (AC) to a lower voltage of Direct Current (DC) that can be used to charge a marine battery. There are two types of battery chargers—manual and automatic.

A manual battery charger must be physically disconnected when the battery has come to a full charge. If not, the battery can be overcharged, and possibly fail. Excess charging current at the end of the charging cycle will heat the electrolyte, resulting in loss of water and active material, substantially reducing battery life.

➡ **As a rule, on manual chargers, when the ammeter on the charger registers half the rated amperage of the charger, the battery is fully charged. This can vary, and it is recommended to use a hydrometer to accurately measure state of charge.**

Automatic battery chargers have an important advantage—they can be left connected (for instance, overnight) without the possibility of overcharging the battery. Automatic chargers are equipped with a sensing device to allow the battery charge to taper off to near zero as the battery becomes fully charged. When charging a low or completely discharged battery, the meter will read close to full rated output. If only partially discharged, the initial reading may be less than full

IGNITION AND ELECTRICAL SYSTEMS

rated output, as the charger responds to the condition of the battery. As the battery continues to charge, the sensing device monitors the state of charge and reduces the charging rate. As the rate of charge tapers to zero amps, the charger will continue to supply a few milliamps of current—just enough to maintain a charged condition.

BATTERY CABLES

Battery cables don't go bad very often, but like anything else, they can wear out. If the cables on your boat are cracked, frayed or broken, they should be replaced.

When working on any electrical component, it is always a good idea to disconnect the negative (-) battery cable. This will prevent potential damage to many sensitive electrical components

Always replace the battery cables with one of the same length, or you will increase resistance and possibly cause hard starting. Smear the battery posts with a light film of dielectric grease, or a battery terminal protectant spray once you've installed the new cables. If you replace the cables one at a time, you won't mix them up.

→**Any time you disconnect the battery cables, it is recommended that you disconnect the negative (-) battery cable first. This will prevent you from accidentally grounding the positive (+) terminal when disconnecting it, thereby preventing damage to the electrical system.**

STARTING CIRCUIT

Description and Operation

♦ See Figure 48

In the early days, all outboard engines were started by simply pulling on a rope wound around the flywheel. As time passed and owners were reluctant to use muscle power, it was necessary to replace the rope starter with some form of power cranking system. Today, many small engines are still started by pulling on a rope, but others have a powered starter motor installed.

The system utilized to replace the rope method was an electric starter motor coupled with a mechanical gear mesh between the starter motor and the powerhead flywheel, similar to the method used to crank an automobile engine.

As the name implies, the sole purpose of the starter motor circuit is to control operation of the starter motor to crank the powerhead until the engine is operating. The circuit includes a relay or magnetic switch to connect or disconnect the motor from the battery. The operator controls the switch with a key switch.

Before you disconnect the cable(s), first turn the ignition to the **OFF** position. This will prevent a draw on the battery which could cause arcing. When the battery cable(s) are reconnected (negative cable last), be sure to check all electrical accessories are all working correctly.

BATTERY STORAGE

If the boat is to be laid up for the winter or for more than a few weeks, special attention must be given to the battery to prevent complete discharge or possible damage to the terminals and wiring. Before putting the boat in storage, disconnect and remove the batteries. Clean them thoroughly of any dirt or corrosion, and then charge them to full specific gravity reading. After they are fully charged, store them in a clean cool dry place where they will not be damaged or knocked over, preferably on a couple blocks of wood. Storing the battery up off the deck, will permit air to circulate freely around and under the battery and will help to prevent condensation.

Never store the battery with anything on top of it or cover the battery in such a manner as to prevent air from circulating around the filler caps. All batteries, both new and old, will discharge during periods of storage, more so if they are hot than if they remain cool. Therefore, the electrolyte level and the specific gravity should be checked at regular intervals. A drop in the specific gravity reading is cause to charge them back to a full reading.

In cold climates, care should be exercised in selecting the battery storage area. A fully-charged battery will freeze at about 60 degrees below zero. A discharged battery, almost dead, will have ice forming at about 19 degrees above zero.

A neutral safety switch is installed into the circuit to permit operation of the starter motor only if the shift control lever is in neutral. This switch is a safety device to prevent accidental engine start when the engine is in gear.

The starter motor is a series wound electric motor which draws a heavy current from the battery. It is designed to be used only for short periods of time to crank the engine for starting. To prevent overheating the motor, cranking should not be continued for more than 30-seconds without allowing the motor to cool for at least three minutes. Actually, this time can be spent in making preliminary checks to determine why the engine fails to start.

Power is transmitted from the starter motor to the powerhead flywheel through a Bendix drive. This drive has a pinion gear mounted on screw threads. When the motor is operated, the pinion gear moves upward and meshes with the teeth on the flywheel ring gear.

When the powerhead starts, the pinion gear is driven faster than the shaft, and as a result, it screws out of mesh with the flywheel. A rubber cushion is built into the Bendix drive to absorb the shock when the pinion meshes with the flywheel ring gear. The parts of the drive must be properly assembled for effi-

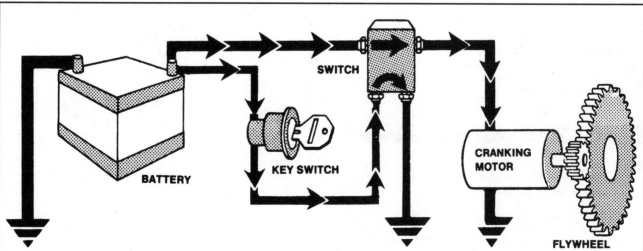

Fig. 48 A typical starting system converts electrical energy into mechanical energy to turn the engine. The components are: Battery, to provide electricity to operate the starter; Ignition switch, to control the energizing of the starter relay or relay; Starter relay or relay, to make and break the circuit between the battery and starter; Starter, to convert electrical energy into mechanical energy to rotate the engine; Starter drive gear, to transmit the starter rotation to the engine flywheel

5-46 IGNITION AND ELECTRICAL SYSTEMS

cient operation. If the drive is removed for cleaning, take care to assemble the parts as illustrated in the accompanying illustrations in this section. If the screw shaft assembly is reversed, it will strike the splines and the rubber cushion will not absorb the shock.

The sound of the motor during cranking is a good indication of whether the starter motor is operating properly or not. Naturally, temperature conditions will affect the speed at which the starter motor is able to crank the engine. The speed of cranking a cold engine will be much slower than when cranking a warm engine. An experienced operator will learn to recognize the favorable sounds of the powerhead cranking under various conditions.

Troubleshooting the Starting System

If the starter motor spins, but fails to crank the engine, the cause is usually a corroded or gummy Bendix drive. The drive should be removed, cleaned, and given an inspection.

1. Before wasting too much time troubleshooting the starter motor circuit, the following checks should be made. Many times, the problem will be corrected.
   - Battery fully charged.
   - Shift control lever in neutral.
   - Main 20-amp fuse located at the base of the fuse cover is good (not blown).
   - All electrical connections clean and tight.
   - Wiring in good condition, insulation not worn or frayed.
2. Starter motor cranks slowly or not at all.
   - Faulty wiring connection
   - Short-circuited lead wire
   - Shift control not engaging neutral (not activating neutral start switch)
   - Defective neutral start switch
   - Starter motor not properly grounded
   - Faulty contact point inside ignition switch
   - Bad connections on negative battery cable to ground (at battery side and engine side)
   - Bad connections on positive battery cable to magnetic switch terminal
   - Open circuit in the coil of the magnetic switch (relay)
   - Bad or run-down battery
   - Excessively worn down starting motor brushes
   - Burnt commutator in starting motor
   - Brush spring tension slack
   - Short circuit in starter motor armature
3. Starter motor keeps running.
   - Melted contact plate inside the magnetic switch
   - Poor ignition switch return action
4. Starter motor picks up speed, put pinion will not mesh with ring gear.
   - Worn down teeth on clutch pinion
   - Worn down teeth on flywheel ring gear
5. Two more areas may cause the powerhead to crank slowly even though the starter motor circuit is in excellent condition
   - A tight or frozen powerhead
   - Water in the lower unit.

Starter Motor

DESCRIPTION & OPERATION

◆ See Figures 49, 50, 51 and 52

As the name implies, the sole purpose of the cranking motor circuit is to control operation of the cranking motor to crank the powerhead until the engine is operating. The circuit includes a relay or magnetic switch to connect or disconnect the motor from the battery. The operator controls the switch with a key switch.

A neutral safety switch is installed into the circuit to permit operation of the cranking motor only if the shift control lever is in neutral. This switch is a safety device to prevent accidental engine start when the engine is in gear.

The cranking motor is a series wound electric motor which draws a heavy current from the battery. It is designed to be used only for short periods of time to crank the engine for starting. To prevent overheating the motor, cranking should not be continued for more than 30-seconds without allowing the motor to cool for at least three minutes. Actually, this time can be spent in making preliminary checks to determine why the engine fails to start.

Power is transmitted from the cranking motor to the powerhead flywheel through a Bendix drive. This drive has a pinion gear mounted on screw threads.

Fig. 49 Typical location of the starter motor (DT55)

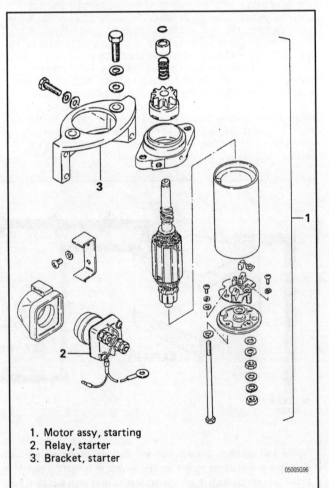

1. Motor assy, starting
2. Relay, starter
3. Bracket, starter

Fig. 50 Typical starter motor and relay assembly

IGNITION AND ELECTRICAL SYSTEMS

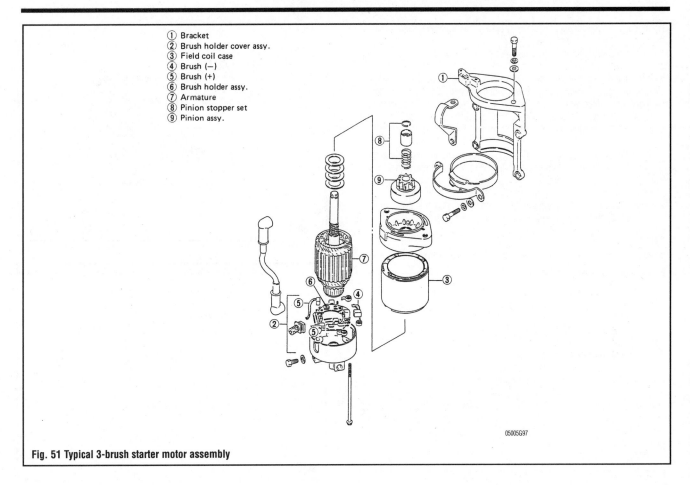

Fig. 51 Typical 3-brush starter motor assembly

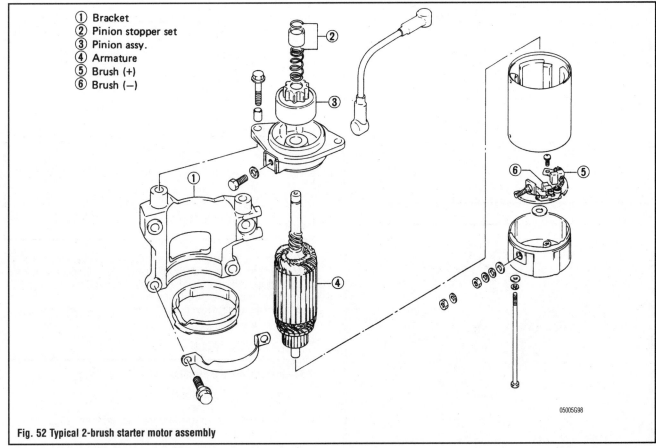

Fig. 52 Typical 2-brush starter motor assembly

5-48 IGNITION AND ELECTRICAL SYSTEMS

When the motor is operated, the pinion gear moves upward and meshes with the teeth on the flywheel ring gear.

When the powerhead starts, the pinion gear is driven faster than the shaft, and as a result, it screws out of mesh with the flywheel. A rubber cushion is built into the Bendix drive to absorb the shock when the pinion meshes with the flywheel ring gear. The parts of the drive must be properly assembled for efficient operation. If the drive is removed for cleaning, take care to assemble the parts as illustrated in the accompanying illustrations in this section. If the screw shaft assembly is reversed, it will strike the splines and the rubber cushion will not absorb the shock.

The sound of the motor during cranking is a good indication of whether the cranking motor is operating properly or not. Naturally, temperature conditions will affect the speed at which the cranking motor is able to crank the engine. The speed of cranking a cold engine will be much slower than when cranking a warm engine. An experienced operator will learn to recognize the favorable sounds of the powerhead cranking under various conditions.

TESTING

♦ See accompanying illustrations

1. Using a multimeter, see if there is any continuity between the commutator and armature core. The tester will indicate infinite resistance if the insulation is in good condition.

2. Using the tester, check for continuity between each pair adjacent commutator segment. If discontinuity is noted at any part of the commutator, replace the whole sub-assembly of the armature.

3. If the surface of the commutator is gummy or otherwise dirty, wipe it off with a cloth dipped in kerosene. If the surface is coarse or in burnt, smooth it with sandpaper. If the surface is grooved deeply, it may be necessary to remove the grooved marks by turning the commutator in a lathe; such turning is often successful in reconditioning the commutator if the extra stock necessary for removal by cutting is available without reducing its diameter to the limit.

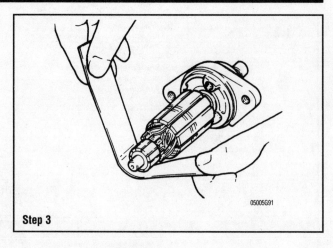
Step 3

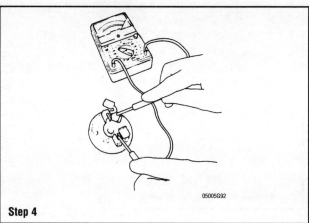
Step 4

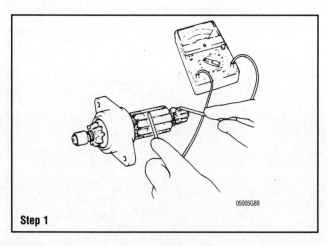
Step 1

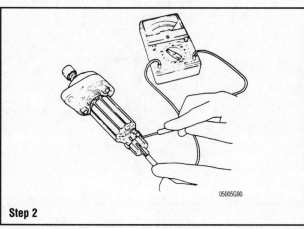
Step 2

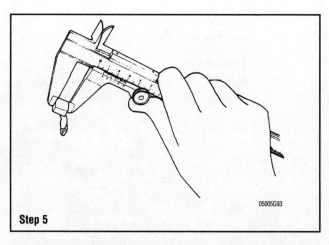
Step 5

4. Using a multimeter, check to be sure that the positive brush holder is insulated from the negative brush holder. The tester should indicate an absence of any continuity between any two brush holders of opposite polarity. If continuity is indicated, replace the brush holder.

5. Check the length of each brush. If the brushes are worn down to the service limit, replace them.

REMOVAL & INSTALLATION

♦ See accompanying illustrations

1. Disconnect the negative battery cable.
2. Remove the engine cover.
3. Disconnect the starter relay electrical relay lead at the bullet connector. Remove the ground wire.

IGNITION AND ELECTRICAL SYSTEMS

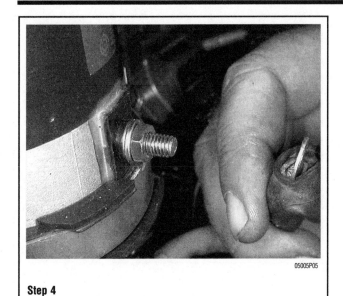
Step 4

Step 7

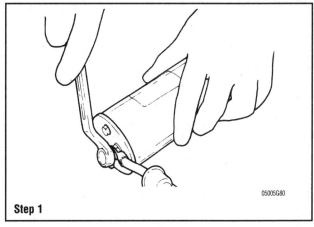
Step 1

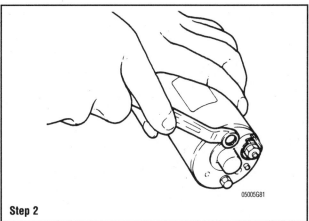
Step 2

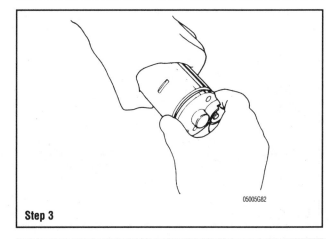
Step 3

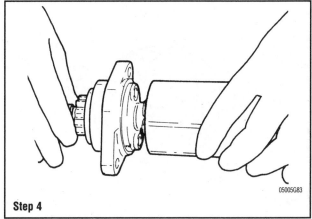
Step 4

4. Disconnect the battery and starter cables.
5. If equipped, remove the starter motor clamp.
6. Remove the mounting bolts holding the starter motor to the mounting bracket.
7. Remove the starter motor.

To install:

8. Install the starter motor onto the mounting bracket. Do not forget to install the clamp if so equipped.
9. Install the mounting bolts and tighten snugly.
10. Connect the starter motor cables. Make sure all connections are tight and free of any corrosion.
11. Install the engine cover.
12. Connect the negative battery cable

OVERHAUL

▶ **See accompanying illustrations**

The following is a typical starter motor overhaul procedure. Some models may vary slightly.

1. Remove the terminal nut and remove the motor cable.
2. Remove the two bolts from the bottom of the motor.
3. Separate the brush holder cover assembly and the two brush springs.
4. Separate the armature from the field coil case.

5-50 IGNITION AND ELECTRICAL SYSTEMS

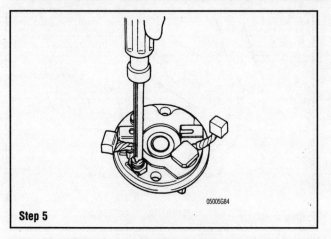
Step 5

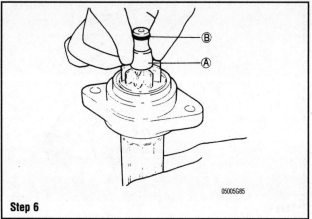
Step 6

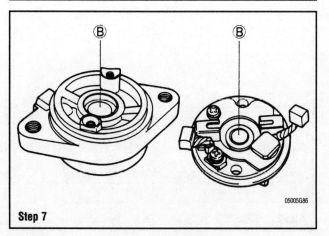
Step 7

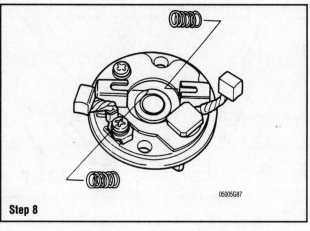
Step 8

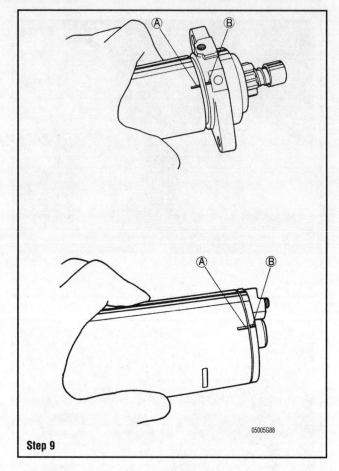
Step 9

5. Remove the screw and nut to remove the brushes.
6. Push down on the pinion stopper "A", remove the stopper ring "B" and take off the pinion.

To assemble:

Assembly is the reverse of disassembly, but the following steps need to be addressed during assembly.

7. Apply water resistant grease to the metal seals and brush holder cover.
8. When setting a brush to each brush holder, be sure to place a brush spring in position before installing the brush.
9. When installing the front cover and the holder cover in the field coil case, align the mark "A" on the field coil case with the mark "B" in both covers. When installing the holder cover, take care not to break the brush.

Starter Motor Relay Switch

DESCRIPTION & OPERATION

♦ See Figure 52

The job of the starter motor relay is to complete the circuit between the battery and starter motor. It does this by closing the starter circuit electromagnetically, when activated by the key switch. This is a completely sealed switch, which meets SAE standards for marine applications. DO NOT substitute an automotive-type relay for this application. It is not sealed and gasoline fumes can be ignited upon starting the powerhead. The relay consists of a coil winding, plunger, return spring, contact disc, and four externally mounted terminals. The relay is installed in series with the positive battery cables mounted to the two larger terminals. The smaller terminals connect to the neutral switch and ground.

To activate the relay, the shift lever is placed in neutral, closing the neutral switch. Electricity coming through the ignition switch goes into the relay coil winding which creates a magnetic field. The electricity then goes on to ground in the powerhead. The magnetic field surrounds the plunger in the relay, which draws the disc contact into the two larger terminals. Upon contact of the termi-

IGNITION AND ELECTRICAL SYSTEMS

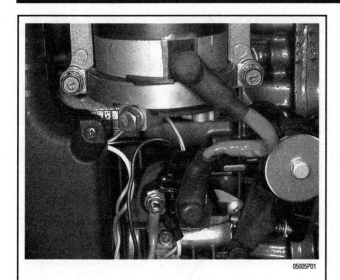

Fig. 53 Typical starter motor relay switch location

nals, the heavy amperage circuit to the starter motor is closed and activates the starter motor. When the key switch is released, the magnetic field is no longer supported and the magnetic field collapses. The return spring working on the plunger opens the disc contact, opening the circuit to the starter.

When the armature plate is out of position or the shift lever is moved into forward or reverse gear, the neutral switch is placed in the open position and the starter control circuit cannot be activated. This prevents the powerhead from starting while in gear.

TESTING

▶ See accompanying illustrations

The relay is usually trouble free. If there are suspected problems with the relay, first test the battery. Then look up the starter circuit in the wiring diagram

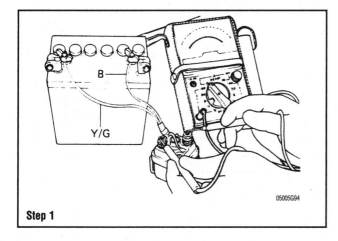

Step 1

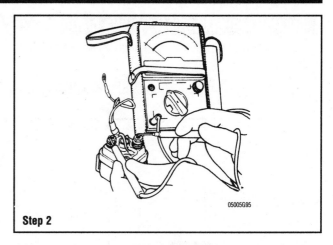

Step 2

for that particular model powerhead. Note that there are two circuits: the heavy amperage starter motor circuit, and the control circuit from the key switch and neutral switch. When the key is turned, listen for a click from the relay. If a click is heard, the control circuit is operating normally. Then test for voltage at the starter motor. If there is voltage at the starter motor, a reading of below 9 volts with no starter action indicates a bad starter or resistance in the circuit. A reading at or near battery voltage indicates that the starter has an open circuit inside.

If no click is heard at the relay, use a jumper wire to jump between the battery terminal of the relay and the "S" terminal. If it now works, the problem is in the control circuit. Using a multimeter, test for voltage at the neutral switch at the control (at both terminals). If there is no voltage, test at the key switch, with the key switch in the start position. Depending upon the specific model, there may be a fuse between the battery and the ignition switch.

1. To test the relay continuity, use a multimeter to check whether continuity exists between the two terminals of the relay with the battery connected.

2. And when the battery is disconnected.

REMOVAL & INSTALLATION

1. Disconnect the negative battery cable.
2. Remove the engine cover.
3. Disconnect the starter relay electrical lead at the bullet connector. Remove the ground wire.
4. Disconnect the battery and starter cables.
5. Remove the starter relay.

To install:

6. Install the starter relay.
7. Connect the battery and starter cables.
8. Connect the starter relay electrical lead at the bullet connector. Install the ground wire.
9. Install the engine cover.
10. Connect the negative battery cable.
11. Check the starting system for proper operation.

5-52 IGNITION AND ELECTRICAL SYSTEMS

IGNITION AND ELECTRICAL WIRING DIAGRAMS

The following diagrams represent the most popular models with the most popular optional equipment.

1988–96 DT2 and 1997 DT2.2 Wiring Diagram

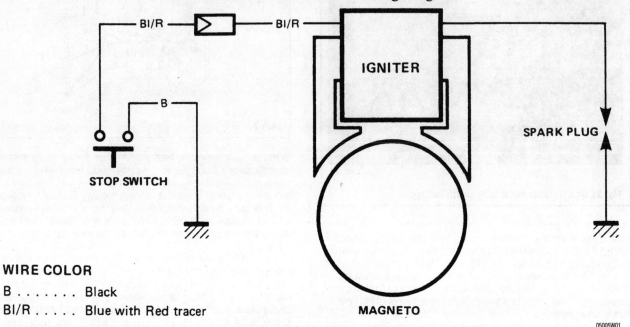

WIRE COLOR

B . . . . Black
Bl/R . . . . . Blue with Red tracer

1988–99 DT4 Wiring Diagram

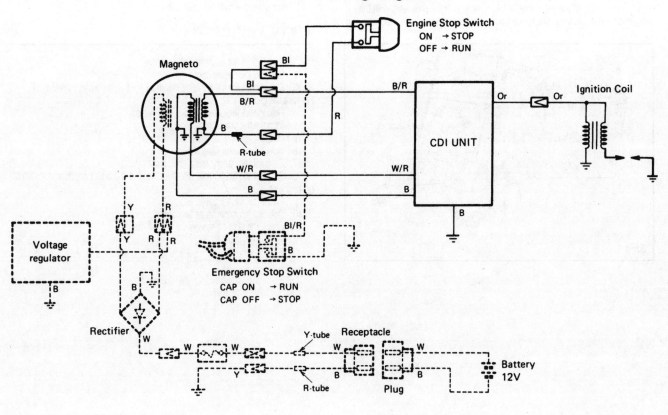

------ Optional

IGNITION AND ELECTRICAL SYSTEMS

1990–98 DT4 Wiring Diagram

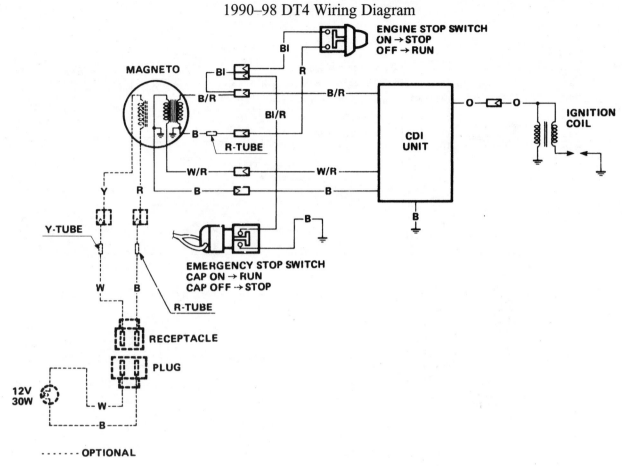

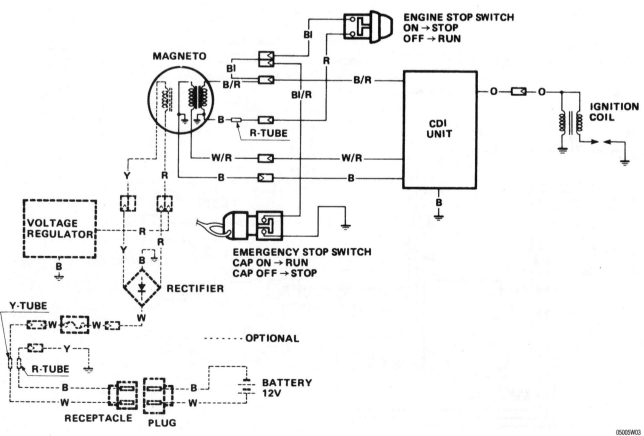

5-54 IGNITION AND ELECTRICAL SYSTEMS

1988 DT8 Wiring Diagram

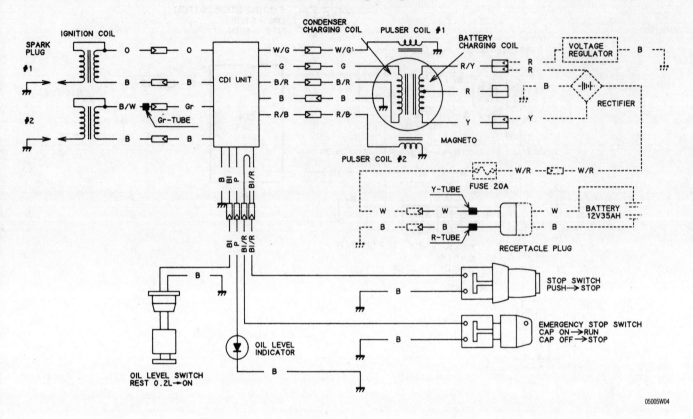

1989–91 DT8 Wiring Diagram

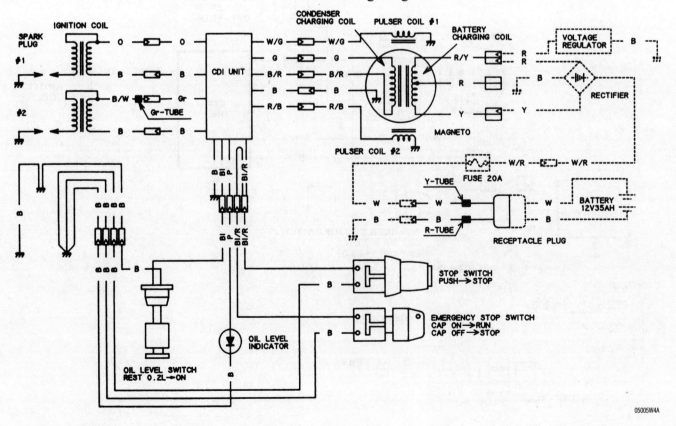

IGNITION AND ELECTRICAL SYSTEMS 5-55

1988–99 DT6 and 1982–97 DT8 Wiring Diagram

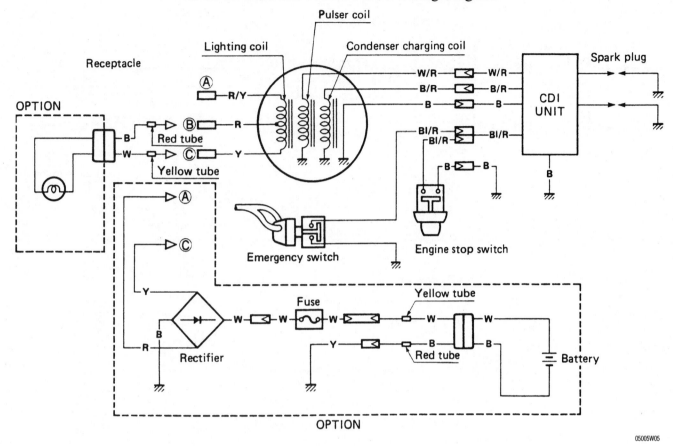

1988–97 DT9.9 and DT15 Wiring Diagram

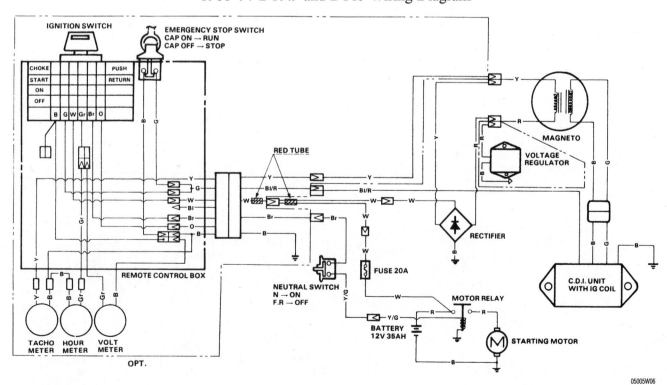

5-56 IGNITION AND ELECTRICAL SYSTEMS

1989–97 DT15C (Manual Start) Wiring Diagram

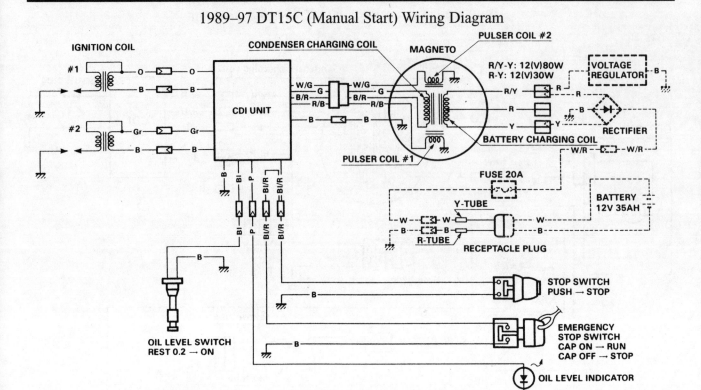

1989–97 DT15C (Electric Start) Wiring Diagram

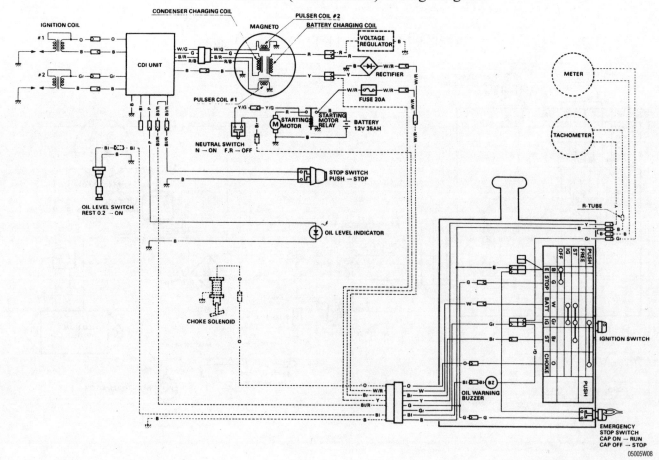

IGNITION AND ELECTRICAL SYSTEMS 5-57

1988 DT20 and DT25 (Manual Start) Wiring Diagram

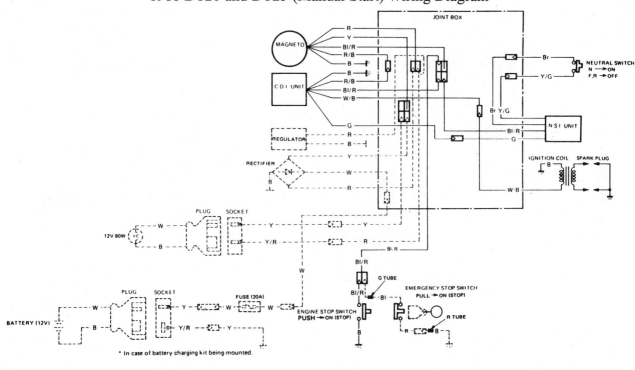

1988 DT20 and DT25 (Electric Start) Wiring Diagram

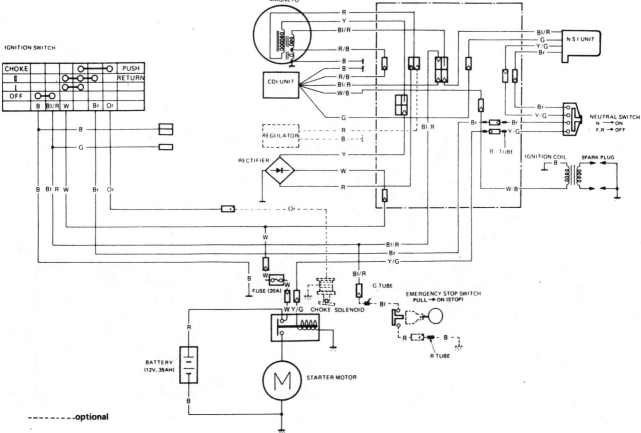

5-58 IGNITION AND ELECTRICAL SYSTEMS

1989 DT25C and DT30C Wiring Diagram

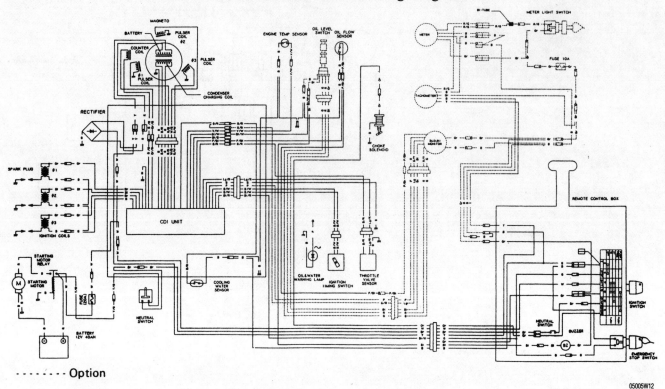

1990 DT25C and DT30C Wiring Diagram

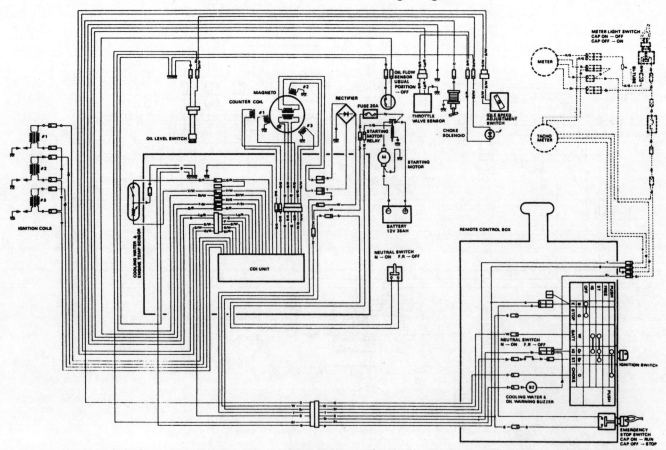

IGNITION AND ELECTRICAL SYSTEMS 5-59

1991 DT25C and DT30C Wiring Diagram

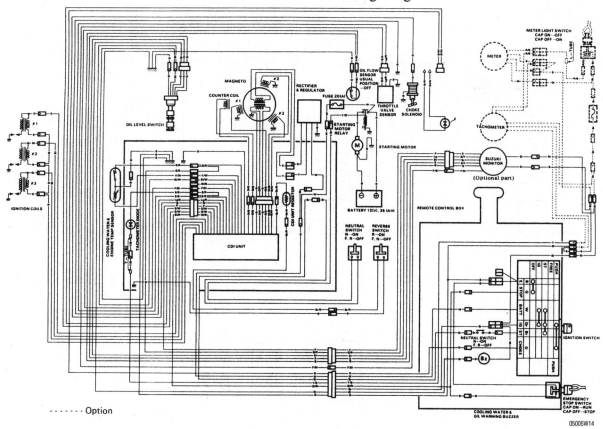

------ Option

1992–99 DT25 and 1992–97 DT30 Wiring Diagram

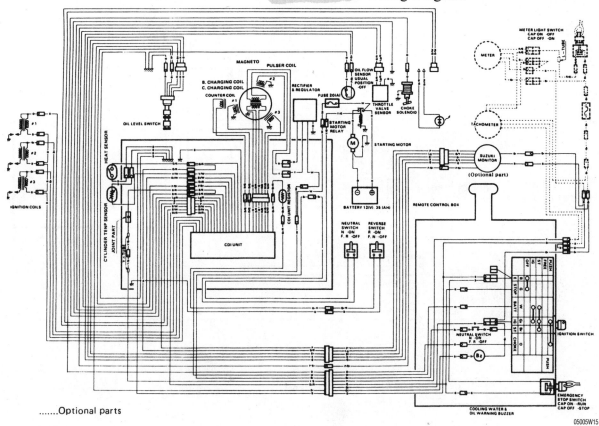

...... Optional parts

5-60 IGNITION AND ELECTRICAL SYSTEMS

1988–89 DT35TC and DT40TC Wiring Diagram

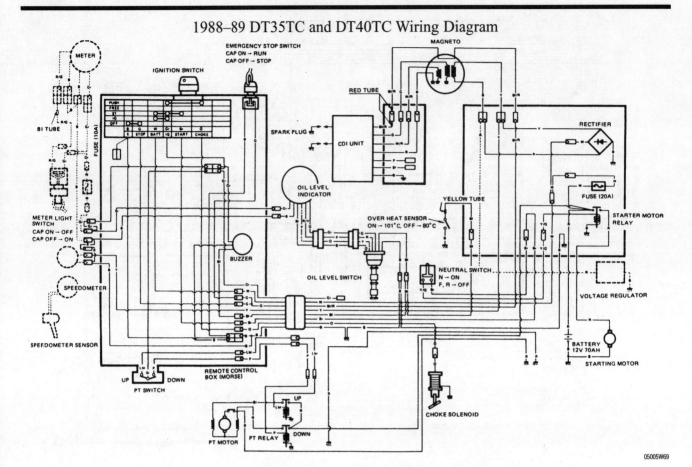

1988–89 DT35C and DT40C Wiring Diagram

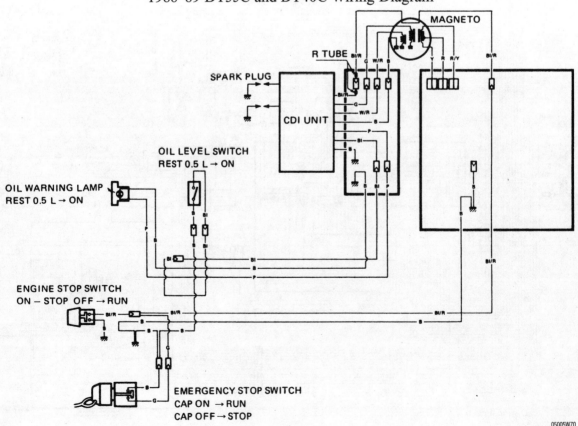

IGNITION AND ELECTRICAL SYSTEMS 5-61

1988–89 DT35CR and DT40CR Wiring Diagram

1990–91 DT40CE Wiring Diagram

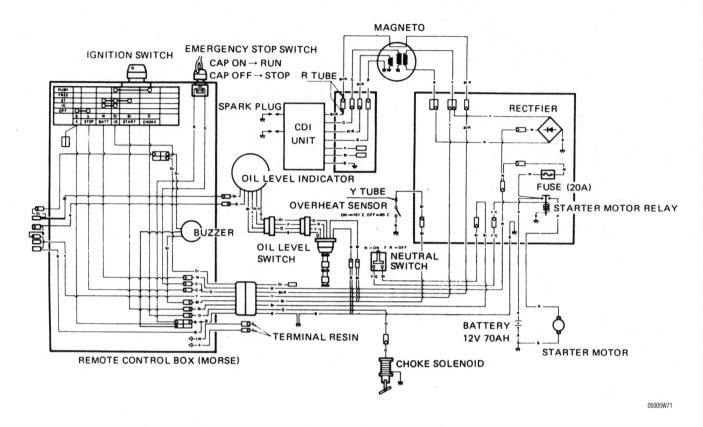

5-62 IGNITION AND ELECTRICAL SYSTEMS

1990–91 DT40CR Wiring Diagram

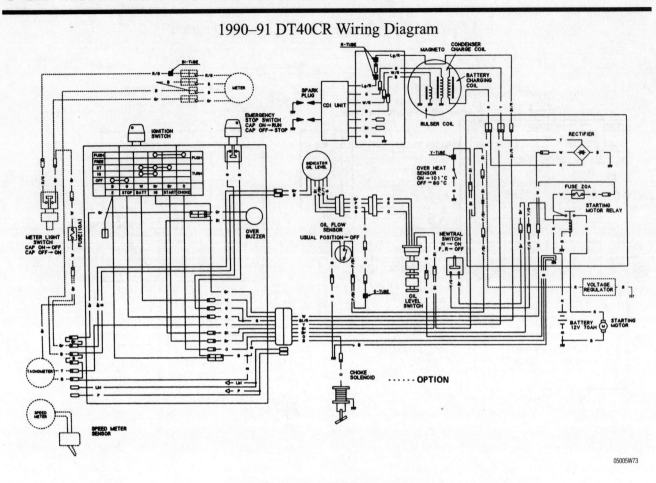

1990–91 DT40TC Wiring Diagram

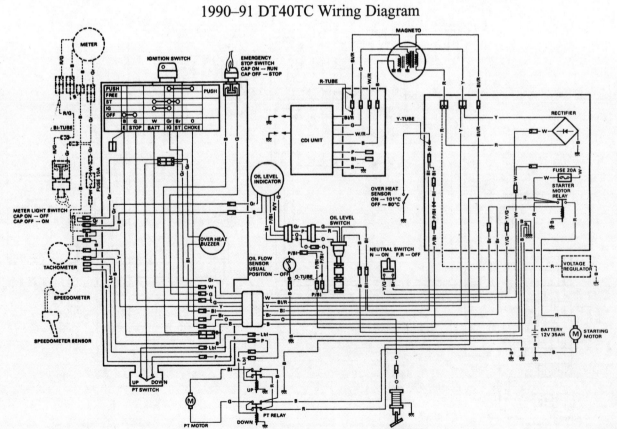

IGNITION AND ELECTRICAL SYSTEMS   5-63

1990–91 DT40C Wiring Diagram

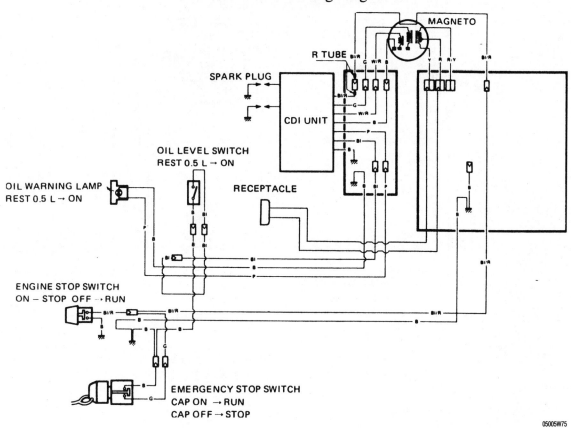

1992–98 DT40C Wiring Diagram

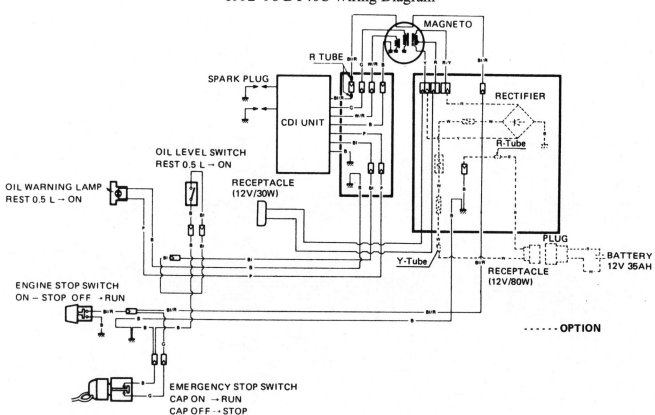

5-64 IGNITION AND ELECTRICAL SYSTEMS

1992–98 DT40CE Wiring Diagram

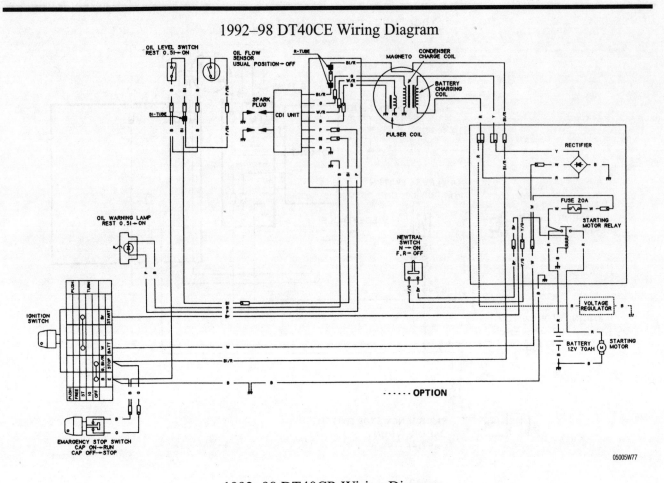

1992–98 DT40CR Wiring Diagram

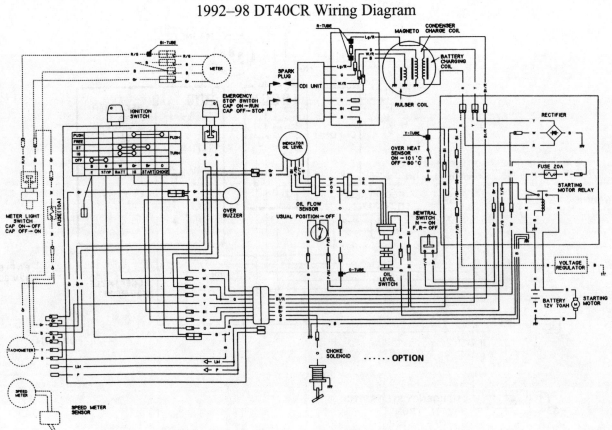

IGNITION AND ELECTRICAL SYSTEMS   5-65

1992–93 DT40TC Wiring Diagram

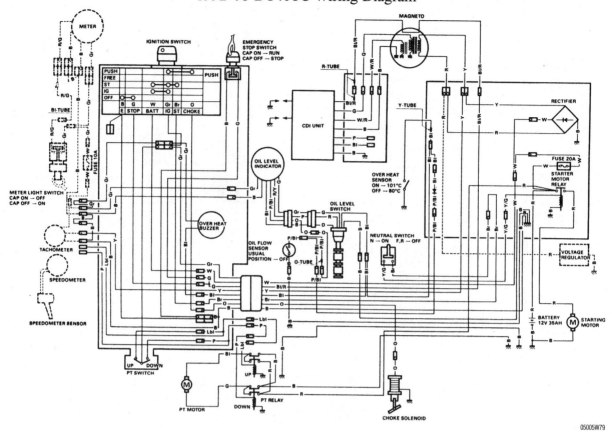

1988 DT55C and DT65C (Australia) Wiring Diagram

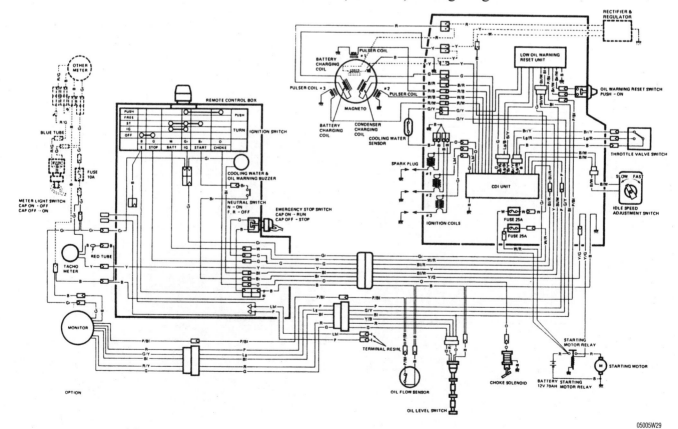

5-66 IGNITION AND ELECTRICAL SYSTEMS

1988 DT55TC and DT65TC (Australia) Wiring Diagram

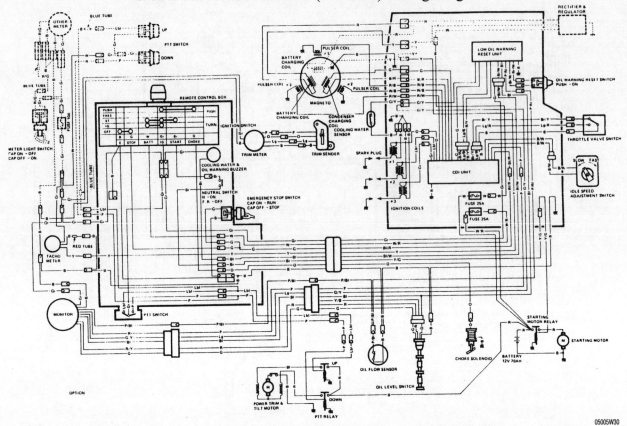

1988 DT55C and DT65C (Except Australia) Wiring Diagram

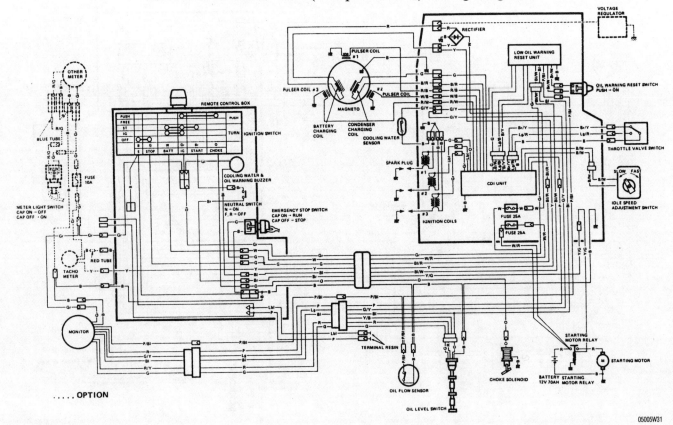

IGNITION AND ELECTRICAL SYSTEMS 5-67

1988 DT55TC and DT65TC (Except Australia) Wiring Diagram

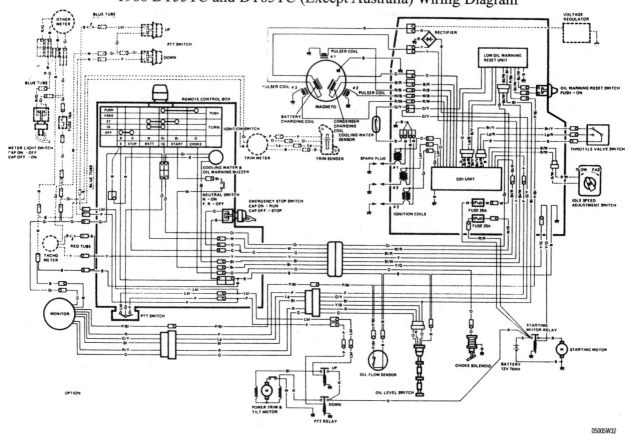

1988 DT55HTC Wiring Diagram

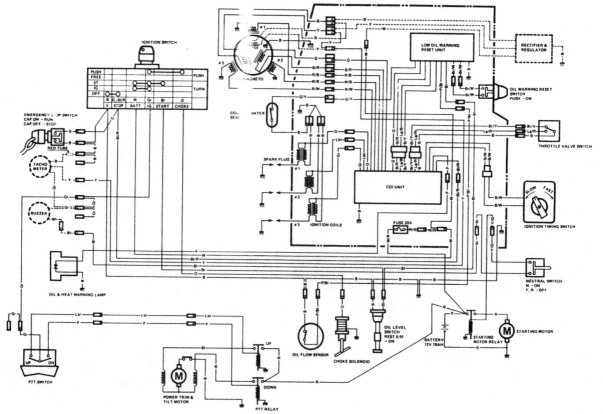

5-68 IGNITION AND ELECTRICAL SYSTEMS

1989 DT55 and DT65 Wiring Diagram

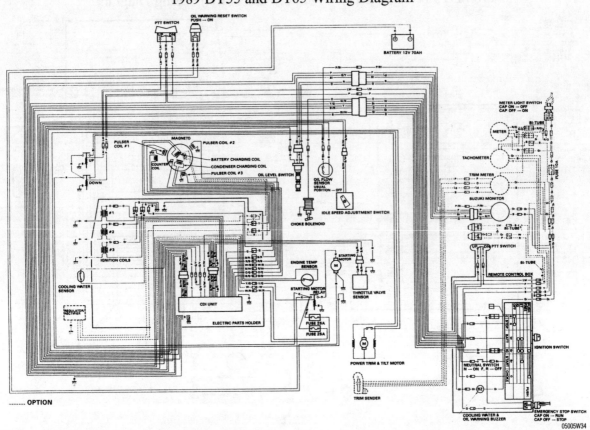

1990–91 DT55 and DT65 Wiring Diagram

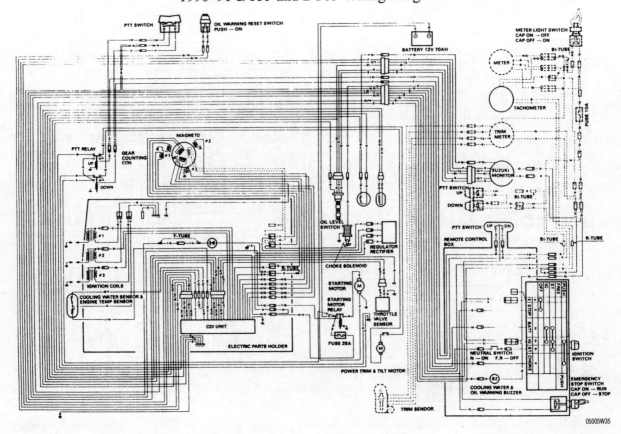

IGNITION AND ELECTRICAL SYSTEMS   5-69

1992 DT55 and DT65 Wiring Diagram

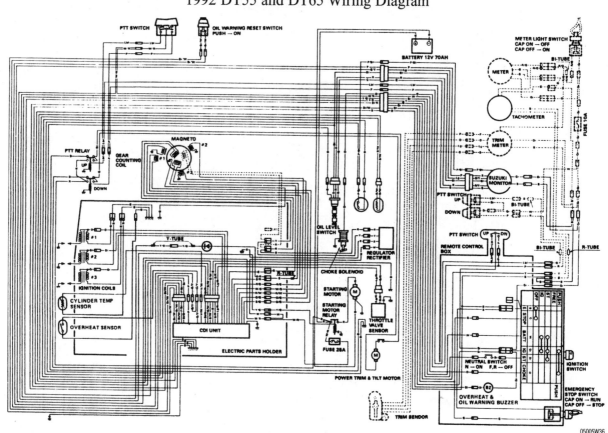

1990–91 DT55HTC Wiring Diagram

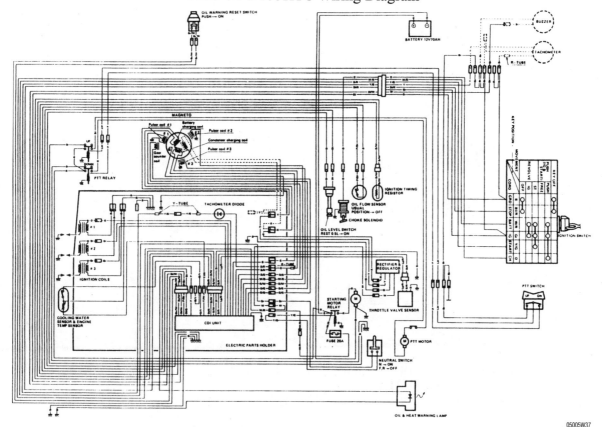

5-70 IGNITION AND ELECTRICAL SYSTEMS

1988 DT75 and DT85 Wiring Diagram

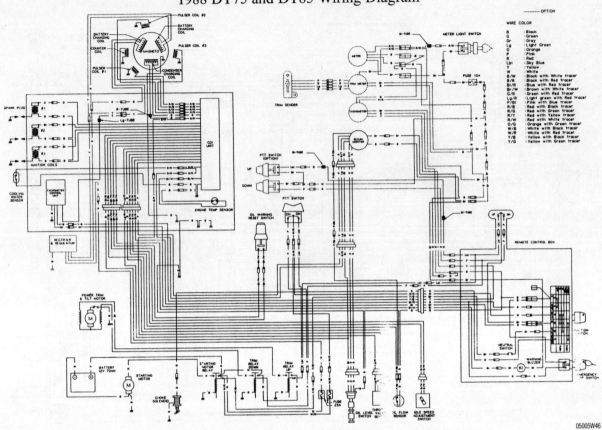

1988 DT75CR and DT85CR Wiring Diagram

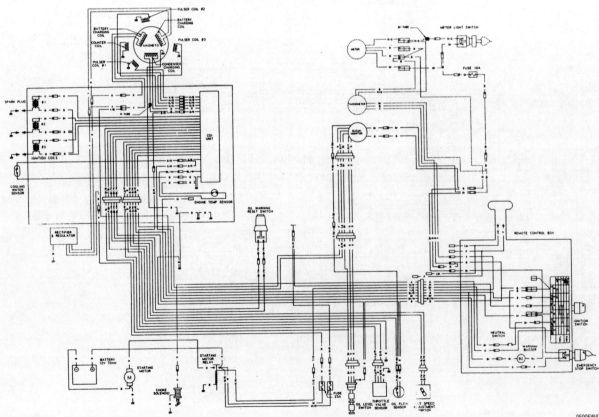

IGNITION AND ELECTRICAL SYSTEMS 5-71

1989 DT75CR and DT85CR Wiring Diagram

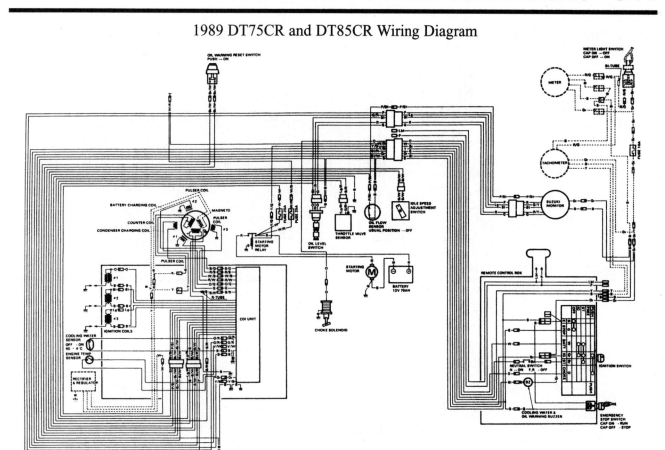

1989 DT75TC and DT85TC Wiring Diagram

------ : Option

NOTE: The tachometer and trim meter are standard equipments for Australian version.

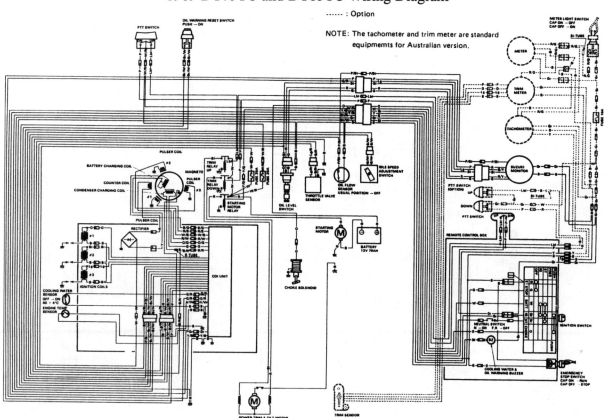

5-72 IGNITION AND ELECTRICAL SYSTEMS

1990–91 DT75 and DT85 Wiring Diagram

1992–94 DT75 and DT85 Wiring Diagram

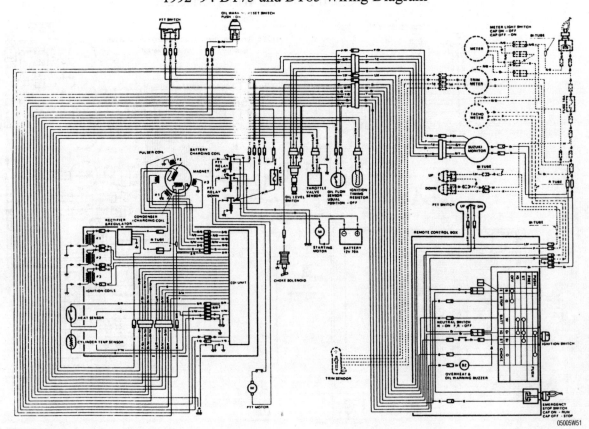

IGNITION AND ELECTRICAL SYSTEMS 5-73

1995–97 DT75 and 1995–99 DT85 Wiring Diagram

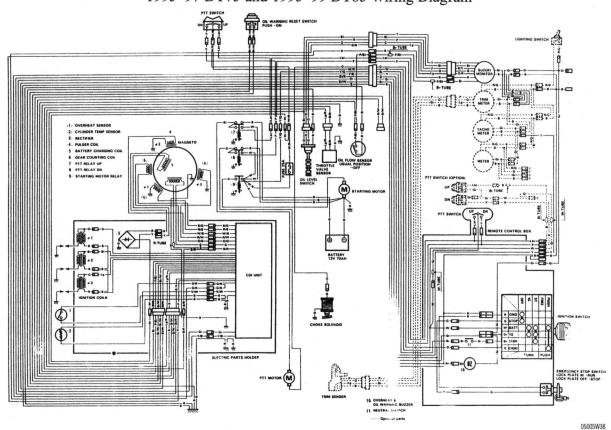

1989–91 DT90 and DT100 Wiring Diagram

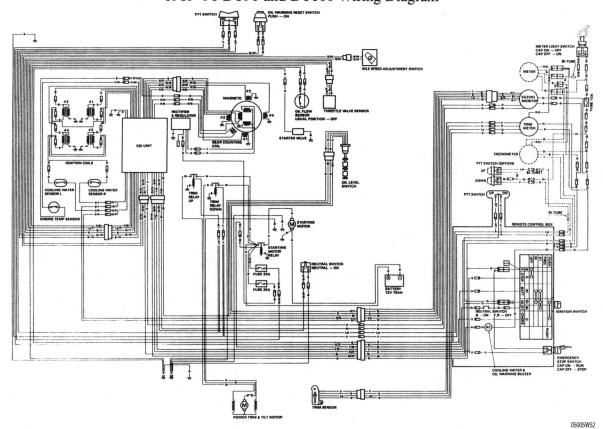

5-74 IGNITION AND ELECTRICAL SYSTEMS

1992–97 DT90 and 1992–99 DT100 Wiring Diagram

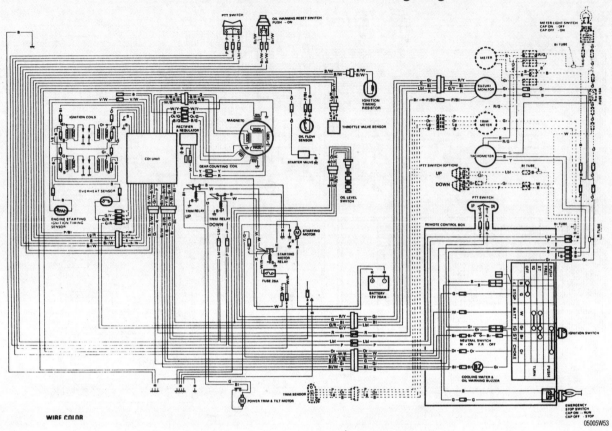

1988–89 DT115 and DT140 Wiring Diagram

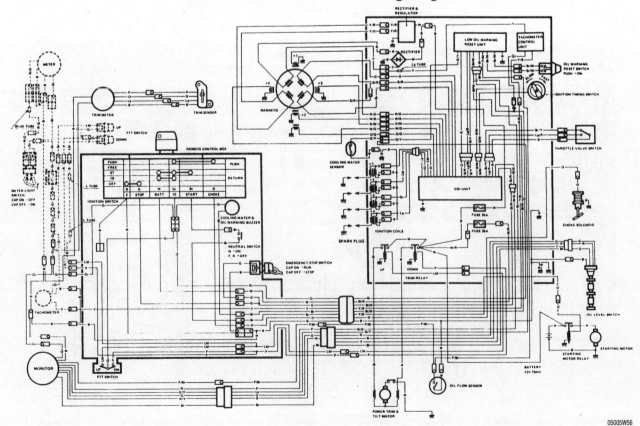

IGNITION AND ELECTRICAL SYSTEMS

1990 DT115 and DT140 Wiring Diagram

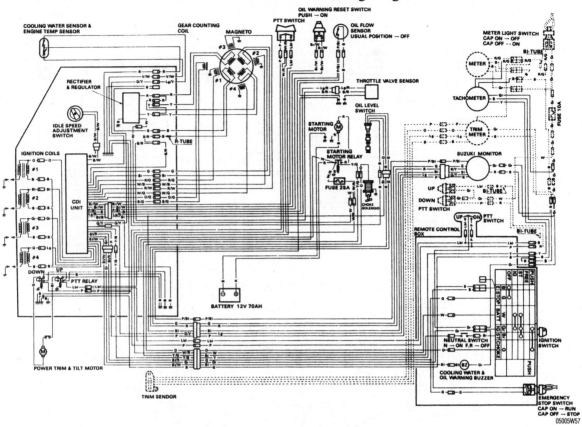

1991–92 DT115 and DT140 Wiring Diagram

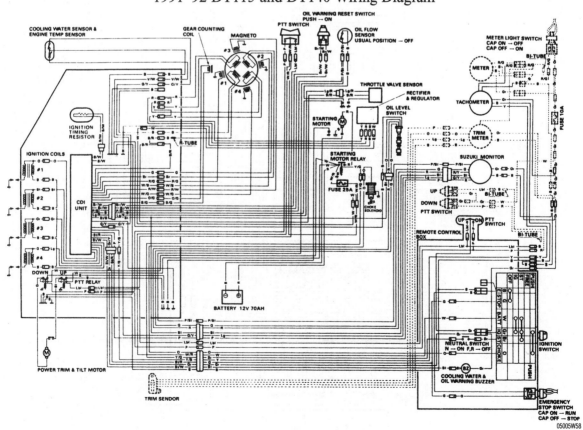

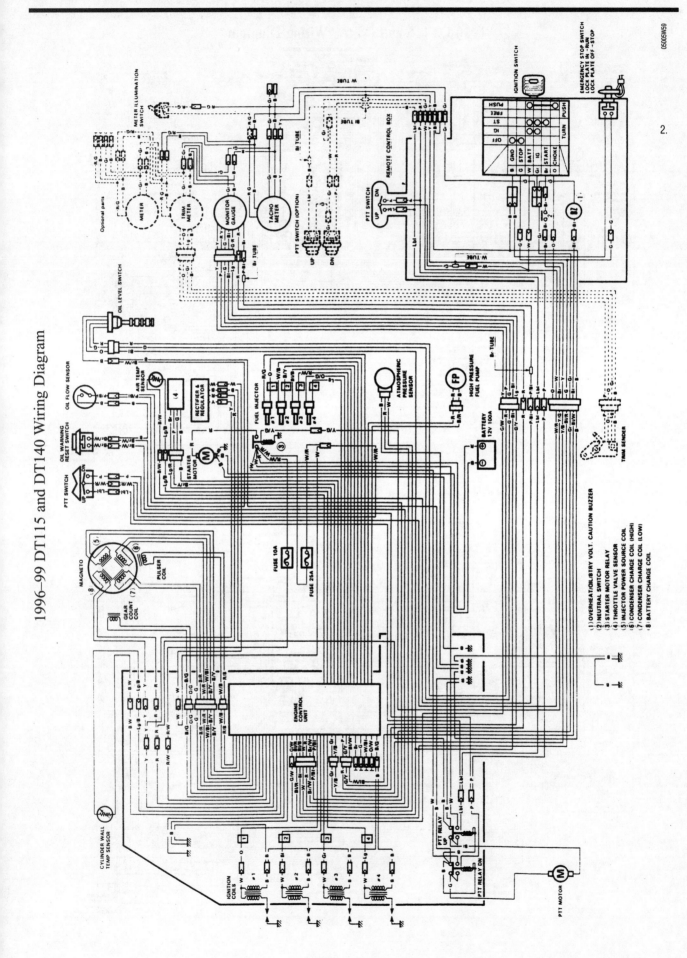

5-76 IGNITION AND ELECTRICAL SYSTEMS

1996–99 DT115 and DT140 Wiring Diagram

IGNITION AND ELECTRICAL SYSTEMS 5-77

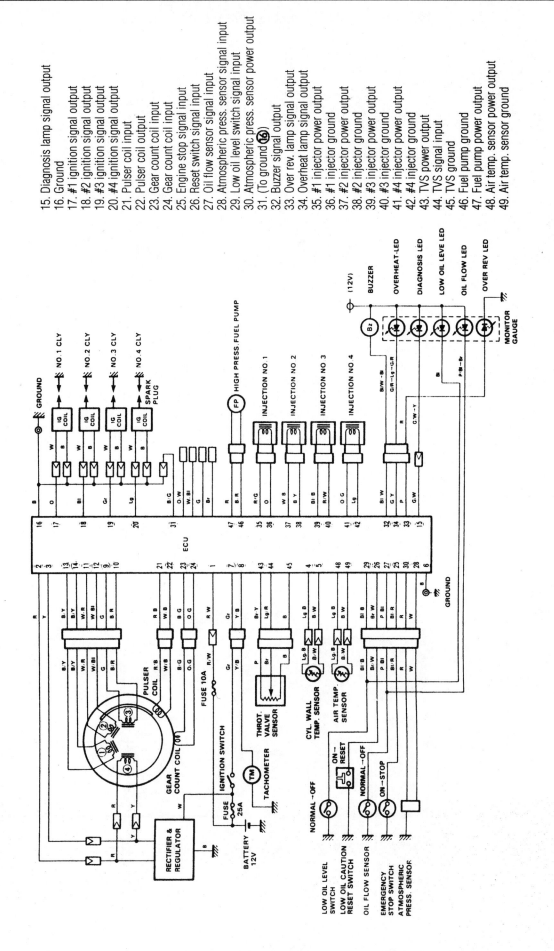

1996–99 DT115 and DT140 Engine Control System Wiring Diagram

5-78 IGNITION AND ELECTRICAL SYSTEMS

1988 DT150, DT175 and DT200 Wiring Diagram

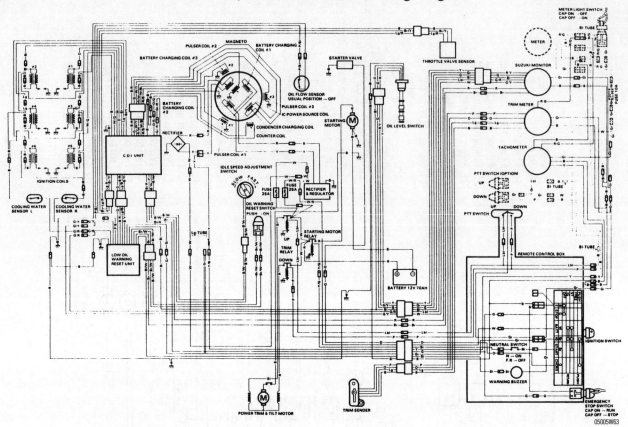

1989–90 DT150, DT175 and DT200 Wiring Diagram

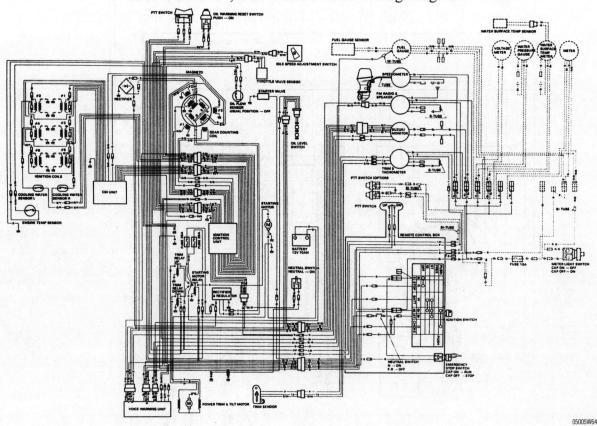

IGNITION AND ELECTRICAL SYSTEMS 5-79

1991–93 DT150, DT175 and DT200 Wiring Diagram

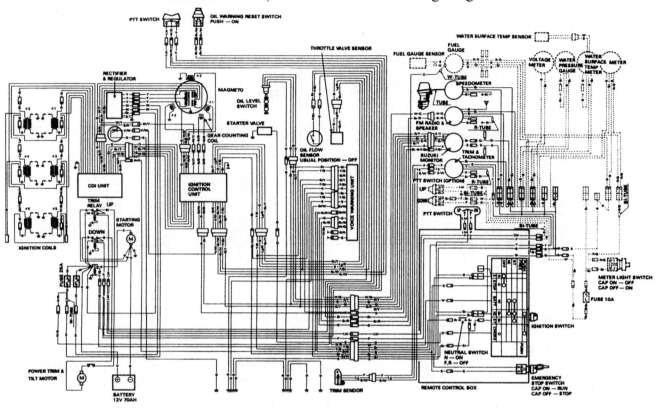

Remove the engine cover.

3. Disconnect the starter relay electrical relay lead at the

1994–99 DT150 and DT200 Wiring Diagram

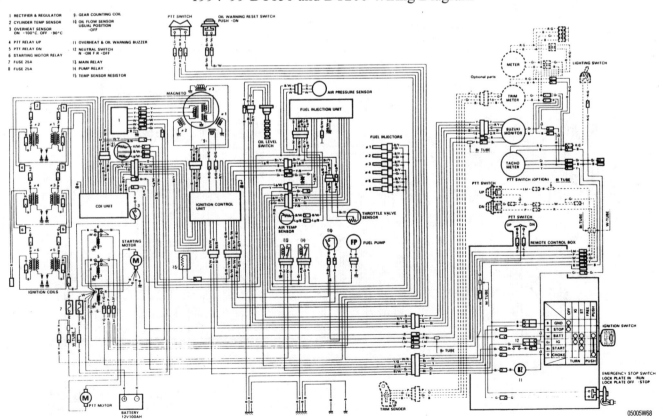

5-80 IGNITION AND ELECTRICAL SYSTEMS

DT150, DT175, DT200 Twin Remote Control Wiring Diagram

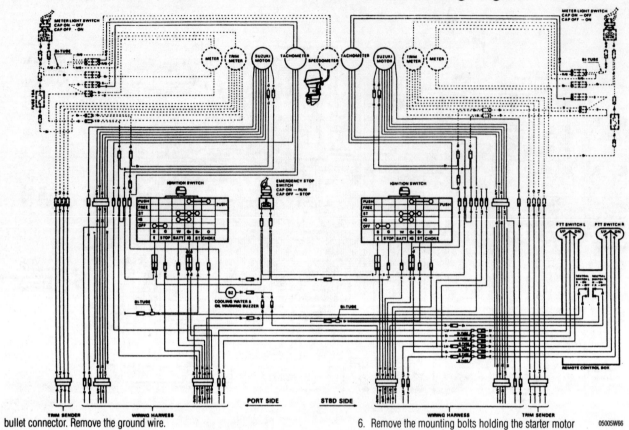

bullet connector. Remove the ground wire.

6. Remove the mounting bolts holding the starter motor

1990–99 DT225 Wiring Diagram

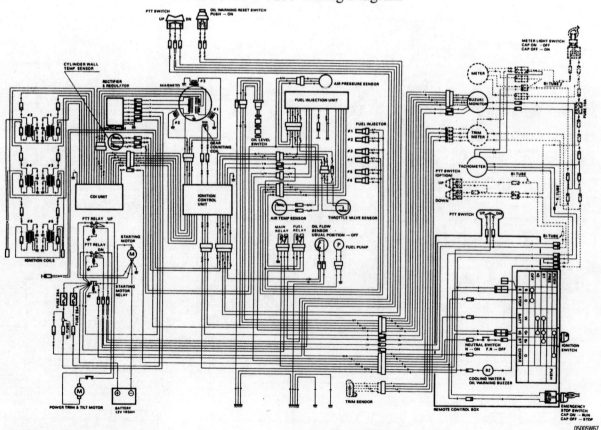

4. Disconnect the battery and starter cables.
5. If equipped, remove the starter motor clamp.

to the mounting bracket.
7. Remove the starter motor.

OIL INJECTION SYSTEM 6-2
DESCRIPTION AND OPERATION 6-2
    OIL PUMP 6-2
    AIR/OIL MIXING VALVE 6-2
TROUBLESHOOTING THE OIL INJECTION
  SYSTEM 6-2
BLEEDING THE OIL INJECTION
  SYSTEM 6-3
    PROCEDURE 6-3
OIL TANK 6-3
    REMOVAL & INSTALLATION 6-3
    CLEANING & INSPECTION 6-4
OIL PUMP 6-5
    REMOVAL & INSTALLATION 6-5
OIL LINES 6-6
    OIL LINE CAUTIONS 6-6
    REMOVAL & INSTALLATION 6-6
AIR/OIL MIXING VALVE 6-7
    REMOVAL & INSTALLATION 6-7
OIL PUMP DISCHARGE RATE 6-7
    TESTING 6-7
    ADJUSTMENT 6-8
OIL PUMP CONTROL ROD 6-8
    ADJUSTMENT 6-8
COOLING SYSTEM 6-11
DESCRIPTION AND OPERATION 6-11
    WATER PUMP 6-11
    THERMOSTAT 6-12
TROUBLESHOOTING THE COOLING
  SYSTEM 6-12
WATER PUMP 6-12
    REMOVAL & INSTALLATION 6-12
    CLEANING & INSPECTION 6-13
THERMOSTAT 6-13
    REMOVAL & INSTALLATION 6-13
    CLEANING & INSPECTION 6-14
**OIL INJECTION WARNING
SYSTEMS 6-14**
DESCRIPTION AND OPERATION 6-14
    LOW OIL LEVEL 6-14
    OIL FLOW 6-14
TROUBLESHOOTING THE OIL INJECTION
  WARNING SYSTEM 6-15
OIL LEVEL SENSOR 6-16
    TESTING 6-16
    REMOVAL & INSTALLATION 6-16
OIL FLOW SENSOR 6-17
    TESTING 6-17
    REMOVAL & INSTALLATION 6-17
    CLEANING & INSPECTION 6-17
**OVERHEAT WARNING
SYSTEM 6-17**
DESCRIPTION AND OPERATION 6-17
TROUBLESHOOTING THE OVERHEAT
  WARNING SYSTEM 6-17
OVERHEAT SENSOR 6-19
    REMOVAL & INSTALLATION 6-19
    TESTING 6-19

SPECIFICATIONS CHARTS
  OIL PUMP DISCHARGE RATE 6-8
TROUBLESHOOTING CHARTS
  OIL INJECTION WARNING
    SYSTEM 6-15
  OVERHEAT WARNING SYSTEM 6-18

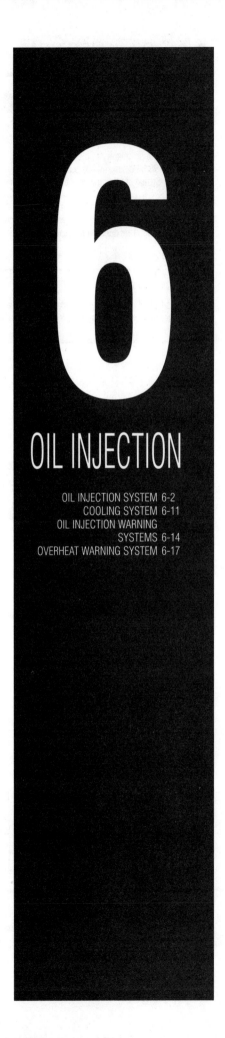

6
OIL INJECTION

OIL INJECTION SYSTEM 6-2
COOLING SYSTEM 6-11
OIL INJECTION WARNING
SYSTEMS 6-14
OVERHEAT WARNING SYSTEM 6-17

OIL INJECTION

OIL INJECTION SYSTEM

Description and Operation

Suzuki is the originator of outboard oil injection, introducing it to the marine industry in 1980. Because of years of experience in motorcycle oil injection technology, Suzuki applied many of those proven principles to their outboard motors.

The oil injection system automatically adjusts the amount of oil to a clean burning 150:1 fuel/oil ratio. Then as engine rpm increases, the ratio changes gradually to a 50:1 mixture at full throttle for greater lubrication.

OIL PUMP

▶ See Figures 1 and 2

The heart of oil injection system is a precision die cast Mikuni oil pump which is gear driven directly by the engine's crankshaft. This mechanical link between engine rpm and the pump means that oil is metered without hesitation in exactly the correct ratio at all throttle openings. To minimize smoke and possible carbon build-up at idle speeds

Once the oil leaves the Mikuni pump, it travels through oil lines to distribution nozzles located downstream of the carburetor or throttle body. The nozzles then inject the oil directly into the incoming air/fuel mixture. This thoroughly mixes the oil, air and fuel before it enters the combustion chamber in order to provide optimum lubrication.

AIR/OIL MIXING VALVE

▶ See Figure 3

On some models, a pre-atomization or mixing valve allows a small amount of crankcase pressure to be injected into the oil lines just before oil reaches the distribution nozzles. This forces thousands of tiny air bubbles into the oil flow before it gets to the combustion chamber. The atomized oil stream is therefore more uniformly distributed throughout the engine for superior lubrication.

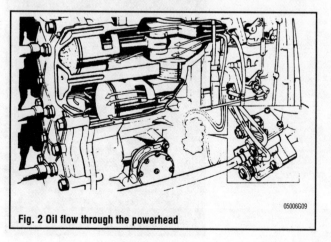

Fig. 1 Exploded view of a typical oil pump
① Oil pump
② Driven gear
③ Retainer
④ Oil pump control rod

Fig. 2 Oil flow through the powerhead

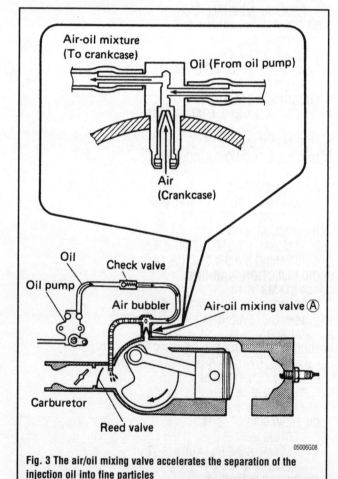

Fig. 3 The air/oil mixing valve accelerates the separation of the injection oil into fine particles

Troubleshooting the Oil Injection System

Like other systems on a 2-stroke engine, the oil injection system emphasizes simplicity. On pre-mix engines, when enough oil accumulated in the crankcase it passed into the combustion chamber where it burned along with the fuel. Since oil is a fuel just like gasoline, it burns as well. Unlike gasoline, oil doesn't burn as efficiently and may produce blue smoke that is seen coming out of the exhaust. If the engine was cold, such as during initial start-up, oil has an even harder time igniting resulting in excessive amounts of smoke.

OIL INJECTION 6-3

Other problems, such oil remaining in the combustion chamber, tend to foul spark plugs and cause the engine to misfire at idle. Most of these problems have been alleviated by the introduction of automatic oil injection systems.

➡ **One of the most common problems with oil injection systems is the use of poor quality injection oil. Poor quality oil tends to gel in the system, clogging lines and filters.**

It is normal for a 2-stroke engine to emit some blue smoke from the exhaust. The blue color of the smoke comes from the burning 2-stroke oil. An excessive amount of blue smoke indicates too much oil being injected into the engine. On Suzuki engines, this is usually caused by an incorrectly adjusted injection control rod.

If the exhaust smoke is white, this is a sign of water entering the combustion chamber. Water may enter as condensation or more seriously may enter through a defective head gasket or cracked head. Usually white smoke from condensation will disappear quickly as the engine warms.

If the exhaust smoke is black, this is a sign of an excessively rich fuel mixture or incorrect spark plugs. The black color of the smoke comes from the fuel burning.

Bleeding the Oil Injection System

PROCEDURE

◆ See Figure 4

1. Place the engine in the full upright position.
2. Remove the engine cover.
3. Connect the outboard to a fuel tank containing a 50:1 pre-mix fuel.
4. Fill the oil tank to the recommended level.

➡ **If the oil pump was removed, fill the oil lines with oil before reconnecting them to the pump fittings.**

5. Place a suitable container under the air bleed screw to catch the 2-stroke oil while bleeding.
6. Remove any access covers and open the air bleed screw on the injection pump 3 turns counterclockwise.
7. Start the engine and maintain engine idle speed during the bleeding process.

➡ **To speed the bleeding process it is possible to disconnect the oil pump link arm and move it to the full output position.**

8. Let the engine idle until the oil leaving the oil pump air bleed screw is free of air bubbles and no air can be seen in the clear plastic oil injection lines.

➡ **On engines using solid color lines, the lines should be disconnected an allowed to drain into a pan.**

9. Stop the engine and retighten the screw.
10. Install the engine cover.

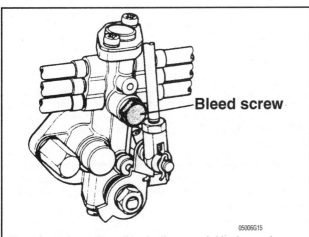

Fig. 4 On engines using solid color lines, an air bleed screw is sometimes used to bleed the system

Oil Tank

REMOVAL & INSTALLATION

◆ See accompanying illustrations

※※ **WARNING**

Proper oil line routing and connections are essential for correct oil injection system operation. The line connections to the powerhead and oil pump look the same but may contain check valves of differing calibrations. Oil lines must be installed between the pump and powerhead correctly and connected to the proper fittings on the intake manifold in order for the system to operate properly.

1. Turn the battery switch off and/or disconnect the negative battery cable.
2. Remove the engine cover.
3. Position a suitable container under the lower cowling to receive oil drained from the tank.
4. Matchmark the oil line for installation reference.
5. Squeeze the oil line to restrict the flow of oil while pulling it free from the fitting.
6. Remove the clamp with a pair of pliers.

Step 5

Step 6

6-4 OIL INJECTION

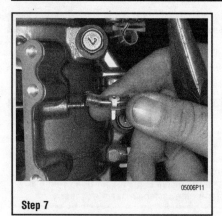
Step 7

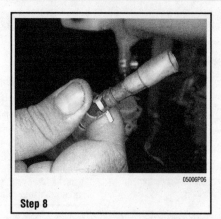
Step 8

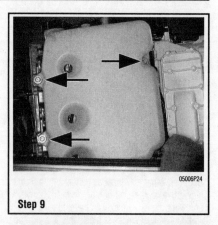
Step 9

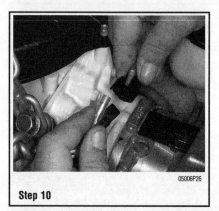
Step 10

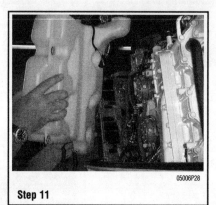
Step 11

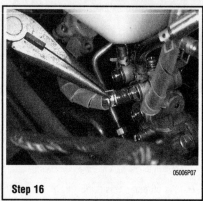
Step 16

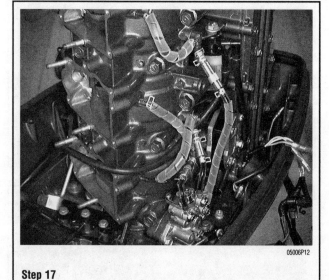
Step 17

7. Slide the clamp up along the oil line to get it out of the way.
8. Cap the end of the oil line to prevent oil leakage and cap the fitting at the oil pump to prevent dirt entering the system.
9. Remove the fasteners attaching the oil tank to the powerhead.
10. Lift the oil tank away from the powerhead slowly and disconnect the oil level sensor electrical harness.
11. Remove the tank from the powerhead.

To install:
12. Position the oil tank on the powerhead after connecting the oil level sensor electrical harness.
13. Install the oil tank and tighten the fasteners securely.
14. Squeeze the oil line to restrict the flow of oil while installing.
15. Remove the caps and install the oil line at its original location.

✴✴ WARNING

Oil lines must be reinstalled with the same type of clamps as removed. The use of a screw type clamp will damage the vinyl line while a tie strap will not provide sufficient clamping pressure.

16. Install the clamp with a pair of pliers and push the clamp up past the rib on the nipple.
17. Ensure all the original oil line protection, is put back in place.
18. Turn the battery switch off and/or disconnect the negative battery cable.
19. Start the powerhead and bleed the air from the oil injection system.
20. Check for proper oil injection system function.
21. Install the engine cover.

CLEANING & INSPECTION

One of the most common problems with oil injection systems is the use of poor quality injection oil. Poor quality oil tends to gel in the system, clogging lines and filters. If this is found to be the case with your system, or if the powerhead has been sitting in storage for a length of time, it is wise to remove the oil tank and clean it with solvent.

While the oil tank is removed, take the opportunity to inspect it for damage and replace it as necessary. The oil tank is the only source of oil for the powerhead. If it should leak, the powerhead will eventually run out of injection oil, with catastrophic and costly results. Remember, there are no parts stores when you are miles out at sea.

OIL INJECTION

Oil Pump

REMOVAL & INSTALLATION

♦ See accompanying illustrations

※※ WARNING

Proper oil line routing and connections are essential for correct oil injection system operation. The line connections to the powerhead and oil pump look the same but may contain check valves of differing calibrations. Oil lines must be installed between the pump and powerhead correctly and connected to the proper fittings on the intake manifold in order for the system to operate properly.

1. Turn the battery switch off and/or disconnect the negative battery cable.
2. Remove the engine cover.
3. Position a suitable container under the lower cowling to receive oil drained from the tank.
4. Remove the oil tank.
5. Disconnect the oil pump control rod at the throttle lever.
6. Mark the exact location of each oil line for installation reference.
7. Remove the oil lines and cap the ends to prevent oil leakage.
8. Cap the fittings at the oil pump to prevent dirt from entering the system.
9. On models equipped with banjo fittings, only remove the fittings if they are to be replaced.
10. Loosen the oil pump mounting hardware.
11. Remove the oil pump from powerhead.
12. Carefully pry the driven gear retainer and gasket loose using a putty knife.
13. Remove the retainer, gasket and driven gear assembly from the powerhead.

To install:
14. Clean the gasket mating surfaces thoroughly.
15. If the pump is to be replaced but the oil lines reused, matchmark the location of all oil lines on the pump for installation reference.
16. If the pump and oil lines are to be replaced, ensure the oil lines are filled with 2-stroke oil prior to starting the powerhead.
17. Install the retainer and driven gear assembly on the powerhead using a new gasket.
18. Position the oil pump on the powerhead and tighten the mounting hardware securely.
19. Connect the oil pump control rod at the throttle lever.
20. If the banjo fittings were removed, always use new gaskets during installation.
21. Position the banjo fittings as illustrated and tighten securely.
22. If the oil lines are to be reused, trim a small amount (¼ in.) off the end prior to connecting to ensure a tight fit.
23. Connect the oil lines to their proper locations.
24. Install the oil tank.
25. Bleed the oil injection system.
26. Connect the negative battery cable and/or turn the battery switch on.
27. Check the oil pump discharge rate.
28. Install the engine cover.

Step 4

Step 5

Step 9

Step 10

Step 11

6-6 OIL INJECTION

Oil Lines

OIL LINE CAUTIONS

1. Do not bend or twist the oil lines when installing.
2. When installing clips, position the tabs toward the inside and make sure they are not in contact with other parts.
3. Check the oil lines, when installed in position, do not come in contact with rods and lever during engine operation.
4. Insall the oil sensor and oil lines with the sensor's arrow mark pointing to the oil pump side.
5. Secure all valves and sensors using their original fasteners.
6. Install hose protectors in their original positions.
7. Extreme caution should be taken not to scratch or damage oil lines.
8. Do not excessively compress an oil line when installing clamps.
9. Always use factory type clamps when installing fuel lines. Never use screw type clamps.
10. When installing the oil tank, ensure oil lines will not be pinched between the tank and the powerhead.

REMOVAL & INSTALLATION

♦ See Figures 5 thru 12

Proper oil line routing and connections are essential for correct oil injection system operation. The line connections to the powerhead and oil pump look the same but may contain check valves of differing calibrations. Oil lines must be installed between the pump and powerhead correctly and connected to the proper fittings on the intake manifold in order for the system to operate properly.

1. Turn the battery switch off and/or disconnect the negative battery cable.
2. Remove the engine cover.
3. Position a suitable container under the lower cowling to receive oil drained from the tank.
4. Remove the oil tank.
5. Mark the exact location of each oil line for installation reference.
6. Remove the clamps with a pair of pliers and push the clamps up along the oil lines.
7. Cap the ends of the oil lines to prevent oil leakage and cap the fitting at the oil pump to prevent dirt entering the system.

To install:

8. If the oil lines are to be reused, trim a small amount (¼ in.) off the end prior to connecting to ensure a tight fit.
9. Ensure the oil lines are filled with 2-stroke oil prior to starting the powerhead.
10. Connect the oil lines to their proper locations.
11. Position the clamps using pliers.

➡ **Do not use screw type clamps as they may damage the oil lines and cause oil leaks.**

12. Install the oil tank.
13. Bleed the oil injection system.
14. Connect the negative battery cable and/or turn the battery switch on.
15. Check the oil pump discharge rate.
16. Install the engine cover.

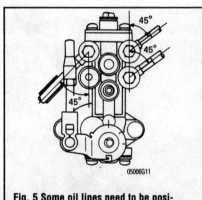

Fig. 5 Some oil lines need to be positioned at specific angles to the oil pump.

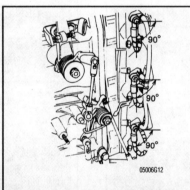

Fig. 6 . . . and to the power head for proper oil flow

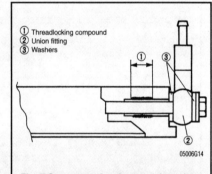

Fig. 7 Some powerheads use outlet unions for oil hose connection. Properly position the unions for leak proof operation

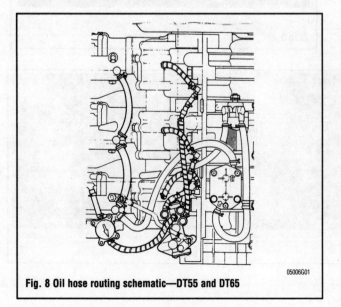

Fig. 8 Oil hose routing schematic—DT55 and DT65

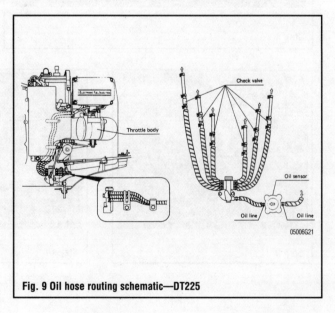

Fig. 9 Oil hose routing schematic—DT225

OIL INJECTION 6-7

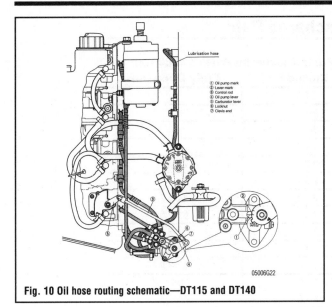

Fig. 10 Oil hose routing schematic—DT115 and DT140

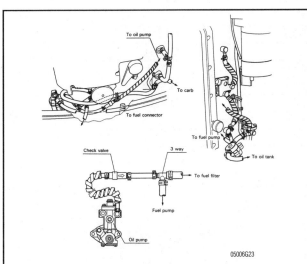

Fig. 11 Single point oil injection system hose routing schematic—DT75 and DT85

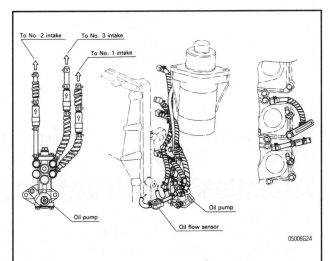

Fig. 12 Multi point oil injection system hose routing schematic—DT75 and DT85

Air/Oil Mixing Valve

REMOVAL & INSTALLATION

1. Remove the engine cover.
2. Position a suitable container to receive oil discharged when the hoses are disconnected.
3. Disconnect and plug the oil lines connected to the valve.
4. Remove the valve from the powerhead.

To install:
5. Clean and inspect the valve. Replace as necessary.
6. Install the valve on the powerhead.
7. Connect the oil lines to the valve and secure them with the clamps, as required.
8. Install the engine cover.

Oil Pump Discharge Rate

TESTING

DT8, DT9.9 and DT15

1. Connect the outboard to a fuel tank containing a 50:1 pre-mix fuel.
2. Disconnect the oil line from the reservoir to the oil pump.
3. Cap the line to prevent oil from spilling during this procedure.
4. Install an engine oil measuring cylinder (09900-21602) or equivalent graduated cylinder on the powerhead.
5. Fill the cylinder with oil and bleed the oil injection system.
6. Install a tachometer, start the engine and maintain an engine speed of 1500 rpm.
7. Select a clean-cut value on the scale of the oil measuring cylinder as a reference point.
8. Run the engine at 1500 rpm for the next 5 minutes.
9. Note the amount of oil discharged by the pump. This is the volume of oil from your reference point to the point at the end of 5 minute period.
10. Compare the discharged oil volume with the "Oil Pump Discharge Rate" chart

➥**Oil pump output test results may vary slightly depending on temperature and testing conditions.**

11. If the oil discharge rate is not within specification, check the injection lines for possible leaks.
12. If the injection lines are dry, the oil pump may be faulty.

Except DT8, DT9.9 and DT15

1. Connect the outboard to a fuel tank containing a 50:1 pre-mix fuel.
2. Disconnect the oil line from the reservoir to the oil pump.
3. Cap the line to prevent oil from spilling during this procedure.
4. Install an engine oil measuring cylinder (09941-8710 or 0990-21602) or equivalent graduated cylinder on the powerhead.
5. Fill the cylinder with oil and bleed the oil injection system.
6. Install a tachometer, start the engine and maintain an engine speed of 1500 rpm.
7. Select a clean-cut value on the scale of the oil measuring cylinder as a reference point.
8. Manually position the oil pump control rod in the fully closed position.
9. Run the engine at 1500 rpm for the next 5 minutes.
10. Note the amount of oil discharged by the pump. This is the volume of oil from your reference point to the point at the end of 5 minute period.
11. Refill the oil measuring cylinder.
12. Select a clean-cut value on the scale of the oil measuring cylinder as a reference point.
13. Manually position the oil pump control rod in the fully open position.
14. Run the engine at 1500 rpm for the next 2 minutes.
15. Note the amount of oil discharged by the pump.
16. Compare the discharged oil volume from the two tests with the "Oil Pump Discharge Rate" chart.

OIL INJECTION

Oil Pump Discharge Rate

Year	Model	Oil Pump Discharge Rate @ 1500 rpm Fully Closed (ml)	Fully Open (ml)
1988-97	8	1.2-1.8	-
1988-97	9.9	1.2-1.8	-
1988-97	15	1.5-2.5	-
1988-99	25, 30	1.0-1.9	1.2-1.7
1988-97	35, 40	1.3-2.3	2.9-4.4
1988-89	55	1.9-3.2	3.4-5.1
1988-98	65	1.9-3.2	4.0-5.9
1988-97	75	2.2-3.7	5.0-7.5
1988-97	85	2.2-3.7	5.9-8.7
1988-99	90, 100	2.5-4.5	6.0-9.0
1988-96	115	3.8-6.8	6.5-7.9
1996-99	115	4.4-7.9	7.6-11.3
1988-99	140	3.8-6.8	8.0-12.0
1996-99	140	4.4-7.9	9.4-14.0
1988-94	150	4.7-8.7	11.5-17.2
1995-97	150	3.4-6.0	8.1-12.1
1998-99	150TC	12.1-14.5	12.9-15.8
1985-97	150STC	3.6-4.8	7.6-9.3
1988-92	175	4.7-8.7	13.3-18.0
1988-97	200	4.7-8.7	13.3-18.0
1998-99	200	5.4-7.8	12.9-15.8
1990-99	225	4.7-8.7	13.3-18.0

➡ Oil pump output test results may vary slightly depending on temperature and testing conditions.

17. If the oil discharge rate is not within specification, check the injection lines for possible leaks.
18. If the injection lines are dry, the oil pump may be faulty.

ADJUSTMENT

▶ See Figure 13

➡ Oil pump control discharge adjustment should only be necessary if the oil pump is removed for service.

Throttle linkage adjustment is an integral step in adjusting the oil control rod.
1. Remove the engine cover
2. Move the throttle to the wide-open position.
3. Ensure that the full-throttle mark on the oil pump control lever aligns with the matchmark on the housing.
4. Close the throttle.
5. Ensure that the closed-throttle mark on the oil pump control lever aligns with the matchmark on the housing.
6. If the marks do not align adjust the control rod or cable as necessary.
7. On rod operated systems, disconnect the control rod from the lever. Loosen the locknut and rotate the rod end as necessary to bring the matchmarks into adjustment. Tighten the locknut and connect the control rod.
8. On cable operated systems, loosen the locknut and rotate the cable adjusting nut as necessary to bring the matchmarks into adjustment. Tighten the locknut.
9. Check the oil pump discharge rate.
10. Install the engine cover.

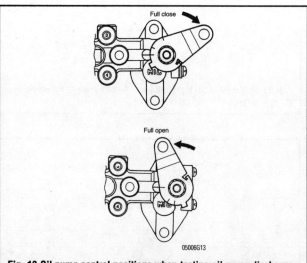

Fig. 13 Oil pump control positions when testing oil pump discharge. Note matchmarks on pump lever and pump body

Oil Pump Control Rod

ADJUSTMENT

Oil pump control rod adjustment should only be necessary if the oil pump is removed for service.

OIL INJECTION 6-9

Except DT55, DT65, 1996-99 DT115 and DT140

♦ See Figures 14, 15 and 16

➡ The throttle linkage must be adjusted prior to performing this procedure. For more information on throttle linkage adjustment, refer to the "Fuel System" section.

1. Remove the control rod from the anchor pins on the oil pump lever and the throttle body lever.
2. Loosen the locknuts on the control rod.
3. Adjust the control rod initial length measured between the center of each hole in the clevis ends. Suzuki provides dimensions for the following models:
   - DT25 and DT30—3.05 in. (77.5mm)
   - DT90, DT100 and V4—4.9 in. (124mm)
   - DT175, 1988-97 DT150 and 1988-93 DT200—5.4 in. (137.5mm)
   - 1984-97 DT200—6.3 in. (113.5mm)
   - 1998-99 DT150—6.6 in. (168.5mm)
   - 1998-99 DT200—6.5 in. (166.0mm)
4. Install the control rod onto the anchor pins on the oil pump lever and the throttle body lever.
5. Move the throttle to the fully closed position.

Fig. 14 Oil pump control rod adjustment— DT90, DT100 and V4

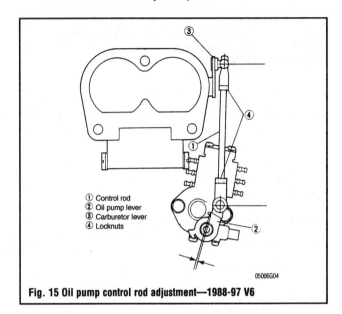

Fig. 15 Oil pump control rod adjustment—1988-97 V6

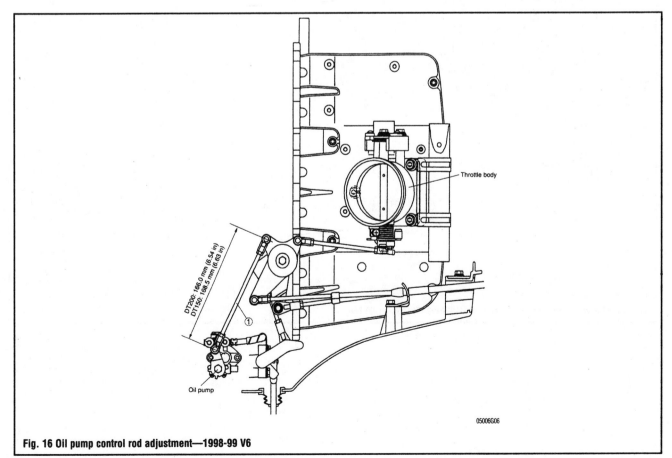

Fig. 16 Oil pump control rod adjustment—1998-99 V6

6-10 OIL INJECTION

6. The clearance between the oil pump lever and the lever stop should be 0.04 in. (1mm).
7. If the clearance is not as specified, readjust the control rod to achieve the proper clearance.
8. Hold the connectors and control in place and tighten the locknuts securely.
9. The control rod clevis ends should be positioned at 90° to each other after tightening the lock nuts.
10. Check for proper oil pump discharge rate.

DT 55 and DT 65

▶ See Figure 17

➥ The throttle linkage must be adjusted prior to performing this procedure. For more information on throttle linkage adjustment, refer to the "Fuel System" section.

1. Remove the engine cover.
2. Loosen the locknuts on the control rod.
3. Move the throttle to the fully closed position.
4. The clearance between the oil pump lever and the lever stop should be 0.04 in. (1mm).
5. If the clearance is not as specified, adjust the control rod to achieve the proper clearance.

1996-99 DT115 and DT140

▶ See Figure 18

➥ The throttle linkage must be adjusted prior to performing this procedure. For more information on throttle linkage adjustment, refer to the "Fuel System" section.

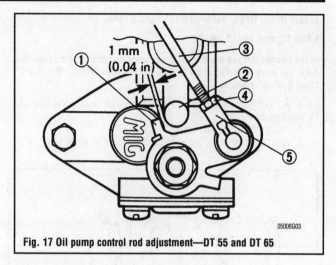

Fig. 17 Oil pump control rod adjustment—DT 55 and DT 65

The oil pump is controlled by a cable on these models.
1. Remove the engine cover.
2. Loosen the locknuts on the cable.
3. Adjust the locknuts until the lever just touches the stopper. This is the fully closed position.
4. Readjust the locknuts to pull the lever open 0.12in (3mm). This is equal to three turns of the locknuts.
5. Tighten the locknuts securely.
6. Ensure the matchmark on the lever aligns with the mark on the pump. If the marks do not align, readjust the cable length.

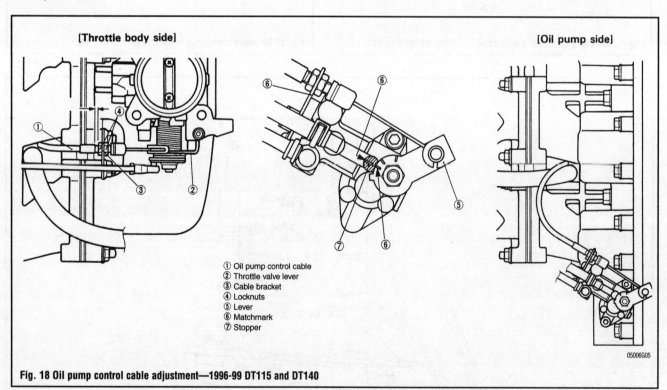

Fig. 18 Oil pump control cable adjustment—1996-99 DT115 and DT140

OIL INJECTION 6-11

COOLING SYSTEM

Description and Operation

♦ See Figures 19, 20 and 21

Water cooling is the most popular method in use to cool outboard powerheads. A "raw-water" type pump delivers seawater to the powerhead, circulating it through the cylinder head(s), the thermostat, the exhaust housing, and back down through the outboard. The water runs down the exhaust cavity and away, either through an exhaust tube or through the propeller hub.

Routine maintenance of the cooling system is quite important, as expensive damage can occur if it overheats. The cooling system is so important, that many outboards covered in this manual incorporate overheat alarm systems and

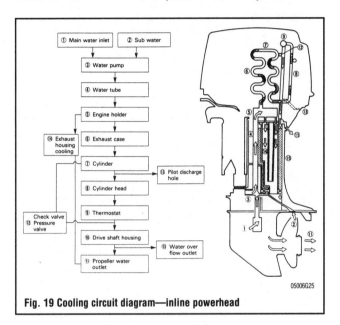

Fig. 19 Cooling circuit diagram—inline powerhead

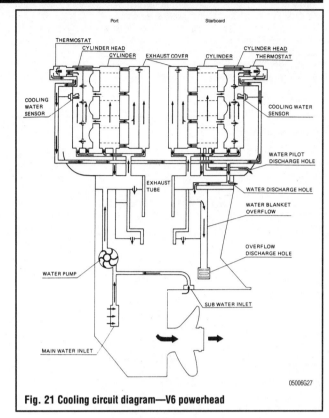

Fig. 21 Cooling circuit diagram—V6 powerhead

speed limiters, in case the engine's operating temperature exceeds predetermined limits.

Poor operating habits can play havoc with the cooling system. For instance, running the engine with the water pickup out of water can destroy the water pump impeller in a matter of seconds. Running in shallow water, kicking up debris that is drawn through the pump, can not only damage the pump itself, but send the debris throughout the entire system, causing water restrictions that create overheating.

WATER PUMP

♦ See Figure 16

The water pumps used on all Suzuki outboards are a displacement type water pump. Water pressure is increased by the change in volume between the impeller and the pump case.

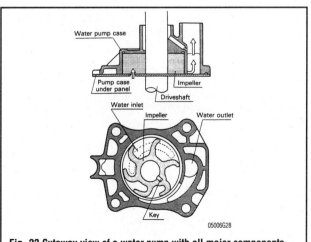

Fig. 22 Cutaway view of a water pump with all major components labeled

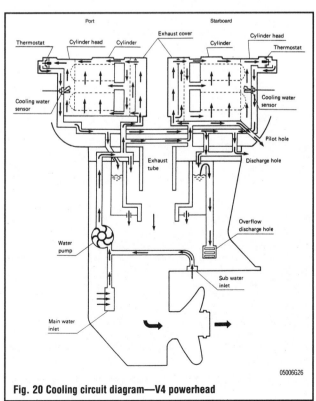

Fig. 20 Cooling circuit diagram—V4 powerhead

6-12 OIL INJECTION

On most outboards, the water pump is mounted on top of the lower unit. A driveshaft key engages a flat on the driveshaft and a notch in the impeller hub. As the driveshaft rotates, the impeller rotates with it.

On the DT2 and DT2.2 the water pump is mounted in a pump case installed on the propeller shaft between the lower unit and propeller. The pump impellers are secured to the propeller shaft by a pin that fits into the propeller shaft and a similar notch in the impeller hub. The propeller on other small displacement models is secured to the drive shaft in the same manner.

THERMOSTAT

▶ See Figure 23

A pellet-type thermostat is used to control the flow of engine water, to provide fast engine warm-up and to regulate water temperatures. A wax pellet element in the thermostat expands when heated and contracts when cooled. The pellet element is connected through a piston to a valve. When the pellet element is heated, pressure is exerted against a rubber diaphragm, which forces the valve to open. As the pellet element is cooled, the contraction allows a spring to close the valve. Thus, the valve remains closed while the water is cold, limiting circulation of water.

As the engine warms, the pellet element expands and the thermostat valve opens, permitting water to flow through the powerhead. This opening and closing of the thermostat permits enough water to enter the powerhead to keep the engine within operating limits.

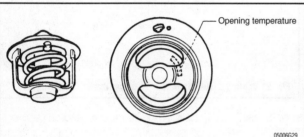

Fig. 23 A pellet-type thermostat is used to control the flow of engine water

Troubleshooting the Cooling System

Water cooling is the most popular method in use on outboard engines today. A "raw-water" pump delivers seawater to the powerhead, circulating it through the cylinder head(s), the thermostat, the exhaust housing, and back down through the outboard. The water runs down the exhaust cavity and away, either through an exhaust tube mounted behind the propeller or, on the larger engines, through the propeller hub.

Routine maintenance of the cooling system is quite important, as expensive damage can occur if it overheats. The cooling system is so important, that many outboards covered in this manual incorporate overheat alarm systems and speed limiters, in case the engine's operating temperature exceeds predetermined limits.

Poor operating habits can play havoc with the cooling system. For instance, running the engine with the water pickup out of water can destroy the water pump impeller in a matter of seconds. Running in shallow water, kicking up debris that is drawn through the pump, can not only damage the pump itself, but send the debris throughout the entire system, causing water restrictions that create overheating.

Symptoms of overheating are numerous and include:
- A "pinging" noise coming from the engine, commonly known as detonation
- Loss of power
- A burning smell coming from the engine
- Paint discoloration on the powerhead in the area of the spark plugs and cylinder heads

If these symptoms occur, immediately seek and correct the cause. If the engine has overheated to the point where paint has discolored, it may be too late to save the powerhead. Powerheads in this state usually require at least partial overhaul.

So what are major causes of overheating? Well the most prevalent cause is lack of maintenance. Other causes which are directly attributable to lack of maintenance or poor operating habits are:
- Fuel system problems causing lean mixture
- Incorrect oil mixture in fuel or a problem with the oil injection system
- Spark plugs of incorrect heat range
- Faulty thermostat
- Restricted water flow through the powerhead due to sand or silt buildup
- Faulty water pump impeller
- Sticking thermostat

Water Pump

REMOVAL & INSTALLATION

Since proper water pump operation is critical to outboard operation, all seals and gaskets should be replaced whenever the water pump is removed. Also, installation of a new impeller each time the water pump is disassembled is good insurance against overheating.

➡ Never turn a used impeller over and reuse it. The impeller rotates with the driveshaft and the vanes take a set in a clockwise direction. Turning the impeller over will cause the vanes to move in the opposite and result in premature impeller failure.

DT2 and DT2.2

➡ The water pumps on the DT2 and DT.2.2 are mounted in a pump case installed on the propeller shaft between the lower unit and propeller. The water pumps can be serviced without removing the lower unit from the drive shaft housing.

1. Remove the propeller.
2. Place a suitable container under the lower unit.
3. Remove the drain screw and drain the lubricant from the unit.
4. Remove the bolts holding the water pump case cover to the lower unit housing and remove the cover.
5. Carefully pry the impeller from the water pump case
6. Remove the impeller drive pin from the propeller shaft.

To install:

7. Insert impeller drive pin in propeller shaft.
8. Install a new impeller in the pump body with a counterclockwise rotating motion.
9. Install the pump case cover
10. Tighten the fasteners securely.
11. Install the propeller.
12. Fill the lower unit with lubricant.
13. Place the outboard in a test tank or move the boat to a body of water.
14. Test the cooling system for proper operation.

Except DT2 and DT2.2

1. Remove the lower unit.
2. Place the lower unit in a suitable holding fixture, keeping the unit upright.
3. As required, remove the water tube from the pump cover .
4. Remove the water pump cover.
5. Slide the impeller off the drive shaft .
6. Remove the impeller drive pin or key from the drive shaft.
7. Carefully pry the pump base plate and gasket assembly free of the lower unit housing.
8. Discard the gasket.

To install:

9. Clean the gasket mating surfaces thoroughly.
10. Install the pump base plate using a new gasket.
11. Install the impeller drive key into the drive shaft .
12. Install a new impeller onto the drive shaft, aligning the impeller slot with the drive shaft key.

➡ Make sure the locating pins are in place prior to installing the pump cover.

OIL INJECTION 6-13

13. Slowly rotate the drive shaft clockwise while sliding the water pump cover down the drive shaft. This will allow the impeller to flex into the housing in the right direction.
14. Coat the water pump cover bolt threads with thread locking compound.
15. Install the lockwashers and bolts, Tightening the bolts securely.
16. As required, install the water tube into the pump cover.
17. Install the lower unit.
18. Place the outboard in a test tank or move the boat to a body of water.
19. Test the cooling system for proper operation.

CLEANING & INSPECTION

➡ **When removing water pump seals, note the direction in which each seal lip faces for proper installation reference.**

Remove the drive shaft grommet, as required and water tube seal from the pump cover. Inspect these rubber parts for wear, hardness or deterioration and replace if necessary.

On models equipped with a water pump cover O-ring, remove the O-ring and replace it.

Check the pump cover for cracks, distortion or melting and replace as required. Clean the pump cover and base plate in solvent and blow dry with compressed air.

1. Throughly remove all gasket residue from the mating surfaces.

Thermostat

REMOVAL & INSTALLATION

♦ **See accompanying illustrations**

1. Locate the thermostat cover on the cylinder head.
2. Remove the cover bolts.
3. Carefully pry the cover from the cylinder head.
4. If the cover does not want to come loose, tap it gently with a plastic hammer.
5. Remove the thermostat cover.
6. Remove the thermostat cover gasket.
7. Remove the thermostat from the cylinder head.

To install:

8. Throughly clean the gasket mating surfaces.
9. Inspect the thermostat bore for signs of corrosion. Outboards used in salt water should be flushed with fresh water after each use to prevent corrosion from forming.
10. Position the thermostat in the cylinder head.

Step 4

Step 5

Step 2

Step 6

6-14 OIL INJECTION

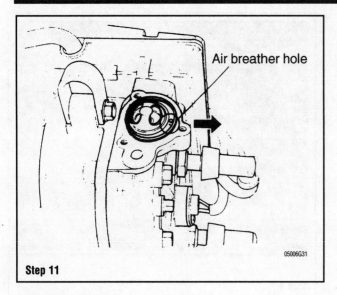

Step 11

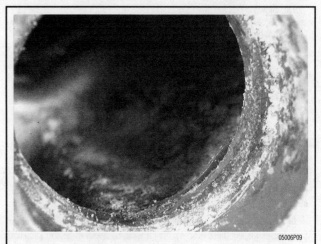

Fig. 24 Corrosion inside the thermostat bore signals a lack of maintenance. Always flush your outboard with fresh water, especially after boating in salt water

11. Point the air breather hole, if equipped, toward the end of the cylinder head (toward the spark plugs).
12. Install the cover using a new gasket.
13. Tighten the cover bolts securely.

CLEANING & INSPECTION

♦ See Figure 24

The cause of a malfunctioning thermostat is often foreign matter stuck to the valve seat. Inspect the thermostat to make sure it is clean and free of foreign matter. If necessary, test the removed thermostat for operation by immersing it in a pot filled with water.

1. With a piece of string pinched by the valve, suspend the thermostat in a pot in such a way that the thermostat floats above the bottom of the pot.
2. Raise the water temperature.
3. If the valve opens (the thermostat releases the string and drops to the bottom of the pot) at the temperature specified on the thermostat, the thermostat is functioning correctly.
4. If the valve does not open (the thermostat remains hanging) at the temperature specified on the thermostat, the thermostat is faulty and should be replaced.

OIL INJECTION WARNING SYSTEMS

Description and Operation

LOW OIL LEVEL

♦ See Figure 25

A low oil level warning light and buzzer are included with most oil injected models. The buzzer and warning light may be installed either in the instrument panel, on the engine case or in the remote control. The light and buzzer serve a dual function. The sending unit, located in the powerhead oil tank, is connected to the warning system through the key switch. When the oil level in the tank drops to a predetermined level, the warning light will change color to alert the operator of a low oil condition. If the operator continues, a buzzer will then sound. After a predetermined time, the engine rpm will be cut to reduce oil consumption and prevent the engine from running out of oil at high speed.

➡ The low oil warning system is equipped with a reset that will allow the operator to run the engine in a low oil condition for approximately 30 minutes. To reset the warning system, remove the engine cover and press the red reset button on the electrical cover with the engine running.

OIL FLOW

♦ See Figures 26 and 27

On some models, the inline oil filter is equipped with an oil flow sensor. If the oil flow is restricted or slows down due to an obstruction within the inline

Fig. 25 As the level of oil in the tank decreases, the float level drops and will illuminate a low oil level lamp at a predetermined level

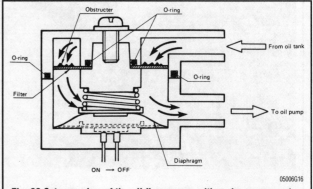

Fig. 26 Cutaway view of the oil flow sensor with major components identified

OIL INJECTION 6-15

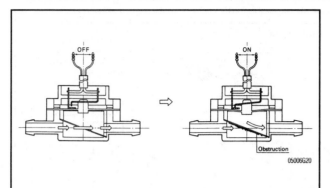

Fig. 27 If the oil flow is restricted or slows down due to an obstruction within the inline oil filter, the integral micro-switch will close

oil filter, the integral micro-switch will close. This will activate the low oil flow and rev limit lights on the monitor. The warning buzzer will sound and the engine speed will automatically decrease.

Troubleshooting the Oil Injection Warning System

When the oil level reaches a predetermined level in the oil tank, the oil level monitor will turn from green to red as the switch in the oil tank senses the low oil condition. If the operator continues to operate the outboard and the oil level drops further, the warning buzzer will begin to sound at 7 second intervals. All the time, the red low oil warning light will be on. After 10 seconds of running, the engine RPM will drop and the rev. limit light on the monitor will come on.

If for some reason full RPM is needed during a low oil condition and the operator does not have oil onboard, the low oil reset button on the engine lower cover can be pressed with the engine running. This button over-rides the rev-limit circuitry, but not the buzzer or light. The operator may continue to run with the buzzer and light on at wide open throttle for approximately 30 minutes.

OIL INJECTION WARNING SYSTEM TROUBLESHOOTING CHART

SYMPTOM	SUZUKI MONITOR					POSSIBLE CAUSE
	BUZZER	OIL	FLOW	WATER	LIMIT	
Key switch-ON Oil light-ON		ON				• Oil level is lower than safety level (1.0 lit).
The buzzer and light do not come on after starting the engine.						• Poor operating or defective cooling water sensor. • Disconnected cooling water sensor lead wires. • Defective buzzer. • Defective monitor. • Malfunction of ignition switch. • Defective reset unit.
While running engine above 3000 RPM, engine RPM drops and buzzer, flow, and over-rev limit light come on.	ON		ON		ON	• Blocked oil filter (clean or replace). • Defective oil flow switch.
Oil level-OK Buzzer ON Over-rev control ON	ON	ON			ON	• Defective oil level switch.
Oil level is less than 1.0 lit. and light, buzzer and over-rev do not come on.						• Defective oil level switch.
Oil is less than 1.0 lit., buzzer and light come on. Over-rev control does not come on.	ON	ON				• Disconnected reset unit lead wires. • Defective CDI unit. • Defective reset unit. • Shorted reset switch.
Oil is less than 1.0 lit. and although over-rev control and lights works, buzzer does not sound.		ON			ON	• Defective buzzer. • Disconnected ignition switch and buzzer lead wires.
While operating engine at over 3000 RPM, engine speed is reduced after the buzzer sounds for 10 seconds.	ON	ON			ON	• Oil level in the oil tank is lower than safety level (1.0 lit).
The buzzer sounds and the oil light stay on even though engine speed is below 3000 RPM.	ON	ON				• Low oil level. • Defective oil level switch.

6-16 OIL INJECTION

Early type

Tester ⊕ lead	Tester ⊖ lead	Float position	Value
Black	Green	No. 1	0 Ω
	Red	No. 2	0 Ω
	Yellow	No. 3	0 Ω
	Orange	No. 4	0 Ω

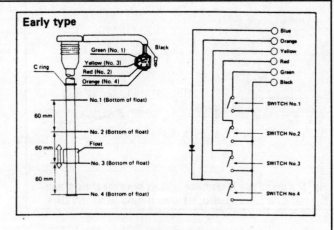

Late type

Tester ⊕ lead	Tester ⊖ lead	Float 1–2 position	Switch 1–2–3 position	Value
Black	Green	No. 1	No. 1	0 Ω
	Red	No. 2	No. 2	0 Ω
	Orange	No. 2	Between No. 2 – No. 3	∞ Ω
		No. 2	No. 3	0 Ω

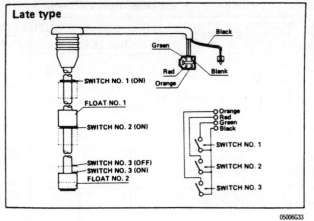

Fig. 28 Testing the oil level sensor

Oil Level Sensor

TESTING

▶ See Figure 28

1. Remove the oil level sensor.
2. Connect a multimeter between the sensor terminals as illustrated and continuity.
3. Continuity should exist between the illustrated terminals when float is at the designated positions.

➡ Some early model sensor use a two position float that should show continuity at the lowest position and no continuity at the upper position.

4. If the sensor does not operate as specified, it may be faulty. Clean the sensor and retest.
5. If the sensor functions properly, check the oil level warning electrical harness for opens or shorts.

REMOVAL & INSTALLATION

▶ See Figures 29, 30 and 31

1. Remove the engine cover.
2. Label and disconnect the oil level sensor electrical harness.
3. Carefully pry the oil level sensor from the oil tank.
4. Carefully remove the oil level sensor from the oil tank.

To install:

5. Place the oil level sensor in the oil tank.

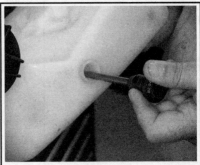

Fig. 29 The oil level sensor is held into the oil tank by a rubber stopper

Fig. 30 The oil level sensor electrical harness runs down the side of the oil tank

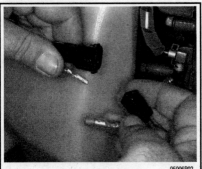

Fig. 31 The oil level sensor uses a multi-pin and a bullet connector

OIL INJECTION 6-17

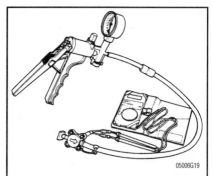

Fig. 32 Connect a hand vacuum pump to the sensor and check the sensor for continuity

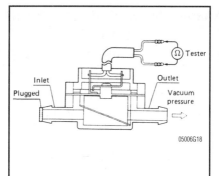

Fig. 33 Continuity should exist when vacuum is applied and should not exist when vacuum is released

Fig. 34 The oil flow sensor is located near the bottom of the oil tank, inline with the oil feed hose

6. Carefully push the oil level rubber stopper into the oil tank.
7. Connect the oil level sensor electrical harness.
8. Install the engine cover.

Oil Flow Sensor

TESTING

♦ See Figures 32, 33 and 34

1. Remove the oil flow sensor.
2. Connect a hand vacuum pump to the outlet side of the sensor and plug the inlet side of the sensor.
3. Check for continuity between the sensor terminals.
4. Continuity should exist when vacuum is applied and should not exist when vacuum is released.
5. If the sensor does not operate as specified, it may be faulty. Clean the sensor and retest.
6. If the sensor functions properly, check the oil flow warning electrical harness for opens or shorts.

REMOVAL & INSTALLATION

1. Remove the engine cover.
2. Remove the oil tank as necessary to gain access to the oil filter.
3. Position a suitable container to receive oil discharged when the hoses are disconnected.
4. Label and disconnect the oil flow sensor electrical harness.
5. Disconnect and plug the oil lines connected to the filter.
6. Remove the filter from the powerhead.

To install:

7. Clean and inspect the filter.
8. Position the filter on the powerhead.
9. Connect the oil lines to the filter and secure them with the clamps.
10. Connect the oil flow sensor electrical harness.
11. Install the oil tank if removed.
12. Install the engine cover.

CLEANING & INSPECTION

♦ See Figure 35

➡ Some oil flow sensors cannot be disassembled for cleaning.

1. Disassemble the oil filter and remove the strainer.
2. Clean the strainer in solvent and blowing it dry using compressed air.
3. Inspect the O-ring, diaphragm and strainer for damage and replace components as necessary.
4. Assemble the oil filter and tighten the screws securely.

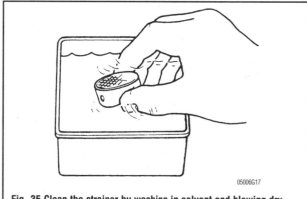

Fig. 35 Clean the strainer by washing in solvent and blowing dry using compressed air

OVERHEAT WARNING SYSTEM

Description and Operation

A overheat warning light and buzzer are included on most models. The buzzer and warning light may be installed either in the instrument panel, on the engine case or in the remote control.

The light and buzzer serve a dual function. The sending unit, usually located in the cylinder head, is connected to the warning system through the key switch. When the powerhead temperature raises to a predetermined temperature, the warning light will illuminate and the warning buzzer will sound to alert the operator of an overheat condition. After a predetermined amount of time, the engine rpm will be cut to reduce the overheat condition and prevent powerhead damage.

Two types of sensors have been used to monitor for overheat conditions. The first monitored cooling water level in the powerhead using a float switch. If the switch sensed water flowing past it, it sent a signal to the control module and the warning light and buzzer would remain off.

The latest style sensor uses a heat sensing bimetal switch installed in the cylinder wall. The switch is designed to turn on at a predetermined cylinder wall temperature.

Troubleshooting the Overheat Warning System

The overheat warning system is activated should cylinder wall temperature become high due to insufficient cooling water. The red temperature lamp will be illuminated and the buzzer will sound.

If the sensor detects overheating when engine speed is more than a predetermined rpm, the rev. limiter will activate and the rev. limit warning lamp will also illuminate.

6-18 OIL INJECTION

OVERHEAT WARNING SYSTEM TROUBLESHOOTING CHART

SYMPTOM	SUZUKI MONITOR					POSSIBLE CAUSE
	BUZZER	OIL	FLOW	WATER	LIMIT	
The buzzer and light do not come on after starting the engine.						• Poor operating or defective cooling water sensor. • Disconnected cooling water sensor lead wires. • Defective buzzer. • Defective monitor. • Malfunction of ignition switch. • Defective reset unit.
The buzzer keeps sounding for 10 to 13 seconds after starting the engine and the water light stays on. • No water coming from the discharge hole. • Water is coming from the discharge hole.	ON			ON		• Blockage of water intake. • Damaged water pump. • Blocked cooling water circuit. • Poor operating or defective cooling water sensor. • Blocked water passage. • Defective cooling water sensor.
While running engine above 3000 RPM, buzzer sounds and cooling water light come on. After 2 or 3 seconds over-rev control comes on.	ON			ON	ON	• Blocked water intake. • Damaged water pump. • Defective cooling water sensor. • Blocked water passage.
Over-rev control does not release despite reset switch having been pushed.						• Defective reset switch. • Defective reset unit.
Oil level-OK Recommended RPM range. No buzzer. Over-rev control-ON					ON	• Defective reset unit. • Defective CDI unit.
Buzzer sounds and cooling water is normal. Over-rev control-ON	ON			ON	ON	• Poor operating or defective cooling water sensor.
Engine speed is reduced after 10 seconds of uneven engine running.					ON	• Engine is over-revving.
Engine smooths out if the throttle is slightly backed-off from the full-open position.					ON/OFF	• Engine is over-revving.
While operating engine at over 3000 RPM, engine speed is reduced after the buzzer sounds for 10 seconds.	ON	ON			ON	• Oil level in the oil tank is lower than safety level (1.0 lit).
The buzzer sounds and the water light is on even though the engine speed has been reduced to below 3000 RPM.	ON			ON		• Blocked water intake. • Damaged water pump. • Defective cooling water sensor. • Blocked water passage.

OIL INJECTION 6-19

Overheat Sensor

REMOVAL & INSTALLATION

Water Level Switch

1. Turn the battery switch off and/or disconnect the negative battery cable.
2. Remove the engine cover.
3. Locate the overheat sensor on the cylinder head.
4. Label and disconnect the switch harness.
5. Remove overheat sensor attaching bolts.
6. Using pliers, grab the tongue of the sensor and twist slightly to remove.

To install:

7. Check the O-ring on the end of the sensor for damage and replace as necessary.
8. Lubricate the O-ring and install the sensor in the cylinder head.
9. Install overheat sensor attaching bolts and tighten them securely.
10. Connect the switch harness.
11. Turn the battery switch on and/or connect the negative battery cable.
12. Test the overheat system for proper operation.
13. Install the engine cover.

Heat Sensing Switch

♦ See Figures 36, 37 and 38

1. Turn the battery switch off and/or disconnect the negative battery cable.
2. Remove the engine cover.
3. Locate the overheat sensor on the cylinder head.
4. Label and disconnect the switch harness.
5. Using pliers, switch grab the switch at the illustrated point with pliers and pull with a slight twist to remove.

To install:

6. Check the O-ring on the end of the sensor for damage and replace as necessary.
7. Lubricate the O-ring.
8. Place a plastic tie-wrap in the sensor bore and push the sensor into the bore.
9. When the sensor tip has reached the bottom of the bore, depress the sensor further and hold in that position.
10. Slowly remove the tie-wrap to release the trapped air at the bottom of the bore.
11. If the sensor is properly installed, it should not protrude more than 0.04 in (1mm) above the surface.

➥If the sensor is not installed properly, it will not be seated at the bottom of the bore and will not give accurate temperature readings. This may lead to an overheat condition.

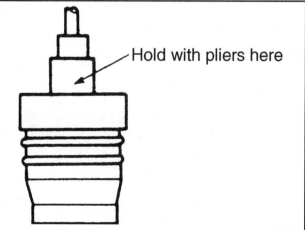

Fig. 36 When removing the heat sensing switch grab the switch at the illustrated point with pliers and pull

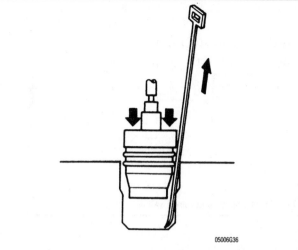

Fig. 37 Place a plastic tie-wrap in the sensor bore and push the sensor into the bore

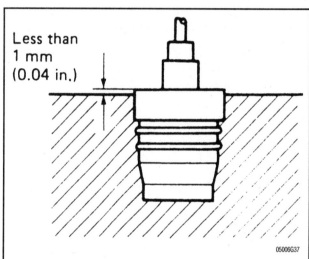

Fig. 38 If the sensor is properly installed, it should not protrude more than 0.04 in (1mm) above the surface

12. Connect the switch harness.
13. Turn the battery switch on and/or connect the negative battery cable.
14. Test the overheat system for proper operation.
15. Install the engine cover.

TESTING

Water Level Switch

♦ See Figure 39

1. Remove the switch from the powerhead.
2. Connect a multimeter between the switch terminals and check for continuity.
3. With the float in the lower position, continuity should exist. With the float in the upper position, there should be no continuity.
4. If the sensor does not operate as specified, it may be faulty. Clean the sensor and retest.
5. Clean the sensor as follows:
    a. Inspect the float to see if it move up and down smoothly.
    b. If the action is stiff, disassemble the switch and flush thoroughly with fresh water to clean.

➥Always remove the switch pin from the left and insert it into the right side of the switch.

6-20 OIL INJECTION

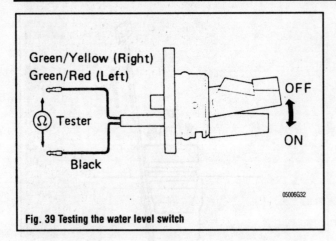

Fig. 39 Testing the water level switch

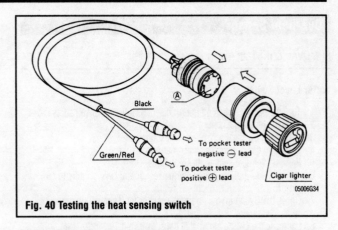

Fig. 40 Testing the heat sensing switch

6. If the sensor functions properly, check the overheat warning electrical harness for opens or shorts.

Heat Sensing Switch

♦ See Figure 40

1. Remove the switch from the powerhead.
2. Connect a multimeter between the switch terminals and check for continuity.
3. With the switch at room temperature, there should be no continuity.
4. Using a cigar lighter from your vehicle, heat the end of the sensor.

➡ **Take care to not touch the sensor with the cigar lighter. Excessive heat will damage the sensor.**

5. As the sensor warms, continuity should exist.
6. If the sensor does not operate as specified, it may be faulty.
7. If the sensor functions properly, check the overheat warning electrical harness for opens or shorts.

ENGINE MECHANICAL 7-2
   THE TWO-STROKE CYCLE 7-2
   FLYWHEEL 7-2
      REMOVAL & INSTALLATION 7-2
      INSPECTION 7-5
   POWERHEAD 7-5
      REMOVAL & INSTALLATION 7-5
      DISASSEMBLY & ASSEMBLY 7-15
POWERHEAD
  RECONDITIONING 7-32
   DETERMINING POWERHEAD
    CONDITION 7-32
      PRIMARY COMPRESSION TEST 7-32
      SECONDARY COMPRESSION
       TEST 7-32
   BUY OR REBUILD? 7-33
   POWERHEAD OVERHAUL TIPS 7-33
      TOOLS 7-34
      CAUTIONS 7-34
      CLEANING 7-34
      REPAIRING DAMAGED
       THREADS 7-34
   POWERHEAD PREPARATION 7-35
   CYLINDER BLOCK AND HEAD 7-35
      GENERAL INFORMATION 7-35
      INSPECTION 7-36
   CYLINDER BORES 7-36
      GENERAL INFORMATION 7-36
      INSPECTION 7-37
      REFINISHING 7-37
   PISTONS 7-37
      GENERAL INFORMATION 7-37
      INSPECTION 7-38
   PISTON PINS 7-39
      GENERAL INFORMATION 7-39
      INSPECTION 7-39
   PISTON RINGS 7-40
      GENERAL INFORMATION 7-40
      INSPECTION 7-41
   CONNECTING RODS 7-41
      GENERAL INFORMATION 7-41
      INSPECTION 7-42
   THE CRANKSHAFT 7-43
      GENERAL INFORMATION 7-43
      INSPECTION 7-44
   BEARINGS 7-44
      GENERAL INFORMATION 7-44
      INSPECTION 7-45
SPECIFICATIONS CHARTS
   TORQUE SPECIFICATIONS 7-46
   ENGINE REBUILDING
    SPECIFICATIONS 7-49

7
POWERHEAD

ENGINE MECHANICAL 7-2
POWERHEAD RECONDITIONING 7-32

POWERHEAD

ENGINE MECHANICAL

The Two-Stroke Cycle

The two-stroke engine can produce substantial power for its size and weight. But why is a two-stroke so much smaller and lighter than a four-stroke? Well, there is no valvetrain. Camshafts, valves and pushrods can really add weight to an engine. A two-stroke engine doesn't use valves to control the air and fuel mixture entering and exiting the engine. There are holes, called ports, cut into the cylinder which allow for entry and exit of the fuel mixture. The two-stroke engine also fires on every second stroke of the piston, which is the primary reason why so much more power is produced than a four-stroke.

Since two-stroke engines discharge approximately one fourth of their fuel unburned, they have come under close scrutiny by environmentalists. Many states have tightened their grip on two-strokes and most manufacturers are hard at work developing new efficient models that can meet the tough emissions standards. Check out your state's regulations before you buy any two-stroke outboard.

The two-stroke engine is able to function because of two very simple physical laws. The first, gases will flow from an area of high pressure to an area of lower pressure. A tire blowout is an example of this principle. The high-pressure air escapes rapidly if the tube is punctured. Second, if a gas is compressed into a smaller area, the pressure increases, and if a gas expands into a larger area, the pressure is decreased. If these two laws are kept in mind, the operation of the two-stroke engine will be easier understood.

Two-stroke engines utilize an arrangement of port openings to admit fuel to the combustion chamber and to purge the exhaust gases after burning has been completed. The ports are located in a precise pattern in order for them to be opened and closed at an exact moment by the piston as it moves up and down in the cylinder. The exhaust port is located slightly higher than the fuel intake port. This arrangement opens the exhaust port first as the piston starts downward and therefore, the exhaust phase begins a fraction of a second before the intake phase.

Actually, the intake and exhaust ports are spaced so closely together that both open almost simultaneously. For this reason, the pistons of most two-stroke engines have a deflector-type top. This design of the piston top serves two purposes very effectively. First, it creates turbulence when the incoming charge of fuel enters the combustion chamber. This turbulence results in more complete burning of the fuel than if the piston top were flat. Second, it forces the exhaust gases from the cylinder more rapidly.

Beginning with the piston approaching top dead center on the compression stroke, the intake and exhaust ports are closed by the piston, the reed valve is open, the spark plug fires, the compressed air/fuel mixture is ignited, and the power stroke begins. The reed valve was open because as the piston moved upward, the crankcase volume increased, which reduced the crankcase pressure to less than the outside atmosphere.

As the piston moves downward on the power stroke, the combustion chamber is filled with burning gases. As the exhaust port is uncovered, the gases, which are under great pressure, escape rapidly through the exhaust ports. The piston continues its downward movement. Pressure within the crankcase increases, closing the reed valves against their seats. The crankcase then becomes a sealed chamber. The air/fuel mixture is compressed ready for delivery to the combustion chamber. As the piston continues to move downward, the intake port is uncovered. A fresh air/fuel mixture rushes through the intake port into the combustion chamber striking the top of the piston where it is deflected along the cylinder wall. The reed valve remains closed until the piston moves upward again.

When the piston begins to move upward on the compression stroke, the reed valve opens because the crankcase volume has been increased, reducing crankcase pressure to less than the outside atmosphere. The intake and exhaust ports are closed and the fresh fuel charge is compressed inside the combustion chamber.

Pressure in the crankcase decreases as the piston moves upward and a fresh charge of air flows through the carburetor picking up fuel. As the piston approaches top dead center, the spark plug ignites the air/fuel mixture, the power stroke begins and one full cycle has been completed.

The exact time of spark plug firing depends on engine speed. At low speed the spark is retarded, fires later than when the piston is at or beyond top dead center. Engine timing is built into the unit at the factory.

At high speed, the spark is advanced, fires earlier than when the piston is at top dead center. On all but the smallest horsepower outboards the timing can be changed adjusted to meet advance and retard specifications.

Because of the design of the two-stroke engine, lubrication of the piston and cylinder walls must be delivered by the fuel passing through the engine. Since gasoline doesn't make a good lubricant, oil must be added to the fuel and air mixture. The trick here is to add just enough oil to the fuel to provide lubrication. If too much oil is added to the fuel, the spark plug can become "fouled" because of the excessive oil within the combustion chamber. If there is not enough oil present with the air/fuel mixture, the piston can "seize" within the cylinder. What usually happens in this case is the piston and cylinder become scored and scratched, from lack of lubrication. In extreme cases, the piston will turn to liquid and eventually disintegrate within the cylinder.

Most two-stroke engines require that the fuel and oil be mixed before being poured into the fuel tank. This is known as "pre-mixing" the fuel. This can become a real hassle. You must be certain that the ratio is correct. Too little oil in the fuel could cause the piston to seize to the cylinder, causing major engine damage and completely ruining your weekend. Most modern two-stroke engines have an oil injection system that automatically mixes the proper amount of oil with the fuel as it enters the engine.

Flywheel

REMOVAL & INSTALLATION

DT2 and DT2.2

♦ See accompanying illustrations

1. Remove the engine cover.
2. Remove the fuel tank assembly.
3. Remove the recoil starter assembly.
4. Remove the starter cup and magneto insulator.
5. Using a flywheel holder (09930–40113 for 1988/89 models; 09930–48720 1990 to present), hold the flywheel and loosen the retaining nut.
6. Using a flywheel rotor remover remove the flywheel. Make sure to keep track of the flywheel key when removing the flywheel.

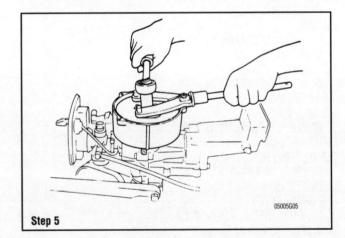
Step 5

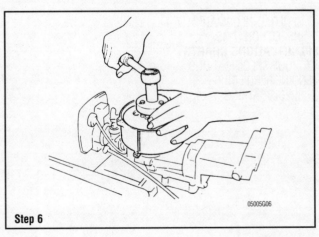
Step 6

POWERHEAD

To install:

7. Before installing the flywheel, thoroughly inspect the crankshaft and flywheel tapers. These surfaces must be absolutely clean and free of oil, grease and dirt. Use solvent and a lint free cloth to clean the surfaces and then blow dry with compressed air.

8. Install the flywheel key, starter cup and flywheel and flywheel bolt. Tighten the bolt to 30–36 ft. lbs. (40–50 Nm.)

9. Install the fuel and engine cover.

DT4, DT6 and DT8

♦ **See accompanying illustrations**

1. Remove the engine cover from the engine.
2. Remove the built-in fuel tank (if equipped).
3. After removing the bolts, remove the recoil starter assembly.
4. Remove the starter cup. If the screws are hard to loosen, use an impact drive to remove them.
5. Use a flywheel holder (09930–40113) to remove the flywheel nut.
6. Use a flywheel holder and flywheel rotor remover (09930–30713) to remove the flywheel.
7. Make sure to remove the flywheel key from the crankshaft.

To install:

8. Install the flywheel key into the keyway on the crankshaft. Make sure the key is seated correctly into the keyway.
9. Install the flywheel onto the crankshaft.
10. Using a flywheel holder, install the flywheel nut and tighten to 32.5 ft. lbs. (45 Nm).
11. Install the starter cup onto the flywheel and tighten the screws.
12. Install the recoil starter.
13. Install the fuel tank.
14. Install the engine cover.

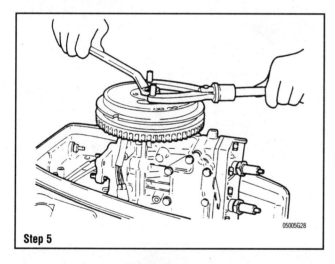
Step 5

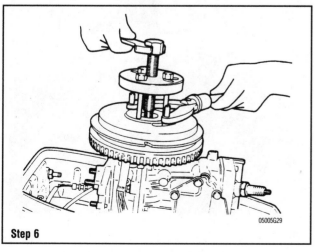
Step 6

DT9.9 and DT15

♦ **See accompanying illustrations**

1. Remove the engine cover from the engine.
2. Remove the two nuts and disconnect the battery/starting motor cables and the neutral switch wire (if equipped).
3. Remove the recoil starter assembly (if equipped).
4. Disconnect the wire lead extending from the stator assembly to the rectifier assembly.
5. Remove the two bolts and remove the starter motor from the engine.
6. Using a flywheel holder (09930–49310), remove the flywheel nut.
7. Using a flywheel holder and the flywheel remover plate (09930–30713), remove the flywheel.

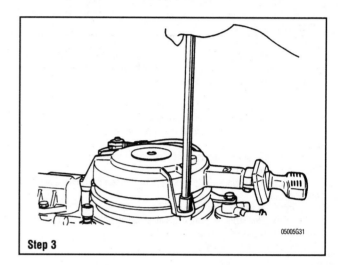
Step 3

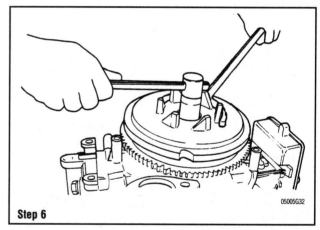
Step 6

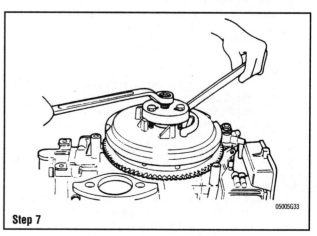
Step 7

7-4 POWERHEAD

Step 8

9. Using a flywheel holder, tighten the flywheel nut to 144.7–151.9 ft. lbs. (200–210 Nm).
10. Connect the stator wire leads to their proper connections.
11. Install the starter pulley onto the flywheel.
12. Install the recoil starter assembly.
13. Connect the battery negative battery cable.
14. Install the engine cover.

DT55, DT65, DT75 and DT85

▶ See accompanying illustrations

1. Remove the engine cover.
2. Remove the electrical junction box cover and disconnect the stator leads.
3. Remove the flywheel cover.
4. Using a flywheel holder (09930–39520) and flywheel & propeller shaft housing remover (09930–39410), remove the flywheel nut.
5. Using the special tools, remove the flywheel from the crankshaft.
6. Make sure to remove the flywheel key before removing the magneto case to prevent tearing the seal.
7. Remove the magneto case from the engine.

To install:
8. Throughly clean the mating surface of the flywheel and crankshaft taper with cleaning solvent. Install the key onto the crankshaft securely.
9. Install the flywheel onto the crankshaft.
10. Using a flywheel holder, tighten the flywheel nut to 144.5–152 ft. lbs. (200–210 Nm).
11. Connect the stator wire leads to their proper connections.

8. If any difficulty is experienced in removing the flywheel, tap the head of the bolt with a hammer. This will usually help in the removal operation.
9. Remove the key from the crankshaft keyway.

To install:
10. Install the key securely into the crankshaft keyway.
11. Install the flywheel onto the crankshaft.
12. Using a flywheel holder, tighten the flywheel nut to 58–65 ft. lbs. (80–90 Nm).
13. Install the starter motor back onto the engine and securely tighten the bolts (if equipped).
14. Reconnect the starter/battery cables and reconnect the neutral safety switch lead wire.
15. Install the recoil starter assembly (if equipped)
16. Connect the stator wires making sure all connections are free from corrosion and are securely fastened.
17. Install the engine cover.

DT20, DT25 and DT30

1. Disconnect the negative battery cable lead to prevent accidental engine start.
2. Remove the engine cover.
3. Disconnect the wire leads in the electrical junction box, leading from the stator assembly.
4. Remove the recoil starter assembly.
5. Using a screwdriver to hold the flywheel, remove the starter pulley bolts and lift of the starter pulley.
6. Using a flywheel holder (09930–48720), remove the flywheel nut.
7. Using a flywheel holder and a flywheel remover (09930–39411), remove the flywheel from the engine.

To install:
8. Throughly clean the mating surface of the flywheel and crankshaft taper with cleaning solvent. Install the key onto the crankshaft securely.
9. Using a flywheel holder, tighten the flywheel nut to 94–108 ft. lbs. (130–150 Nm).
10. Install the starter pulley onto the flywheel.
11. Install the recoil starter assembly.
12. Connect the battery negative battery cable.
13. Install the engine cover.

DT35 and DT40

1. Disconnect the negative battery cable lead to prevent accidental engine start.
2. Remove the engine cover.
3. Disconnect the wire leads in the electrical junction box and CDI/control unit holder, leading from the stator assembly.
4. Remove the recoil starter assembly.
5. Using a flywheel holder (09930–39520), remove the flywheel nut.
6. Using a flywheel holder and a flywheel remover (09930–39410), remove the flywheel and key.

To install:
7. Throughly clean the mating surface of the flywheel and crankshaft taper with cleaning solvent. Install the key onto the crankshaft securely.
8. Install the flywheel onto thew crankshaft.

Step 5

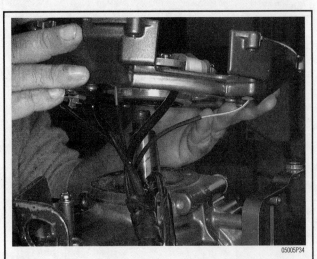
Step 7

POWERHEAD

12. Connect the battery negative battery cable.
13. Install the engine cover.

DT115, DT140 and V4

1. Remove the engine cover.
2. Remove the electrical junction box cover and disconnect the stator leads.
3. Remove the flywheel cover.
4. Using a flywheel holder (09930–48720), remove the flywheel nut.
5. Using a flywheel holder, flywheel remover (09930–39411) and flywheel bolts (09930–39420), remove the flywheel.
6. Remove the flywheel from the crankshaft.

To install:

7. Throughly clean the mating surface of the flywheel and crankshaft taper with cleaning solvent. Install the key onto the crankshaft securely.
8. Install the flywheel onto the crankshaft.
9. Using a flywheel holder, tighten the flywheel nut to 181–188 ft. lbs. (250–260 Nm).
10. Connect the stator wire leads to their proper connections.
11. Connect the battery negative battery cable.
12. Install the engine cover.

V6

1. Remove the engine cover.
2. Remove the electrical junction box cover and disconnect the stator leads.
3. Remove the flywheel cover.
4. Using a flywheel holder (09930–48720), remove the flywheel nut.
5. Using a flywheel holder, flywheel remover (09930–39411) and flywheel bolts (09930–39420), remove the flywheel.

To install:

6. Throughly clean the mating surface of the flywheel and crankshaft taper with cleaning solvent. Install the key onto the crankshaft securely.
7. Install the flywheel onto the crankshaft.
8. Using a flywheel holder, tighten the flywheel nut to 181–188 ft. lbs. (250–260 Nm).
9. Connect the stator wire leads to their proper connections.
10. Connect the battery negative battery cable.
11. Install the engine cover.

INSPECTION

Check the flywheel carefully for cracks or fractures.

✳✳ CAUTION

A cracked or chipped flywheel must be replaced. A damaged flywheel may fly apart at high rpm, throwing metal fragments over a large area. Do not attempt to repair a damaged flywheel.

Check tapered bore of flywheel and crankshaft taper for signs of fretting or working.
On electric start models, check the flywheel teeth for excessive wear or damage.
Check crankshaft and flywheel nut threads for wear or damage.
Replace flywheel, crankshaft and/or flywheel nut as required.

Powerhead

REMOVAL & INSTALLATION

When removing any powerhead, it is a good idea to make a sketch or take an instant picture of the location, routing and positioning of electrical harnesses, brackets and component locations for installation reference.

➡ Sometimes when attempting to remove the powerhead it won't come loose from the adapter. The gasket may hold the powerhead. Rock the powerhead back and forth or give it a gentile nudge with a pry bar. If the gasket breaks loose and the powerhead still will not come loose, then the driveshaft is seized to the crankshaft at the splines.

The following procedures assume that the outboard has been removed from the boat and placed on a suitable work stand. If the powerhead is being removed with the outboard still mounted on the boat and the powerhead is equipped with an electric starter, disconnect first the negative, then the positive battery cables to prevent accidental starting.

On some powerheads it will be necessary to remove attached components if the powerhead is to be overhauled. Refer to the specific sections covering these components for removal and installation information.

DT2 and DT2.2

♦ **See accompanying illustrations**

1. Remove the engine covers.
2. Turn the fuel shutoff to the **OFF** position. Disconnect and plug the fuel line.
3. Remove the fuel tank.
4. Remove the rewind starter assembly.
5. Remove the starter cup and flywheel insulator.
6. Using a flywheel holder loosen and remove the flywheel nut.
7. Remove the flywheel using a flywheel puller.

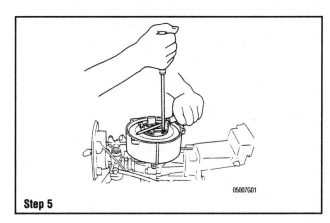

Step 5

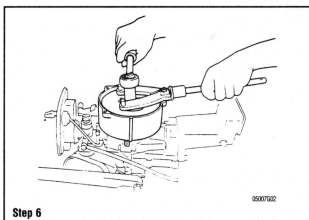

Step 6

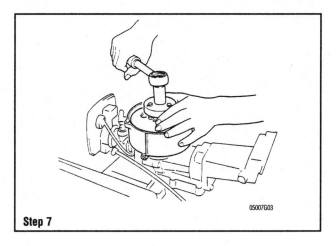

Step 7

7-6 POWERHEAD

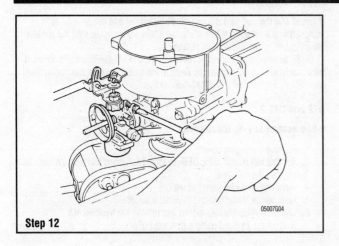
Step 12

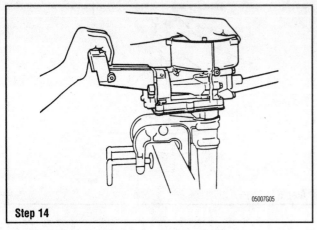
Step 14

8. Disconnect the spark plug lead.
9. Label and disconnect the stator lead wires.
10. Remove the choke knob
11. Remove the throttle link knob and the control panel.
12. Remove the carburetor and fuel shut-off valve.
13. Remove the bolts holding the powerhead to the driveshaft housing.
14. Remove the powerhead.

➡ **If the powerhead will not come off, rotate the propeller to free the powerhead from the driveshaft.**

15. Remove and discard the powerhead mounting gasket.

To install:

16. Clean the powerhead mounting and driveshaft housing gasket surfaces thoroughly.
17. Lightly coat driveshaft splines with marine grease.
18. Install a new powerhead new mounting gasket.
19. Install the powerhead, rotating the propeller as required to align driveshaft and crankshaft splines.
20. Coat powerhead mounting bolt threads with thread locking compound.
21. Install the powerhead mounting bolts and tighten to specification.
22. Install the carburetor and fuel shut-off valve.
23. Install the control panel, the choke knob and the throttle link knob.
24. Connect the stator lead wires.
25. Connect the spark plug lead.
26. Install the flywheel.
27. Using a flywheel holder tighten the flywheel nut.
28. Install the starter cup and flywheel insulator.
29. Install the rewind starter assembly.
30. Install the fuel tank.
31. Connect the fuel line and turn the fuel shutoff to the **ON** position.
32. Start the engine and make adjustments as necessary.
33. Check engine for proper operation.
34. Install the engine covers.

DT4

1. Remove the engine cover.
2. Remove the fuel tank.
3. Remove the hand rewind starter.
4. Using a flywheel holder loosen and remove the flywheel nut.
5. Remove the flywheel using the flywheel puller.
6. Label and disconnect the stator and CDI unit electrical leads.
7. Remove the ignition coil and CDI unit.
8. Disconnect the throttle cable from the carburetor.
9. Loosen the hose clamp, disconnect and plug the fuel hose.
10. Remove the silencer.
11. Remove the carburetor.
12. Remove the fuel pump.
13. Remove the powerhead mounting bolts.
14. Remove the powerhead.

➡ **If the powerhead will not come off, rotate the propeller to free the powerhead from the driveshaft.**

15. Remove and discard the powerhead mounting gasket.

To install:

16. Clean the powerhead mounting and driveshaft housing gasket surfaces thoroughly.
17. Lightly coat the driveshaft splines with marine grease.
18. Install a new powerhead new mounting gasket.
19. Install the powerhead, rotating the propeller as required to align driveshaft and crankshaft splines.
20. Coat powerhead mounting bolt threads with silicone sealer.
21. Install the powerhead mounting bolts and tighten to specification.
22. Install the fuel pump.
23. Install the carburetor.
24. Install the silencer.
25. Connect the throttle cable top the carburetor and adjust it to specification.
26. Install the ignition coil and CDI unit.
27. Connect the stator and CDI unit electrical leads.
28. Install the flywheel. Tighten the flywheel nut to specification.
29. Install the hand rewind starter.
30. Install the fuel tank.
31. Unplug and connect the fuel hose. Using a new hose clamp, fasten the hose properly.
32. Start the engine and make adjustments as necessary.
33. Check engine for proper operation.
34. Install the engine cover.

DT6 and DT8

♦ **See accompanying illustrations**

1. Remove the engine cover.
2. Remove the silencer cover.
3. Loosen the hose clamp, disconnect and plug the fuel hose.
4. Disconnect the choke knob.

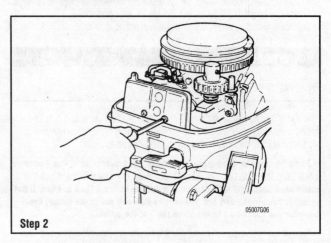
Step 2

POWERHEAD

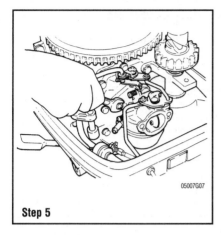
Step 5

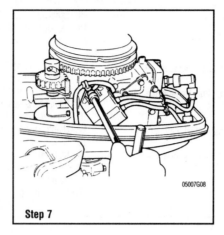
Step 7

Step 8

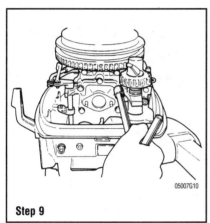
Step 9

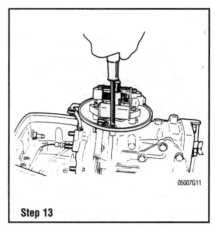
Step 13

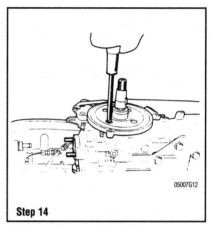
Step 14

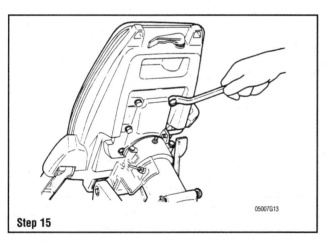
Step 15

5. Remove the carburetor.
6. Label and disconnect the stator and CDI unit electrical leads.
7. Remove the ignition coil and CDI unit.
8. Remove the fuel pump with the fuel filter still attached.
9. Remove the hand rewind starter.
10. Disconnect the throttle cable from the carburetor.
11. Using a flywheel holder loosen and remove the flywheel nut.
12. Remove the flywheel using the flywheel puller.
13. Remove the stator.
14. Remove the upper oil seal housing.
15. Remove the powerhead mounting bolts
16. Remove the powerhead.

➡ If the powerhead will not come off, rotate the propeller to free the powerhead from the driveshaft.

17. Remove and discard the powerhead mounting gasket.

To install:

18. Clean the powerhead mounting and driveshaft housing gasket surfaces thoroughly.
19. Lightly coat the driveshaft splines with marine grease.
20. Install a new powerhead new mounting gasket.
21. Install the powerhead, rotating the propeller as required to align driveshaft and crankshaft splines.
22. Coat powerhead mounting bolt threads with silicone sealer.
23. Install the powerhead mounting bolts and tighten to specification.
24. Install the upper oil seal housing.
25. Install the stator.
26. Install the flywheel.
27. Using a flywheel holder install and tighten the flywheel nut specification.
28. Connect the throttle cable to the carburetor.
29. Install the hand rewind starter.
30. Install the fuel pump using a new fuel filter.
31. Install the ignition coil and CDI unit.
32. Install the carburetor.
33. Connect the choke knob.
34. Unplug and connect the fuel hose. Using a new hose clamp, fasten the hose properly.
35. Install the silencer cover.
36. Start the engine and make adjustments as necessary.
37. Check engine for proper operation.
38. Install the engine cover.

7-8 POWERHEAD

DT9.9 and DT15

♦ See accompanying illustrations

1. Remove the engine cover.
2. On electric start models, disconnect the electrical cables between the battery and starter.
3. On oil injected models, disconnect and plug the oil line. Remove the oil tank.
4. Disconnect the neutral switch.
5. Loosen the neutral starter interlock locknut and disconnect the interlock cable from the throttle limiter.
6. Loosen the throttle cable locknuts and disconnect the cable from the control lever.
7. Remove the silencer cover.
8. Remove the recoil starter assembly.
9. Label and disconnect the CDI unit electrical leads.
10. Remove the CDI unit.
11. Label and disconnect the neutral start switch electrical lead.
12. Remove the neutral start switch.
13. Label and disconnect the stator electrical leads.
14. Remove the rectifier assembly.
15. Remove the starter motor relay.
16. Remove the starter motor.
17. Using a flywheel holder loosen and remove the flywheel nut.
18. Remove the flywheel using the flywheel puller.
19. Remove the flywheel key.
20. Disconnect the stator electrical leads
21. Remove the throttle control lever.
22. Disconnect and plug the hoses from the fuel filter.

Step 2

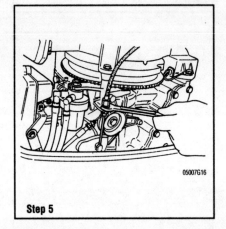
Step 5

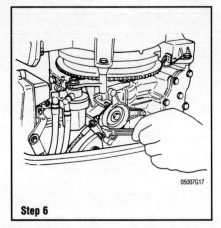
Step 6

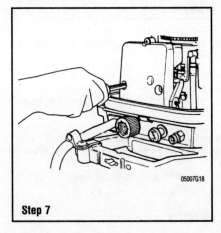
Step 7

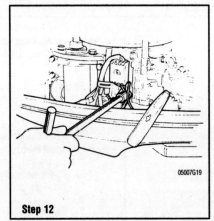
Step 12

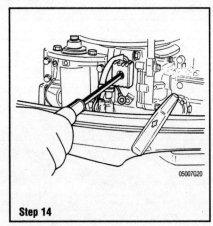
Step 14

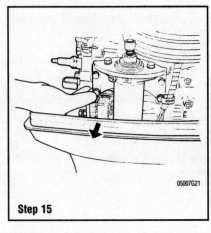
Step 15

Step 16

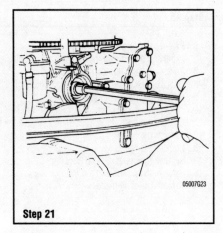
Step 21

POWERHEAD 7-9

Step 23

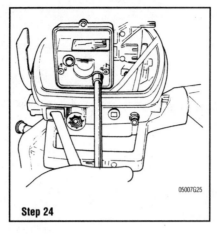
Step 24

Step 28

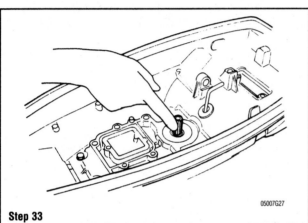
Step 33

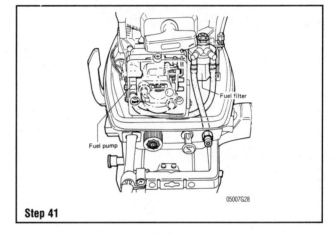
Step 41

23. Remove the fuel filter.
24. Remove the silencer case and carburetor.
25. Remove the fuel pump.
26. Remove the nut on the backside of the starter switch assembly.
27. Unclamp the starter cable clamp and remove the grommet from the cable. Pull the starter cable out.
28. Remove the stator assembly.
29. Remove the bolts holding the powerhead to the driveshaft housing.
30. Remove the powerhead.

➡ **If the powerhead will not come off, rotate the propeller to free the powerhead from the driveshaft.**

31. Remove and discard the powerhead mounting gasket.

To install:
32. Clean the powerhead mounting and driveshaft housing gasket surfaces thoroughly.
33. Lightly coat the driveshaft splines with marine grease.
34. Install a new powerhead new mounting gasket.
35. Install the powerhead, rotating the propeller as required to align driveshaft and crankshaft splines.
36. Coat powerhead mounting bolt threads with silicone sealer.
37. Install the powerhead mounting bolts and tighten to specification.
38. Install the stator assembly.
39. Install the starter cable, grommet and starter cable clamp.
40. Install the nut on the backside of the starter switch assembly.
41. Install the fuel pump and fuel filter, routing the hoses properly.
42. Install the silencer case and carburetor.
43. Connect the hoses to the fuel filter.
44. Install the throttle control lever.
45. Install the flywheel key and flywheel.
46. Using a flywheel holder tighten the flywheel nut.
47. Install the starter motor.
48. Install the starter motor relay.
49. Install the rectifier assembly.
50. Connect the stator electrical leads.
51. Install the neutral start switch.
52. Connect the neutral start switch electrical lead.
53. Install the CDI unit.
54. Connect the CDI unit electrical leads.
55. Install the recoil starter assembly.
56. Install the silencer cover.
57. Connect the cable to the control lever and adjust to specification.
58. Connect the interlock cable to the throttle limiter and adjust to specification.
59. Connect the neutral switch.
60. On electric start models, connect the electrical cables between the battery and starter.
61. On oil injected models, install the oil tank and properly connect the oil lines.
62. Start the engine and make adjustments as necessary.
63. Check engine for proper operation.
64. Install the engine cover.

DT20, DT25 and DT30

1. Remove the engine cover.
2. Remove the recoil starter.
3. Remvoe the electric parts holder and rectifier cover.
4. Disconnect the starter cable from the neutral start lever.
5. Disconnect the negative, then the positive battery cables.
6. Disconnect the spark plug leads.
7. Remove the CDI unit from the electric parts holder.
8. Remove the electric parts holder.
9. Disconnect and plug the oil hose from the oil tank.
10. Remove the oil tank.
11. Disconnect the throttle valve sensor, choke solenoid and idle speed adjustment switch electrical leads.
12. Disconnect the fuel hose from the engine under cover.

7-10 POWERHEAD

13. Remove the wiring harness stopper and disconnect the ignition coil lead wires.
14. Remove the CDI unit and the wiring harness assembly.
15. Remove the silencer cover.
16. Disconnect the choke lever.
17. Remove the bolts holding the powerhead to the driveshaft housing.
18. Remove the powerhead.

➡ If the powerhead will not come off, rotate the propeller to free the powerhead from the driveshaft.

19. Remove and discard the powerhead mounting gasket.

To install:

20. Clean the powerhead mounting and driveshaft housing gasket surfaces thoroughly.
21. Lightly coat the driveshaft splines with marine grease.
22. Install a new powerhead new mounting gasket.
23. Install the powerhead, rotating the propeller as required to align driveshaft and crankshaft splines.
24. Coat powerhead mounting bolt threads with silicone sealer.
25. Install the powerhead mounting bolts and tighten to specification.
26. Connect the choke lever.
27. Install the silencer cover.
28. Install the CDI unit and the wiring harness assembly.
29. Install the wiring harness stopper and connect the ignition coil lead wires.
30. Connect the fuel hose to the engine under cover.
31. Connect the throttle valve sensor, choke solenoid and idle speed adjustment switch electrical leads.
32. Install the oil tank.
33. Connect the oil hose to the oil tank.
34. Install the electric parts holder.
35. Install the CDI unit from the electric parts holder.
36. Connect the spark plug leads.
37. Connect the battery cables, positive side first.
38. Connect the starter cable to the neutral start lever. Adjust to specification
39. Install the electric parts holder and rectifier cover.
40. Install the recoil starter.
41. Start the engine and make adjustments as necessary.
42. Check engine for proper operation.
43. Install the engine cover.

DT35 and DT40

1. Disconnect the negative battery cable, then the positive battery cable.
2. Remove the engine cover.
3. Disconnect the neutral start interlock cable from the throttle limiter.
4. Label and disconnect the oil tank electrical leads.
5. Disconnect and plug the oil tank hose.
6. Remove the oil tank.
7. Remove the recoil starter assembly.
8. Remove the silencer cover.
9. Disconnect the oil pump control rod from the carburetor.
10. Remove the choke knob.
11. Disconnect and plug the fuel lines.
12. Remove the carburetor.
13. Disconnect the throttle control link rods and remove the throttle lever.
14. Disconnect the starter motor, starter relay and neutral switch electrical leads.
15. Remove the starter motor assembly.
16. Loosen the bolt in the electrical parts holder and disconnect the black ground wire.
17. Label and disconnect all electrical leads in the electrical parts holder.
18. Disconnect and plug the hoses from the fuel filter.
19. Remove the fuel filter.
20. Remove the flywheel.

21. Remove the bolts holding the powerhead to the driveshaft housing.
22. Remove the powerhead.

➡ If the powerhead will not come off, rotate the propeller to free the powerhead from the driveshaft.

23. Remove and discard the powerhead mounting gasket.

To install:

24. Clean the powerhead mounting and driveshaft housing gasket surfaces thoroughly.
25. Lightly coat the driveshaft splines with marine grease.
26. Install a new powerhead new mounting gasket.
27. Install the powerhead, rotating the propeller as required to align driveshaft and crankshaft splines.
28. Coat powerhead mounting bolt threads with silicone sealer.
29. Install the powerhead mounting bolts and tighten to specification.
30. Install the flywheel.
31. Install the fuel filter.
32. Connect the hoses to the fuel filter.
33. Connect all electrical leads in the electrical parts holder.
34. Connect the black ground wire and tighten the bolt in the electrical parts holder.
35. Install the starter motor assembly.
36. Connect the starter motor, starter relay and neutral switch electrical leads.
37. Install the throttle lever and connect the throttle control link rods.
38. Install the carburetor.
39. Adjust the link rods to specification
40. Connect the fuel lines
41. Install the choke knob.
42. Connect and adjust the oil pump control rod at the carburetor.
43. Install the silencer cover.
44. Install the recoil starter assembly.
45. Install the oil tank.
46. Connect the oil tank hose.
47. Connect the oil tank electrical leads.
48. Connect and adjust the neutral start interlock cable at the throttle limiter.
49. Connect the negative battery cable, then the positive battery cable.
50. Start the engine and make adjustments as necessary.
51. Check engine for proper operation.
52. Install the engine cover.

DT50

♦ See accompanying illustrations

1. Disconnect the negative battery cable, then the positive battery cable.
2. Remove the engine cover.
3. Remove the silencer cover

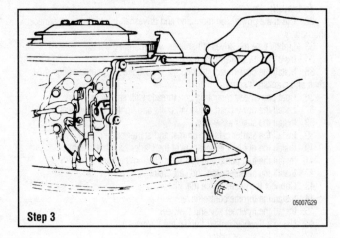

Step 3

POWERHEAD

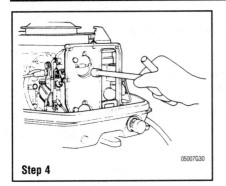
Step 4

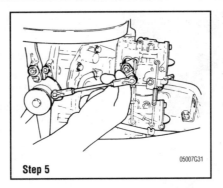
Step 5

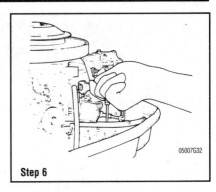
Step 6

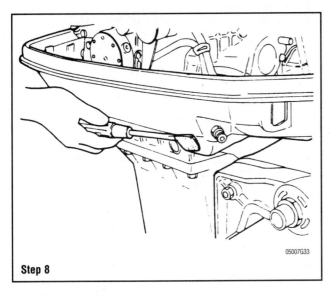
Step 8

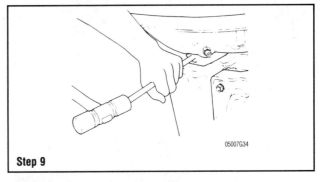
Step 9

4. Remove the silencer case.
5. Disconnect the carburetor throttle rod.
6. Disconnect and plug the fuel hoses.
7. Remove the carburetors and choke solenoid.
8. Remove the grommet and loosen the clutch shaft double nuts.
9. Disconnect the clutch rod from the clutch shaft by driving the clutch rod out using a drift.
10. Remove the bolts holding the powerhead to the driveshaft housing.
11. Remove the powerhead.

➡ **If the powerhead will not come off, rotate the propeller to free the powerhead from the driveshaft.**

12. Remove and discard the powerhead mounting gasket.

To install:

13. Clean the powerhead mounting and driveshaft housing gasket surfaces thoroughly.
14. Lightly coat the driveshaft splines with marine grease.
15. Install a new powerhead new mounting gasket.
16. Install the powerhead, rotating the propeller as required to align driveshaft and crankshaft splines.
17. Coat powerhead mounting bolt threads with silicone sealer.
18. Install the powerhead mounting bolts and tighten to specification.
19. Connect the clutch rod to the clutch shaft by driving the clutch rod in using a drift.
20. Tighten the clutch shaft double nuts to specification.
21. Install the grommet.
22. Install the carburetors and choke solenoid.
23. Connect the fuel hoses.
24. Connect and adjust the carburetor throttle rod.
25. Install the silencer cover and case.
26. Connect the negative battery cable, then the positive battery cable.
27. Start the engine and make adjustments as necessary.
28. Check engine for proper operation.
29. Install the engine cover.

DT55 and DT65

♦ **See accompanying illustrations**

1. Disconnect the negative battery cable, then the positive battery cable.
2. Remove the engine cover.
3. Label and disconnect the electrical leads to the oil level switch.
4. Disconnect and plug the oil hose.
5. Remove the oil tank.
6. Remove the power trim and tilt motor relay cover, then remove the relays from the cylinder.
7. Remove the electrical parts holder cover. Label and disconnect all electrical connectors in the electrical parts holder.
8. Remove the CDI unit.
9. Disconnect the electrical leads from the starter motor and relay.
10. Remove the flywheel.
11. Disconnect and plug the fuel hose.
12. Remove the clutch link from the throttle control arm and the clutch shaft side arm.
13. Disconnect the water outlet hose from the engine lower cover.
14. Remove the bolts securing the lower under cover on each side of the outboard.

Step 14

7-12 POWERHEAD

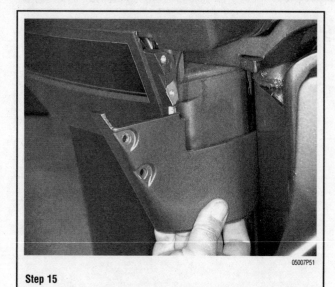
Step 15

Step 16

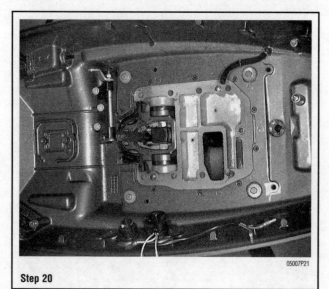
Step 20

15. Remove both under covers.
16. Remove the bolts holding the powerhead to the driveshaft housing.
17. Remove the powerhead.

➡ If the powerhead will not come off, rotate the propeller to free the powerhead from the driveshaft.

18. Remove and discard the powerhead mounting gasket.

To install:

19. Clean the powerhead mounting and driveshaft housing gasket surfaces thoroughly.
20. Lightly coat the driveshaft splines with marine grease.
21. Install a new powerhead new mounting gasket.
22. Install the powerhead, rotating the propeller as required to align driveshaft and crankshaft splines.
23. Coat powerhead mounting bolt threads with silicone sealer.
24. Install the powerhead mounting bolts and tighten to specification.
25. Install both under covers and tighten the bolts securely.
26. Connect the water outlet hose to the engine lower cover.
27. Install the clutch link on the throttle control arm and the clutch shaft side arm.
28. Connect the fuel hose.
29. Install the flywheel.
30. Connect the electrical leads to the starter motor and relay.
31. Install the CDI unit.
32. Install the electrical parts holder cover.
33. Connect all electrical connectors in the electrical parts holder.
34. Install the power trim and tilt relays.
35. Install the power trim and tilt motor relay cover.
36. Install the oil tank.
37. Connect the oil hose.
38. Connect the electrical leads to the oil level switch.
39. Connect the negative battery cable, then the positive battery cable.
40. Start the engine and make adjustments as necessary.
41. Check engine for proper operation.
42. Install the engine cover.

DT75 and DT85

♦ See accompanying illustrations

1. Remove the engine cover.
2. Remove the lower front cover.

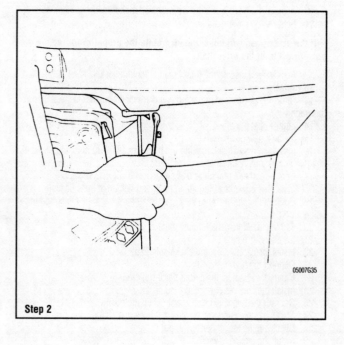
Step 2

POWERHEAD 7-13

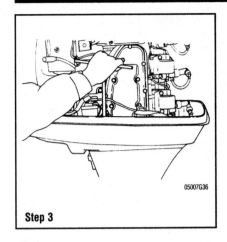
Step 3

Step 4

Step 5

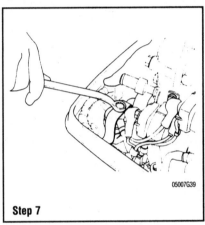
Step 7

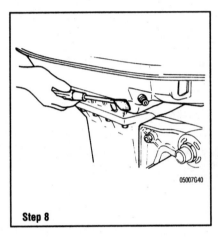
Step 8

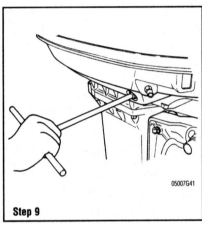
Step 9

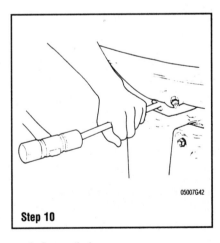
Step 10

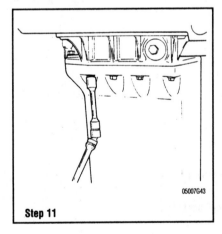
Step 11

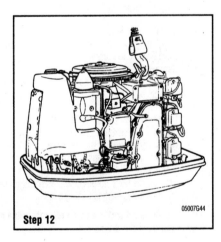
Step 12

3. Remove the lower rear cover.
4. Disconnect the negative, then the positive battery cable.
5. Remove the battery cable clamp, disconnect the cable grommet from the lower support housing and remove the cables.
6. Disconnect and plug the fuel hose.
7. Remove the fuel filter.
8. Remove the grommet on the starboard side of the driveshaft housing.
9. Remove the clutch shaft nuts.
10. Disconnect the clutch shaft rod by driving it out of its bore with a drift.
11. Remove the bolts holding the powerhead to the driveshaft housing.
12. Remove the powerhead using a hoist.

➡If the powerhead will not come off, rotate the propeller to free the powerhead from the driveshaft.

13. Remove and discard the powerhead mounting gasket.
14. Remove the aligning dowel pins from the bottom of the powerhead.
To install:
15. Clean the powerhead mounting and driveshaft housing gasket surfaces thoroughly.

7-14 POWERHEAD

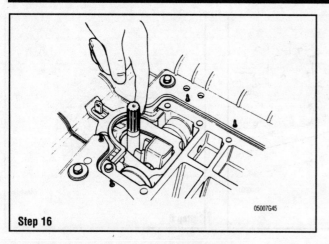
Step 16

16. Lightly coat the driveshaft splines with marine grease.
17. Install a new powerhead new mounting gasket.
18. Install the powerhead, rotating the propeller as required to align driveshaft and crankshaft splines.
19. Coat powerhead mounting bolt threads with silicone sealer.
20. Install the powerhead mounting bolts and tighten to specification.
21. Install the clutch shaft rod by driving it into its bore with a drift.
22. Install the clutch shaft nuts.
23. Install the grommet on the starboard side of the driveshaft housing.
24. Install the fuel filter.
25. Connect the fuel hose.
26. Install the battery cables and secure them with the cable clamps.
27. Install the cable grommet in the lower support housing.
28. Connect the negative battery cable, then the positive battery cable.
29. Start the engine and make adjustments as necessary.
30. Check engine for proper operation.
31. Install the engine cover.

DT115 and DT140

1. Remove the engine cover.
2. Label and disconnect the oil level switch electrical leads.
3. Disconnect and plug the oil hose.
4. Remove the oil tank.
5. Remove the idle speed adjustment switch, if equipped and electrical parts holder cover.
6. Disconnect the negative, then the positive battery cables.
7. Disconnect the power trim and tilt electrical leads.
8. Remove the electrical parts holder assembly.
9. Disconnect and plug the fuel hose.
10. Remove the lower engine covers.
11. Disconnect the water outlet hose.
12. Remove the bolts holding the powerhead to the driveshaft housing.
13. Remove the powerhead using a hoist.

➡ If the powerhead will not come off, rotate the propeller to free the powerhead from the driveshaft.

14. Remove and discard the powerhead mounting gasket.
15. Remove the aligning dowel pins from the bottom of the powerhead.

To install:
16. Clean the powerhead mounting and driveshaft housing gasket surfaces thoroughly.
17. Lightly coat the driveshaft splines with marine grease.
18. Install a new powerhead new mounting gasket.
19. Install the powerhead, rotating the propeller as required to align driveshaft and crankshaft splines.
20. Coat powerhead mounting bolt threads with silicone sealer.
21. Install the powerhead mounting bolts and tighten to specification.
22. Connect the water outlet hose.
23. Install the lower engine covers.
24. Connect the fuel hose.
25. Install the electrical parts holder assembly.
26. Connect the power trim and tilt electrical leads.
27. Connect the positive, then the negative battery cables.
28. Install the idle speed adjustment switch, if equipped and electrical parts holder cover.
29. Install the oil tank.
30. Connect and plug the oil hose.
31. Label and Connect the oil level switch electrical leads.
32. Install the engine cover.

V4

1. Remove the engine top cover.
2. Remove the relay cover under the starter.
3. Remove the starter cable from the starter.
4. Disconnect the starter relay from the relay holder.
5. Remove the relay holder and bolt.
6. Disconnect the negative battery cable and the power trim and tilt motor cable from the engine.
7. Remove the flywheel cover.
8. Remove the power trim and tilt motor relays together with the relay holder.
9. Remove the cover from the electrical parts holder.
10. Label and disconnect all electrical leads inside the electrical parts holder.
11. Remove the bolts securing the electrical parts holder and remove it.
12. Label and disconnect the ground wire from each cylinder head.
13. Label and disconnect the oil level switch wire leads.
14. Remove the oil tank.
15. Disconnect and plug the fuel hose from the fuel filter.
16. Disconnect the throttle valve sensor electrical lead, the oil sensor electrical lead and the starter valve electrical lead.
17. Remove the lower front under cover.
18. Remove the lower rear under cover.
19. Disconnect the water outlet hose.
20. Remove the bolts holding the powerhead to the driveshaft housing.
21. Remove the clutch connector rod pin.
22. Attach a hoist to the engine hooks and lift the powerhead slightly.

➡ If the powerhead will not come off, rotate the propeller to free the powerhead from the driveshaft.

23. Disconnect the clutch lever rod from the clutch shaft.
24. Disconnect the upper clutch rod from the clutch shaft.
25. Remove the powerhead using a hoist.
26. Remove and discard the powerhead mounting gasket.
27. Remove the aligning dowel pins from the bottom of the powerhead.

To install:
28. Clean the powerhead mounting and driveshaft housing gasket surfaces thoroughly.
29. Lightly coat the driveshaft splines with marine grease.
30. Install a new powerhead new mounting gasket.
31. Install the powerhead, rotating the propeller as required to align driveshaft and crankshaft splines.
32. Coat powerhead mounting bolt threads with silicone sealer.
33. Install the powerhead mounting bolts and tighten to specification.
34. Connect the upper clutch rod to the clutch shaft.
35. Connect the clutch lever rod to the clutch shaft.
36. Install the clutch connector rod pin.
37. Connect the water outlet hose.
38. Install the lower rear under cover.
39. Install the lower front under cover.
40. Connect the throttle valve sensor electrical lead, the oil sensor electrical lead and the starter valve electrical lead.
41. Connect the fuel hose to the fuel filter.
42. Install the oil tank.
43. Connect the oil level switch wire leads.
44. Connect the ground wire from each cylinder head.
45. Install the electrical parts holder.
46. Connect all electrical leads inside the electrical parts holder.
47. Install the cover on the electrical parts holder.
48. Install the power trim and tilt motor relays together with the relay holder.
49. Install the flywheel cover.
50. Connect the negative battery cable and the power trim and tilt motor cable to the engine.
51. Install the relay holder and bolt.

POWERHEAD 7-15

52. Connect the starter relay to the relay holder.
53. Install the starter cable to the starter.
54. Install the relay cover under the starter.
55. Start the engine and make adjustments as necessary.
56. Check engine for proper operation.
57. Install the engine cover.

V6

1. Remove the engine cover.
2. Disconnect the negative, then the positive battery cables.
3. Remove the cover from the electrical parts holder.
4. Disconnect the battery cable and the power trim and tilt motor electrical leads from the powerhead.
5. Disconnect and plug the fuel hose.
6. Remove the cover from the electrical parts holder
7. Disconnect all electrical leads inside the electrical parts holder, then remove the electrical parts holder.
8. Remove the bolts holding the powerhead to the driveshaft housing.
9. Remove the lower front under cover.
10. Remove the lower rear under cover.
11. Disconnect the water outlet hose.
12. Remove the bolts holding the powerhead to the driveshaft housing.
13. Remove the upper to lower clutch rod clevis pin.
14. Attach a hoist to the engine hooks and lift the powerhead slightly.

➡ **If the powerhead will not come off, rotate the propeller to free the powerhead from the driveshaft.**

15. Disconnect the clutch lever rod connector from the clutch shaft.
16. Disconnect the upper clutch rod from the clutch shaft.
17. Remove the powerhead using a hoist.
18. Remove and discard the powerhead mounting gasket.
19. Remove the aligning dowel pins from the bottom of the powerhead.

To install:

20. Clean the powerhead mounting and driveshaft housing gasket surfaces thoroughly.
21. Lightly coat the driveshaft splines with marine grease.
22. Install a new powerhead new mounting gasket.
23. Install the powerhead, rotating the propeller as required to align driveshaft and crankshaft splines.
24. Install the upper to lower clutch rod clevis pin.
25. Coat powerhead mounting bolt threads with silicone sealer.
26. Install the powerhead mounting bolts and tighten to specification.
27. Connect the water outlet hose.
28. Install the lower rear under cover.
29. Install the lower front under cover.
30. Install the electrical parts holder, then connect all electrical leads inside the electrical parts holder.
31. Install the cover on the electrical parts holder.
32. Connect the fuel hose.
33. Connect the battery cable and the power trim and tilt motor electrical leads to the powerhead.
34. Connect the positive, then the negative battery cables.
35. Start the engine and make adjustments as necessary.
36. Check engine for proper operation.
37. Install the engine cover.

DISASSEMBLY & ASSEMBLY

DT2 and DT2.2

♦ **See accompanying illustrations**

1. Loosen the cylinder head nuts in several stages using a criss-cross pattern.
2. Remove the nuts and lift the cylinder head from the cylinder block.
3. Remove and discard the cylinder head gasket.
4. Loosen the crankcase bolts in several stages using a criss-cross pattern.
5. Carefully pry apart and separate the crankcase halves.

➡ **If the halves resist coming apart, tap them lightly with a plastic hammer. Do not pry heavily on the crankcase halves, as severe damage may occur.**

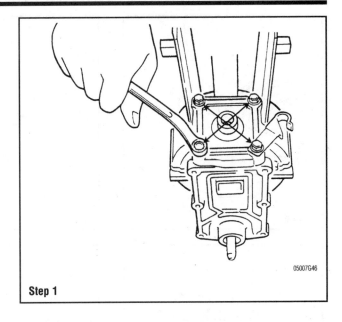

Step 1

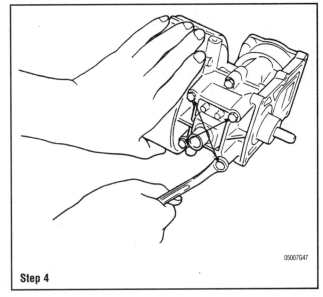

Step 4

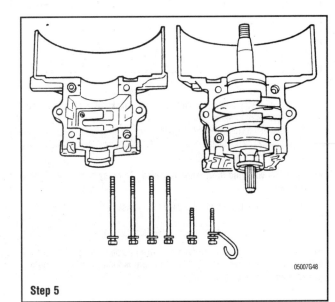

Step 5

7-16 POWERHEAD

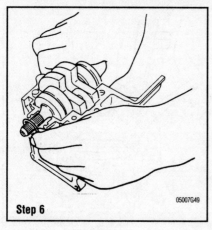
Step 6

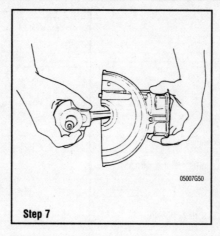
Step 7

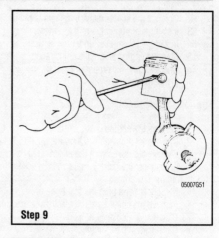
Step 9

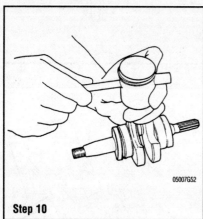
Step 10

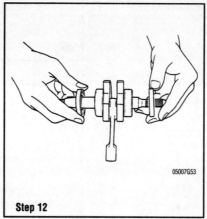
Step 12

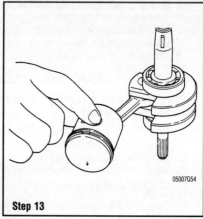
Step 13

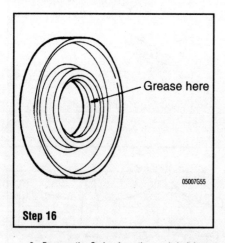
Step 16

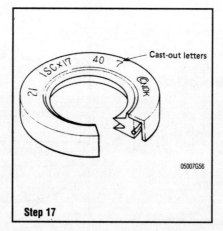
Step 17

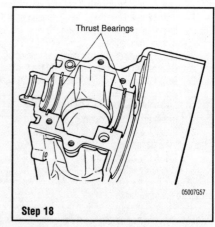
Step 18

6. Remove the O-ring from the crankshaft lower end.
7. To remove the rotating assembly, hold the crankshaft while sliding the cylinder block away from the piston.
8. Remove the crankshaft thrust rings.
9. Carefully pry the piston pin retainer from its groove on the piston. Then repeat the procedure for the other side.

➡ The circlip will tend to fly when removed. Be ready to catching it as it comes free.

9. Piston pin retainers are only good for one usage, discard the retainers after removing them.
10. Push the piston pin from it bore in the piston using a brass drift if necessary.

➡ The piston pin should slide smoothly from its bore. If opposition is felt, either the bore is out of round or the pin is bent. Inspect the components and replace as necessary.

11. Remove the piston and needle bearing from the connecting rod.
12. Slide the top and bottom crankshaft oil seals from the crankshaft.

To assemble:
13. Install the piston and needle bearing on the connecting rod. Make sure the arrow on the piston points to the splined portion of the crankshaft (downward).
14. Push the piston pin into the piston bore using a brass drift if necessary.
15. Install new piston pin retainers
16. Lubricate the seals with marine grease prior to installation.
17. Install the top and bottom crankshaft oil seals with the numbers facing the ends of the crankshaft.
18. Install the crankshaft thrust rings.
19. Lubricate the rotating assembly with 2-stroke oil prior to installation.

POWERHEAD 7-17

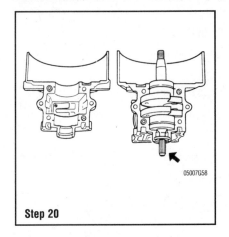
Step 20

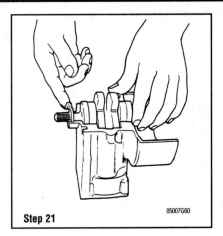
Step 21

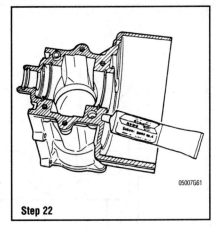
Step 22

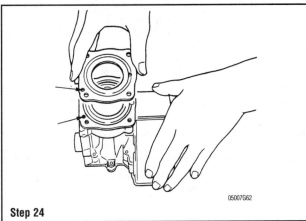
Step 24

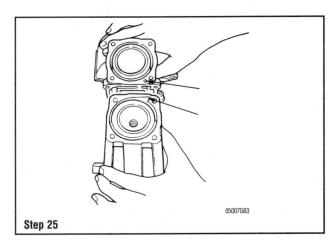
Step 25

does not turn freely, determine the cause and correct prior to assembling the powerhead.

DT4

▶ See Figures 1 and 2

1. Loosen the clips and disconnect the lubrication hose from each fitting.
2. Remove the mounting bolts and remove the rewind starter mounting base.
3. Remove the screws and remove the water jacket cover and gasket.
4. Loosen the cylinder head bolts in several stages using a criss-cross pattern.
5. Remove the cylinder head.
6. Loosen the crankcase bolts in several stages using a criss-cross pattern.
7. Carefully pry apart and separate the crankcase halves.
8. To remove the rotating assembly, hold the crankshaft while sliding the cylinder block away from the piston.

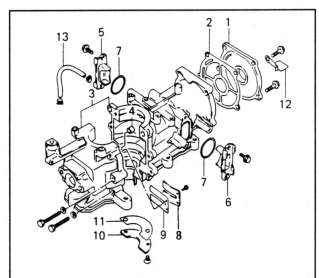

20. Install the rotating assembly into the cylinder block, making sure the crankshaft faces the correct way. The splined end of the crankshaft should face driveshaft housing side.
21. Apply marine grease to the O-ring and the splined portion of the crankshaft. Install the O-ring.
22. Apply a bead of Suzuki Bond No. 4 (99000-31030), or equivalent sealer to the crankcase halves (shaded areas in the illustration).
23. Tighten the crankcase bolts to specification in several stages using a criss-cross pattern.
24. Install a new cylinder head gasket noting the position of the cooling water hole.
25. Install the cylinder head matching the waterway on the head to the water hole on the cylinder block.
26. Tighten the cylinder head nuts to specification in several stages using a criss-cross pattern.
27. Rotate the crankshaft several turns to check for binding. If crankshaft

1. Cover, cylinder head
2. Gasket, cylinder head cover
3. Cylinder
4. Dowel pin
5. Cover, port, R
6. Cover, port, L
7. O ring
8. Stopper, reed valve
9. Reed, valve
10. Cover, water jacket
11. Gasket, water jacket
12. Cover, lower drain hole
13. Hose, lubrication

Fig. 1 Exploded view of the powerhead—DT4

7-18 POWERHEAD

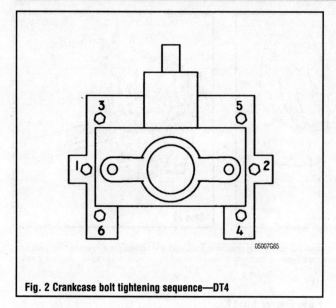

Fig. 2 Crankcase bolt tightening sequence—DT4

9. Carefully pry the piston pin retainer from its groove on the piston. Then repeat the procedure for the other side.

➡ **The circlip will tend to fly when removed. Be ready to catching it as it comes free.**

10. Piston pin retainers are only good for one usage, discard the retainers after removing them.
11. Push the piston pin from it bore in the piston using a brass drift if necessary.

➡ **The piston pin should slide smoothly from its bore. If opposition is felt, either the bore is out of round or the pin is bent. Inspect the components and replace as necessary.**

12. Remove the piston from the connecting rod.
13. Slide the top and bottom crankshaft oil seals from the crankshaft.

To assemble:
14. Install the piston on the connecting rod. Make sure the arrow on the piston points to the splined portion of the crankshaft (downward).
15. Push the piston pin into the piston bore using a brass drift if necessary.
16. Install new piston pin retainers
17. Lubricate the seals with marine grease prior to installation.
18. Install the top and bottom crankshaft oil seals with the numbers facing the ends of the crankshaft.
19. Install the crankshaft thrust washer.
20. Lubricate the rotating assembly with 2-stroke oil prior to installation.
21. Install the rotating assembly into the cylinder block, making sure the crankshaft faces the correct way. The splined end of the crankshaft should face driveshaft housing side.
22. The crankshaft thrust washer should fit snugly into the groove in the cylinder block.
23. Apply marine grease to the splined portion of the crankshaft.
24. Apply a bead of Suzuki Bond No. 4 (99000-31030), or equivalent sealer to the crankcase halves (shaded areas in the illustration).
25. Tighten the crankcase bolts to specification in several stages using a criss-cross pattern.
26. Install a new cylinder head gasket.
27. Install the cylinder head and tighten the cylinder head bolts to specification in several stages using the correct torque sequence.
28. Rotate the crankshaft several turns to check for binding. If crankshaft does not turn freely, determine the cause and correct prior to assembling the powerhead.

DT6 and DT8

◆ See accompanying illustrations

1. Loosen the cylinder head bolts in several stages using a criss-cross pattern.
2. Remove the cylinder head.
3. Loosen the inlet case bolts in several stages using a criss-cross pattern.
4. Remove the inlet case.
5. Remove the reed valve assembly.

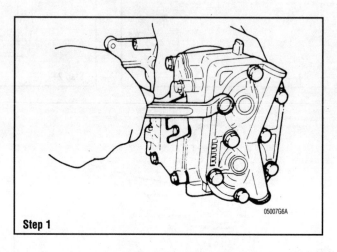
Step 1

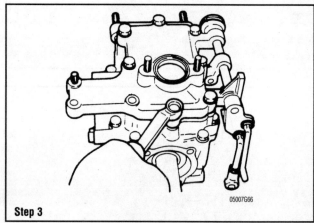
Step 3

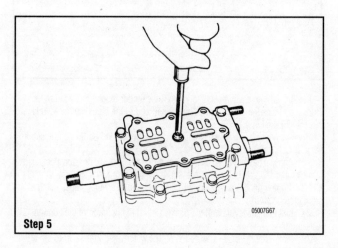
Step 5

POWERHEAD 7-19

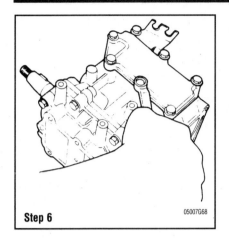
Step 6

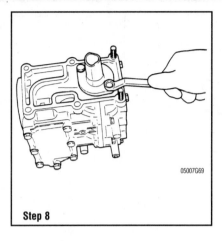
Step 8

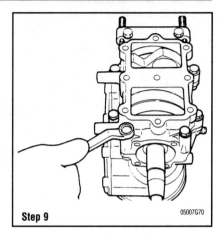
Step 9

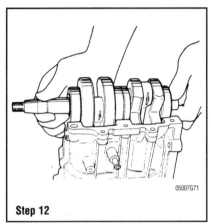
Step 12

Step 13

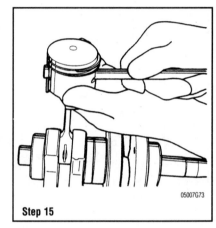
Step 15

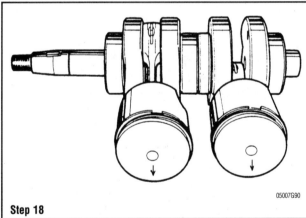
Step 18

6. Loosen the exhaust cover bolts in several stages using a criss-cross pattern.
7. Remove the exhaust cover.
8. Remove the lower oil seal housing.
9. Loosen the crankcase bolts in several stages using a criss-cross pattern.
10. Separate the crankcase from the cylinder block.
11. Carefully pry apart and separate the crankcase halves.
12. Remove the rotating assembly by carefully lifting the crankshaft from the cylinder block.
13. Carefully pry the piston pin retainer from its groove on the piston. Then repeat the procedure for the other side.

➡ The circlip will tend to fly when removed. Be ready to catching it as it comes free.

14. Piston pin retainers are only good for one usage, discard the retainers after removing them.
15. Push the piston pin from it bore in the piston using a brass drift if necessary.

➡ The piston pin should slide smoothly from its bore. If opposition is felt, either the bore is out of round or the pin is bent. Inspect the components and replace as necessary.

16. Remove the piston from the connecting rod.
17. Slide the top and bottom crankshaft oil seals from the crankshaft.

To assemble:
18. Install the piston on the connecting rod. Make sure the arrow on the pistons point to the exhaust port side.
19. Push the piston pin into the piston bore using a brass drift if necessary.
20. Install new piston pin retainers.
21. Lubricate the seals with marine grease prior to installation.
22. Install the top and bottom crankshaft oil seals with the numbers facing the ends of the crankshaft.
23. Install the crankshaft upper bearing, making sure the clearance between (A) and (B) as illustrated is not greater than 0.02 in. (0.5mm). If the clearance is greater, add a shim (09160-35001) to decrease the distance.
24. Lubricate the rotating assembly with 2-stroke oil prior to installation.
25. Install the rotating assembly into the cylinder block, making sure the crankshaft faces the correct way. The splined end of the crankshaft should face driveshaft housing side.

7-20 POWERHEAD

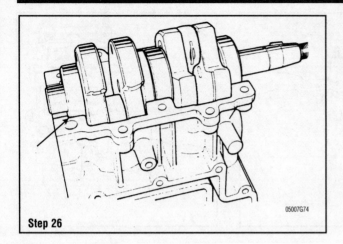
Step 26

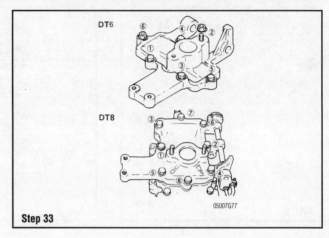
Step 33

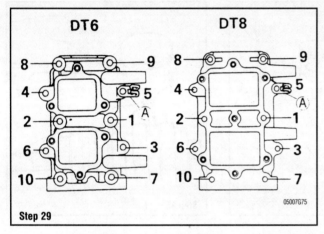
Step 29

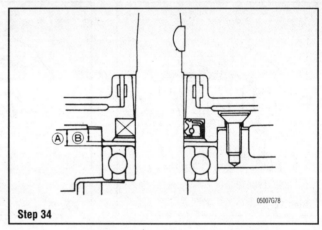
Step 34

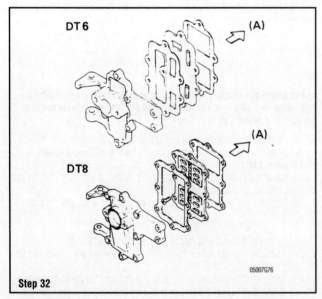
Step 32

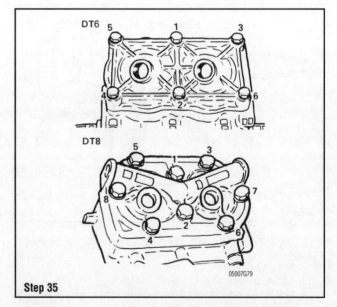
Step 35

26. The crankshaft lower bearing should fit snugly into the groove in the cylinder block.
27. Apply marine grease to the splined portion of the crankshaft.
28. Apply a bead of Suzuki Bond No. 4 (99000-31030), or equivalent sealer to the crankcase halves (shaded areas in the illustration).
29. Install the crankcase and tighten the bolts to specification in several stages using the proper tightening sequence.
30. Install the lower oil seal housing.
31. Install the exhaust cover.

32. Install the reed valve assembly making sure the reed valves face the crankshaft.
33. nstall the inlet case and tighten the bolts to specification in several stages using the proper tightening sequence.
34. Install the cylinder head using a new gasket and tighten the bolts to specification in several stages using the proper tightening sequence.
35. Rotate the crankshaft several turns to check for binding. If crankshaft does not turn freely, determine the cause and correct prior to assembling the powerhead.

POWERHEAD 7-21

DT9.9 and DT15

♦ See accompanying illustrations

1. Loosen the inlet case bolts in several stages using a criss-cross pattern.
2. Remove the inlet case (1), inner piece (2) and reed valve assembly (3).
3. Remove the exhaust plate and gasket from the cylinder.
4. Remove the exhaust cover attaching bolts.
5. Insert a pry bar between the exhaust cover and cylinder block and pry the cover to remove it. Discard the exhaust cover gasket.
6. Remove the exhaust plate and discard the gasket.
7. Loosen the cylinder head bolts in several stages using a criss-cross pattern.
8. Remove the cylinder head. If the cylinder head is hard to remove, use a plastic hammer to drive the dead at the placed indicated in the illustration.
9. Loosen the crankcase bolts in several stages using a criss-cross pattern. Then remove the crankcase.
10. Remove the rotating assembly by carefully lifting the crankshaft from the cylinder block.
11. Carefully pry the piston pin retainer from its groove on the piston. Then repeat the procedure for the other side.

➡ The circlip will tend to fly when removed. Be ready to catching it as it comes free.

12. Piston pin retainers are only good for one usage, discard the retainers after removing them.
13. Push the piston pin from it bore in the piston using a brass drift if necessary.

➡ The piston pin should slide smoothly from its bore. If opposition is felt, either the bore is out of round or the pin is bent. Inspect the components and replace as necessary.

14. Remove the piston from the connecting rod.

To assemble:

15. Install the piston on the connecting rod. Make sure the arrow on the pistons point to the exhaust port side.
16. Push the piston pin into the piston bore using a brass drift if necesary.
17. Install new piston pin retainers.
18. Lubricate the seals with marine grease prior to installation.
19. Install the top and bottom crankshaft oil seals with the numbers facing the ends of the crankshaft.
20. Lubricate the rotating assembly with 2-stroke oil prior to installation.
21. Install the rotating assembly into the cylinder block, making sure the crankshaft faces the correct way. The splined end of the crankshaft should face driveshaft housing side.

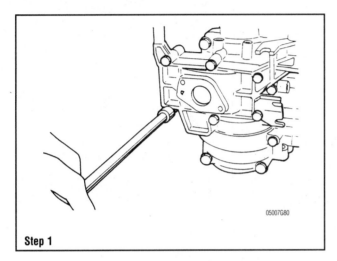
Step 1

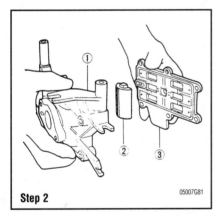
Step 2

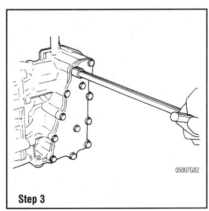
Step 3

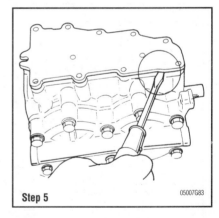
Step 5

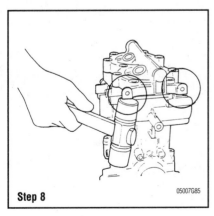
Step 8

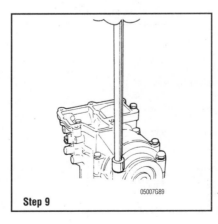
Step 9

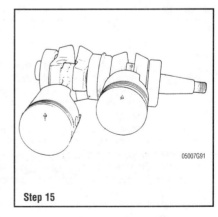
Step 15

7-22 POWERHEAD

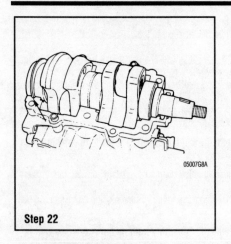
Step 22

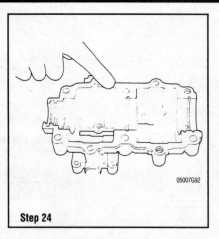
Step 24

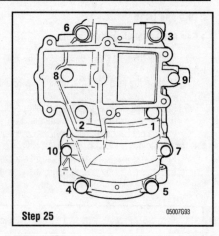
Step 25

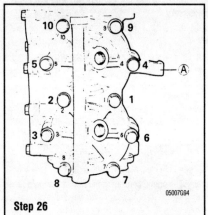
Step 26

Step 29

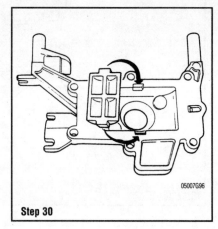
Step 30

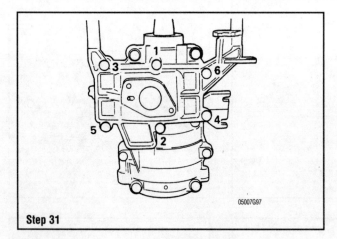
Step 31

22. The flange of the lower oil seal should fit snugly into the groove in the cylinder block.
23. Apply marine grease to the splined portion of the crankshaft.
24. Apply a bead of Suzuki Bond No. 4 (99000-31030), or equivalent sealer to the crankcase halves (shaded areas in the illustration).
25. Install the crankcase and tighten the bolts to specification in several stages using the proper tightening sequence.
26. Install the cylinder head using a new gasket and tighten the bolts to specification in several stages using the proper tightening sequence.
27. Install the exhaust plate using a new gasket.
28. Install the exhaust cover and tighten the attaching bolts securely.
29. Install the reed valve assembly making sure the reed valves face the crankshaft.
30. Install the inlet case piece with the arched surface downward and fit the tangs into the notches of the inlet case.
31. Install the inlet case and tighten the bolts to specification in several stages using the proper tightening sequence.

DT25 and DT30

♦ See Figures 3 and 4

1. Remove all components still attached to the powerhead until the unit is down to the bare cylinder block assembly
2. Remove the thermostat cover and thermostat.
3. Loosen the inlet case bolts in several stages using a criss-cross pattern.
4. Remove the inlet case.
5. Remove the reed valve assembly.
6. Remove the exhaust cover attaching bolts.

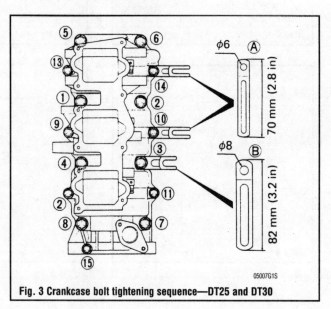

Fig. 3 Crankcase bolt tightening sequence—DT25 and DT30

POWERHEAD 7-23

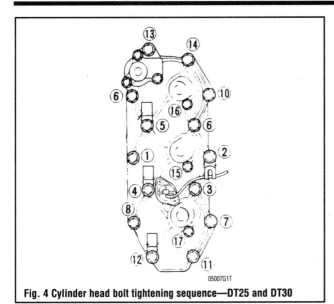

Fig. 4 Cylinder head bolt tightening sequence—DT25 and DT30

7. Insert a pry bar between the exhaust cover and cylinder block and pry the cover to remove it. Discard the exhaust cover gasket.
8. Remove the lower oil seal housing attaching bolts.
9. Insert a pry bar between the oil seal housing and cylinder block and pry the cover to remove it. Discard the oil seal housing gasket.
10. Loosen the cylinder head bolts in several stages using a criss-cross pattern.
11. Insert a pry bar between the cylinder head and cylinder block and pry the head to remove it. Discard the cylinder head gasket.
12. Loosen the cylinder head cover bolts.
13. Insert a pry bar between the cylinder head cover and the cylinder head. Pry the head cover to remove it.
14. Loosen the crankcase bolts.
15. Insert a pry bar between crankcase and the cylinder block. Pry the crankcase to remove it.
16. Remove the rotating assembly by carefully lifting the crankshaft from the cylinder block.
17. Carefully pry the piston pin retainer from its groove on the piston. Then repeat the procedure for the other side.

➡ **The circlip will tend to fly when removed. Be ready to catching it as it comes free.**

18. Piston pin retainers are only good for one usage, discard the retainers after removing them.
19. Push the piston pin from it bore in the piston using a brass drift if necessary.

➡ **The piston pin should slide smoothly from its bore. If opposition is felt, either the bore is out of round or the pin is bent. Inspect the components and replace as necessary.**

20. Remove the piston from the connecting rod.

To assemble:

21. Install the piston on the connecting rod. Make sure the arrow on the pistons point to the exhaust port side.
22. Push the piston pin into the piston bore using a brass drift if necessary.
23. Install new piston pin retainers.
24. Lubricate the seals with marine grease prior to installation.
25. Install the top and bottom crankshaft oil seals with the numbers facing the ends of the crankshaft.
26. Lubricate the rotating assembly with 2-stroke oil prior to installation.
27. Install the rotating assembly into the cylinder block, making sure the crankshaft faces the correct way. The splined end of the crankshaft should face driveshaft housing side.
28. Ensure the crankshaft C-rings are fitted into their grooves in the crankcase and the flange of the oil seals fit snugly into their grooves. Check that the bearing stopper pins are resting snugly in their cutaways.
29. Apply marine grease to the splined portion of the crankshaft.

30. Apply a bead of Suzuki Bond No. 4 (99000-31030), or equivalent sealer to the crankcase halves (shaded areas in the illustration).
31. Ensure the crankcase locating pins are installed prior to installing the crankcase.
32. Install the crankcase and tighten the bolts to specification in several stages using the proper tightening sequence.
33. Install the cylinder head using a new gasket. Tighten the cylinder head and cylinder head cover bolts to specification in several stages using the proper tightening sequence.
34. Install the exhaust cover and tighten attaching bolts securely.
35. Install the reed valve assembly and tighten attaching bolts securely.
36. Install the inlet case and tighten the bolts to specification in several stages using the proper tightening sequence.
37. Install the thermostat cover and thermostat.
38. Install all previously remove all components to the powerhead.

DT35 and DT40

♦ **See accompanying illustrations**

1. Remove all components still attached to the powerhead until the unit is down to the bare cylinder block assembly
2. Remove the thermostat cover and thermostat.
3. Loosen the inlet case bolts in several stages using a criss-cross pattern.
4. Remove the inlet case.
5. Remove the reed valve assembly.
6. Remove the exhaust cover attaching bolts.
7. Insert a pry bar between the exhaust cover and cylinder block and pry the cover to remove it. Discard the exhaust cover gasket.
8. Remove the lower oil seal housing attaching bolts.
9. Insert a pry bar between the oil seal housing and cylinder block and pry the cover to remove it. Discard the oil seal housing gasket.
10. Loosen the cylinder head bolts in several stages using the sequence defined by the numbers punched into the bolt heads. Start with bolt number 13 and work toward bolt number 1.
11. Insert a pry bar between the cylinder head and cylinder block and pry the head to remove it. Discard the cylinder head gasket.
12. Loosen the cylinder head cover bolts.
13. Insert a pry bar between the cylinder head cover and the cylinder head. Pry the head cover to remove it.
14. Loosen the crankcase bolts.
15. Insert a pry bar between crankcase and the cylinder block. Pry the crankcase to remove it.
16. Remove the rotating assembly by carefully lifting the crankshaft from the cylinder block.
17. Carefully pry the piston pin retainer from its groove on the piston. Then repeat the procedure for the other side.

➡ **The circlip will tend to fly when removed. Be ready to catching it as it comes free.**

18. Piston pin retainers are only good for one usage, discard the retainers after removing them.
19. Push the piston pin from it bore in the piston using a brass drift if necessary.

➡ **The piston pin should slide smoothly from its bore. If opposition is felt, either the bore is out of round or the pin is bent. Inspect the components and replace as necessary.**

20. Remove the piston from the connecting rod.

To assemble:

21. Install the piston on the connecting rod. Make sure the arrow on the pistons point to the exhaust port side.
22. Push the piston pin into the piston bore using a brass drift if necessary.
23. Install new piston pin retainers.
24. Lubricate the seals with marine grease prior to installation.
25. Install the top and bottom crankshaft oil seals with the numbers facing the ends of the crankshaft.
26. Lubricate the rotating assembly with 2-stroke oil prior to installation.
27. Install the rotating assembly into the cylinder block, making sure the crankshaft faces the correct way. The splined end of the crankshaft should face driveshaft housing side.

7-24 POWERHEAD

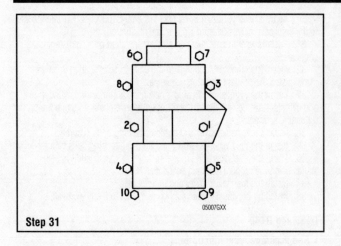
Step 31

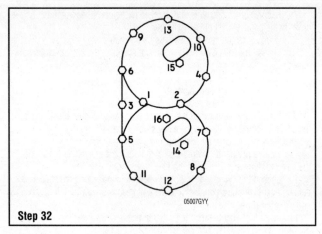
Step 32

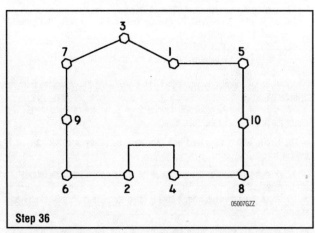
Step 36

37. Install the thermostat cover and thermostat.
38. Install all previously removed components to the powerhead.

DT50

▶ See accompanying illustrations

1. Remove all components still attached to the powerhead until the unit is down to the bare cylinder block assembly
2. Loosen the cylinder head bolts in several stages using a criss-cross pattern.
3. Remove the cylinder head.
4. Remove the valve and spring from the cylinder block.
5. Loosen the inlet case bolts in several stages using a criss-cross pattern.
6. Remove the inlet case.

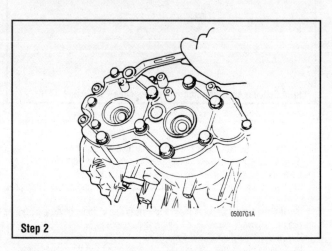
Step 2

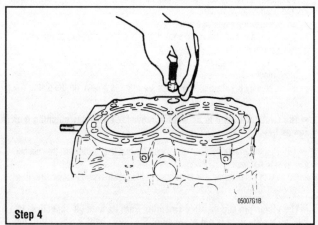
Step 4

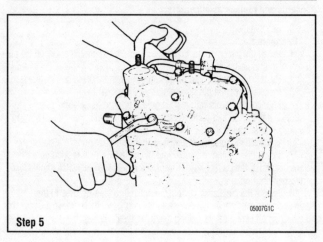
Step 5

28. The flange of the middle and lower oil seals should fit snugly into the groove in the cylinder block.
29. Apply marine grease to the splined portion of the crankshaft.
30. Apply a bead of Suzuki Bond No. 4 (99000-31030), or equivalent sealer to the crankcase halves (shaded areas in the illustration).
31. Install the crankcase and tighten the bolts to specification in several stages using the proper tightening sequence.
32. Install the cylinder head using a new gasket. Tighten the cylinder head and cylinder head cover bolts to specification in several stages using the proper tightening sequence.
33. Install the lower oil seal housing and tighten attaching bolts securely.
34. Install the exhaust cover and tighten attaching bolts securely.
35. Install the reed valve assembly and tighten attaching bolts securely.
36. Install the inlet case and tighten the bolts to specification in several stages using the proper tightening sequence.

POWERHEAD 7-25

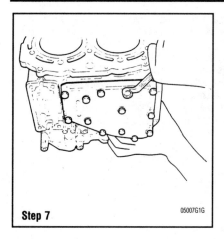
Step 7

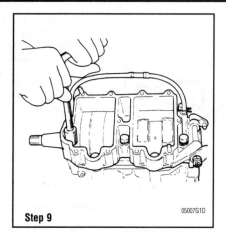
Step 9

Step 11

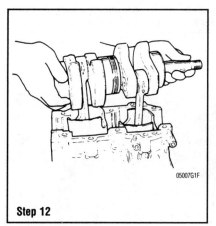
Step 12

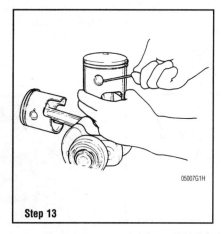
Step 13

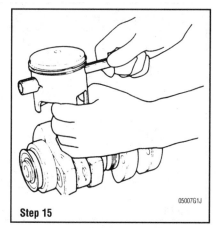
Step 15

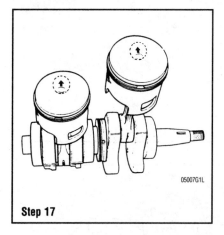
Step 17

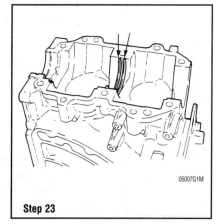
Step 23

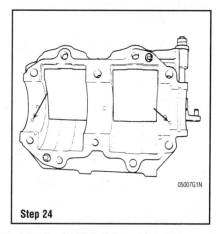
Step 24

7. Loosen the exhaust cover bolts in several stages using a criss-cross pattern.
8. Remove the exhaust cover.
9. Loosen the crankcase bolts in several stages using a criss-cross pattern.
10. Separate the crankcase from the cylinder block.
11. Remove the lower oil seal housing.
12. Remove the rotating assembly by carefully lifting the crankshaft from the cylinder block.
13. Carefully pry the piston pin retainer from its groove on the piston. Then repeat the procedure for the other side.

➡ **The circlip will tend to fly when removed. Be ready to catching it as it comes free.**

14. Piston pin retainers are only good for one usage, discard the retainers after removing them.
15. Push the piston pin from it bore in the piston using a brass drift if necessary.

➡ The piston pin should slide smoothly from its bore. If opposition is felt, either the bore is out of round or the pin is bent. Inspect the components and replace as necessary.

16. Remove the piston from the connecting rod.

To assemble:
17. Install the piston on the connecting rod. Make sure the arrow on the pistons point to the exhaust port side.
18. Push the piston pin into the piston bore using a brass drift if necessary.
19. Install new piston pin retainers.
20. Lubricate the seals with marine grease prior to installation.
21. Install the top and bottom crankshaft oil seals with the numbers facing the ends of the crankshaft.
22. Lubricate the rotating assembly with 2-stroke oil prior to installation.
23. Ensure the two thrust rings are properly installed prior to installing the rotating assembly.
24. Make sure there is no remaining O-ring compound in the illustrated area.

7-26 POWERHEAD

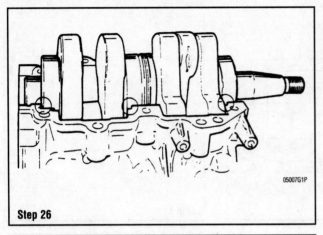
Step 26

25. Install the rotating assembly into the cylinder block, making sure the crankshaft faces the correct way. The splined end of the crankshaft should face driveshaft housing side.
26. Ensure the bearing stopper pins rest snugly in the cutaways.
27. Lubricate the seals with marine grease and install.
28. Install the crankcase and tighten the bolts to specification in several stages using the proper tightening sequence.
29. Install the exhaust cover and tighten the bolts securely.
30. Install the inlet case and tighten the bolts to specification in several stages using the proper tightening sequence.
31. Install the valve and spring into the cylinder block.
32. Install the cylinder head and tighten the bolts to specification in several stages using the proper tightening sequence.
33. Install all components previously removed.

DT55 and DT65

▶ See Figures 5 and 6

1. Remove all components still attached to the powerhead until the unit is down to the bare cylinder block assembly
2. Remove the thermostat cover and thermostat.
3. Loosen the inlet case bolts in several stages using a criss-cross pattern.
4. Remove the inlet case.

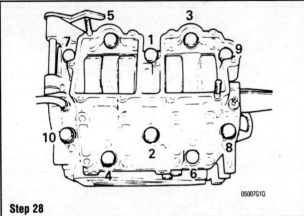
Step 28

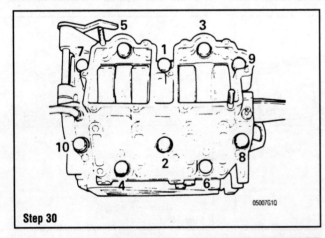
Step 30

Fig. 5 Crankcase bolt tightening sequence—DT55 and DT65

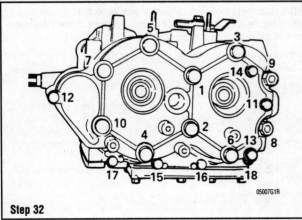
Step 32

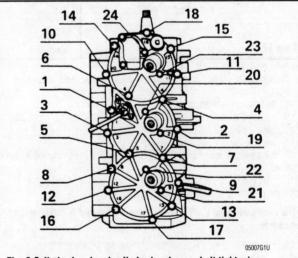
Fig. 6 Cylinder head and cylinder head cover bolt tightening sequence—DT55 and DT65

POWERHEAD 7-27

5. Remove the reed valve assembly.
6. Remove the exhaust cover attaching bolts.
7. Insert a pry bar between the exhaust cover and cylinder block and pry the cover to remove it. Discard the exhaust cover gasket.
8. Remove the lower oil seal housing attaching bolts.
9. Insert a pry bar between the oil seal housing and cylinder block and pry the cover to remove it. Discard the oil seal housing gasket.
10. Loosen the cylinder head bolts in several stages using a criss-cross pattern.
11. Insert a pry bar between the cylinder head and cylinder block and pry the head to remove it. Discard the cylinder head gasket.
12. Loosen the cylinder head cover bolts.
13. Insert a pry bar between the cylinder head cover and the cylinder head. Pry the head cover to remove it.
14. Loosen the crankcase bolts.
15. Insert a pry bar between crankcase and the cylinder block. Pry the crankcase to remove it.
16. Remove the rotating assembly by carefully lifting the crankshaft from the cylinder block.
17. Carefully pry the piston pin retainer from its groove on the piston. Then repeat the procedure for the other side.

➡The circlip will tend to fly when removed. Be ready to catching it as it comes free.

18. Piston pin retainers are only good for one usage, discard the retainers after removing them.
19. Push the piston pin from it bore in the piston using a brass drift if necessary.

➡The piston pin should slide smoothly from its bore. If opposition is felt, either the bore is out of round or the pin is bent. Inspect the components and replace as necessary.

20. Remove the piston from the connecting rod.

To assemble:
21. Install the piston on the connecting rod. Make sure the arrow on the pistons point to the exhaust port side.
22. Push the piston pin into the piston bore using a brass drift if necessary.
23. Install new piston pin retainers.
24. Lubricate the seals with marine grease prior to installation.
25. Install the top and bottom crankshaft oil seals with the numbers facing the ends of the crankshaft.
26. Lubricate the rotating assembly with 2-stroke oil prior to installation.
27. Install the rotating assembly into the cylinder block, making sure the crankshaft faces the correct way. The splined end of the crankshaft should face driveshaft housing side.
28. Ensure the crankshaft C-rings are fitted into their grooves in the crankcase and the flange of the oil seals fit snugly into their grooves.
29. Check that the bearing stopper pins are resting snugly in their cutaways.
30. Apply marine grease to the splined portion of the crankshaft.
31. Apply a bead of Suzuki Bond No. 4 (99000-31030), or equivalent sealer to the crankcase halves (shaded areas in the illustration).
32. Ensure the crankcase locating pins are installed prior to installing the crankcase.
33. Install the crankcase and tighten the bolts to specification in several stages using the proper tightening sequence.
34. Install the cylinder head using a new gasket. Tighten the cylinder head and cylinder head cover bolts to specification in several stages using the proper tightening sequence.
35. Install the exhaust cover and tighten attaching bolts securely.
36. Install the reed valve assembly and tighten attaching bolts securely.
37. Install the inlet case and tighten the bolts to specification in several stages using the proper tightening sequence.
38. Install the thermostat cover and thermostat.
39. Install all previously remove all components to the powerhead.

DT75 and DT85

♦ See Figures 7, 8, 9 and 10

1. Remove all components still attached to the powerhead until the unit is down to the bare cylinder block assembly
2. Remove the thermostat cover and thermostat.

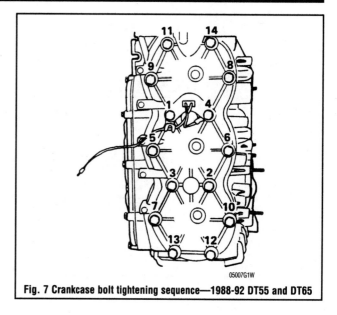
Fig. 7 Crankcase bolt tightening sequence—1988-92 DT55 and DT65

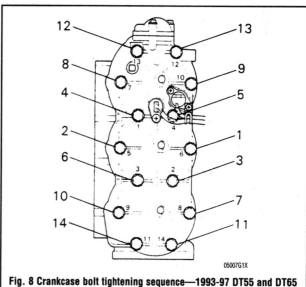
Fig. 8 Crankcase bolt tightening sequence—1993-97 DT55 and DT65

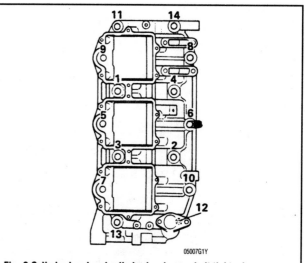
Fig. 9 Cylinder head and cylinder head cover bolt tightening sequence—1988-92 DT55 and DT65

7-28 POWERHEAD

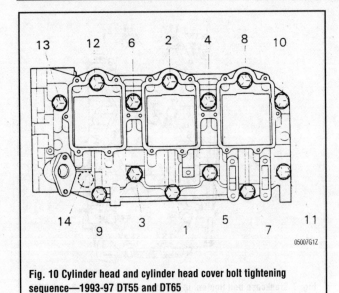

Fig. 10 Cylinder head and cylinder head cover bolt tightening sequence—1993-97 DT55 and DT65

3. Loosen the inlet case bolts in several stages using a criss-cross pattern.
4. Remove the inlet case.
5. Remove the reed valve assembly.
6. Remove the exhaust cover attaching bolts.
7. Insert a pry bar between the exhaust cover and cylinder block and pry the cover to remove it. Discard the exhaust cover gasket.
8. Remove the lower oil seal housing attaching bolts.
9. Insert a pry bar between the oil seal housing and cylinder block and pry the cover to remove it. Discard the oil seal housing gasket.
10. Loosen the cylinder head bolts in several stages using a criss-cross pattern.
11. Insert a pry bar between the cylinder head and cylinder block and pry the head to remove it. Discard the cylinder head gasket.
12. Loosen the cylinder head cover bolts.
13. Insert a pry bar between the cylinder head cover and the cylinder head. Pry the head cover to remove it.
14. Loosen the crankcase bolts.
15. Insert a pry bar between crankcase and the cylinder block. Pry the crankcase to remove it.
16. Remove the rotating assembly by carefully lifting the crankshaft from the cylinder block.
17. Carefully pry the piston pin retainer from its groove on the piston. Then repeat the procedure for the other side.

➡ The circlip will tend to fly when removed. Be ready to catching it as it comes free.

18. Piston pin retainers are only good for one usage, discard the retainers after removing them.
19. Push the piston pin from it bore in the piston using a brass drift if necessary.

➡ The piston pin should slide smoothly from its bore. If opposition is felt, either the bore is out of round or the pin is bent. Inspect the components and replace as necessary.

20. Remove the piston from the connecting rod.

To assemble:
21. Install the piston on the connecting rod. Make sure the arrow on the pistons point to the exhaust port side.
22. Push the piston pin into the piston bore using a brass drift if necessary.
23. Install new piston pin retainers.
24. Lubricate the seals with marine grease prior to installation.
25. Install the top and bottom crankshaft oil seals with the numbers facing the ends of the crankshaft.
26. Lubricate the rotating assembly with 2-stroke oil prior to installation.
27. Install the rotating assembly into the cylinder block, making sure the crankshaft faces the correct way. The splined end of the crankshaft should face driveshaft housing side.
28. Ensure the crankshaft C-rings are fitted into their grooves in the crankcase and the flange of the oil seals fit snugly into their grooves. Check that the bearing stopper pins are resting snugly in their cutaways.
29. Apply marine grease to the splined portion of the crankshaft.
30. Apply a bead of Suzuki Bond No. 4 (99000-31030), or equivalent sealer to the crankcase halves (shaded areas in the illustration).
31. Ensure the crankcase locating pins are installed prior to installing the crankcase.
32. Install the crankcase and tighten the bolts to specification in several stages using the proper tightening sequence.
33. Install the cylinder head using a new gasket. Tighten the cylinder head and cylinder head cover bolts to specification in several stages using the proper tightening sequence.
34. Install the exhaust cover and tighten attaching bolts securely.
35. Install the reed valve assembly and tighten attaching bolts securely.
36. Install the inlet case and tighten the bolts to specification in several stages using the proper tightening sequence.
37. Install the thermostat cover and thermostat.
38. Install all previously removed components to the powerhead.

DT115 and DT140

▶ See Figures 11 and 12

1. Remove all components still attached to the powerhead until the unit is down to the bare cylinder block assembly.
2. Remove the silencer cover and then the silencer case.
3. Loosen the inlet case bolts in several stages using a criss-cross pattern.
4. Remove the inlet case.
5. Remove the reed valve assembly.
6. Remove the exhaust cover attaching bolts.
7. Insert a pry bar between the exhaust cover and cylinder block and pry the cover to remove it. Discard the exhaust cover gasket.
8. Loosen the cylinder head bolts in several stages using a criss-cross pattern.
9. Insert a pry bar between the cylinder head and cylinder block and pry the head to remove it. Discard the cylinder head gasket.
10. Loosen the crankcase bolts.
11. Insert a pry bar between crankcase and the cylinder block. Pry the crankcase to remove it.
12. Remove the rotating assembly by carefully lifting the crankshaft from the cylinder block.
13. Carefully pry the piston pin retainer from its groove on the piston. Then repeat the procedure for the other side.

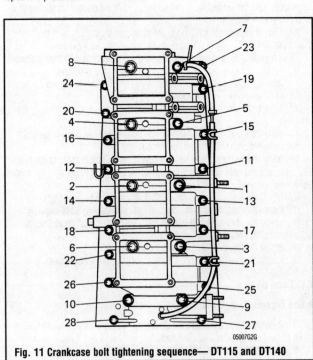

Fig. 11 Crankcase bolt tightening sequence— DT115 and DT140

POWERHEAD 7-29

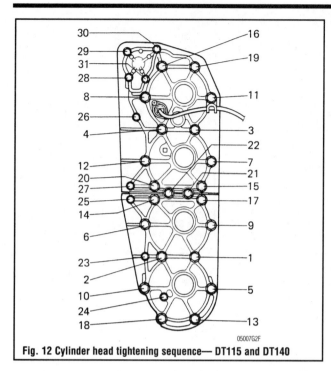

Fig. 12 Cylinder head tightening sequence— DT115 and DT140

➡ **The circlip will tend to fly when removed. Be ready to catching it as it comes free.**

14. Piston pin retainers are only good for one usage, discard the retainers after removing them.
15. Push the piston pin from it bore in the piston using a brass drift if necessary.

➡ **The piston pin should slide smoothly from its bore. If opposition is felt, either the bore is out of round or the pin is bent. Inspect the components and replace as necessary.**

16. Remove the piston from the connecting rod.

To assemble:

17. Install the piston on the connecting rod. Make sure the arrow on the pistons point to the exhaust port side.
18. Push the piston pin into the piston bore using a brass drift if necessary.
19. Install new piston pin retainers.
20. Lubricate the seals with marine grease prior to installation.
21. Install the top and bottom crankshaft oil seals with the numbers facing the ends of the crankshaft.
22. Lubricate the rotating assembly with 2-stroke oil prior to installation.
23. Firmly insert crankshaft locating pins.
24. Install the rotating assembly into the cylinder block, making sure the crankshaft faces the correct way. The splined end of the crankshaft should face driveshaft housing side.
25. Firmly fit the bearing races onto the locating pins with punch mark stamped on the circumference of the bearings directed upwards.
26. Check if the projection of spacer is in the hole of crankcase correctly.
27. Make sure that the flange of the upper oil seal fit snugly into the groove provided in crankcase.
28. Check to be sure the bearing stopper pin is resting snugly in the cutaway.
29. Also make sure that the C-ring is fitted into groove in the crankcase.
30. Make sure that the flange of oil seal housing fits snugly into the groove provided in crankcase.
31. Apply marine grease to the splined portion of the crankshaft.
32. Apply a bead of Suzuki Bond No. 4 (99000-31030), or equivalent sealer to the crankcase halves (shaded areas in the illustration).
33. Install the crankcase and tighten the bolts to specification in several stages using the proper tightening sequence.
34. Install the cylinder head using a new gasket. Tighten the cylinder head and cylinder head cover bolts to specification in several stages using the proper tightening sequence.
35. Install the exhaust cover and tighten attaching bolts securely.
36. Install the reed valve assembly and tighten attaching bolts securely.
37. Install the inlet case and tighten the bolts to specification in several stages using the proper tightening sequence.
38. Install all previously removed components to the powerhead.

V4

▶ **See accompanying illustrations**

1. Remove all components still attached to the powerhead until the unit is down to the bare cylinder block assembly.
2. Remove the exhaust cover bolts.
3. Using a pry bar, remove the exhaust cover and plate.
4. Remove the silencer cover.
5. Position the power unit with the case and silencer facing downwards on a level surface so that it will not fall over.

➡ **Place a thick rubber sheet under the silencer to prevent scratching the surface.**

6. Remove the cylinder head bolts.
7. Tap on the cylinder head with a plastic hammer to remove.
8. Remove the oil seal housing.
9. Remove the hexagon cylinder bolts.
10. Remove the crankcase nuts.
11. Pulling upward carefully and slowly, remove each cylinder head.
12. Carefully pry the piston pin retainer from its groove on the piston. Then repeat the procedure for the other side.

➡ **The circlip will tend to fly when removed. Be ready to catching it as it comes free.**

13. Piston pin retainers are only good for one usage, discard the retainers after removing them.
14. Push the piston pin from it bore in the piston using a brass drift if necessary.

➡ **The piston pin should slide smoothly from its bore. If opposition is felt, either the bore is out of round or the pin is bent. Inspect the components and replace as necessary.**

15. Remove the piston from the connecting rod.
16. Place the crankcase on a wooden stand.
17. Remove the inlet cases with the reed valves.
18. Remove the clutch shaft.
19. Remove the crankcase bolts and lift the rotating assembly out from the crankcase.

To assemble:

20. Lubricate the seals with marine grease prior to installation.
21. Install the top and bottom crankshaft oil seals with the numbers facing the ends of the crankshaft.
22. Lubricate the rotating assembly with 2-stroke oil prior to installation.

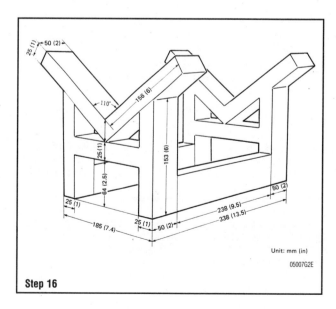

Step 16

7-30 POWERHEAD

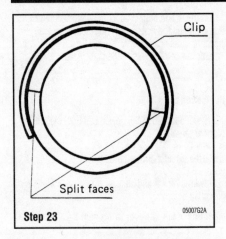
Step 23

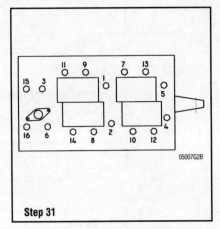
Step 31

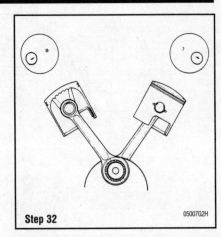
Step 32

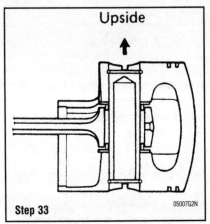
Step 33

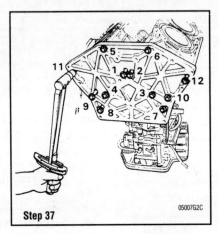
Step 37

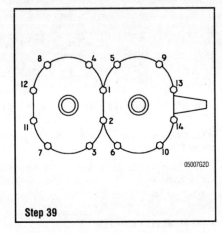
Step 39

23. Install the middle bearing on the crankshaft with the clip covering both split faces of the bearing.
24. Install the rotating assembly into the cylinder block, making sure the crankshaft faces the correct way. The splined end of the crankshaft should face driveshaft housing side.
25. Firmly fit the bearing races onto the locating pins with punch mark stamped on the circumference of the bearings directed upwards.
26. Ensure the flanges of the upper oil seal fit snugly into the groove provided in crankcase.
27. Check to be sure that the under bearing stopper pin is resting snugly in the cutaway.
28. Make the "C" ring is fitted into groove in the crankcase.
29. Check to be sure that the end gap of the seal rings face up.
30. Apply marine grease to the splined portion of the crankshaft.
31. Install the crankcase and tighten the crankcase nuts to specification using the proper tightening sequence.

➡ **After the crankcase has been assembled, turn the crankshaft to see whether it makes any abnormal noise. If the crankshaft makes any abnormal noise, disassemble the crankcase to find the trouble.**

32. Install the piston on the connecting rod. Make sure the arrow on the pistons point to the exhaust port side.
33. Push the piston pin into the piston bore using a brass drift if necessary.
34. Install new piston pin retainers.
35. Install a new cylinder head gasket and hold it in place with a light film of grease.
36. Install each cylinder head onto the cylinder block and snug each bolt to hold the cylinders in place.
37. Install the cylinder/crankcase assembly plate (09912-68720) and tighten the bolts to the specified torque using the proper sequence.
38. Tighten the hexagon cylinder bolts to specification using a criss-cross tightening sequence.
39. Install the cylinder head and tighten the cylinder head bolts to specification using the proper tightening sequence.

40. Install the oil seal housing and tighten the bolts to specification.
41. Install the exhaust cover and plate. Tighten the exhaust cover bolts to specification.
42. Install all components previously removed from the cylinder block.

V6

♦ See accompanying illustrations

1. Remove all components still attached to the powerhead until the unit is down to the bare cylinder block assembly.
2. Remove the exhaust cover bolts.
3. Using a pry bar, remove the exhaust cover and plate.
4. Remove the silencer cover.
5. Position the power unit with the case and silencer facing downwards on a level surface so that it will not fall over.

➡ **Place protective padding under the inlet case to prevent scratching the surface.**

6. Remove the cylinder head bolts.
7. Tap on the cylinder head with a plastic hammer to remove.
8. Remove the oil seal housing.
9. Remove the hexagon cylinder bolts.
10. Pulling upward carefully and slowly, remove each cylinder head.
11. Carefully pry the piston pin retainer from its groove on the piston. Then repeat the procedure for the other side.

➡ **The circlip will tend to fly when removed. Be ready to catching it as it comes free.**

12. Piston pin retainers are only good for one usage, discard the retainers after removing them.
13. Push the piston pin from it bore in the piston using a brass drift if necessary.

➡ **The piston pin should slide smoothly from its bore. If opposition is felt, either the bore is out of round or the pin is bent. Inspect the components and replace as necessary.**

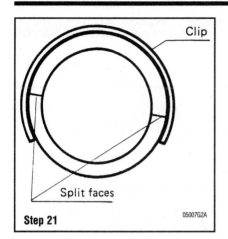
Step 21

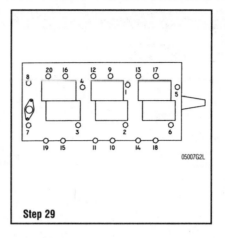
Step 29

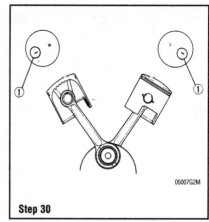
Step 30

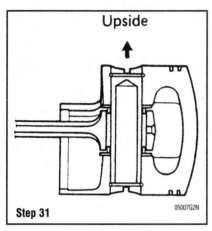
Step 31

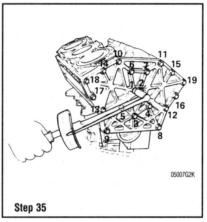
Step 35

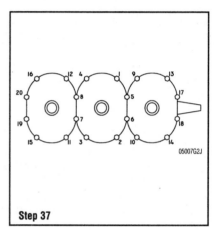
Step 37

14. Remove the piston from the connecting rod.
15. Place the crankcase on a fabricated wooden stand.
16. Remove the inlet cases with the reed valves.
17. Remove the crankcase bolts and lift the rotating assembly out from the crankcase.

To assemble:
18. Lubricate the seals with marine grease prior to installation.
19. Install the top and bottom crankshaft oil seals with the numbers facing the ends of the crankshaft.
20. Lubricate the rotating assembly with 2-stroke oil prior to installation.
21. Install the middle bearing on the crankshaft with the clip covering both split faces of the bearing.
22. Install the rotating assembly into the cylinder block, making sure the crankshaft faces the correct way. The splined end of the crankshaft should face driveshaft housing side.
23. Firmly fit the bearing races onto the locating pins with punch mark stamped on the circumference of the bearings directed upwards.
24. Ensure the flanges of the upper oil seal fit snugly into the groove provided in crankcase.
25. Check to be sure that the under bearing stopper pin is resting snugly in the cutaway.
26. Make the "C" ring is fitted into groove in the crankcase.
27. Check to be sure that the end gap of the seal rings face up.
28. Apply marine grease to the splined portion of the crankshaft.

29. Install the crankcase and tighten the crankcase nuts to specification using the proper tightening sequence.

➡ **After the crankcase has been assembled, turn the crankshaft to see whether it makes any abnormal noise. If the crankshaft makes any abnormal noise, disassemble the crankcase to find the trouble.**

30. Install the piston on the connecting rod. Make sure the arrow on the pistons point to the exhaust port side.
31. Push the piston pin into the piston bore using a brass drift if necessary.
32. Install new piston pin retainers.
33. Install a new cylinder head gasket and hold it in place with a light film of grease.
34. Install each cylinder head onto the cylinder block and snug each bolt to hold the cylinders in place.
35. Install the cylinder/crankcase assembly plate (09912-68720) and tighten the bolts to the specified torque using the proper sequence.
36. Using a special socket (09911-78730), tighten the cylinder bolts to specification using a criss-cross tightening sequence.
37. Install the cylinder head and tighten the cylinder head bolts to specification using the proper tightening sequence.
38. Install the oil seal housing and tighten the bolts to specification.
39. Install the exhaust cover and plate. Tighten the exhaust cover bolts to specification.
40. Install all components previously removed from the cylinder block.

7-32 POWERHEAD

POWERHEAD RECONDITIONING

Determining Powerhead Condition

Anything that generates heat and/or friction will eventually burn or wear out (for example, a light bulb generates heat, therefore its life span is limited). With this in mind, a running powerhead generates tremendous amounts of both; friction is encountered by the moving and rotating parts inside the powerhead and heat is created by friction and combustion of the fuel. However, the powerhead has systems designed to help reduce the effects of heat and friction and provide added longevity. The oil injection system combines oil with the fuel to reduce the amount of friction encountered by the moving parts inside the powerhead, while the cooling system reduces heat created by friction and combustion. If either system is not maintained, a break-down will be inevitable. Therefore, you can see how regular maintenance can affect the service life of your powerhead.

There are a number of methods for evaluating the condition of your powerhead. A secondary compression test can reveal the condition of your pistons, piston rings, cylinder bores and head gasket(s). A primary compression test can determine the condition of all engine seals and gaskets. Because the 2-stroke powerhead is a pump, the crankcase must be sealed against pressure created on the down stroke of the piston and vacuum created when the piston moves toward top dead center. If there are air leaks into the crankcase, insufficient fuel will be brought into the crankcase and into the cylinder for normal combustion.

PRIMARY COMPRESSION TEST

Because the 2-stroke powerhead is a pump, the crankcase must be sealed against pressure created on the down stroke of the piston and vacuum created when the piston moves toward top dead center. If there are air leaks into the crankcase, insufficient fuel will be brought into the crankcase and into the cylinder for normal combustion.

➡ **If it is a very small leak, the powerhead will run poorly, because the fuel mixture will be lean and cylinder temperatures will be hotter than normal.**

Air leaks are possible around any seal, O-ring, cylinder block mating surface or gasket. Always replace O-rings, gaskets and seals when service work is performed.

If the powerhead is running, soapy water can be sprayed onto the suspected sealing areas. If bubbles develop, there is a leak at that point. Oil around sealing points and on ignition parts under the flywheel indicates a crankcase leak.

The base of the powerhead and lower crankshaft seal is impossible to check on an installed powerhead. When every test and system have been checked out and the bottom cylinder seems to be effecting performance, then the lower seal should be tested.

Adapter plates available from tool manufacturers to seal the inlet, exhaust and base of the powerhead. Adapter plates can also be manufactured by cutting metal block off plates from pieces of plate steel or aluminum. A pattern made from the gaskets can be used for an accurate shape. Seal these plates using rubber or silicone gasket making compound.

1. Install adapter plates over the intake ports and the exhaust ports to completely seal the powerhead.

➡ **When installing the adapter plates, make sure to leave the water jacket holes open.**

2. Into one adapter, place an air fitting which will accept a hand air pump.
3. Using the hand pump (or another regulated air source), pressurize the crankcase to five pounds of pressure.
4. Spray soapy water around the lower seal area and other sealed areas watching for bubbles which indicate a leaking point.
5. Turn the powerhead upside down and fill the water jacket with water. If bubbles show up in the in the water when a positive pressure is applied to the crankcase, there may be cracks or corrosion holes in the cooling system passages. These holes can cause a loss of cooling system effectiveness and lead to overheating.
6. After the pressure test is completed, pull a vacuum to stress the seals in the opposite direction and watch for a pressure drop.
7. Note the leaking areas and replace the seals or gaskets.

SECONDARY COMPRESSION TEST

♦ See Figure 13

The actual pressure measured during a secondary compression test is not as important as the variation from cylinder to cylinder. On multi-cylinder powerheads, a variation of 15 psi or more is considered questionable. On single cylinder powerheads, a drop of 15 psi from the normal compression pressure you established when it was new is cause for concern (you did do a compression test on it when it was new, didn't you?).

➡ **If the powerhead been in storage for an extended period, the piston rings may have relaxed. This will often lead to initially low and misleading readings. Always run an engine to operating temperature to ensure that the reading you get is accurate.**

1. Disable the ignition system by removing the lanyard clip. If you do not have a lanyard, take a wire jumper lead and connect one end to a good engine ground and the other end to the metal connector inside the spark plug boot, using one jumper for each plug wire. Never simply disconnect all the plug wires.

✳✳ CAUTION

Removing all the spark plugs and cranking over the powerhead can lead to an explosion if raw fuel/oil sprays out of the plug holes. A plug wire could spark and ignite this mix outside of the combustion chamber if it isn't grounded to the engine.

2. Remove all the spark plugs and be sure to keep them in order. Carefully inspect the plugs, looking for any inconsistency in coloration and for any sign of water or rust near the tip.
3. Thread the compression gauge into the No. 1 spark-plug hole, taking care to not crossthread the fitting.
4. Open the throttle to the wide open throttle position and hold it there.

➡ **Some engines allow only minimal opening if the gearshift is in neutral, to guard against over-revving.**

5. Crank over the engine an equal number of times for each cylinder you test, zeroing the gauge for each cylinder.
6. If you have electric start, count the number of seconds you count. On manual start, pull the starter rope four to five times for each cylinder you are testing.
7. Record your readings from each cylinder. When all cylinders are tested,

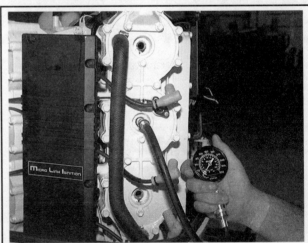

Fig. 13 The actual pressure measured during a secondary compression test is not as important as the variation from cylinder to cylinder

POWERHEAD 7-33

compare the readings and determine if pressures are within the 15 psi criterion.

8. If compression readings are lower than normal for any cylinders, try a "wet" compression test, which will temporarily seal the piston rings and determine if they are the cause of the low reading.

9. Using a can of fogging oil, fog the cylinder with a circular motion to distribute oil spray all around the perimeter of the piston. Retest the cylinder.

   a. If the compression rises noticeably, the piston rings are sticking. You may be able to cure the problem by decarboning the powerhead.

   b. If the dry compression was really low and no change is evident during the wet test, the cylinder is dead. The piston and/or are worn beyond specification and a powerhead overhaul or replacement is necessary.

10. If two adjacent cylinders on a multi-cylinder engine give a similarly low reading then the problem may be a faulty head gasket. This should be suspected if there was evidence of water or rust on the spark plugs from these cylinders.

Buy or Rebuild?

▶ See Figures 14 and 15

Now that you have determined that your powerhead is worn out, you must make some decisions. The question of whether or not a powerhead is worth rebuilding is largely a subjective matter and one of personal worth. Is the powerhead a popular one, or is it an obsolete model? Are parts available? Is the outboard it's being put into worth keeping? Would it be less expensive to buy a new powerhead, have your powerhead rebuilt by a pro, rebuild it yourself or buy a used powerhead? Or would it be simpler and less expensive to buy another outboard? If you have considered all these matters and more and have still decided to rebuild the powerhead, then it is time to decide how you will rebuild it.

➡ **The editors at Seloc® feel that most powerhead machining should be performed by a professional machine shop. Don't think of it as wasting money, rather, as insurance that the job has been done right the first time. There are many expensive and specialized tools required to perform such tasks as boring and honing a powerhead. Even inspecting the parts requires expensive micrometers and gauges to properly measure wear and clearances. Also, a machine shop can deliver to you clean and ready to assemble parts, saving you time and aggravation. Your maximum savings will come from performing the removal, disassembly, assembly and installation of the powerhead and purchasing or renting only the tools required to perform the above tasks. Depending on the particular circumstances, you may save 40 to 60 percent of the cost doing these yourself.**

A complete rebuild or overhaul of a powerhead involves replacing or reconditioning all of the moving parts (pistons, rods, crankshaft, etc.) with new or remanufacturerd ones and machining the non-moving wearing surfaces of the block and heads. Unfortunately, this may not be cost effective. For instance, your crankshaft may have been damaged or worn, but it can be machined for a minimal fee.

So, as you can see, you can replace everything inside the powerhead, but, it is wiser to replace only those parts which are really needed and, if possible, repair the more expensive ones.

Powerhead Overhaul Tips

▶ See Figure 16

Most powerhead overhaul procedures are fairly standard. In addition to specific parts replacement procedures and specifications for your individual powerhead, this section is also a guide to acceptable rebuilding procedures. Examples of standard rebuilding practice are given and should be used along with specific details concerning your particular powerhead.

Competent and accurate machine shop services will ensure maximum performance, reliability and powerhead life. In most instances it is more profitable for the do-it-yourself mechanic to remove, clean and inspect the component, buy the necessary parts and deliver these to a shop for actual machine work.

Much of the assembly work (crankshaft, bearings, pistons, connecting rods and other components) is well within the scope of the do-it-yourself mechanic's

Fig. 14 The question of whether or not a powerhead is worth rebuilding is largely a subjective matter and one of personal worth. This powerhead is not worth much in its present condition

Fig. 15 A burned piston like this one will be replaced during an overhaul. The condition which caused the hole in the top of the piston must be identified and corrected or the same thing will happen again

Fig. 16 Much of the assembly work (crankshaft, bearings, pistons, connecting rods and other components) is well within the scope of the average do-it-yourself mechanic's tools and abilities

7-34 POWERHEAD

tools and abilities. You will have to decide for yourself the depth of involvement you desire in a powerhead repair or rebuild.

TOOLS

The tools required for a powerhead overhaul or parts replacement will depend on the depth of your involvement. With a few exceptions, they will be the tools found in an average do it yourselfer's tool kit. More in-depth work will require some or all of the following:
- A dial indicator (reading in thousandths) mounted on a universal base
- Micrometers and telescope gauges
- Jaw and screw-type pullers
- Scraper
- Ring groove cleaner
- Piston ring expander and compressor
- Ridge reamer
- Cylinder hone or glaze breaker
- Plastigage®
- Powerhead stand

The use of most of these tools is illustrated in this section. Many can be rented for a one-time use from a local parts store or tool supply house.

Occasionally, the use of special tools is necessary. See the information on Special Tools and the Safety Notice in the front of this book before substituting another tool.

CAUTIONS

Aluminum is extremely popular for use in powerheads, due to its low weight. Observe the following precautions when handling aluminum parts:
- Never hot tank aluminum parts, the caustic hot tank solution will eat the aluminum
- Remove all aluminum parts (identification tag, etc.) from powerhead parts prior to hot tanking
- Always coat threads lightly with oil or anti-seize compounds before installation, to prevent seizure
- Never overtighten bolts or spark plugs especially in aluminum threads

When assembling the powerhead, any parts that will be exposed to frictional contact must be prelubed to provide lubrication at initial start-up. Any product specifically formulated for this purpose can be used.

When semi-permanent (locked, but removable) installation of bolts or nuts is desired, threads should be cleaned and coated with Loctite® or another similar, commercial non-hardening sealant.

CLEANING

Before the powerhead and its components are inspected, they must be thoroughly cleaned. You will need to remove any varnish, oil sludge and/or carbon deposits from all of the components to insure an accurate inspection. A crack in the block or cylinder head can easily become overlooked if hidden by a layer of sludge or carbon.

Most of the cleaning process can be carried out with common hand tools and readily available solvents or solutions. Carbon deposits can be chipped away using a hammer and a hard wooden chisel. Old gasket material and varnish or sludge can usually be removed using a scraper and/or cleaning solvent. Extremely stubborn deposits may require the use of a power drill with a wire brush. Always follow any safety recommendations given by the manufacturer of the tool and/or solvent. You should always wear eye protection during any cleaning process involving scraping, chipping or spraying of solvents.

➡️**If using a wire brush, use extreme care around any critical machined surfaces (such as the gasket surfaces, bearing saddles, cylinder bores, etc.). USE OF A WIRE BRUSH IS NOT RECOMMENDED ON ANY ALUMINUM COMPONENTS.**

An alternative to the mess and hassle of cleaning the parts yourself is to drop them off at a local machine shop. They will, more than likely, have the necessary equipment to properly clean all of the parts for a nominal fee.

✳✳ CAUTION

Always wear eye protection during any cleaning process involving scraping, chipping or spraying of solvents.

Remove any plugs or pressed-in bearings and carefully wash and degrease all of the powerhead components including the fasteners and bolts. Small parts should be placed in a metal basket and allowed to soak. Use pipe cleaner type brushes and clean all passageways in the components.

Use a ring expander to remove the rings from the pistons. Clean the piston ring grooves with a ring groove cleaner or a piece of broken ring. Scrape the carbon off of the top of the piston. You should never use a wire brush on the pistons. After preparing all of the piston assemblies in this manner, wash and degrease them again.

REPAIRING DAMAGED THREADS

▶ See Figures 17, 18, 19, 20 and 21

Several methods of repairing damaged threads are available. Heli-Coil®, Keenserts® and Microdot® are among the most widely used. All involve basically the same principle—drilling out stripped threads, tapping the hole and installing a prewound insert—making welding, plugging and oversize fasteners unnecessary.

Two types of thread repair inserts are usually supplied: a standard type for most inch coarse, inch fine, metric course and metric fine thread sizes and a spark lug type to fit most spark plug port sizes. Consult the individual tool manufacturer's catalog to determine exact applications. Typical thread repair kits will contain a selection of prewound threaded inserts, a tap (corresponding to the outside diameter threads of the insert) and an installation tool. Spark plug inserts usually differ because they require a tap equipped with pilot threads and a combined reamer/tap section. Most manufacturers also supply blister-packed thread repair inserts separately in addition to a master kit containing a variety of taps and inserts plus installation tools.

Before attempting to repair a threaded hole, remove any snapped, broken or damaged bolts or studs. Penetrating oil can be used to free frozen threads. The offending item can usually be removed with locking pliers or using a screw/stud extractor. After the hole is clear, the thread can be repaired, as shown in the series of accompanying illustrations and in the kit manufacturer's instructions.

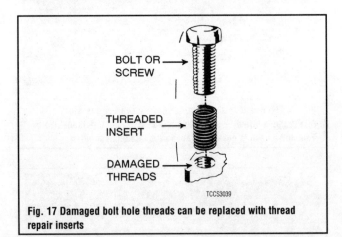

Fig. 17 Damaged bolt hole threads can be replaced with thread repair inserts

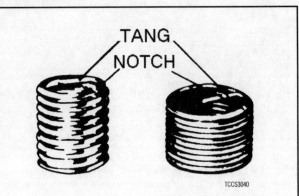

Fig. 18 Standard thread repair insert (left) and a spark plug thread insert

POWERHEAD 7-35

Fig. 19 Drill out the damaged threads with the specified size bit. Be sure to drill completely through the hole or to the bottom of a blind hole

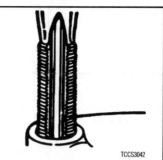

Fig. 20 Using the kit, tap the hole to receive the thread insert. Keep the tap well oiled and back it out frequently to avoid clogging the threads

Fig. 21 Screw the insert onto the installer tool until the tang engages the slot. Thread the insert into the hole until it is ¼–½ turn below the top surface, then remove the tool and break of the tang using a punch

Powerhead Preparation

To properly rebuild a powerhead, you must first remove it from the outboard, then disassemble and inspect it. Ideally you should place your powerhead on a stand. This affords you the best access to the components. Follow the manufacturer's directions for using the stand with your particular powerhead.

Now that you have the powerhead on a stand, it's time to strip it of all but the necessary components. Before you start disassembling the powerhead, you may want to take a moment to draw some pictures, fabricate some labels or get some containers to mark and hold the various components and the bolts and/or studs which fasten them. Modern day powerheads use a lot of little brackets and clips which hold wiring harnesses and such and these holders are often mounted on studs and/or bolts that can be easily mixed up. The manufacturer spent a lot of time and money designing your outboard and they wouldn't have wasted any of it by haphazardly placing brackets, clips or fasteners. If it's present when you disassemble it, put it back when you assemble it, you will regret not remembering that little bracket which holds a wire harness out of the path of a rotating part.

You should begin by unbolting any accessories attached to the powerhead. Remove any covers remaining on the powerhead. The idea is to reduce the powerhead to the bare necessities (cylinder head(s), cylinder block, crankshaft, pistons and connecting rods), plus any other 'in block' components.

Cylinder Block and Head

GENERAL INFORMATION

▶ See Figures 22 thru 27

The cylinder block is made of aluminum and may have cast-in iron cylinder liners. It is the major part of the powerhead and care must be given to this part when service work is performed. Mishandling or improper service procedures

Fig. 22 The cylinder block . . .

Fig. 23 . . . crankcase half . . .

Fig. 24 . . . and cylinder head make up the major components of the cylinder assembly

Fig. 25 In this cylinder, an exhaust port can be seen above the level of the piston. The inlet port is on the opposite side of the cylinder wall, below the piston

Fig. 26 The cylinder block and crankcase half are machined to fit together perfectly. They provide a cradle for the spinning crankshaft

Fig. 27 To seal the ends of the cylinder assembly around the crankshaft, O-rings are installed around the end caps and neoprene seals are installed inside the cap and seal against the crankshaft

7-36 POWERHEAD

performed on this assembly may make scrap out of an otherwise good casting. The cylinder assembly casting and other major castings on the outboard are expensive and need to be cared for accordingly.

There are three parts to the cylinder assembly, the cylinder block, the cylinder head and the crankcase half. The cylinder block and crankcase half are married together and line bored to receive the crankshaft bearings, reed blocks and on some powerheads sealing rings. After this operation they are treated as one casting.

➡Remember that anything done to the mating surfaces during service work will change the inner bore diameter for the main bearings, reed blocks and sealing rings and possibly prevent the block and crankcase mating surfaces from sealing.

The only service work allowed on the mating surface is a lapping operation to remove nicks from the service. Carefully guard this surface when other service work is being performed. The different sealing materials used to seal the mating surfaces are sealing strips, sealing compound and Loctite®.

Since the 2-stroke powerhead operates like a pump with one inlet and one outlet for each cylinder, special sealing features must be designed into the cylinder assembly to seal each individual cylinder in a multi-cylinder powerhead. Each inlet manifold must be completely sealed both for vacuum and pressure. One way of doing this internally is with a labyrinth seal, which is located between two adjacent cylinders next to the crankshaft. It may be of aluminum or brass, formed in the assembly and machined with small circular grooves running very close to a machined area on the crankshaft. The tolerance is so close that fuel residue puddling in the seal effectively completes the seal between the cylinder block and crankcase halves against the crankshaft. Crankcase pressures are therefore retained to each individual cylinder. No repair of the labyrinth seal is made. If damage has occurred to the seal, the main bearings have allowed the crankshaft to run out and rub.

Another method of internal sealing between the crankcases is with seal rings. These rings are installed in grooves in the crankshaft. When the crankshaft is installed, the sealing rings mate up to and seal against the web in the cylinder block crankcase halves and crankshaft. Sealing rings of different thickness are available for service work. The side tolerance is close, so puddled fuel residue will effectively complete the seal between crankcases and crankshaft.

To seal the ends of the cylinder assembly around the crankshaft, O-rings are installed around the end caps and neoprene seals are installed inside the cap and seal against the crankshaft.

INSPECTION

♦ See Figures 28, 29 and 30

Everytime the cylinder head is removed, the cylinder head and cylinder block deck should be checked for warping. Do this with a straight edge or a surface block. If the cylinder head or cylinder block deck are warped, the surface should

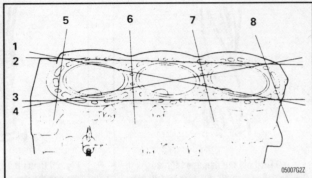

Fig. 29 When inspecting components for warping, check in multiple directions

Fig. 30 To help prevent bolts from seizing due to corrosion, coat threads with a good antiseize compound

be machined flat by a competent machine shop. Minor warpage may be cured by using emery paper in a figure eight motion on a surface block until the surface is true.

Inspect the cylinder head and cylinder block for cracks and damage to the bolt holes caused by galvanic corrosion. On models which do not use a cylinder head, check the cylinder dome for holes or cracks caused by overheating and pre-ignition. The spark plug threads may also be damaged by overtorquing the spark plug.

Quite often the small bolts around the cylinder block sealing area are seized by corrosion. If white powder is evident around the bolts, stop. Galvanic corrosion is probably seizing the shank of the bolt and possibly the threads as well. Putting a wrench on them may just twist the head off, creating one big mess. Know the strength of the bolt and stop before it breaks. If it does break, don't reach for an easy out, it won't work.

A good way to service these seized bolts is with localized heat (from a heat gun, not a torch) and a good penetrating oil. Heat the aluminum casting, not the bolt. This releases the bolt from the corrosive grip by creating clearance between the bolt, the corrosion and the aluminum casting. Be careful because too much heat will melt the casting. Many bolts can be released in this way, preventing drilling out the total bolt and heli-coiling the hole or tapping the hole for an oversize bolt.

To help prevent bolts from seizing due to corrosion, coat threads with a good antiseize compound.

Cylinder Bores

GENERAL INFORMATION

The purpose of the cylinder bore is to help lock in combustion gases, provide a guided path for the piston to travel within, provide a lubricated surface for

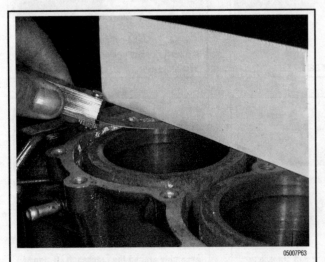

Fig. 28 Everytime the cylinder head is removed, the cylinder head and cylinder block deck should be checked for warping using a straight edge and a feeler gauge

the piston rings to seal against and transfer heat to the cooling system. These functions are carried out through all engine speeds. To function properly the cylinder has to have a true machined surface and must have the proper finish installed on it to retain lubricant.

INSPECTION

♦ See Figures 31 and 32

The roundness of the cylinder diameter and the straightness of the cylinder wall should be inspected carefully. Micrometer readings should be taken at several points to determine the cylinder condition. Start at the bottom using an outside micrometer or dial bore gauge. By starting at the bottom, below the area of ring travel, cylinder bore diameter can be determined and a determination can be made if the powerhead is standard or has been bored oversize. Take the second measurement straight up from the first in the area of the ports and note that the cylinder is larger here. This is the area where the rings ride and it has worn slightly. Take the third measurement within a half inch of the top of the cylinder, straight up from where the second measurement was taken. These three measurements should be repeated with the measuring instrument turned 90° clockwise.

After the readings are taken, you will have enough information to access the cylinder condition. This will tell you if the rings can simply be replaced or if the cylinder will need to be overbored. While measuring the cylinder, you should also be noting if there is a cross-hatched pattern on the cylinder walls. Also note any scuffing or deep scratches.

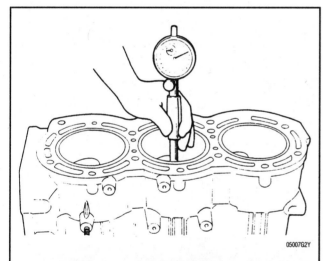

Fig. 31 The roundness of the cylinder diameter and the straightness of the cylinder wall should be inspected using a dial bore gauge

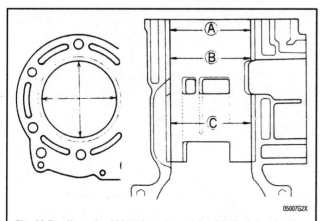

Fig. 32 Readings should be taken at several points to determine the cylinder condition. Start at the bottom and work your way to the top

REFINISHING

If the cylinder is out of round, worn beyond specification, scored or deeply scratched, reboring will be necessary. If the cylinder is within specification, it can be deglazed with a flex hone and new rings installed.

➡ Some cylinders are chrome plated and require special service procedures. Consult a qualified machine shop when dealing with chrome plated cylinders.

Almost all engine block refinishing must be performed by a machine shop. If the cylinders are not to be rebored, then the cylinder glaze can be removed with a ball hone. When removing cylinder glaze with a ball hone, use a light or penetrating type oil to lubricate the hone. Do not allow the hone to run dry as this may cause excessive scoring of the cylinder bores and wear on the hone. If new pistons are required, they will need to be installed to the connecting rods. This should be performed by a machine shop as the pistons must be installed in the correct relationship to the rod or engine damage can occur.

When deglazing, it is important to retain the factory surface of the cylinder wall. The cross-hatched patter on the cylinder wall is used to retain oil and seal the rings. As the piston rings move up and down the wall, a glaze develops. The hone is used to remove this glaze and reestablish the basket weave pattern. The pattern and the finish is has a satin look and makes an excellent surface for good retention of 2-stroke oil on the cylinder wall.

There is nothing magic about the crosshatch angle but there should be one similar to what the factory used. (approximately 20–40°). Too steep an angle or too flat a pattern is not acceptable and as it is not good for ring seating. Since the hone reverses as it is being pushed down and pulled up the cylinder wall, many different angles are created. Multiple criss-crossing angles are the secret for longevity of the cylinder and the rings. The pattern allows 2-stroke oil to flow under the piston ring bearing surface and prevents a metal-to-metal contact between the cylinder wall and piston ring. The satin finish is necessary to prevent early break-in scuffing and to seat the ring correctly.

After the cylinder hone operation has been completed, one very important job remains. The grit that was developed in the machining process must be thoroughly cleaned up. Grit left in the powerhead will find its way into the bearings and piston rings and become embedded into the piston skirts, effectively grinding away at these precision parts. Relate this to emery paper applied to a piece of steel or steel against a grinding stone. The effect is removal of material from the steel. Grit left in the powerhead will damage internal components in a very short time.

Wiping down the cylinder bores with an oil or solvent soaked rag does not remove grit. Cleaning must be thorough so that all abrasive grit material has been removed from the cylinders. It is important to use a scrub brush and plenty of soapy water. Remember that aluminum is not safe with all cleaning compounds, so use a mild dish washing detergent that is designed to remove grease. After the cylinder is thought to be clean, use a white paper towel to test the cylinder. Rub the paper towel up and down on the cylinder and look for the presence of gray color on the towel. The gray color is grit. Re-scrub the cylinder until it is perfectly clean and passes the paper towel test. When the cylinder passes the test, immediately coat it with 2-stroke oil to prevent rust from forming.

➡ Rust forms very quickly on clean, oil free metal. Immediately coat all clean metal with 2-stroke oil to prevent the formation of rust.

Pistons

GENERAL INFORMATION

♦ See Figures 33, 34, 35 and 36

Piston are the moveable end of a cylinder. The cylinder bore provides a guided path for the piston allowing a small clearance between the piston skirt and cylinder wall. This clearance allows for piston expansion and controls piston rock within the cylinder.

Modern piston design is such that the head of the piston directs incoming fuel toward the top of the cylinder and outgoing exhaust to the exhaust port in the cylinder wall. This design is called a deflector type piston head. The deflector dome deflects the incoming fuel upward to the spark plug end of the cylinder, partially cooling the cylinder and spark plug tip. It also purges the spent

7-38 POWERHEAD

Fig. 33 A hole placed in the side of the piston, commonly referred to as the piston boss, is used to mount the piston to the piston pin

Fig. 34 The piston has machined grooves in which the rings are installed. They are carried along with the piston as it travels up and down the cylinder wall

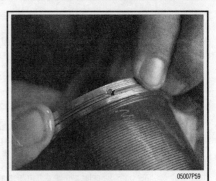

Fig. 35 There is one small pin in each ring groove to prevent the ring from rotating

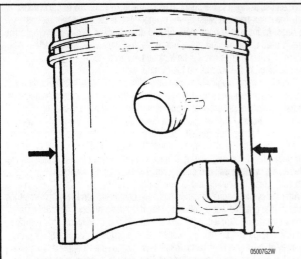

Fig. 36 Piston diameter should be measured at a specific position on the piston which the manufacturer will specify

INSPECTION

♦ See Figures 37 and 38

The piston needs to be inspected for damage. Check the head for erosion caused by excessive heat, lean mixtures and out of specification timing/synchronization. Examine the ring land area to see if it is flat and not rounded over. Also look for burned through areas caused by preignition. Check the skirt for scoring caused by a break through of the oil film, excessive cylinder wall temperatures, incorrect timing/synchronization or inadequate lubrication.

To measure the piston diameter, place an outside micrometer on the piston skirt at the specified location. All pistons in a given powerhead should read the same. Check the specifications for placement of the micrometer when measuring pistons. Generally there is a specific place on the piston. This is especially true of barrel shaped pistons that are larger in the middle than they are at the top and bottom.

If the piston looks reasonably good after cleaning, take a close look at the ring lands. Wear may develop on the bottom of the ring lands. This wear is usually uneven, causing the ring to push on the higher areas and loads the ring unevenly when inertia is the greatest. Such uneven support of the ring will cause ring breakage and the piston will need to be replaced.

When installing a new ring in the groove, measure the ring side clearance against specification. Also check the see if the ring pins are there and that they have not loosened. Measure the skirt to see if the piston is collapsed.

gases from the cylinder. In essence, the incoming fuel charge is chasing out the exhaust gases from the cylinder.

Not all piston designs are of the deflector head type. Other pistons have a small convex crown on the piston head. In this case, port design aids in directing the incoming fuel upward. The piston head bears the brunt of the combustion force and heat. Most of the heat is transferred from the piston head through the rings to the cylinder wall and then on to the cooling system.

The piston design can be round, cam ground or barrel shaped. The cam ground design allows for expansion of the piston in a controlled manner. As the piston heats up, expansion take place and the piston moves out along the piston pin becoming more round as it warms up. Barrel shaped pistons rock very slightly in the bore which helps to keep the rings free.

The piston has machined ring grooves in which the rings are installed. They are carried along with the piston as it travels up and down the cylinder wall. There is one small pin in each ring groove to prevent the ring from rotating. The piston skirt is the bearing area for thrust and rides on the cylinder wall oil film. The side thrust of the piston is dependent upon piston pin location. If the pin is in the center of the piston, then there will be more thrust. If the pin is offset a few thousandths of an inch from the center of the piston, there will be less thrust. A used piston will have one side of the piston skirt show more signs of wear than the opposite side. The side showing wear is the major thrust side.

Thrust is caused by the pendulum action of the rod following the crankshaft rotation, which pulls the rod out from under the piston. The combustion pressure therefore pushes and thrusts the piston skirt against the cylinder wall. Some heat is also transferred at this point. The other skirt receives only minor pressure. Some pistons have small grooves circling the skirts to retain oil in the critical area between the skirt and the cylinder wall.

Fig. 37 This piston is severely scored from lack of lubrication and should not be reused

POWERHEAD 7-39

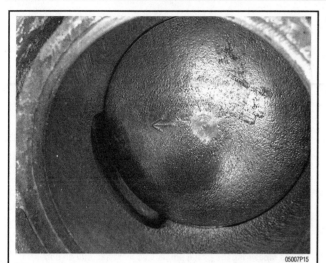

Fig. 38 Pistons should be installed with the arrow facing the exhaust port

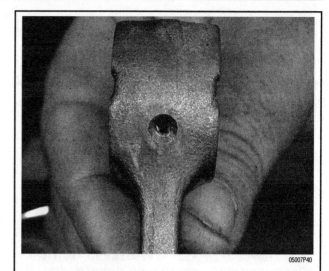

Fig. 40 . . . a similar hole in the connecting rod also oils the pin

Piston Pins

GENERAL INFORMATION

♦ See Figures 39, 40, and 41

A hole placed in the side of the piston, commonly referred to as the piston boss, is used to mount the piston to the piston pin. The combustion pressure is transferred to the piston pin and connecting rod bearing, then on to the crankshaft where it is converted to rotary motion. The pin is fitted to the piston bosses. The piston pin is the inner bearing race for the bearing mounted in the small end of the connecting rod. This transfers the combustion pressures into the connecting rod and allows the rod to swing with a pendulum-like action.

Piston pins are secured into both piston bosses. All have retainers and in addition some use a press fit to secure the pin. There are some models which use a slip fit. These may require special installation techniques.

Another type of pin fitting is loose on one side and tight on the other. This type aids in removal of the pin without collapsing the piston. With this design, always press on the pin from the loose boss side. The piston is marked on the inside of the piston skirt with the word "loose" to identify the loose boss.

Fig. 41 This piston uses a floating pin design. Once the retainers are removed, the pin should slide out easily

Always press with the loose side up and press the pin all the way through and out. When installing, press with the loose side up.

In all pressing operations, set the piston in a cradle block to support the piston. Some pistons require heating to expand the piston bosses so the pin can be pressed out without collapsing the piston. Other pistons just have a slip fit.

INSPECTION

♦ See Figures 42, 43, 44, 45 and 46

Check the piston pin retainer grooves for evidence of the retainers moving as they may have been distorted. Always replace the retainers once they have been removed. If there is evidence of wear in any of these areas, the piston should be replaced.

Inspect piston pin for wear in the bearing area. Rust marks caused by water will leave a needle bearing imprint. Chatter marks on the pin indicate that the piston pin should be replaced. If these marks are not too heavy, they may possibly be cleaned with emery paper for loose needle bearings or crocus cloth for caged bearings.

If the piston pin checks out visually, measure its outside diameter and compare that measurement with the inside diameter of the piston pin bore. Proper clearance is vital to providing enough lubrication.

Fig. 39 The holes in the bottom of this piston pin bore provide oiling to the piston pin . . .

7-40 POWERHEAD

Fig. 42 Measuring the piston pin bore inside diameter. This reading will be compared with the piston pin outside diameter to determine pin-to-bore clearance

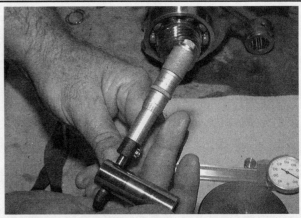

Fig. 43 Measuring the piston pin outside diameter with an outside micrometer at the point where the pin aligns with the piston pin bore . . .

Fig. 44 . . . and also at the point where the pin aligns with the connecting rod bore

Fig. 45 Typical caged bearing design used to support piston pins

Fig. 46 The small end bore in the connecting rod must be perfectly round to prevent bearing troubles

Piston Rings

GENERAL INFORMATION

▶ See Figure 47

The piston ring seals the piston to the cylinder bore, just as other seals are used on the crankshaft and lower unit. To perform correctly, the rings must conform to the cylinder wall and maintain adequate pressure to insure their sealing action at required operating speeds and temperatures. There are different designs used throughout the outboard industry. A given manufacturer will select a ring design that meets the operating requirements of the powerhead. This may be a standard ring, a pressure back (Keystone) ring or a combination of rings.

The functions of the piston ring include sealing the combustion gases so they cannot pass between the piston and the cylinder wall into the crankcase upsetting the pulse and maintaining an oil film in conjunction with the cylinder wall finish throughout the ring travel area. The rings also transfer heat picked up by the piston during combustion. This heat is transferred into the cylinder wall and thus to the cooling system. There are either two or three rings per piston, which perform these functions.

➡ An oil control ring is not used on 2-stroke engines.

All piston rings used are of the compression type. This means that they are for sealing the clearance between piston and cylinder wall. They are not allowed to rotate on the piston as automotive piston rings do. They are prevented from rotating by a pin n the piston ring groove. If the ring was allowed to turn, a ring end could snap into the cylinder port and become broken. The ring ends are specially machined to compensate for the pin. As the rings warm up in a running powerhead they expand, thereby requiring a specific end gap between the ring ends for expansion. This ring gap decreases upon warm-up, effectively limiting blowing gases (from the combustion process) from going into the crankcase. The rings ride in a piston ring groove with minimal side clearance, which gives them support as they move up and down the cylinder wall. With this support, combustion gas pressure and oil effectively seal the piston ring

Fig. 47 The piston ring seals the piston to the cylinder bore

POWERHEAD

against the ring land and the cylinder wall. As long as the oil mix is correct and temperatures remain where they should, the rings will provide service for many hours of operation.

INSPECTION

▶ See Figures 48, 49, 50, 51 and 52

One of the first indications of ring trouble is the loss of compression and performance. When compression has been lost or lowered because of the ring not sealing, the ring is either broken or stuck with carbon, gum or varnish. Improper oil mixing and stale gasoline provide the carbon, gum and varnish which cause the rings to stick. Low octane fuel, improperly adjusted timing/synchronization and lean fuel mixtures can damage the ring land, causing the ring to stick or break.

➡ **Running the outboard out of the water for even a few seconds can have damaging effects on the rings, pistons, cylinder walls and water pump.**

To determine if the rings fit the cylinder and piston, two measurements are taken; ring gap and ring side clearance. To determine these measurements, the ring is pushed into the cylinder bore using the piston skirt, so it will be square. Position each ring, one at a time, at the bottom of the cylinder (the smallest diameter) and using a feeler gauge, measure the expansion space between the ring ends. This is known as the ring gap measurement. Compare this measurement against the specifications. If the measurement is too small, the ring must be filed to increase the gap. If it is too large, either the bore is too large or the ring is not correct for the powerhead.

After ring end gap has been determined, position each ring in the piston ring groove and using a feeler gauge, measure between the ring and the piston ring land. Compare this measurement against the specifications. If the measurement is too small, the ring groove may be compressed. Inspect the ring groove and ring land condition. If it is too large the ring may not correct for the powerhead.

Connecting Rods

GENERAL INFORMATION

The connecting rod transfers the combustion pressure from the piston pin to the crankshaft, changing the vertical motion into rotary motion. In doing so, the connecting rod swings back and forth on the piston pin like a pendulum while it is traveling up and down. It goes down by combustion pressure and goes up by flywheel momentum and/or other power strokes on a multi-cylinder powerhead. The connecting rod can be of aluminum on smaller horsepower fishing outboards or of steel on larger horsepower models.

Most connecting rod designs use a steel liner with needle bearings in the large end and a pressed-in needle bearing in the small end.

The steel rod is a bearing race at both the large and small ends of the rod. It is hardened to withstand the rolling pressures applied from the loose or caged needle bearings. Unlike many connecting rod designs, these rods do not use two piece caps. The connecting rod big end is one piece. This requires the crankshaft to be pressed together to form a rotating assembly with the connecting rods.

The connecting rods are mist lubricated. Some of the rods have a trough design in the shank area. Oil holes may be drilled into the bearing area at both ends of this trough. Oil mist that falls out of the fuel will settle into the rod trough and collect. As the rod moves in and out, the oil is sloshed back and

Fig. 48 To determine ring gap, use a feeler gauge to measure the expansion space between the ring ends with the ring installed in the cylinder

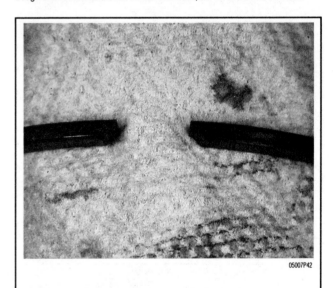

Fig. 49 Some rings are square . . .

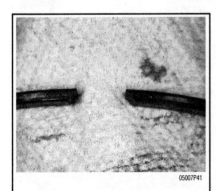

Fig. 50 . . . while other rings have a notched shape

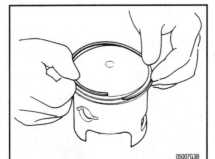

Fig. 51 When installing rings, place the ring in the groove and work it around the piston using a spiral motion until the ring is properly seated

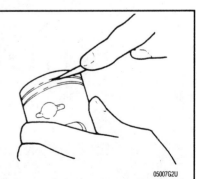

Fig. 52 Decarbon the piston rings using a ring groove cleaner or a broken piece of piston ring

7-42 POWERHEAD

forth in the trough and out the oil holes into the rod or piston pin bearings. This provides sufficient lubrication for these bearings. When the rod is equipped with oil holes, the oil holes have to be placed in the upward position toward the tapered end of the crankshaft when reassembled.

INSPECTION

▶ See Figures 53 thru 58

Damage to the connecting rod can be caused by lack of lubrication and will result in galling of the bearing and eventual seizing to the crankshaft. Over speeding of the powerhead may also cause the upper shank area of the rod to stretch and break near the piston pin.

Steel rods are inspected in the bearing areas, much like you would inspect a roller bearing. Look for scoring, pit marks, chatter marks, rust and color change. A blue color indicates overheating of the bearing surface. Minor rust marks or scoring may be cleaned up using crocus cloth for caged needle bearings or emery paper for loose needle bearings. A piece of round stock, cut with a slot in one end to accept a small piece of emery paper and mounted in a drill motor, can be used to clean up the rod ends.

The rod also needs to be checked to see if it is bent or has a twist in it. To do this, remove the piston and place the rod on a surface plate or a piece of flat glass (automotive widow). Using a flash light behind the rod and looking from in front of the rod, check for any light which can be seen under the rod ends. If light can be seen shining under the rod ends, the rod is bent and it must be replaced. You can also use a .002 feeler gauge. See if it will start under the machined area of the rod. If it will, the rod is bent. Examine the rod bolts and studs for damage and replace the nuts where used. Always reinstall the rod back on the same journal from which it was removed. The needle bearings, rod bear-

Fig. 55 . . . the rotating assembly is then lifted into position and the pistons are then aligned with the bores . . .

Fig. 56 . . . the pistons are slipped into their bores . . .

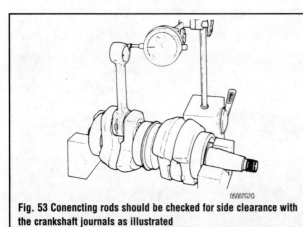

Fig. 53 Conencting rods should be checked for side clearance with the crankshaft journals as illustrated

Fig. 54 Rotating assembly installation starts by compressing the piston rings with clamps . . .

Fig. 57 . . . and the clamps are removed . . .

POWERHEAD 7-43

Fig. 58 . . . the crankcase can then be torqued into place

ing surface and crankshaft journal are all mated to each other once the powerhead has been run.

→When installing the connecting rod, the long sloping side must be installed toward the exhaust side of the cylinder assembly and if there is a hole in the connecting rod, position the oil hole upward. Some piston designs are marked with the word "UP". This side should be placed toward the tapered end of the crankshaft.

The Crankshaft

GENERAL INFORMATION

♦ See Figures 59, 60 and 61

The crankshaft is used to convert vertical motion received from the mounted connecting rod into rotary motion, which turns the driveshaft. It mounts the flywheel, which imparts a momentum to smooth out pulses between power

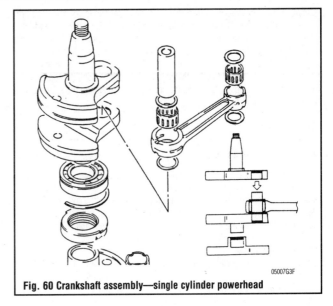
Fig. 60 Crankshaft assembly—single cylinder powerhead

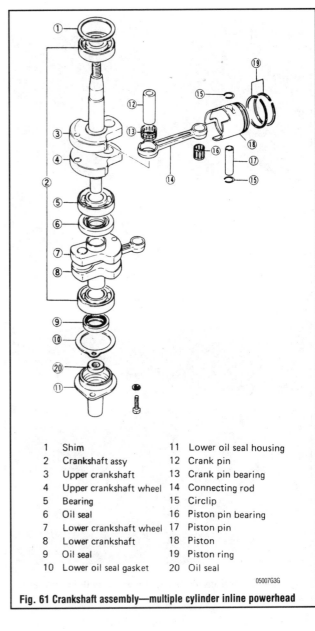

1	Shim	11	Lower oil seal housing
2	Crankshaft assy	12	Crank pin
3	Upper crankshaft	13	Crank pin bearing
4	Upper crankshaft wheel	14	Connecting rod
5	Bearing	15	Circlip
6	Oil seal	16	Piston pin bearing
7	Lower crankshaft wheel	17	Piston pin
8	Lower crankshaft	18	Piston
9	Oil seal	19	Piston ring
10	Lower oil seal gasket	20	Oil seal

Fig. 61 Crankshaft assembly—multiple cylinder inline powerhead

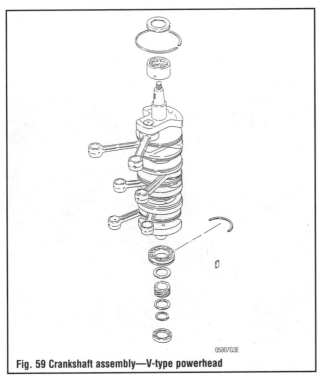
Fig. 59 Crankshaft assembly—V-type powerhead

strokes. It also provides sealing surfaces for the upper and lower seals and provides a surface for the labyrinth seal to hold oil against and a groove in which sealing rings are installed to seal pressures into each crankcase. Mounted main bearings control the axial movement of the crankshaft as it accomplishes these functions. The crankshaft bearing journals are case hardened to be able to withstand the stresses applied by the floating needle bearings used for connecting rod and main bearings. In essence, the crankshaft journals are the inner bearing races for the needle bearings.

INSPECTION

♦ See Figures 62 and 63

Pressure from the power stoke applied to the crankshaft rod journal by the needle bearings has a tendency to wear the journal on one side. During crankshaft inspection, the journals should also be measured with a micrometer to determine if they are round and straight. They should also be inspected for scoring, pitting, rust marks, chatter marks and discoloration caused by heat.

Check the sealing surfaces for grooves worn in by the upper and lower crankshaft seals. Take a look at the splined area which receives the driveshaft. Inspect the side of the splines for wear. This wear can be caused by lack of lubrication or improper lubricant applied during a seasonal service. An exhaust housing/lower unit that has received a sudden impact can be warped and this can also cause spline damage in the crankshaft.

The crankshaft cannot be repaired because of the case hardening and the possibility of changing the metallurgical properties of the material during the welding and machine operation. Also, there are no oversized bearings available. Repairs are limited to cleaning up the journal surface with 320 emery paper when loose needle bearings are run on the journal. Where caged roller bearings are run, the journal may be polished with crocus cloth.

The tapered end of the crankshaft has a spline or a keyway and key which times the flywheel to the crankshaft. Inspect the spline or key and keyway for damage. The crankshaft taper should be clean and free of scoring, rust and lubrication. The taper must match the flywheel hub. If someone has hit the flywheel with a heavy hammer or has used an improper puller to remove the flywheel, the flywheel and hub may be warped. Place the flywheel on the tapered end of the crankshaft and check the fit. If there is any rocking indication a distorted hub, replace the flywheel. Always use a puller which pulls from the bolt pattern or threaded inner hub of the flywheel. Never use a puller on the outside of the flywheel.

The taper is used to lock the flywheel hub to the crankshaft. When mounting the flywheel to the crankshaft the taper on the crankshaft an in the flywheel hub must be cleaned with a fast evaporating solvent. No lubrication on the crankshaft taper or flywheel hub should be done. The flywheel nut must be torqued to specification to obtain a press fit between the flywheel hub and the crankshaft taper. If the nut is not brought to specifications, the flywheel may spin on the crankshaft causing major damage.

The flywheel key is for alignment purposes and sets the flywheel's relative position to the crankshaft. Check the key for partial shearing on the side. If there is any indication of shearing, replace the key. Also check the keyway in the flywheel and crankshaft for damage. If there is damage which will allow incorrect positioning of the flywheel, the powerhead timing will be off.

Bearings

GENERAL INFORMATION

♦ See Figure 64

Needle bearings are used to carry the load which is applied to the piston and rod. This load is developed in the combustion process and the bearings reduce the friction between the crankshaft and the connecting rod. They roll with little effort and at times have been referred to as anti-friction bearings, as they reduce friction by reducing the surface area that is in contact with the crankshaft and the connecting rod. These needle bearings are of two types, loose and caged. When loose bearings are used, there can be upwards to 32 loose bearings floating between the rod journal of the crankshaft and the connecting rod. These bearings are aided in rolling by the movement of the crank pin journal and the connecting rod pendulum action. The surface installed on the journal and rod encourages needle rotation because of its relative roughness. If the journal and rod surface was polished with crocus cloth, the loose needle bearings would

Fig. 62 Crankshaft seals should always be lubricated prior to installation

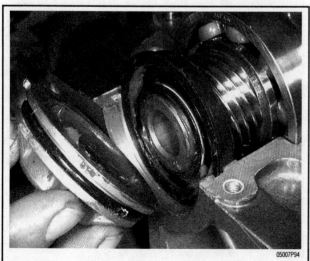

Fig. 63 Most cranshafts use two types of seals, an O-ring and sealing ring

Fig. 64 Typical caged bearing assembly

POWERHEAD 7-45

have the tendency to scoot, wearing both surfaces. So, journals and rods which uses the loose needle bearings are cleaned up sing 320 grit emery paper.

Caged needle bearings used a reduced number of needles and the needles are kept separated and are encouraged to roll by the cage. The cage also controls end movement of the bearings. Because of the cage, the journal and rod surfaces can be smoother, so these surfaces are polished with crocus cloth.

Main bearings are used to mount and control the axial movement of the crankshaft. They are either ball, needle or split race needle bearings. The split race needle bearings are held together with a ring and are sandwiched between the crankcase and cylinder assembly. The split race bearings are commonly used as center main bearings, as this is the only type of bearing that can be easily installed in this location. The ball bearings may be mounted as top or bottom mains on the crankshaft.

The bearing is made up of three parts—the inner race, needle and the outer race. In most industrial applications, the outer or inner race of a needle bearing assembly is held in a fixed position by a housing or shaft. The connecting rod needle bearings in the outboard powerhead have the same basic parts, but differ in that both inner and outer races are in motion. The outer race—the connecting rod—is swinging like a clock pendulum. The inner race—the crankshaft—is rotating and the needle bearing is floating between the two races.

INSPECTION

When the powerhead is disassembled and inspection of the parts is made, then by necessity along with examining the needle bearings, the crankshaft main bearing journal, rod journal and connecting rod bearing surfaces are also examined. The surfaces of all three of these parts can give a tremendous amount of information and the examination will determine if the parts are reusable.

Surfaces should be examined for scoring, pitting, chatter marks, rust marks, spalling and discoloration from overheating of the bearing surfaces. Minor scoring or pitting and rust marks may be cleaned up and the surfaces brought back to a satisfactory condition. This is done using crocus cloth for caged needle bearings and 320 grit emery paper for loose needle bearings. This is not a metal removing process, rather just a clean up of the surfaces.

Needle bearings are used as main bearings and are inspected for the same conditions as listed above. There are no oversized bearings available for rod or main bearings. Because of the hardness of the crankshaft (a bearing race), it should not be turned or welded up in order to bring it back to standard size. The welding process may stress the metallurgical properties of the crankshaft, developing cracks.

The caged rod bearings and split race main bearings are inspected for the same condition as loose needle bearings, plus the cage is examined for wear, cracks and breaks.

Ball bearings are used for top and bottom main bearings in some powerheads. These may be pressed onto the crankshaft or pressed into the end cap. To examine these bearings, wash, dry, oil and check them on the crankshaft or in the bearing cap. Turn the bearing by hand and feel if there is any roughness or catching. Try to wobble the bearing by grasping the outer race, (inner race) checking for looseness of the bearing. Replace the bearing if any of these conditions are found. If the bearing is pressed off (out) the bearing will probably be damaged and should be replaced.

If new or used bearings are contaminated with grit or dirt particles at the time of installation, abrasion will naturally follow. Many bearing failures are due to the introduction of foreign material into the internal parts of the bearing during assembly. Misalignment of the rod cap,, torque of the rod bolts and lack of proper lubrication also cause failures. Bearing failure is usually detected by a gradual rise in operating noise, excessive looseness (axial) in the bearing and shaft deflection. Keep the work area clean and use needle bearing grease or multipurpose grease to hold the bearings in place. This grease will dissipate quickly as the fuel mixture comes in contact with it. Do not use a wheel bearing or chassis grease as this will cause damage to the bearings. Oil the ball bearings with 2-stroke oil upon installation. Remember to keep them clean.

Exploded view of a typical powerhead showing major components

Engine Torque Specifications

Component	Diameter	Standard (ft. lbs.)	Metric (Nm)
Conventional bolt/nut			
	5mm	1.5-3	2-4
	6mm	3-5	4-7
	8mm	7-11.5	10-16
	10mm	16-25.5	22-35
Stainless steel bolt/nut			
	5mm	1.5-3	2-4
	6mm	4.5-7	6-10
	8mm	11-14.5	15-20
	10mm	24.5-29.5	34-41
Clamp bracket			
DT4		17.5-19	24-26
DT9.9, DT15, DT35, DT40		2-2.5	20-25
DT15C		16.5-18	23-25
DT25C, DT30C, DT55, DT65, DT75, DT85 DT90, DT100, DT150, DT175, DT200, DT225		29.5-32.5	41-45
Clamp bracket shaft			
DT115, DT140, DT150, DT175, DT200		31	43
Crankcase			
DT2, DT4, DT9.9, DT15, DT25C, DT30C, DT35, DT40, DT55, DT65		6-8.5	8-12
DT15C		14.5-19	20-26
DT35, DT40, DT55, DT65, DT75, 85		33.5-39	46-54
DT115, DT140	8mm	16.5	23
DT115, DT140	10mm	36	50
Crankcase			
DT150, DT175, DT200	6mm	10	14
	8mm	14.5-19	20-26
	12mm	54	75
DT225	8mm	14.5-19	20-26
	10mm	33.5-39	46-54
Cylinder head			
DT6, DT8, DT9.9, DT15, DT35, DT40		14.5-19	20-26
DT15C, DT55, DT65		15-18	21-25
DT25C, DT30C		6-8.5	8-12
DT75, DT85, DT90, DT100		33.5-39.5	46-54
DT90, DT100		20-23	28-32
DT115, DT140	8mm	14.5-19	20-26
	10mm	33.5-39.5	46-54
DT150, DT175, DT200	8mm	21.5	30
	10mm	54	75
DT225	8mm	20-23	28-32
	12mm	33.5-39.5	46-54
Cylinder head cover		6-8.5	8-12

POWERHEAD 7-47

Engine Torque Specifications

Component	Diameter	Standard (ft. lbs.)	Metric (Nm)
Drive shaft housing			
DT55, DT65		1.5-2	15-20
DT75, 85		6-8.5	8-12
DT90, DT100, DT150, DT175, DT200, DT225		24.5-29.5	34-41
DT115, DT140		13	18
Engine holder			
DT4, DT9.9, DT15, DT25C, DT30C, DT35, DT40, DT90, DT100		11-14.5	15-20
DT55, DT65, DT75, DT85		13-20	18-28
DT200	8mm	11-14.5	15-20
	10mm	24.5-29.5	34-41
DT225	8mm	11-14.5	15-20
	10mm	24.5-29.5	34-41
Exhaust cover			
DT6, DT8		3-5	4-7
Except DT6, DT8		6-8.5	8-12
Exhaust tube			
DT35, DT40, DT55, DT65, DT75, DT85, DT90, DT100, DT225		6-8.5	8-12
DT115, DT140	6mm	6	8
DT115, DT140	8mm	13	18
DT150, DT175, DT200		14.5	20
Exhaust tube housing			
DT150, DT175, DT200		14.5	20
Except DT150, DT175, DT200		6-8.5	8-12
Flywheel			
DT2		29-36	40-50
DT4		32.5	45
DT6, DT8		43.5-50.5	60-70
DT9.9, DT15		58-65	80-90
DT20		72.5-79.5	100-110
DT25C, DT30C		94-108.5	130-150
DT35, DT40, DT55, DT65		20-21	200-210
DT75, 85		144.5-152	200-210
DT90, DT100 DT115, DT140, DT150, DT175, DT200, DT225		180-188	250-260
Gear case			
DT4, DT6, DT8, DT9.9, DT15, DT35, DT40		11-14.5	15-20
DT55, DT65	8mm	11-14.5	15-20
DT55, DT65	10mm	24.5-29.5	34-41
DT75, 85	8mm	11-14.5	15-20
DT75, 85	10mm	24.5-29.5	34-41
DT90, DT100, DT150, DT175, DT200, DT225		36-43.5	50-60
DT115, DT140		40	55

7-48 POWERHEAD

Engine Torque Specifications

Component	Diameter	Standard (ft. lbs.)	Metric (Nm)
Gearcase mounting		11-14.5	15-20
Lower mounting			
DT9.9, DT15, DT15C		11-14.5	15-20
DT25C, DT30C		18-19.5	25-27
DT35, DT40		33.5-39	46-54
DT55, DT65		5.6-6.4	56-64
DT75, 85		40.5-46.5	5.6-6.4
DT90, DT100, DT115, DT140 DT150, DT175, DT200, DT225		65-72.5	90-100
DT150, DT175, DT200		68.5	95
Pinion			
DT25C, DT30C		19.5-21.5	27-30
DT35, DT40, DT55, DT65		21.5-29	30-40
DT75, 85		50-58	70-80
DT90, DT100		58-72.5	80-100
DT115, DT140		68.5	95
DT150, DT175, DT200, DT225		101.5-108.5	140-150
Power unit mounting			
DT115, DT140		32	44
DT150, DT175, DT200	8mm	16.5	23
DT150, DT175, DT200	10mm	32	44
Propeller			
DT9.9, DT15, DT35, DT40, DT75, DT85, DT115, DT140, DT150, DT175, DT200		36-43.5	50-60
DT15C		12.5-14.5	17-20
DT25C, DT30C		21-22.5	29-31
Propeller shaft bearing			
DT4, DT15C, DT25C, DT30C		4.5-7	6-10
DT35, DT40, DT55, DT65, DT75, DT85		11-12.5	15-20
DT115, DT140		13	18
DT150, DT175, DT200		137.5	190
Propeller shaft housing stopper			
DT90, DT100		108.5-123	150-170
DT150, DT175, DT200, DT225		130-144.5	180-200
Propeller shaft		36-45	50-62
PT unit		11-14.5	15-20
PTT unit			
DT75, 85		3.5-5	5-7
DT115, DT140			
Tilt Cylinder		94	130
Trim Cylinder		58	80

POWERHEAD 7-49

Engine Torque Specifications

Component	Diameter	Standard (ft. lbs.)	Metric (Nm)
Union fuel filter		18-25.5	25-35
Upper mount			
DT9.9, DT15, DT15C		11-14.5	15-20
DT25C, DT30C, DT35, DT40, DT55, DT65		24.5-29.5	34-41
DT90, DT100, DT115, DT140, DT150, DT175, DT200, DT200, DT225		40.5-46.5	56-64
Water pump case			
DT35, DT40		4.5-7	6-10
Except DT35, DT40		11-14.5	15-20

Engine Rebuilding Specifications—DT2

Component	Standard (in.)	Standard Metric (mm)	Service Limit Standard (in.)	Service Limit Metric (mm)
Crankshaft deflection	0.0012	0.03		
Conrod deflection	0.12	3		
Cylinder head distortion	0.0012	0.03		
Cylinder distortion	0.0012	0.03		
Piston diameter	1.6118-1.6124	40.940-40.955		
Cylinder bore	1.6142-1.6148	41.000-41.015		
Piston to cylinder clearance	0-0020-0.0026	0.052-0.067	0.0058	0.147
Piston diameter measuring position ①	0.6	15		
Cylinder measuring position ②	0.9	22		
Wear on cylinder bore	0.004	0.1		
Piston pin diameter	0.4723-0.4724	11.996-12.000	0.4717	11.98
Piston pin hole diameter	0.4724-0.4727	11.998-12.006	0.4736	12.03
Piston ring end gap				
1st	0.004-0.010	0.10-0.25	0.024	0.6
2nd	0.004-0.012	0.10-0.30	0.024	0.6
Maximum reed stop opening	0.15-0.17	3.8-4.2		
Reed seat clearance	0.008	0.2		

① From piston skirt end
② From cylinder top surface

7-50 POWERHEAD

Engine Rebuilding Specifications—DT2.2

Component	Standard Standard (in.)	Metric (mm)	Service Limit Standard (in.)	Metric (mm)
Crankshaft runout	0.000-0.002	0.00-0.05	0.0039	0.10
Conrod deflection	0.12	3.0		
Cylinder head distortion	0.002	0.05		
Cylinder distortion	0.002	0.05		
Piston diameter	1.6905-1.6911	42.940-42.955		
Cylinder bore	1.6929-1.6935	43.000-43.015		
Piston to cylinder clearance	0.0020-0.0026	0.052-0.067	0.0058	0.147
Piston diameter measuring position ①	0.6	15		
Cylinder measuring position ②	0.9	22		
Wear on cylinder bore	0.004	0.10		
Piston pin diameter	0.4723-0.4724	11.996-12.000	0.4717	11.980
Piston pin hole diameter	0.4725-0.4728	12.002-12.010	0.4736	12.030
Piston ring end gap	0.004-0.012	0.10-0.30		
Maximum reed stop opening	0.15-0.17	3.8-4.2		
Reed to seat clearance	0.008	0.20		

① From piston skirt end
② From cylinder top surface

Engine Rebuilding Specifications—DT4

Component	Standard Standard (in.)	Metric (mm)	Service Limit Standard (in.)	Metric (mm)
Crankshaft deflection	0.002	0.05		
Conrod deflection	0.16	4.0		
Cylinder head distortion	0.0012	0.030		
Cylinder distortion	0.0012	0.030		
Piston diameter	1.9961-1.9967	49.940-49.955		
Cylinder bore	1.9685-1.9691	50.000-50.015		
Piston to cylinder clearance	0.0020-0.0026	0.052-0.067	0.0058	0.147
Piston diameter measuring position ①	0.7	19		
Cylinder measuring position ②	4.3	110		
Wear on cylinder bore	0.004	0.10		
Piston pin diameter	0.4722-0.4724	11.995-12.000	0.4717	11.980
Piston pin hole diameter	0.4725-0.4728	12.002-12.010	0.4736	12.030
Piston ring end gap	0.006-0.014	0.15-0.35	0.028	0.70
Maximum reed stop opening	0.19-0.20	4.8-5.2		
Reed seat clearance	0.008	0.20		

① From piston skirt end
② From cylinder top surface

POWERHEAD 7-51

Engine Rebuilding Specifications—DT6 and DT8

Component	Standard (in.)	Standard Metric (mm)	Service Limit Standard (in.)	Service Limit Metric (mm)
Crankshaft deflection	0.002	0.05		
Conrod deflection	0.16	4.0		
Cylinder head distortion	0.0012	0.030		
Cylinder distortion	0.0012	0.030		
Piston to cylinder clearance	0-0020-0.0026	0.052-0.067		
Piston diameter measuring position ①	0.7	18		
Cylinder measuring position ②	0.9	23		
Wear on cylinder bore	0.004	0.10		
Piston pin diameter	0.4723-0.4724	11.996-12.000	0.4714	11.98
Piston pin hole diameter	0.4725-0.4728	12.002-12.010	0.474	12.03
Piston ring end gap	0.006-0.0138	0.15-0.35	0.031	0.80
Maximum opening for reed being bowed	0.16-0.181	4.1-4.5		
Reed-to-seat clearance	0.008	0.20		

① From piston skirt end
② From cylinder top surface

Engine Rebuilding Specifications—DT9.9 and DT15

Component	Standard (in.)	Standard Metric (mm)	Service Limit Standard (in.)	Service Limit Metric (mm)
Crankshaft deflection	0.002	0.05		
Conrod deflection	0.16	4.0		
Cylinder head distortion	0.0012	0.030		
Cylinder distortion	0.0012	0.030		
Piston to cylinder clearance	0.0020-0.0026	0.052-0.067		
Piston diameter measuring position ①	0.8	21		
Cylinder measuring position ②	1.1	28		
Wear on cylinder bore	0.004	0.10		
Piston pin diameter	0.5510-0.5512	13.995-14.000	0.5504	13.980
Piston pin hole diameter	0.5511-0.5514	13.998-14.006	0.5524	14.030
Piston ring end gap	0.008-0.016	0.20-0.40	0.031	0.80
Maximum opening for reed being				
DT9.9	0.15-0.17	3.8-4.2		
DT15	0.217-0.232	5.5-5.9		
Reed-to-seat clearance	0.008	0.20		

① From piston skirt end
② From cylinder top surface

Engine Rebuilding Specifications—DT15C

Component	Standard Standard (in.)	Metric (mm)	Service Limit Standard (in.)	Metric (mm)
Crankshaft deflection	0.002	0.05		
Conrod deflection	0.2	5.0		
Cylinder head distortion	0.0012	0.030		
Cylinder distortion	0.0012	0.030		
Piston diameter	2.3207-2.3213	58.945-58.960		
Cylinder bore	2.3228-2.3234	59.000-59.015		
Piston to cylinder clearance	0.0017-0.0024	0.042-0.062	0.0056	0.142
Piston diameter measuring position ①	0.8	20		
Cylinder measuring position ②	1.1	28		
Wear on cylinder bore	0.004	0.10		
Piston pin diameter	0.5510-0.5512	13.995-14.000	0.5504	13.980
Piston pin hole diameter	0.5511-0.5513	13.998-14.004	0.5524	14.030
Piston ring end gap	0.004-0.010	0.10-0.25	0.031	0.8
Maximum reed stop opening	0.22-0.23	5.5-5.9		
Reed seat clearance	0.008	0.20		

① From piston skirt end
② From cylinder top surface

Engine Rebuilding Specifications—DT20

Component	Standard Standard (in.)	Metric (mm)	Service Limit Standard (in.)	Metric (mm)
Crankshaft deflection	0.002	0.05		
Conrod deflection	0.2	5.0		
Cylinder head distortion	0.0012	0.030		
Cylinder distortion	0.0012	0.030		
Piston to cylinder clearance	0.0024-0.0035	0.060-0.090		
Piston diameter measuring position ①	0.98	25		
Cylinder measuring position ②	1.18	30		
Wear on cylinder bore	0.004	0.10		
Piston pin diameter	0.6297-0.6299	15.995-16.000	0.5504	13.980
Piston pin hole diameter	0.6298-0.6302	15.998-16.006		
Piston ring end gap	0.0059-0.0138	0.15-0.35	0.028	0.7
Maximum reed stop opening	0.433	11.0		
Reed seat clearance	0.008	0.20		

① From piston skirt end
② From cylinder top surface

POWERHEAD 7-53

Engine Rebuilding Specifications—DT25C and DT30C

Component	Standard Standard (in.)	Standard Metric (mm)	Service Limit Standard (in.)	Service Limit Metric (mm)
Crankshaft deflection	0.002	0.05		
Conrod deflection	0.2	5.0		
Cylinder head distortion	0.0012	0.030		
Cylinder distortion	0.0012	0.030		
Piston diameter	2.4378-2.4384	61.920-61.935		
Cylinder bore	2.4409-2.4415	62.000-62.015		
Piston to cylinder	0.0028-0.0034	0.072-0.087	0.0066	0.167
Piston diameter measuring position ①	0.9	23		
Cylinder measuring position ②	1.1	28		
Wear on cylinder bore	0.004	0.10		
Piston pin diameter	0.6297-0.6299	15.995-16.000	0.6291	15.980
Piston pin hole diameter	0.6300-0.6303	16.002-16.010	0-6311	16.030
Piston ring end gap	0.006-0.014	0.15-0.35	0.031	0.8
Maximum reed stop opening	0.15	3.8		
Reed seat clearance	0.008	0.20		

① From piston skirt end
② From cylinder top surface

Engine Rebuilding Specifications—DT35 and DT40

Component	Standard Standard (in.)	Standard Metric (mm)	Service Limit Standard (in.)	Service Limit Metric (mm)
Crankshaft deflection				
Upper/Lower	0.002	0.05		
Middle	0.003	0.07		
Conrod deflection	0.2	5.0		
Cylinder head distortion	0.0012	0.030		
Cylinder distortion	0.0012	0.030		
Piston diameter	3.1060-3.1066	78.893-78.908		
Cylinder bore	3.11027-3.1108	79.000-79.015		
Piston to cylinder clearance	0.0039-0.0045	0.099-0.114	0.0079	0.194
Piston diameter measuring position ①	1.1	28.0		
Cylinder measuring position ②	1.38	35.0		
Wear on cylinder bore	0.004	0.10		
Piston pin diameter	0.7872-0.7874	19.995-20.000	0.7866	19.980
Piston pin hole diameter	0.7873-0.7876	19.998-20.006	0.7B87	20.032
Piston ring end gap	0.008-0.016	0.2-0.4	0.031	0.8
Maximum reed stop opening	0.35-0.36	8.9-9.3		
Reed seat clearance	0.008	0.20		

① From piston skirt end
② From cylinder top surface

POWERHEAD

Engine Rebuilding Specifications—DT55 and DT65

Component	Standard (in.)	Standard Metric (mm)	Service Limit Standard (in.)	Service Limit Metric (mm)
Crankshaft deflection	0.002	0.05		
Conrod deflection	0.2	5		
Cylinder head distortion	0.0012	0.030		
Cylinder distortion	0.0012	0.031		
Piston diameter	2.8698-2.8704	72.893-72.908		
Cylinder bore	2.8740-2.8746	73.000-73.015		
Piston to cylinder	0.0039-0.0045	0.099-0.114	0.0076	0.194
Piston diameter measuring position ①	1.1	28.0		
Cylinder measuring position ②	1.38	35.0		
Wear on cylinder bore	0.004	0.10		
Piston pin diameter	0.7872-0.7874	19.995-20.000	0.7795	19.98
Piston pin hole diameter	0.7873-0.7876	19.998-20.006	0.7886	20.03
Piston ring end gap	0.008-0.016	0.20-0.40	0.031	0.8
Maximum reed stop opening	0.37-0.38	9.4-9.8		
Reed seat clearance	0.008	0.2		

① From piston skirt end
② From cylinder top surface

Engine Rebuilding Specifications—DT75 and DT85

Component	Standard (in.)	Standard Metric (mm)	Service Limit Standard (in.)	Service Limit Metric (mm)
Crankshaft deflection	0.002	0.05		
Conrod deflection	0.2	5.0		
Cylinder head distortion	0.0012	0.030		
Cylinder distortion	0.0012	0.030		
Piston to cylinder clearance	0.0053-0.0065	0.135-0.165		
Piston diameter measuring position ①	1.38	35.0		
Cylinder measuring position ②	1.77	45.0		
Wear on cylinder bore	0.004	0.10		
Piston pin diameter	0.7872-0.7874	19.995-20.000	0.7866	19.98
Piston pin hole diameter	0.7873-0.7876	19.998-20.006	0.7892	20.046
Piston ring end gap	0.008-0.016	0.2-0.4	0.031	0.8
Maximum opening for reed being bowed	0.30-0.31	7.6-7.9		
Reed-to-seat clearance	0.008	0.20		

① From piston skirt end
② From cylinder top surface

POWERHEAD 7-55

Engine Rebuilding Specifications—DT90 and DT100

Component	Standard (in.)	Standard Metric (mm)	Service Limit Standard (in.)	Service Limit Metric (mm)
Crankshaft runout	0.002	0.05		
Conrod deflection	0.2	5.0		
Cylinder head distortion	0.0012	0.030		
Cylinder distortion	0.0012	0.030		
Piston diameter	3.3018-3.3024	83.865-83.880		
Cylinder bore	3.3071-3.3077	84.000-84.015		
Piston to cylinder clearance	0-0051-0.0055	0.13-0 14	0.0087	0.22
Piston diameter measuring position ①	1.0	25.0		
Cylinder measuring position ②	1.0	25.0		
Piston pin diameter	0.7872-0.7874	19.995-20.000	0.7866	19.980
Piston pin hole diameter	0.7875-0.7878	20.002-20.010	0.7886	20.030
Piston ring end gap	0.008-0.016	0.20-0.40	0.031	0.80
Maximum reed stop opening	0.413	10.5		
Reed seat clearance	0.008	0.20		

① From piston skirt end
② From cylinder top surface

Engine Rebuilding Specifications—DT115 and DT140

Component	Standard (in.)	Standard Metric (mm)	Service Limit Standard (in.)	Service Limit Metric (mm)
Crankshaft runout	0.000-0.002	0.00-0.05	0.0039	0.10
Conrod deflection	0.2	5.0		
Cylinder head distortion				
1988-95	0.0012	0.030		
1996-99	0.0039	0.10		
Piston diameter	3.3024-3.3030	83.880-83.895		
Cylinder bore	3.3071-3.3130	84.000-84.015		
Piston to cylinder clearance	0.0045-0.0049	0.115-0.125	0.0081	0.205
Piston diameter measuring position ①	1.0	25		
Cylinder measuring position ②	1.2	30		
Wear on cylinder bore	0.004	0.10		
Piston pin diameter	0.8659-0.8661	21.995-22.000	0.8653	21.980
Piston pin hole diameter	0.8662-0.8665	22.002-22.010	0.8673	22.030
Piston ring end gap	0.008-0.016	0.20-0.40	0.031	0.80
Maximum reed stop opening	0.31-0.33	7.9-8.3		
Reed seat clearance	0.008	0.20		

① From piston skirt end
② From cylinder top surface

Engine Rebuilding Specifications—DT150, DT175 and DT200

Component	Standard (in.)	Standard Metric (mm)	Service Limit Standard (in.)	Service Limit Metric (mm)
Crankshaft deflection	0.002	0.05		
Conrod deflection	0.2	5.0		
Cylinder head distortion	0.0012	0.030		
Cylinder distortion	0.0012	0.030		
Piston diameter	3.3018-3.3024	83.865-83.880		
Cylinder bore	3.3071-3.3077	84.000-84.015		
Piston to cylinder clearance	0.0051-0.0055	0.130-0.140	0.0087	0.220
Piston diameter measuring position ①	1.3	32		
Cylinder measuring position ②	1.6	40		
Wear on cylinder bore	0.004	0.10		
Piston pin diameter	0.8659-0.8661	21.995-22.000	0.8654	21.980
Piston pin hole diameter	0.8662-0.8665	22.002-22.010	0.8673	22.030
Piston ring end gap	0.008-0.016	0.20-0.40	0.031	0.80
Maximum reed stop opening				
DT150	0.24	6.0		
DT175-DT200	0.29	7.4		
Reed seat clearance	0.008	0.20		

① From piston skirt end

② From cylinder top surface

Engine Rebuilding Specifications—DT225

Component	Standard (in.)	Standard Metric (mm)	Service Limit Standard (in.)	Service Limit Metric (mm)
Crankshaft deflection	0.002	0.05		
Conrod deflection	0.2	5.0		
Cylinder head distortion	0.0012	0.030		
Cylinder distortion	0.0012	0.030		
Piston diameter	3.3035-3.3041	83.910-83.925		
Cylinder bare	3.3071-3.3080	84.000-84.023		
Piston to cylinder clearance	0.0030-0.0045	0.075-0.113		
Piston diameter measuring position ①	1.3	32.0		
Cylinder measuring position ②	1.6	40.0		
Wear on cylinder bore	0.004	0.10		
Piston pin diameter	0.8659-0.8661	21.995-22.000	0.8654	21.980
Piston pin hole diameter	0.8662-0.8665	22.001-22.006	0.8673	22.030
Piston ring end gap	0.008-0.016	0.2-0.4	0.031	0.8
Maximum reed stop opening	0.26	6.6		
Reed seat clearance	0.008	0.20		

① From piston skirt end

② From cylinder top surface

LOWER UNIT 8-2
GENERAL INFORMATION 8-2
SHIFTING PRINCIPLES 8-2
   STANDARD ROTATING UNIT 8-2
   COUNTERROTATING UNIT 8-2
TROUBLESHOOTING THE LOWER
  UNIT 8-2
PROPELLER 8-2
   REMOVAL & INSTALLATION 8-2
LOWER UNIT—NO REVERSE GEAR 8-3
   REMOVAL & INSTALLATION 8-3
LOWER UNIT—WITH REVERSE
  GEAR 8-3
   REMOVAL & INSTALLATION 8-3
LOWER UNIT OVERHAUL 8-6
DT2 AND DT2.2 (FORWARD ONLY) 8-6
   DISASSEMBLY 8-6
   CLEANING & INSPECTION 8-7
   ASSEMBLY 8-8
   SHIMMING PROCEDURE 8-9
DT4 8-9
   DISASSEMBLY 8-9
   CLEANING & INSPECTION 8-9
   ASSEMBLY 8-10
   SHIMMING PROCEDURE 8-11
DT6 AND DT8 8-11
   DISASSEMBLY 8-11
   INSPECTION & CLEANING 8-13
   ASSEMBLY 8-14
   SHIMMING PROCEDURE 8-14
DT9.9 AND DT15 8-14
   DISSASSEMBLY 8-14
   CLEANING & INSPECTION 8-15
   ASSEMBLY 8-16
   SHIMMING PROCEDURE 8-17
DT20, DT25 AND DT30 8-19
   DISASSEMBLY 8-19
   CLEANING & INSPECTION 8-20
   ASSEMBLY 8-20
   SHIMMING PROCEDURE 8-22
DT35 AND DT40 8-24
   DISASSEMBLY 8-24
   CLEANING & INSPECTION 8-24
   ASSEMBLY 8-24
   SHIMMING PROCEDURE 8-26
DT55 AND DT65 8-28
   DISASSEMBLY 8-28
   CLEANING & INSPECTION 8-30
   ASSEMBLY 8-31
   SHIMMING PROCEDURE 8-32
DT75 AND DT85 8-35
   DISASSEMBLY 8-35
   CLEANING & INSPECTION 8-36
   ASSEMBLY 8-38
   SHIMMING PROCEDURE 8-39
DT115 AND DT140 8-44
   DISASSEMBLY 8-44
   CLEANING & INSPECTION 8-44
   ASSEMBLY 8-45
   SHIMMING PROCEDURE 8-45

V4 AND V6 8-46
   DISASSEMBLY 8-46
   CLEANING & INSPECTING 8-47
   ASSEMBLY 8-47
   SHIMMING PROCEDURE 8-50
JET DRIVE 8-51
DESCRIPTION AND OPERATION 8-51
MODEL IDENTIFICATION AND SERIAL
  NUMBERS 8-51
JET DRIVE ASSEMBLY 8-51
   REMOVAL & INSTALLATION 8-51
   ADJUSTMENT 8-54
   DISASSEMBLY 8-56
   CLEANING & INSPECTING 8-56
   ASSEMBLING 8-57

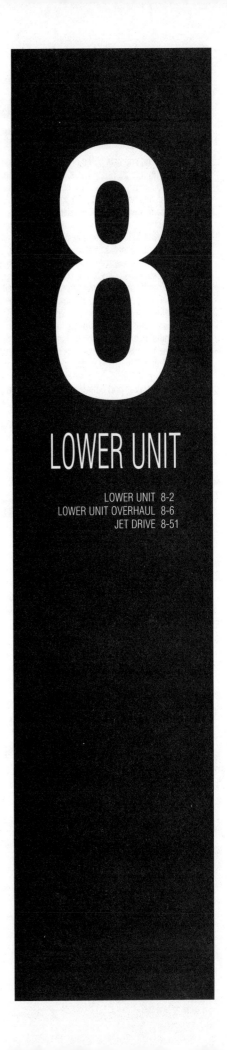

8
LOWER UNIT

LOWER UNIT 8-2
LOWER UNIT OVERHAUL 8-6
JET DRIVE 8-51

8-2 LOWER UNIT

LOWER UNIT

General Information

The lower unit consists of the driveshaft, water pump, pinion gear, bearings, forward and reverse gears, propeller shaft. The shifting mechanism and the lower unit housing. The housing is bolted to the exhaust housing, which places the driveshaft and water tube(s) through the center of the exhaust housing. The water tube carries water from the water pump to the powerhead.

The driveshaft splines insert into the crankshaft which then transmits power from the powerhead to the gearcase. A pinion gear on the driveshaft takes the power from the driveshaft and transfers the power to the propeller shaft. The powerhead driveshaft rotates in a clockwise direction continuously while the engine is running, propeller shaft direction is controlled by the gearcase shifting assembly.

The lower unit can be removed from the engine without removing the engine from it's mounting on the boat. The lower units in this chapter differ somewhat in their design and construction and require different servicing procedures.

The lower unit is normally trouble free until water enters the gearcase, the operator shifts incorrectly, or the oil is not changed regularly and corrosion enters the unit. Constant maintenance is required to prevent these problems. Because the unit is normally underwater, extra care must be taken to prevent problems.

Shifting the unit in and out of gear needs to be quick and positive to prevent rounding over the shift clutch dogs. Slow engagement will damage the parts. This problem is evident when the unit jumps out of gear.

Following the recommended oil change schedule allows the oil to be drained and checked for contamination, especially water intrusion. A milky looking oil is a sign of water has entered the gearcase and it must then be pressure tested to check the seals and the leaks repaired to prevent damage to the bearings and gears.

Shifting Principles

STANDARD ROTATING UNIT

Non-Reverse Type

Some of the smaller engines equipped with a neutral, but no reverse-gear, utilize a spring-loaded clutch to shaft between neutral and forward gear.

Reverse Type

On Suzuki outboard engines equipped with a reverse gear, a sliding-type clutch engages the chosen gear in the gearcase housing. This clutch when engaged, creates a direct connection that then moves the power flow from the pinion to the propeller shaft.

Power flow in the lower unit goes through the driveshaft into the pinion gear, which constantly turns the forward and reverse gears in opposite directions. The clutch dog is part of the shaft mechanism and is splined to the propeller shaft. The clutch dog is held in the central position (neutral) between the forward and reverse gears. When the shift shaft (rod) is moved, the shift cam (shifter) moves the follower (shifter shaft), which in turn, moves the clutch dog into mesh with the selected gear. Power is then transmitted from the gear through the clutch dog into the propeller shaft, and finally to the propeller.

V4 and V6 Smooth-Shift

1. This design provides easy shifting, positive engagement and extended durability of the gears and shifting mechanism. Its compact design is also partly responsible for the slim design of the lower unit housing with its smooth water flow and low forward water intake.

COUNTERROTATING UNIT

As mentioned earlier in this section a single design shifting mechanism is employed on both the standard and counter-rotating units, with the counter-rotating shift mechanism having the shift rod being turned 180° from the standard shift mechanism.

The main physical differences lie in the shifting mechanism (mirror image) and the propeller shaft. The counterrotating unit has a shoulder machined into it for the forward bevel gear tapered roller bearing. Another difference regards nomenclature: what would be the forward gear on a standard unit becomes the reverse gear on a counterrotating unit and what would normally be the reverse gear on a standard unit, becomes the forward gear on a counterrotating unit. The pinion gear remains the same and driveshaft rotation remains the same as on a standard lower unit.

Mirror image shifting mechanisms produce counter rotation of the propeller shaft. This type lower unit consists of the same major identical components as the standard unit.

On a standard lower unit, the cam on the shift shaft is located on the starboard side of the shifter. Therefore, when the rod is rotated counterclockwise, the clutch shifter is pulled forward and the forward gear is engaged.

On a counterrotating lower unit, the cam on the shift rod is located on the port side of the shifter. Therefore, when the rod is rotated counterclockwise, the clutch shifter is pushed back and the gear in the aft end of the housing (which normally is the reverse gear) is engaged. In this manner, the rotation of the propeller shaft is reversed. The same logic applies to the selection of reverse gear.

➡ **Counter-rotational shifting is accomplished without modification to the shift cable at the shift box. The normal setup is essential for correct shifting. The only special equipment the counterrotating unit requires is the installation of a left-hand propeller.**

Troubleshooting the Lower Unit

Once a season the lower unit needs to be dropped and lubrication on the driveshaft splines renewed. An extreme pressure moly lube is applied directly to the splines. Seals or an O-ring are used around the driveshaft to retain the lubrication and keep the water and exhaust from the splined joint area. This is a mandatory service to keep the splines from rusting together. Exhaust pressure and water are both present at the joint seal. If a seal failure occurs, water washes the lubricant from the splines and rusting will occur. The rust can be so severe that the two shafts will be rusted together. This will first be known there is an attempt to drop the lower unit. it will not separate from the powerhead.

If rust has seized the joint, the drive shaft will have to be cut with a saw or cutting torch. Some driveshafts use stainless steel and others a good grade of steel. The best preventive service is not to let it happen. Service the spline joint each season along with a water pump impeller replacement. When water has been found in the lower unit oil, the unit should be pressure tested with air to 16–18 psi and the submerged under water or sprayed with a mixture of soapy water. Bubbles will come out of the areas that are leaking. Also, the unit should be subjected to 3–5 inches of vacuum. This will test the seals in the other direction. The pressure vacuum should hold for a few minutes.

After the shafts and housings are cleaned and inspected, special attention should be paid to the area's that are leaking. It is common for grit in the water to wear thew shafts at the seal contact points or for corrosion to eat away at the housing.

If the gearshift jumps out of gear, check the detent balls, spring, clutch dog, gears and shifter.

If the gearshift won't shift, check the pivot pin, lever or cable adjustment, shift rod connection, gearcase components and driveshaft. Replace any damaged or worn parts.

If the gearcase is seized, check the gearcase for lubricant. If lubricant is present, drain and disassemble the gearcase. Inspect all components for damage or corrosion. Replace broken or corroded components. Check for a distorted gearcase housing.

Propeller

REMOVAL & INSTALLATION

DT2, DT2.2, DT4, DT6 and DT8

♦ See accompanying illustrations

1. Disconnect the spark plug lead(s) to prevent accidental starting of the engine.

LOWER UNIT   8-3

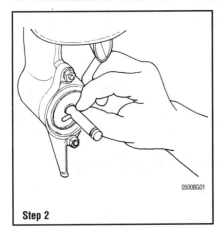
Step 2

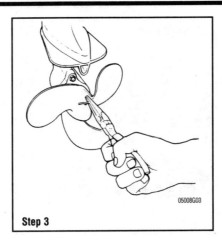
Step 3

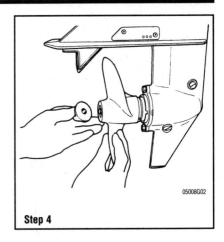
Step 4

2. On DT2 models, remove and discard the cotter pin. Remove the propeller from the shaft.
3. On DT4 and DT6 models, remove and discard the cotter pin.
4. Then, unscrew the propeller nut and remove the propeller from the shaft.
5. Remove drive pin from propeller with an appropriate punch.
6. Clean propeller shaft thoroughly.
7. Inspect the pin engagement slot in the propeller hub and shaft for wear or damage. Replace as necessary.

To install:

8. Clean propeller shaft thoroughly.
9. Lubricate the propeller shaft with waterproof marine grease.
10. Inspect the pin engagement slot in the propeller hub and shaft for wear or damage.
11. Install the propeller on the shaft
12. Install new drive and cotter pins. Bend the ends of the cotter pin over completely.
13. Install the propeller nut.
14. Connect the spark plug lead(s).

Except DT2, DT2.2, DT4, DT6 and DT8

1. Disconnect the spark plug leads to prevent accidental starting of the engine.
2. On models equipped with a cotter pin, remove it. Discard the cotter pin as it must not be reused.
3. On models equipped with a lockwasher, straighten the tab lockwasher.
4. Remove the propeller nut from the shaft.
5. Remove the tab lockwasher (if used) and propeller nut spacer from the shaft.
6. Remove the propeller and bushing stopper from the shaft.

To install:

7. Throughly clean and inspect the propeller shaft for damage.
8. Lubricate propeller shaft with waterproof marine grease.
9. Install the propeller on the shaft.
10. If a tab lockwasher is used, check washer tab condition and replace as required.
11. If a cotter pin is used, install a new one and bend the ends over completely.
12. Tighten the propeller nut to the correct torque:
   • DT9.9–DT200: 36–43.5 ft. lbs. (50–60 Nm)
   • DT15C: 12.5–14.5 ft. lbs. (17–20 Nm)
   • DT25C, DT30C: 21–22.5 ft. lbs. (29–31 Nm)

Lower Unit—No Reverse Gear

REMOVAL & INSTALLATION

DT2

1. Remove the engine cover and disconnect the spark plug lead to prevent any accidental engine starting during the lower unit removal.
2. Place the shift lever in **FORWARD**.
3. Remove the propeller.
4. Place a suitable container under the gearcase. Remove the drain screw and drain the lubricant from the unit.

➡ If the lubricant is creamy in color or metallic particles are found, the gearcase must be completely disassembled to determine and correct the cause of the problem.

5. Wipe a small amount of lubricant on a finger and rub the finger and thumb together. Check for the presence of metallic particles in the lubricant. Note the color of the lubricant. A white or creamy color indicates water in the lubricant. Check the drain container for signs of water separation from the lubricant.
6. Remove the bolt cover.
7. Remove the 2 nuts holding the gearcase to the driveshaft housing
8. Carefully separate then remove the gearcase from the driveshaft housing.
9. Install the gearcase in an appropriate holding fixture.

✲✲ CAUTION

Do not grease the top of the driveshaft. This may excessively preload the driveshaft and crankshaft when the mounting fasteners are tightened and cause a premature failure of the power head or gearcase.

10. Lightly lubricate the driveshaft splines with waterproof marine grease.
11. Apply a thin but uniform coat of a silicone sealer to the gearcase and driveshaft housing mating surfaces.
12. Wipe the driveshaft housing bolt threads with thread locking compound or equivalent.

To install:

13. Position the gearcase under the driveshaft housing. Align the driveshaft splines with the crankshaft.

✲✲ CAUTION

Do not rotate the flywheel counterclockwise. This can damage the water pump impeller and cause the engine to overheat.

14. Seat the gearcase against the driveshaft housing, rotating the flywheel clockwise as required until the driveshaft and crankshaft engage.
15. Install the gearcase nuts and lockwashers and tighten securely.
16. Install the mounting nut cover.
17. Install the propeller.
18. Reconnect the spark plug lead and refill the gearcase with the proper type and quantity of lubricant.

Lower Unit—With Reverse Gear

REMOVAL & INSTALLATION

♦ See accompanying illustrations

DT4, DT6 and DT8

1. Remove the engine cover and disconnect the spark plug lead(s) as a safety precaution to prevent any accidental starting of the engine during lower unit removal.
2. Place the shift lever in **FORWARD**.
3. Remove the propeller.

8-4 LOWER UNIT

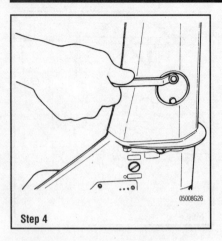
Step 4

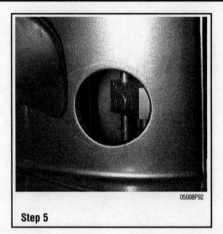
Step 5

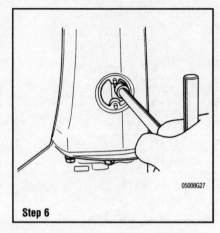
Step 6

4. On DT6, DT8 models, perform the following, remove the bolts holding the shift rod adjusting cover and remove the cover
5. If it is equipped with a rubber cover, remove that to access the shift rod
6. Loosen the shift rod connector bolt
7. Remove the bolts holding the gearcase to the driveshaft housing.

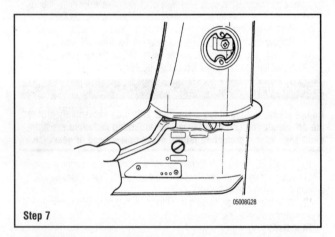
Step 7

8. Tilt the driveshaft housing up and carefully separate it from the gearcase.
9. Remove the gearcase from the driveshaft housing.
10. Install the gearcase in an appropriate holding fixture.
11. Place a suitable container under the gearcase. Remove the drain screw and drain the lubricant from the unit.

➡ If the lubricant is creamy in color or metallic particles are found, the gearcase must be completely disassembled to determine and correct the cause of the problem.

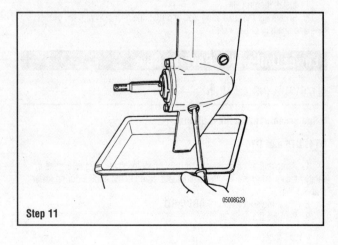
Step 11

12. Wipe a small amount of lubricant on a finger and rub the finger and thumb together. Check for the presence of metallic particles in the lubricant. Note the color of the lubricant. A white or creamy color indicates water in the lubricant. Check the drain container for signs of water separation from the lubricant.

✶✶ CAUTION

Do not grease the top of the driveshaft. This may excessively preload the driveshaft and crankshaft when the mounting fasteners are tightened and cause a premature failure of the power head or gearcase.

To install:

13. Make sure the shift rod is in the **NEUTRAL** position and lightly lubricate the driveshaft splines with waterproof marine grease.
14. Apply a thin but uniform coat of a silicone sealer to the gearcase and driveshaft housing mating surfaces.
15. Apply a small amount of thread locking compound to the driveshaft housing bolts.
16. Position the gearcase under the driveshaft housing. Align the driveshaft splines with the crankshaft, insert the water tube into the water pump case and fit the upper shift rod into the shift rod connector.

✶✶ CAUTION

Do not rotate the flywheel counterclockwise. This can damage the water pump impeller and cause the engine to overheat.

17. Seat the gearcase against the driveshaft housing, rotating the flywheel clockwise as required until the driveshaft and crankshaft engage.
18. Install the gearcase fasteners and tighten to specifications.
19. On DT6 and DT8 models, perform the following:
   • Make sure the engine shift lever is in **FORWARD** and install the shift rod connector bolt.
   • Place the shift lever in **NEUTRAL** and make sure the propeller rotates freely, then shift back into **FORWARD** and make sure the propeller will only rotate clockwise. If the propeller does not rotate as indicated, loosen the shift rod connector bolt and readjust the position of the connector or rods as required.
20. When connecting the clutch rod, equalize the distance of forward and reverse shifts by means of the clutch rod connector.
21. After adjusting the clutch rod, replace the clutch adjusting hole cover gasket (if equipped) or install the rubber cover.
22. Apply silicone sealant to the mating surfaces of the driveshaft housing and the gearcase and secure the housing to the gearcase. A thin uniform coat of silicone is all that is need for sealing these surfaces.
23. Apply a coat of waterproof marine grease to the splines of the driveshaft.
24. Apply a thread locking compound to the gearcase bolts and tighten to 11–14.5 ft. lbs. (15–20 Nm).
25. Install the propeller.
26. Refill the gearcase with proper type and quantity of lubricant.
27. Reconnect the spark plug lead(s).

LOWER UNIT 8-5

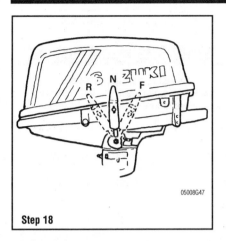
Step 18

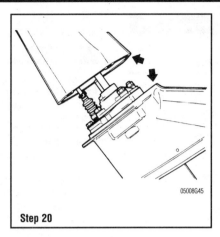
Step 20

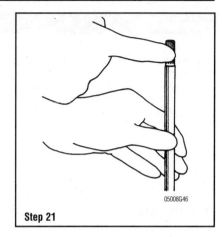
Step 21

DT9.9 to DT35 and DT40

The gearcase used on these models varies primarily in shape, size and component location. This is a basic procedure that covers all models listed. Your model may vary slightly from the one shown, but has the same basic components and all models listed are serviced in essentially the same way.

1. Remove the engine cover and disconnect the spark plug leads to prevent accidental starting of the engine during lower unit removal.
2. Shift the outboard into **FORWARD**.
3. Remove the propeller.
4. Loosen the clutch rod locknut.
5. Loosen the clutch rod turnbuckle and separate the clutch rod from the shift rod.
6. Remove the bolts and lockwashers securing the gearcase to the driveshaft housing.
7. Carefully separate the gearcase from the driveshaft housing.
8. Install the gearcase in an appropriate holding fixture.
9. Place a suitable container under the gearcase. Remove the vent and drain the lubricant from the unit.

➡ If the lubricant is creamy in color or metallic particles are found, the gearcase must be completely disassembled to determine and correct the cause of the problem.

10. Wipe a small amount of the drained lubricant on a finger and rub the finger and thumb together. Check for the presence of metallic particles in the lubricant. Note the color of the lubricant. A white or creamy color indicates water in the lubricant. Check the drain container for signs of water separation from the lubricant.

✱✱ CAUTION

Do not grease the top of the driveshaft. This may excessively preload the driveshaft and crankshaft when the mounting fasteners are tightened and cause a premature failure of the power head or gearcase.

To install:

11. Pull the lower shift rod upward as far as possible to make sure it is in the **FORWARD** position and lightly lubricate the driveshaft splines with waterproof marine grease.
12. Apply a thin but uniform coat of a silicone sealer to the gearcase and driveshaft housing mating surfaces.
13. Make sure the locating dowels are in place.
14. Position the gearcase under the driveshaft housing.
15. Align the driveshaft splines with the crankshaft, insert the water tube into the water pump case and the water pump seal tube.

➡ On 1989-on models, make sure to connect the clutch rod to the shift shaft with chamfered side of the lower nut facing upward

16. Attach the shift rod into the clutch rod.

✱✱ CAUTION

Do not rotate the flywheel counterclockwise. This can damage the water pump impeller.

17. Seat the gearcase in place, rotating the flywheel clockwise as required until the driveshaft and crankshaft engage.
18. Install the gearcase bolts and lockwashers and tighten to specification.
19. Make sure the engine shift lever is in **FORWARD** and tighten the shift rod connector or upper shift rod turnbuckle. Place the shift lever in **NEUTRAL** and make sure the propeller rotates freely, then shift back into **FORWARD** and make sure the propeller will only rotate clockwise. If the propeller does not rotate as indicated, loosen the shift rod connector or upper shift rod turnbuckle and readjust the position of the connector/turnbuckle or rods as required. When shift pattern is correct, tighten the shift rod connector locknut securely.
20. Install the propeller.
21. Reconnect the spark plug leads and refill the gearcase with proper type and quantity of lubricant.
22. Install the engine cover.

DT55 and DT65

1. Remove the engine cover and disconnect the spark plug leads to prevent accidental starting of the engine during lower unit removal.
2. Place a container under the gearcase. Remove the vent and drain plugs. Drain the lubricant from the unit.

➡ If the lubricant is creamy in color or metallic particles are found, the gearcase must be completely disassembled to determine and correct the cause of the problem.

3. Wipe a small amount of lubricant on a finger and rub the finger and thumb together. Check for the presence of metallic particles in the lubricant. Note the color of the lubricant. A white or creamy color indicates water in the lubricant. Check the drain container for signs of water separation from the lubricant.
4. Remove the propeller as described in this chapter.
5. Place the outboard in **NEUTRAL**.
6. Remove the nut from the clutch shaft.
7. Disconnect the clutch shaft from the clutch rod.
8. Remove the bolts and lockwashers securing the gearcase to the driveshaft housing.
9. Remove the gearcase from the driveshaft housing and mount it in a suitable holding fixture.

✱✱ CAUTION

Do not grease the top of the driveshaft. This may excessively preload the driveshaft and crankshaft when the mounting bolts are tightened and cause a premature failure of the power head or gearcase.

To install:

10. Lightly lubricate the driveshaft splines and the O-ring seal around the driveshaft bearing with waterproof marine grease.
11. Apply a thin but uniform coat of a silicone sealer to the gearcase and driveshaft housing mating surfaces.
12. Make sure the locating dowels are in place.
13. Position the gearcase under the driveshaft housing.
14. Align the driveshaft splines with the crankshaft, insert the water tube into the water pump case and the water pump seal tube.
15. Guide the clutch rod into the clutch shaft.

8-6 LOWER UNIT

❈❈❈ CAUTION

Do not rotate the flywheel counterclockwise. This can damage the water pump impeller and cause the engine to overheat.

16. Seat the gearcase in place, rotating the flywheel clockwise as required until the driveshaft and crankshaft engage.
17. Apply a silicone sealer to the mounting bolt threads and tighten to specifications.
18. Connect the clutch shaft onto the clutch rod and install the nut. Tighten the nut securely.
19. Install the propeller.
20. Reconnect the spark plug leads and refill the gearcase with proper type and quantity of lubricant. Install the engine cover.

DT75, DT85, DT115, DT140, V4 and V6 models

1. Remove the engine cover.
2. Disconnect the spark plug leads to prevent accidental starting of the engine during lower unit removal.
3. Place a container under the gearcase. Remove the vent and drain plugs.
4. Drain the lubricant from the unit.

➡ **If the lubricant is creamy in color or metallic particles are found, the gearcase must be completely disassembled to determine and correct the cause of the problem.**

5. Wipe a small amount of lubricant on a finger and rub the finger and thumb together. Check for the presence of metallic particles in the lubricant. Note the color of the lubricant. A white or creamy color indicates water in the lubricant. Check the drain container for signs of water separation from the lubricant.
6. Remove the propeller.
7. Place the outboard in **NEUTRAL** and remove the cover at the front of the clutch rod.
8. Use needlenose pliers and remove the clutch rod connector cotter pin. Discard the cotter pin.
9. Remove the connector pin.
10. On V6 models, remove the trim tab bolt access cap on the top surface of the driveshaft housing. Remove the bolt and lockwasher securing the trim tab and remove the trim tab.
11. On all other models, remove the bolt and lockwasher securing the trim tab and remove the trim tab.
12. Remove the bolts and lockwashers securing the gearcase to the driveshaft housing.
13. Remove the gearcase from the driveshaft housing and mount it in a suitable holding fixture.

❈❈❈ CAUTION

Do not grease the top of the driveshaft. This may excessively preload the driveshaft and crankshaft when the mounting bolts are tightened and cause a premature failure of the power head or gearcase.

To install:

14. Lightly lubricate the driveshaft splines with waterproof marine grease or equivalent.
15. Apply a thin but uniform coat of a silicone sealer to the gearcase and driveshaft housing mating surfaces.
16. Make sure the locating dowels are in place.
17. Position the gearcase under the driveshaft housing.
18. Align the driveshaft splines with the crankshaft, insert the water tube into the water pump case and the water pump seal tube.
19. Guide the clutch rod into the clutch rod connector.

❈❈❈ CAUTION

Do not rotate the flywheel counterclockwise. This can damage the water pump impeller.

20. Seat the gearcase in place, rotating the flywheel clockwise as required until the driveshaft and crankshaft engage.
21. On DT115 and DT140 models, apply thread locking compound to the mounting bolt threads and tighten securely.
22. On all other models, apply a silicone sealer to the mounting bolt threads and tighten to specifications.
23. Install the trim tab and the bolt and lockwasher. Tighten the bolt securely. On V6 models, install the trim cap over the bolt hole in the driveshaft housing.
24. Connect the clutch rod into the clutch rod connector and insert the connector pin through both parts.
25. Install a new cotter pin. Bend the ends over completely.
26. Install the clutch rod cover.
27. Install the propeller.
28. Reconnect the spark plug leads and refill the gearcase case with proper type and quantity of lubricant.
29. Install the engine cover.

LOWER UNIT OVERHAUL

DT2 and DT2.2 (Forward Only)

DISASSEMBLY

♦ **See accompanying illustrations**

1. Remove the gearcase.
2. Secure the gearcase in a suitable holding fixture or a vise with protective jaws. If protective jaws are not available, position the gearcase upright with the skeg between wooden blocks in the vise.
3. Remove the water pump case cover.
4. Remove the water pump impeller.
5. Remove the pin.
6. Carefully pry the water pump case free from the gearcase housing.

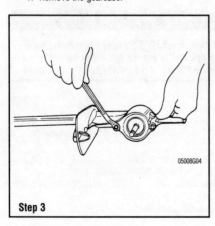

Step 3

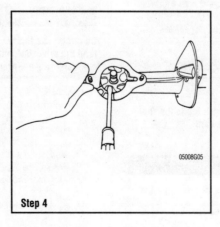

Step 4

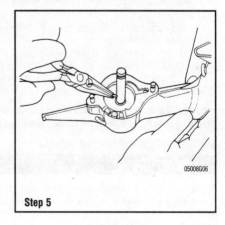

Step 5

LOWER UNIT 8-7

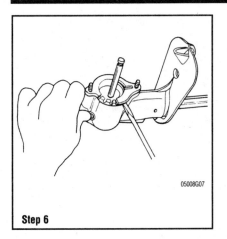
Step 6

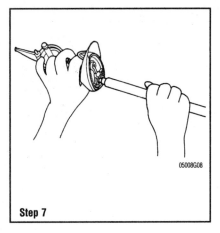
Step 7

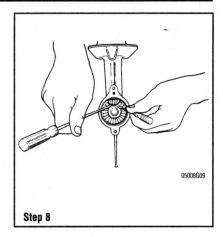
Step 8

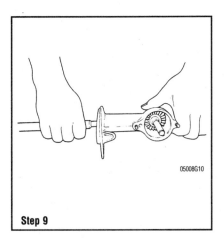
Step 9

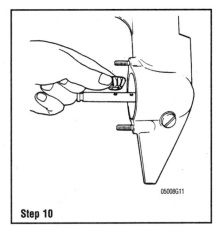
Step 10

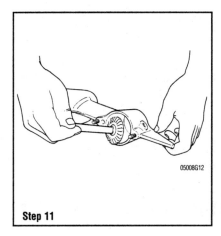
Step 11

7. Draw out the water tube and driveshaft seal pipe.
8. Use a pair of screwdrivers to pry the E-clip from the pinion gear.
9. Take out the driveshaft.
10. Remove the pinion gear and shim(s) from the gearcase housing.
11. Pull out the propeller shaft and gear assembly.
12. Remove the shim(s) from the gearcase housing.
13. Remove the driveshaft seal pipe bushing and oil seal.

CLEANING & INSPECTION

♦ See Figures 1 and 2

1. Clean and inspect all parts.
2. Inspect the driveshaft pinion and propeller shaft gear for broken teeth and wear.
3. Inspect the water pump impeller for wear and damage.
4. Inspect the driveshaft housing oil seal and water pump case oil seal lips for tears, nicks or other damage and wear.
5. Inspect the propeller for damage or distortion.
6. Check the driveshaft and propeller shaft bearings for wear and damage. Make sure there is no damage to these bearings and that they spin smoothly with no rattles or noise.
7. Inspect the cooling water passage to make sure it is clear of debris.
8. If inspection of the driveshaft bearings indicates replacement is required, proceed as follows:
   - Remove the driveshaft snap ring with snap-ring pliers.
   - Remove the shim(s) from the top driveshaft bearing.
   - Insert a bearing remover handle into the seal pipe bore.
   - Insert the bearing remover into the gearcase bore and attach to the remover
   - Pull the 2 driveshaft bearings and spacer from the gearcase.

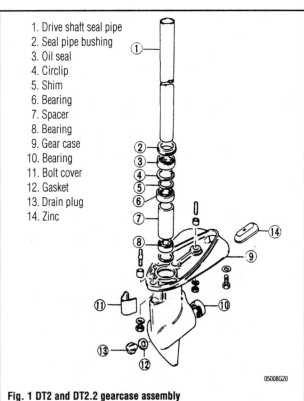

1. Drive shaft seal pipe
2. Seal pipe bushing
3. Oil seal
4. Circlip
5. Shim
6. Bearing
7. Spacer
8. Bearing
9. Gear case
10. Bearing
11. Bolt cover
12. Gasket
13. Drain plug
14. Zinc

Fig. 1 DT2 and DT2.2 gearcase assembly

8-8 LOWER UNIT

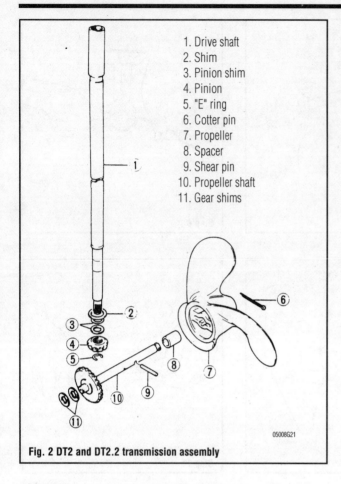

Fig. 2 DT2 and DT2.2 transmission assembly

1. Drive shaft
2. Shim
3. Pinion shim
4. Pinion
5. "E" ring
6. Cotter pin
7. Propeller
8. Spacer
9. Shear pin
10. Propeller shaft
11. Gear shims

3. Install the propeller shaft shims and propeller shaft in the gearcase housing.
4. Position the pinion gear and shim(s) in the housing under the driveshaft bore, insert the driveshaft and engage the pinion gear. Install the E-clip on the end of the driveshaft to retain the pinion gear.
5. Install the driveshaft seal pipe and water tube.
6. Remove the water pump case bearing and oil seal.
7. Install a new bearing with an installer (09914–79510).
8. Install a new oil seal and coat its lips with waterproof marine grease.
9. Before installing the water pump case to the gearcase, apply waterproof marine grease to the oil seal.
10. Also apply waterproof marine grease to the O-ring.
11. Install the water pump. Be sure to position the water pump impeller in the direction shown in the illustration. Note that the impeller vanes curve back relative to the direction of rotation. Don't forget to install the impeller pin.

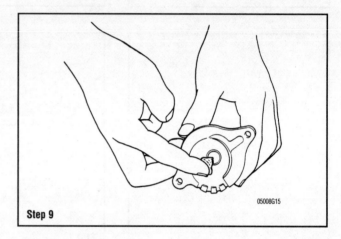

Step 9

ASSEMBLY

♦ See accompanying illustrations

1. If the driveshaft bearings are replaced:
   • Install new bearings with the spacer between them using an appropriate bearing installer.
   • Install the shim(s) on the top driveshaft bearing.
   • Install the driveshaft snap ring with pliers.
2. When installing new oil seals, make sure each seal is in its specific location.

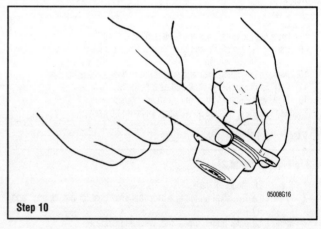

Step 10

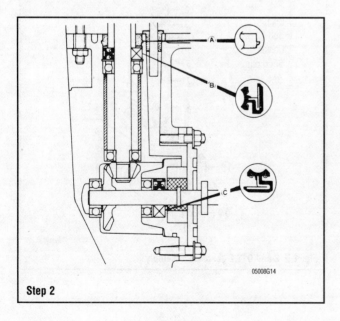

Step 2

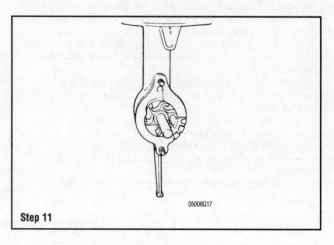

Step 11

LOWER UNIT 8-9

12. Apply waterproof marine grease to the propeller shaft and then fit the propeller onto the greased shaft.
13. After installing the gearcase, fill with the correct amount of SAE 90 hypoid gear oil: 1.35 oz. (40cc).
14. Check gearcase lubricant level after engine has been run. Change the lubricant after 10 hours of operation (break-in period).

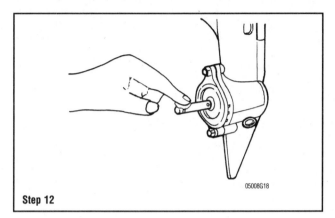
Step 12

ment exceeds this specification, remove the snap ring and exchange the shim(s) as required to bring the clearance within specifications. Shim stock is available in three sizes: 0.004 in. (0.1 mm); 0.008 in. (0.2 mm); 0.020 in. (0.5 mm).
2. Once the driveshaft bearing shimming is correct, install a new oil seal and driveshaft bushing. Coat the seal lips with waterproof marine grease.
3. Install the propeller shaft shims and propeller shaft back into the gearcase.

DT4

DISASSEMBLY

1. Secure the gearcase in a holding fixture or a vise with protective jaws. If protective jaws are not available, Place the skeg between two wooden blocks in a vise.
2. Remove the 2 bearing housing bolts. Remove the bearing housing and propeller shaft assembly .

➥**Insert a screwdriver into the propeller shaft bearing housing at the section marked with an "O". Gently pry with the screwdriver to separate the housing section.**

3. Remove the propeller shaft from the bearing housing.
4. Remove the water pump.
5. After removing the bolt, detach the shift rod guide stopper.
6. Pull out the shift rod assembly from the gearcase.
7. Use a screwdriver to carefully pry loose the driveshaft bearing housing along with the driveshaft.
8. Reach into the propeller shaft bore and remove the pinion gear and the pinion thrust washers .
9. Remove the forward gear and its shims and thrust washer .
10. Remove the bolt on the side of the gearcase, then remove the water filter.
11. Remove the reverse gear and bearing housing from the propeller shaft assembly.
12. Use a screwdriver to remove the clutch dog spring from around the clutch dog shifter.
13. Using a small drift, drive the pin out of the clutch dog shifter.
14. Slide the clutch dog off the propeller shaft.
15. Remove the push pin and return spring from the end of the propeller shaft

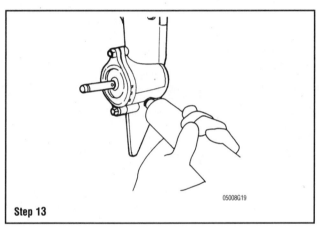
Step 13

CLEANING & INSPECTION

♦ **See Figures 4 and 5**

1. Clean and inspect all parts.
2. Inspect for abnormal or excessive wear on the following parts:
   • Forward gear bearings
   • Pinion gear
   • Reverse gear
   • Driveshaft
   • Propeller shaft

➥**Make sure to closely check all the engagement dogs on the gears and clutch dog shifter.**

   • Inspect the clutch push rod and clutch cam for excessive wear and damage.
   • Check the water pump impeller for wear and damage.
   Perform a gearcase pressure check.
   • While rotating the driveshaft and propeller shaft, apply the pressure (14.2 psi) through the oil drain plug using the oil leakage tester (09950–69511) and air pump assembly (09821–00004). Once the correct pressure has been reached, watch the pressure, if it falls, then there is a sealing problem and you will need to find out which seal is leaking. Use a mixture of soapy water and spray the area around each seal to check for air bubbles escaping. This will indicate which seal is leaking and needs to be replaced.

➥**First apply a low pressure of 2.8–5.7 psi to set the sealing lip of the seals. Then apply the full pressure to the seals. This check should be done after reassembly.**

Be sure that the coolant passages are clean and free of signs of corrosion. Also make sure that the water tube is clear.
Inspect the splines of the driveshaft for signs of wear and damage.

SHIMMING PROCEDURE

♦ **See Figure 3**

1. Temporarily install the driveshaft with the pinion gear and shims. Pull upward on driveshaft. It should not move more than 0.004in. (0.1 mm). If move-

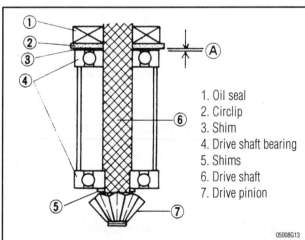

1. Oil seal
2. Circlip
3. Shim
4. Drive shaft bearing
5. Shims
6. Drive shaft
7. Drive pinion

Fig. 3 Temporarily install the driveshaft with the pinion gear and shims. Pull upward on driveshaft. It should not move more than 0.004in. (0.1 mm)

8-10 Lower Unit

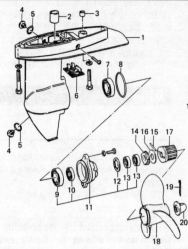

1. Case set, gear
2. Bush
3. Pin
4. Plug, drain
5. Gasket, drain plug
6. Filter, water
7. Bearing, gear
8. O ring, bearing housing
9. Bearing, reverse gear
10. Bearing, propeller shaft
11. Housing, prop shaft bearing
12. Washer, oil seal stopper
13. Oil seal
14. Protector, oil seal
15. Shear pin
16. Washer
17. Bush, propeller
18. Propeller
19. Cotter pin
20. Nut

Fig. 4 DT4 gearcase assembly

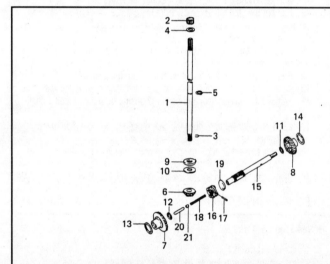

1. Shaft, drive
2. Bush, drive shaft upper
3. Ring, drive shaft snap
4. Thrust washer
5. Circlip
6. Pinion
7. Gear, forward
8. Gear, reverse
9. Thrust washer
10. Washer
11. Thrust washer REV
12. Thrust washer FWD
13. Shim, FWD
14. Shim, REV
15. Shaft, propeller
16. Shifter, clutch dog
17. Pin, clutch dog
18. Spring
19. Spring, dog
20. Rod, push
21. Pin, push

Fig. 5 DT4 transmission assembly

Check the clutch return spring for strength by measuring its free length. The spring may be re-used for assembly if the free length is within limits.
- Standard: 2.7 in.(69 mm)
- Service limit: 2.6 in. (67 mm)

ASSEMBLY

Before reassembly, liberally coat the forward gear, driveshaft and propeller shaft with outboard motor gear oil.

➡ **Make sure to note the direction of assembly on the clutch dog shifter. The end with the "F" is meant to face the forward gear. Be sure to mount the shifter correctly on the propeller shaft.**

Care must be used when installing the clutch dog shifter on the propeller shaft.
- First, insert the return spring into the propeller shaft and slide the clutch dog shifter onto the propeller shaft
- Then install the spring pin into the slots provided on both the propeller shaft and the clutch dog shifter and finally push the pin and push rod in the propeller shaft

➡ **After connecting the dog shifter to the push pin, check to be sure that the spring pin is all the way in, with its driven end flush with the surface of the dog shifter. Fit the spring snugly into the groove on the dog shifter so that the return spring does not come out.**

➡ **Be sure that the forward and reverse gears are separated and installed correctly. Both gears look alike, but the forward gear has an oil pocket located in the area around the dogs.**

1. Install the water filter and bolt. Tighten the bolt securely.
2. Fit the forward gear shims over the shaft at the rear of the gear and install into the prop shaft bore.
3. Insert the pinion gear with shims into the prop shaft bore. Fit the gear into the driveshaft bore and mesh it with the forward gear.
4. Apply a silicone sealer to the mating surfaces of the driveshaft bearing housing and gearcase housing mating surfaces.
5. Insert the driveshaft into the gearcase housing with a rotating motion and engage the pinion gear. Install the pinion gear circlip onto the driveshaft.
6. Lubricate the driveshaft bearing housing oil seal lip with waterproof marine grease. Install the housing on the driveshaft and seat into the gearcase housing.
7. Install the shift rod assembly.
8. Install the shift rod guide stopper and bolt. Tighten the bolt securely.
9. Install the water pump.
10. Check pinion gear depth and forward gear backlash.
11. Insert propeller shaft into housing bore and engage the forward gear.
12. Coat the bearing housing outer edges (front and rear) with waterproof marine grease.
13. Carefully install bearing housing on propeller shaft.

LOWER UNIT 8-11

14. Check propeller shaft thrust clearance.
15. When proper propeller shaft thrust clearance has been established, coat the bearing housing bolt threads with silicone sealer. Install housing (use installer 09914–79610) and tighten bolts to specifications.
16. Install the gearcase. Fill with recommended type and quantity of lubricant.
17. Check gearcase lubricant level after engine has been run. Change the lubricant after 10 hours of operation (break-in period).

SHIMMING PROCEDURE

Measurement Of Thrust Play

1. Using a dial indicator and stand, push the propeller shaft inward towards the gearcase. Keep the shaft in this position and set the dial indicator on the shaft with the indicator rod depressed 0.78 in. (2 mm).
2. Keeping the shaft pushed in, zero the dial indicator gauge, then pull the propeller shaft outwards until the dial indicator reads its maximum: 0.0039–0.0079 in. (0.1–0.2 mm).
3. To measure the thrust play on the drive shaft, set up the dial indicator on the drive shaft. Again, press in the shaft and zero the dial indicator and then pull out and read the dial indicator for the maximum thrust play: 0.008–0.016 in. (0.2–0.4 mm).

Pinon And Forward Gear Tooth Contact Pattern Adjustment

♦ See accompanying illustrations

Coat the entire surface of the forward gear lightly with Prussian Blue or equivalent, and install the propeller shaft and all related parts into the gearcase. It is unnecessary to install the propeller shaft housing.
While pushing the propeller shaft in, hold it firmly by hand so it does not turn. Slowly rotate the drive shaft about 5 turns, and remove the forward gear from the gearcase. Check the Prussian Blue for proper the gear tooth contact pattern.
1. The forward gear runs with a localized tooth contact, and from this it can be determined whether the tooth bearing position is correct or not.
2. This is the correct gear tooth bearing position. It is advisable to obtain this tooth contact pattern by means of shim adjustments.

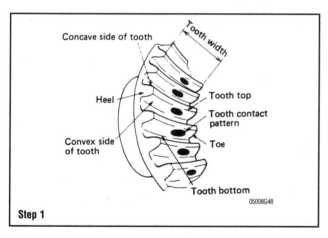
Step 1

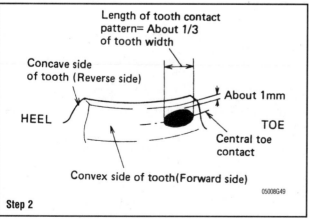
Step 2

3. This shows a top side toe contact. In this case, decrease the forward gear shim thickness and increase the pinion gear shim thickness.

➡ **The top side toe contact will result in a chipped forward gear tooth or damage to a tooth bottom of the pinion gear. Avoid this gear tooth bearing position.**

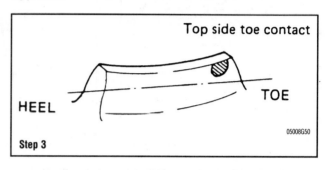
Step 3

4. If the tooth contact pattern is as shown, the bottom side toe contact is produced. In this case, increase the forward gear shim thickness and decrease the pinion gear thickness and check the tooth contact pattern.

➡ **If this gear tooth bearing position is not correct, the pinion gear tooth may be chipped, this condition must be corrected.**

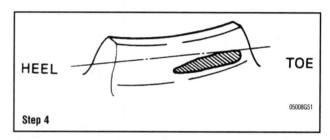
Step 4

DT6 and DT8

DISASSEMBLY

♦ See accompanying illustrations

1. Remove the gearcase.
2. Secure the gearcase in a suitable holding fixture or a vise with protective jaws. If protective jaws are not available, position the gearcase upright with the skeg between wooden blocks in a vise.
3. Remove the water pump.

➡ **If bearing housing is corroded in prop shaft bore and cannot be removed easily by hand, tap side of housing cap with a rubber mallet to rotate cap ears, then pry bearing housing off.**

4. Remove the 2 bearing housing bolts. Remove the bearing housing and propeller shaft assembly.

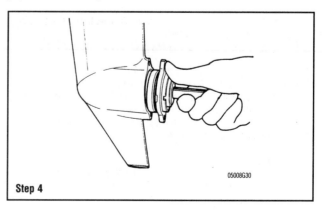
Step 4

8-12 LOWER UNIT

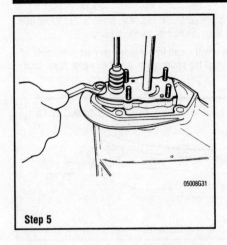
Step 5

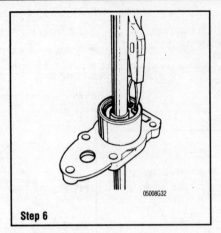
Step 6

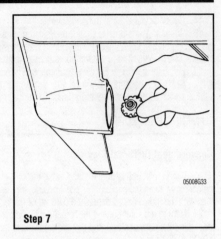
Step 7

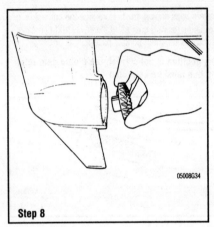
Step 8

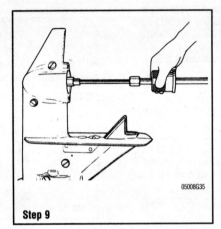
Step 9

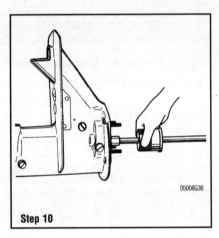
Step 10

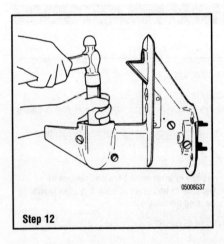
Step 12

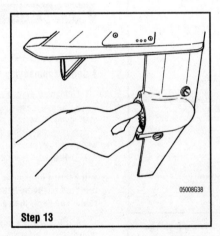
Step 13

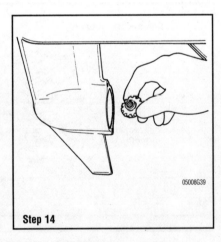
Step 14

5. Unbolt the driveshaft bearing housing and remove the bearing housing, the driveshaft and the shift rod assembly.

6. Remove the snap-ring from the driveshaft bearing housing and pull the driveshaft out.

7. Remove the pinion gear from the gearcase.

8. And then remove the forward gear and shim(s).

9. If necessary, obtain a bearing remover (09913–69911) and slide shaft (09930–30102) to remove the forward gear bearing.

10. Use the same remover/slide hammer combination to remove the gearcase driveshaft needle roller bearing.

11. Clean and inspect all parts.

To install:

12. Install new forward gear and driveshaft bearings with installer (09914–79610) or equivalent.

13. Fit the forward gear shim(s) over the shaft at the rear of the gear and install the gear into the gearcase.

14. Insert the pinion gear with shim(s) into the prop shaft bore. Make sure that the pinion gear meshes with the forward gear.

15. Lubricate the driveshaft bearing housing oil seal lip with waterproof marine grease. Install the housing on the driveshaft and seat into the gearcase housing. Install housing snap ring (if used) and make sure it fits properly into its groove.

16. Insert the driveshaft into the gearcase housing with a rotating motion and engage the pinion gear.

17. Fit the adjusting shim (which was removed in disassembly) to the forward end of the driveshaft. Fit the bearing housing to the gearcase. Insert the serrated end of the driveshaft into the pinion gear.

18. Adjust the shifting cam and shifting rod to the specified length:

LOWER UNIT   8-13

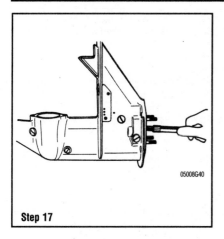
Step 17

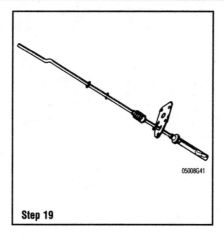
Step 19

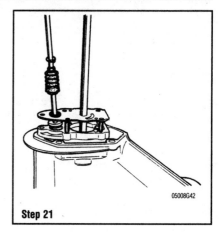
Step 21

- Short shaft: 17.23 in. (438 mm)
- Long shaft: 22.24 in. (565 mm)

19. Insert the shifting rod into the gearcase and retain the rod with the shifting cam stop screw. Check to be sure that the shifting rod is still capable of smooth movement.
20. Install shift cam stop screw.

➡ **The propeller shaft must be out of the gearcase when installing or removing the clutch rod.**

21. Install the shifting rod guide in the sequence illustrated. Make sure to apply waterproof marine grease to the O-ring.
22. Install the water pump.
23. Check pinion gear depth and forward gear backlash.
24. Remove and discard the bearing housing bearing and oil seal and replace with a new seal.
25. Install a new housing bearing with installer (09914–79610) or equivalent. Coat seal lips with waterproof marine grease.
26. Install a new bearing housing O-ring. Lubricate housing O-ring and end of propeller shaft pushrod with waterproof marine grease.
27. Insert propeller shaft into housing bore and engage the forward gear.
28. Coat bearing housing outer edges (front and rear) with waterproof marine grease.
29. Carefully install bearing housing on propeller shaft.
30. Check propeller shaft thrust clearance as described in this chapter.
31. When proper propeller shaft thrust clearance has been established, coat the bearing housing bolt threads with Silicone sealer. Install housing and tighten bolts securely.
32. Fill the gearcase with the recommended type and quantity of lubricant.
33. Check gearcase lubricant level after engine has been run. Change the lubricant after 10 hours of operation (break-in period).

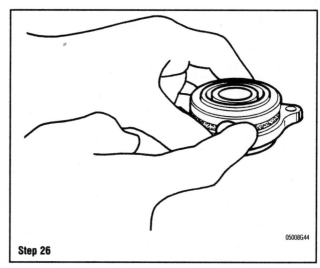
Step 26

INSPECTION & CLEANING

♦ **See Figures 6 and 7**

Wash all parts completely and dry them using compressed air. Inspect each part and service it, as necessary, or replace the part if it does not meet specification. Parts to be inspected and items to be checked are as follows:
- All bearings for wear and damage

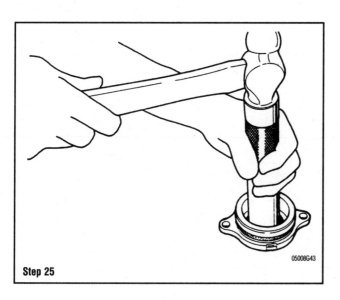
Step 25

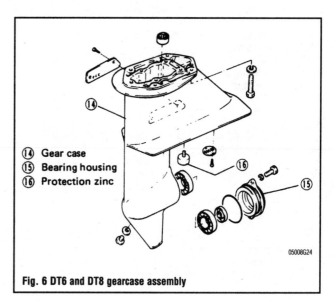
⑭ Gear case
⑮ Bearing housing
⑯ Protection zinc

Fig. 6 DT6 and DT8 gearcase assembly

8-14 LOWER UNIT

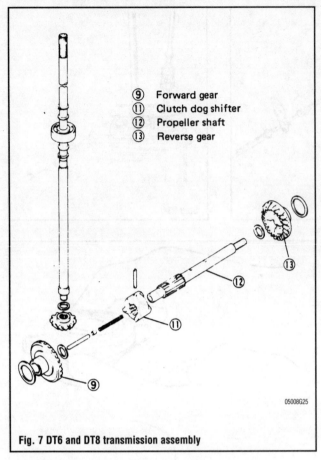

- ⑨ Forward gear
- ⑪ Clutch dog shifter
- ⑫ Propeller shaft
- ⑬ Reverse gear

Fig. 7 DT6 and DT8 transmission assembly

- Propeller shaft and driveshaft for wear at oil seal contact points
- All gear teeth for damage or wear
- Dogs on clutch dog shifter for wear and damage
- Dogs on the forward and reverse gears for damage and wear
- Shifting cam and pushrod for wear
- O-rings and oil seals for cracks, tears and wear
- Propeller for nicks, bent blades or other damage and wear
- Cooling circuit for clogging or other obstructions
- Gearcase for rusting, pitting and distortion.

ASSEMBLY

1. Install new forward gear and driveshaft bearings with installer (09914-79610) or equivalent.
2. Fit the forward gear shim(s) over the shaft at the rear of the gear and install into the prop shaft bore.
3. Insert the pinion gear with shim(s) into the prop shaft bore. Fit the gear into the driveshaft bore and mesh it with the forward gear.
4. Lubricate the driveshaft bearing housing oil seal lip with waterproof marine grease. Install the housing on the driveshaft and seat into the gearcase housing. Install housing snap ring (if used) and make sure it fits properly into its groove.
5. Insert the driveshaft into the gearcase housing with a rotating motion and engage the pinion gear.
6. Lubricate the shift rod O-ring with waterproof marine grease and insert shift rod assembly into gearcase.
7. Install shift cam stop screw.
8. Install the water pump.
9. Check pinion gear depth and forward gear backlash.
10. Remove and discard the bearing housing bearing and oil seal.
11. Install a new housing bearing with installer (09914–79610) or equivalent. Coat seal lips with waterproof marine grease.
12. Install a new bearing housing O-ring. Lubricate housing O-ring and end of propeller shaft pushrod with waterproof marine grease.

13. Insert propeller shaft into housing bore and engage the forward gear.
14. Coat bearing housing outer edges (front and rear) with waterproof marine grease.
15. Carefully install bearing housing on propeller shaft.
16. Check propeller shaft thrust clearance.
17. When proper propeller shaft thrust clearance has been established, coat the bearing housing bolt threads with Silicone sealer. Install housing and tighten bolts securely.
18. Install the gearcase. Fill with recommended type and quantity of lubricant.
19. Check gearcase lubricant level after engine has been run. Change the lubricant after 10 hours of operation (break-in period).

SHIMMING PROCEDURE

Adjust the forward gear in such a way that the tooth contact pattern produced by rolling this gear against the pinion gear will extend from the center of the tooth toward the toe. This is accomplished by increasing or decreasing the shim thickness.

Be sure to produce the specified amount of backlash between the pinion gear and forward gear. To measure the backlash, hold the pinion steady with your hand and move the forward gear back and forth. Check the backlash at the heel. Backlash should measure 0.0039–0.0079 in. (0.1–0.2 mm).

Be sure that there is no backlash between the reverse gear and the pinion gear, although a small amount of backlash is permitted. To eliminate the backlash, if any, increase the thickness of the shim behind the reverse gear.

By means of the shim on the inner side of the forward gear and reverse gear, adjust the thrust play of the propeller shaft to this specification: 0.002–0.020 in. (0.05–0.50 mm).

DT9.9 and DT15

DISASSEMBLY

♦ See accompanying illustrations

1. Remove the gearcase.
2. Secure the gearcase in a suitable holding fixture or a vise with protective jaws. If protective jaws are not available, position the gearcase upright with the skeg between wooden blocks in a vise.
3. Remove the 2 bearing housing bolts.

➡ **If the bearing housing is corroded in the prop shaft bore and cannot be removed easily by hand, carefully tap the side of the housing cap with a rubber mallet to rotate the cap ears, then pry housing off.**

4. Attach the special tools sliding shaft (09930–30102) and propeller shaft housing remover (09950–59320) to the propeller shaft splined end.
5. Using the special tool, remove the bearing housing and propeller shaft assembly.
6. Remove the water pump.

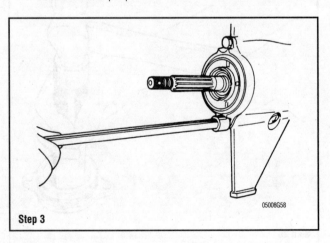

Step 3

LOWER UNIT 8-15

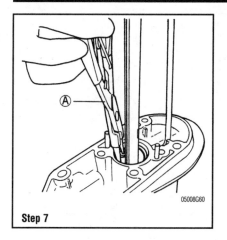
Step 7

Step 8

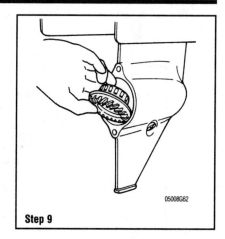
Step 9

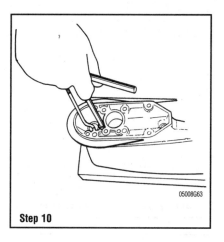
Step 10

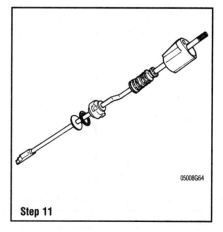
Step 11

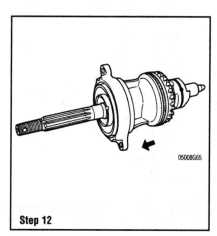
Step 12

7. Using closing-type snap ring pliers, remove the gearcase circlip.
8. Remove the pinion, thrust washers, shims, and thrust bearing and pull out the driveshaft from the gearcase.
9. Then remove the forward gear and shim(s). Be sure to remove all shims and reinstall the same amount of shims.
10. Remove the shift rod collar nut.
11. Then remove the shift rod assembly.
12. Remove the reverse gear and bearing housing from the propeller shaft assembly that had been already removed.
13. Disassemble the propeller shaft and clutch assembly.
14. After removing the clutch shifter dog spring from around the shifter, use a drift to drive out the pin and then slide the clutch dog shifter off the propeller shaft.
15. Remove the return spring from the end of the propeller shaft.

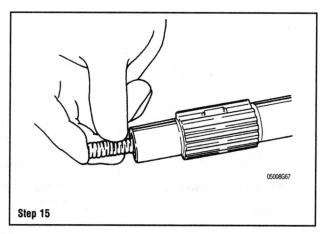
Step 15

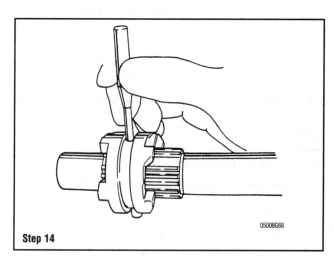
Step 14

CLEANING & INSPECTION

♦ See Figures 8 thru 13

Wash all parts completely and dry them using compressed air. Inspect each part and service it, as necessary, or replace the part if it does not meet specification. Parts to be inspected and items to be checked are as follows:
- All bearings for wear and damage
- Propeller shaft and driveshaft for wear at oil seal contact points
- All gear teeth for damage or wear
- Dogs on clutch dog shifter for wear and damage
- Dogs on the forward and reverse gears for damage and wear
- Shifting cam and pushrod for wear
- Perform a gearcase pressure test to check the seals. Use the oil leakage tester (09950–69510) and air pump assembly (09821–00004) and pressurize

8-16 LOWER UNIT

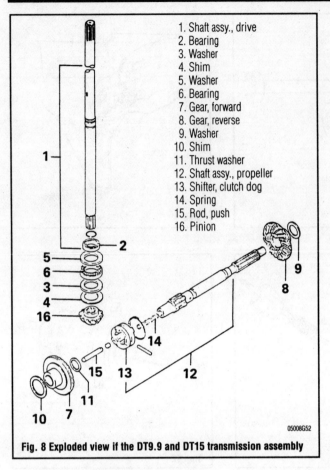

1. Shaft assy., drive
2. Bearing
3. Washer
4. Shim
5. Washer
6. Bearing
7. Gear, forward
8. Gear, reverse
9. Washer
10. Shim
11. Thrust washer
12. Shaft assy., propeller
13. Shifter, clutch dog
14. Spring
15. Rod, push
16. Pinion

Fig. 8 Exploded view if the DT9.9 and DT15 transmission assembly

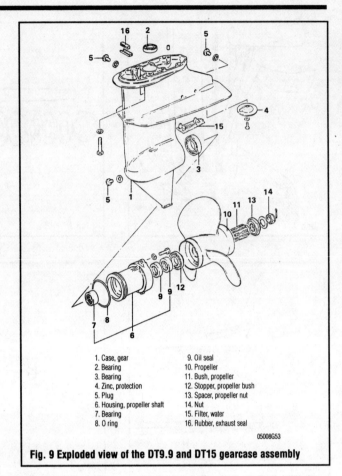

1. Case, gear
2. Bearing
3. Bearing
4. Zinc, protection
5. Plug
6. Housing, propeller shaft
7. Bearing
8. O ring
9. Oil seal
10. Propeller
11. Bush, propeller
12. Stopper, propeller bush
13. Spacer, propeller nut
14. Nut
15. Filter, water
16. Rubber, exhaust seal

Fig. 9 Exploded view of the DT9.9 and DT15 gearcase assembly

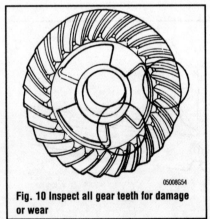

Fig. 10 Inspect all gear teeth for damage or wear

Fig. 11 Dogs on clutch dog shifter for wear and damage

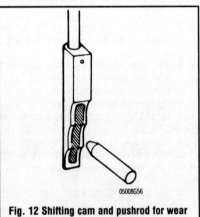

Fig. 12 Shifting cam and pushrod for wear

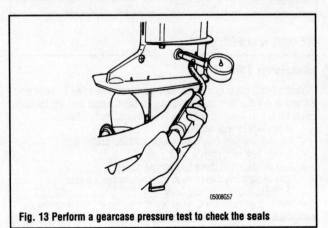

Fig. 13 Perform a gearcase pressure test to check the seals

the gearcase to 14.22 psi (1.0 kg/cm). With the gearcase pressurized, spray soapy water onto the seals and check for escaping air bubbles.
- O-rings and oil seals for cracks, tears and wear
- Propeller for nicks, bent blades or other damage and wear
- Cooling circuit for clogging or other obstructions
- Gearcase for rusting, pitting and distortion.

ASSEMBLY

▶ See accompanying illustrations

1. Reassemble the clutch dog shifter. This shifter has the letter "F" on the end meant to face the forward gear.
2. First, insert the return spring and the clutch dog shifter into the propeller shaft, the spring pin into the slots provided on both the propeller shaft and the dog shifter, and finally push the pin and the push rod into the propeller shaft.

LOWER UNIT 8-17

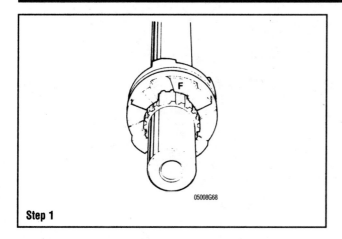
Step 1

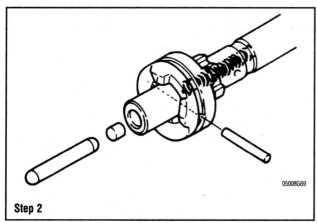
Step 2

13. Coat bearing housing outer edges (front and rear) with waterproof marine grease.
14. Carefully install bearing housing on propeller shaft.
15. Check propeller shaft thrust clearance.
16. When proper propeller shaft end play/thrust clearance has been established, coat the bearing housing bolt threads with a silicone sealer.
17. Use the special tool (09922–59510) to install housing and tighten bolts to 4.5–7.0 ft. lbs. (6–10 Nm).
18. Install the water pump.
19. Install the gearcase. In assembling the gearcase assembly and extension case to the driveshaft housing, apply silicone sealant to both joining faces and assemble after aligning the dowel pins. Tighten the bolts to 11.0–14.5 ft. lbs. (15–20 Nm).
20. Fill with recommended type and quantity of lubricant.
21. Check gearcase lubricant level after engine has been run. Change the lubricant after 10 hours of operation (break-in period).

SHIMMING PROCEDURE

Transmission Gear Adjustment

THRUST WASHER & SHIM MOUNTING POSITION

♦ See Figures 14 and 15

If the lower unit has been rebuilt or any of the components have been replaced, then shimming for correct gear contact and backlash will need to be performed.
Initial selection-shim adjustment may be required.

3. Fit the forward gear shim(s) over the shaft at the rear of the gear and install gear/shim assembly into the prop shaft bore.
4. Insert the shift rod assembly into the gearcase bore and install the collar nut.
5. Insert the pinion gear into the prop shaft bore. Fit the gear into the driveshaft bore and mesh it with the forward gear.
6. Reassemble the bearing, shims and thrust washer to the driveshaft in the same order as noted during removal.
7. Insert the driveshaft into the gearcase housing with a rotating motion and engage the pinion gear.
8. Install a new driveshaft bearing snap ring. Make sure the snap ring fits properly into its groove.
9. Install rubber exhaust seal.
10. Check pinion gear depth and forward gear backlash.
11. Install a new bearing housing O-ring. Lubricate housing O-ring with waterproof marine grease.
12. Insert propeller shaft into housing bore and engage the forward gear.

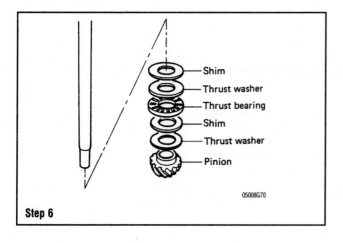
Step 6

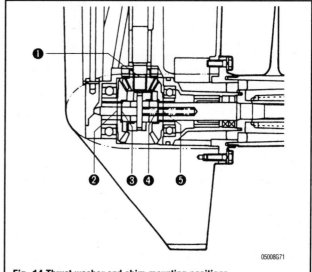
Fig. 14 Thrust washer and shim mounting positions

	Numerical index/Item	Available Thickness	Design specification thickness
①	Pinion gear back-up shim (mm)	0.7, 0.8, 0.9, 1.0, 1.1, 1.2, 1.3	1.0
②	Forward gear back-up shim (mm)	0.7, 0.8, 0.9, 1.0, 1.1, 1.2, 1.3	1.0
③	Forward gear thrust washer (mm)	1.5	1.5
④	Reverse gear thrust washer (mm)	1.1, 1.2, 1.3, 1.4, 1.5, 1.6, 1.7, 1.8, 1.9	1.5
⑤	Reverse gear back-up shim (mm)	0.7, 0.8, 0.9, 1.0, 1.1, 1.2, 1.3	1.0

Fig. 15 Shim specification table

8-18 LOWER UNIT

Forward Gear & Pinion Gear

♦ See Figure 16

The following steps need to be performed prior to adjustment.

1. Install forward gear bearing, back-up shim, forward gear, and forward gear thrust washer. Install thrust washer, thrust bearing, thrust washer, pinion back-up shim and pinion gear.

➥When installing the forward gear back-up shim, choose the shim thinner than the design specification for calculating adjustment.

2. Slide drive shaft assembly down into the gearcase and install the pinion gear on the driveshaft splines.
3. Install the pinion gear circlip.
4. Prior to checking the tooth contact pattern, be sure a slight amount of backlash exists between the pinion gear and forward gear by slightly rotating the driveshaft or forward gear by hand. If there is no backlash, reduce back-up shim thickness. If there is too much backlash, increase the back-up shim thickness.

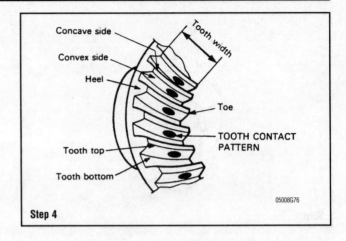
Step 4

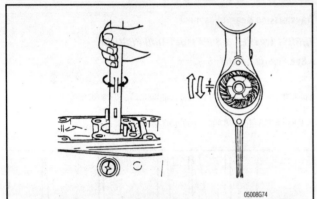

Fig. 16 Prior to checking the tooth contact pattern, be sure a slight amount of backlash exists between the pinion gear and forward gear by slightly rotating the driveshaft or forward gear by hand

5. This is the optimum tooth contact. Doing a shim adjustment may be necessary to obtain this contact pattern.
6. An example of incorrect top side toe contact. To correct this condition, decrease the forward gear shim thickness and slightly increase the pinion gear shim thickness.

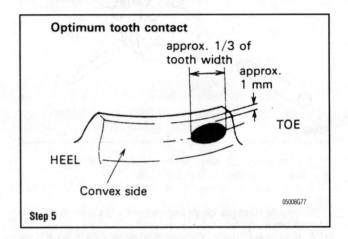
Step 5

CHECKING AND ADJUSTING TOOTH CONTACT (FORWARD/PINION GEARS)

♦ See accompanying illustrations

Check tooth contact pattern using the following procedure.
1. Too correctly assess tooth contact, smear a light coating of Prussian Blue compound on the convex surface of the forward gear.
2. Install the propeller shaft and bearing housing assembly (minus the reverse gear and internal components).
3. Push the propeller shaft inward and hold it in that position. Rotate the driveshaft clockwise 5–6 times by hand.
4. Then pull out the propeller shaft and bearing housing and then check the gear tooth contact pattern.

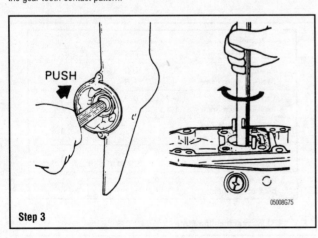
Step 3

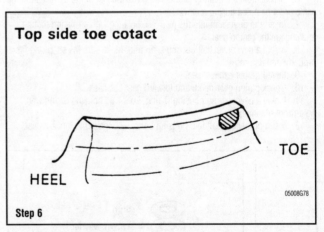
Step 6

➥Do not set the tooth contact in this position (top side toe contact). Damage and chipping of the pinion and forward gear may result.

7. An example of incorrect bottom side toe contact. To correct this condition, increase the forward gear shim thickness and slightly decrease the pinion gear shim thickness.

➥Do not set the tooth contact in this position (bottom side toe contact). Damage and chipping of the pinion gear may result.

LOWER UNIT **8-19**

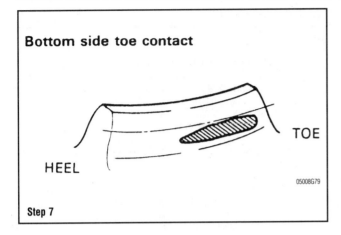

Step 7

8. After adjusting the tooth contact pattern, install the propeller shaft, bearing housing assembly, reverse gear and all related shims and washers.
9. Recheck the amount of backlash by slightly rotating the drive shaft by hand. Backlash should not be less than when checked at the start of this entire procedure.
10. If backlash is less, reduce the reverse gear back-up shim thickness.

CHECKING PROPELLER SHAFT THRUST PLAY

▶ See Figure 17

To perform the thrust play measurement, obtain the following special tool
- Gear adjusting gauge (09951–09510).

After adjusting all the gear positions, measure the propeller shaft thrust play and if it is not within specification: 0.008–0.016 in. (0.2–0.4 mm), you must make a shim adjustment.

→Maintain the forward gear thrust washer at standard thickness (1.5 mm) and adjust only the reverse gear thrust washer with shims.

1. Set the gear adjusting gauge to the propeller shaft.
2. Push the propeller shaft inward.
3. Hold the shaft in and set the dial gauge pointer to zero.
4. Slowly pull the propeller shaft outward and read the maximum thrust play measurement on the dial.
   - If the play is larger than specified, increase the reverse gear thrust washer thickness.
   - If the play is smaller than specified, reduce the reverse gear thrust washer thickness.

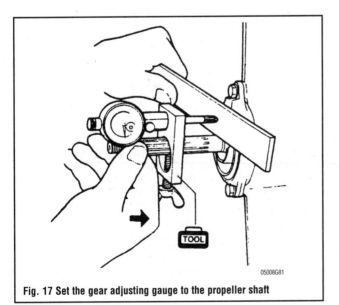

Fig. 17 Set the gear adjusting gauge to the propeller shaft

DT20, DT25 and DT30

DISASSEMBLY

▶ See Figure 18 and 19

1. Remove the propeller bearing housing bolts.
2. Obtain the following special tools:
   - Sliding hammer (09930–30102)
   - Propeller shaft remover attachment (09950–59310).
3. Attach the special tools to the propeller shaft and using the slide hammer attachment, remove the propeller bearing housing assembly and propeller shaft assembly.
4. Remove the water pump assembly.
5. Remove the key, detach the pump case lower plate and gasket.
6. Fit the drive shaft holder (09921–29610) to the splined end of the driveshaft.
7. Pad the sides of the gearcase, hold the pinion nut with a wrench, and turn the driveshaft with the special tool, loosening the pinion nut securing the pinion gear to the driveshaft.
8. Remove the two bolts located on the driveshaft bearing housing.
9. To separate the driveshaft bearing housing from the gearcase, use two 6 mm jacking bolts to separate the two components. Make sure to screw the jacking bolts equally to prevent damage to the bearing housing and gearcase.
10. Pull out the driveshaft with the driveshaft bearing housing.
11. Remove the pinion gear from the gearcase.
12. Remove the forward gear with its shim and bearing.

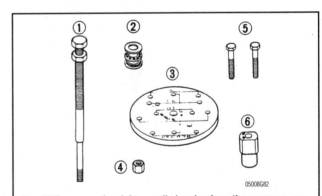

Fig. 18 To remove the pinion needle bearing from the gearcase, use the following special tools

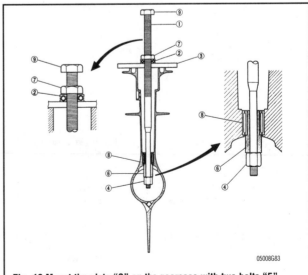

Fig. 19 Mount the plate "3" on the gearcase with two bolts "5"

8-20 LOWER UNIT

13. Take out the pinion shim.
14. Remove the screw holding the shift rod guide and pull out the shift rod assembly.
15. Using a pair of snap ring pliers, remove the circlip holding the collar preload spring and remove the drive shaft, preload spring, spring collar and washers.
16. Remove the dog spring from around the clutch dog shifter.
17. Remove the push rod from the end of the propeller shaft.
18. Using a drift, push out the pin from the clutch dog shifter.
19. Remove the clutch dog shifter, push pin and return spring from the propeller shaft.
20. To remove the pinion needle bearing from the gearcase, use the following special tools:
    - Remover shaft "1" (09951–49910)
    - Bearing "2" (09951–69910)
    - Plate "3" (09951–39914)
    - Nut "4" (09951–29910)
    - Bolts "5" (01107–08408)
    - Attachment "6" (09951–19610)
21. Remove the four stud bolts on the gearcase.
22. Mount the plate "3" on the gearcase with two bolts "5".
23. Have the remover shaft "1" threaded through the plate so that the bearing "2" stays between the plate and turning nut "7", and put it in the gearcase. Then, install the attachment "6" and hold it with the nut "4" at the end of the remover shaft.
24. Remove the pinion needle bearing "8" by turning the lower nut "7" clockwise with the upper nut "9" held tight.

CLEANING & INSPECTION

♦ See Figure 20

Wash all parts completely and dry them using compressed air. Inspect each part and service it, as necessary, or replace the part if it does not meet specification. Parts to be inspected and items to be checked are as follows:
- All bearings for wear and damage
- Propeller shaft and driveshaft for wear at oil seal contact points
- All gear teeth for damage or wear
- Dogs on clutch dog shifter for wear and damage
- Dogs on the forward and reverse gears for damage and wear
- Shifting cam and pushrod for wear
- Perform a gearcase pressure test to check the seals. Use the oil leakage tester (09950–69511) and air pump assembly (09821–00004) and pressurize the gearcase to 14.22 psi (1.0 kg/cm). With the gearcase pressurized, spray soapy water onto the seals and check for escaping air bubbles.
- O-rings and oil seals for cracks, tears and wear
- Propeller for nicks, bent blades or other damage and wear
- Cooling circuit for clogging or other obstructions
- Gearcase for rusting, pitting and distortion.

ASSEMBLY

♦ See Figures 21, 22 and 23

Drive the pinion needle bearing into the gearcase using the following special tools:

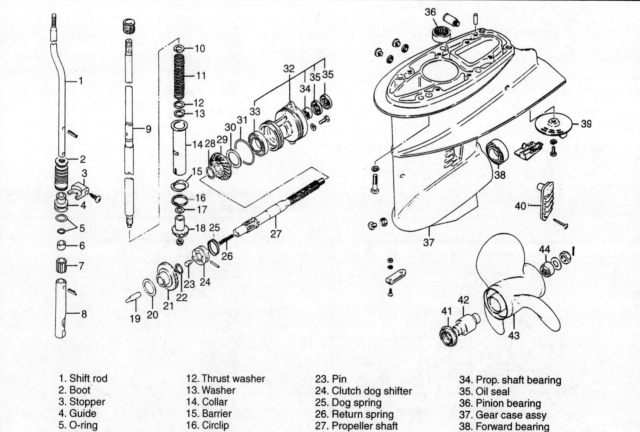

1. Shift rod
2. Boot
3. Stopper
4. Guide
5. O-ring
6. Magnet spacer
7. Magnet
8. Cam
9. Drive shaft assy
10. Washer
11. Spring
12. Thrust washer
13. Washer
14. Collar
15. Barrier
16. Circlip
17. Shim
18. Pinion gear
19. Push rod
20. Shim
21. Forward gear
22. Thrust washer
23. Pin
24. Clutch dog shifter
25. Dog spring
26. Return spring
27. Propeller shaft
28. Thrust washer
29. Reverse gear
30. Shim
31. O-ring
32. Prop. shaft housing
33. Reverse bearing
34. Prop. shaft bearing
35. Oil seal
36. Pinion bearing
37. Gear case assy
38. Forward bearing
39. Trim tab
40. Water filter
41. Prop. bush stopper
42. Bush
43. Propeller
44. Spacer

Fig. 20 Exploded view if the DT25 and DT30 lower unit assembly

LOWER UNIT   8-21

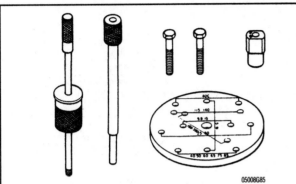

Fig. 21 Drive the pinion needle bearing into the gearcase using the following special tools

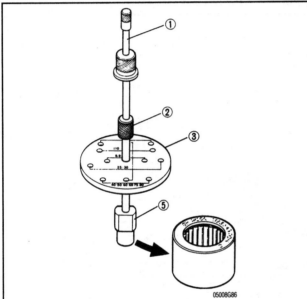

Fig. 22 Thread the slide shaft "1" into the top of the installer shaft "2" and have the installer shaft "2" inserted through the plate "3", then install the attachment "5" at the shaft end with the pinion needle bearing attached to it

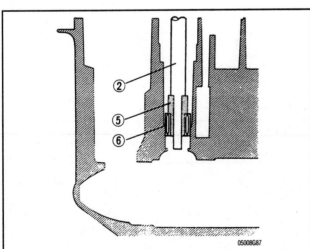

Fig. 23 Lay the gearcase down, and insert the installer shaft into it, making sure of the position of the pinion needle bearing "6" that is being installed

- "1" Rotor remover slide shaft (09930–30102)
- "2" Installer shaft (09951–59910)
- "3" Pinion needle bearing plate (09951–39914)
- "4" Bolts (01107–08408)
- "5" Attachment (09951–19610)

➡ Make sure to thoroughly clean out the inside of the gearcase before installation.

   1. Thread the slide shaft "1" into the top of the installer shaft "2" and have the installer shaft "2" inserted through the plate "3", then install the attachment "5" at the shaft end with the pinion needle bearing attached to it.

❊❊❊ CAUTION

When installing the pinion needle bearing, make sure to have the stamped marks on the bearing facing up.

   2. Lay the gearcase down, and insert the installer shaft into it, making sure of the position of the pinion needle bearing "6" that is being installed.
   3. Set the gearcase upright and install the plate on the gearcase with the two bolts "4", then install the pinion needle bearing by hitting the installer shaft lightly until the coupler touches the plate.

❊❊❊ CAUTION

If the pinion needle bearing does not install smoothly, it might be misaligned. Make sure to realign the bearing before proceeding or damage to the bearing and gearcase may result. Do not use force to set the bearing, just use repeated light hammering strokes on the special tool until the bearing is seated in the gearcase.

   4. Before installing the preload spring, be sure to put the end of the spring into the notch that is provided in the inner race of the drive shaft tapered bearing.
   5. Be sure to not miss the driveshaft lockwasher when putting the collar on the preload spring. Fit the onto the collar, making the tongue of the washer show out of the slip provided in the top of the collar, and then put on the collar. After this, insert the thrust washer at the near side.
   6. After putting on the preload spring collar, push down the collar all the way to compress the preload spring inside, and then install the circlip to hold the collar in place.
   7. Install the shift rod and shift rod guide assembly.
   8. Install the pinion shim.
   9. Install the forward gear, shim and bearing into the gearcase.
   10. Install the pinion shim into the gearcase.
   11. Install the driveshaft and driveshaft bearing housing into the gearcase and tighten the retaining bolts.
   12. Install the water pump assembly.
   13. Install the pinion gear into the gearcase, making sure to mesh it with the forward gear.
   14. Using the special tool to turn the drive shaft, apply a small amount of thread locking compound to the driveshaft threads and tighten the pinion nut to 19.5–21.5 ft. lbs. (27–30 Nm). Make sure to pad the gearcase to prevent the wrench from damaging it.
   15. The clutch dog shifter is marked for easy assembly. The end with the letter "F" must face the forward gear. Be sure to mount the clutch dog shifter correctly on the propeller shaft.
   16. Insert the return spring and push pin into the propeller shaft.
   17. Bring the push pin to the slot, making the hole of the push pin visible in the slot; then slide the dog shifter over the slot aligning the pin hole to the hole in the shaft. Insert the drive spring pin through the dog shifter and hole of the push pin.
   18. Check to be sure that the pin is all the way in, with its driven end flush with the surface of the clutch dog shifter.
   19. Fit the spring snugly into the groove on the clutch dog shifter to retain the pin in place.
   After fitting the dog shifter and connecting pin on the propeller shaft, install the reverse gear, using the forward and reverse thrust washers and reverse shim that were removed during disassembly.
   Use the following special tools to install the bearing housing into the gearcase.

- Propeller shaft housing installer (09922–59410)
- Bearing installer handle (09922–59420)

8-22 LOWER UNIT

20. With the special tool installed, drive the bearing housing into the gearcase until it is seated firmly.

21. Tighten the propeller shaft housing bolts to 11.0–14.5 ft. lbs. (15–20 Nm).

SHIMMING PROCEDURE

Transmission Gear Adjustment

▶ See Figure 24

PINION GEAR TO FORWARD GEAR BACKLASH

▶ See Figure 25

1. To measure the backlash, hold the driveshaft steady and move the forward gear back and forth.
2. Check the backlash at the heel of the forward gear.
3. If the amount of backlash is larger than 0.004–0.008 in. (0.1–0.2 mm), increase the thickness of each of the pinion gear back-up shim(s).
4. If the backlash is smaller, decrease the thickness of each shim.

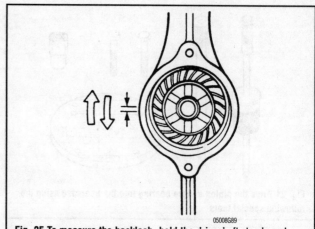

Fig. 25 To measure the backlash, hold the driveshaft steady and move the forward gear back and forth

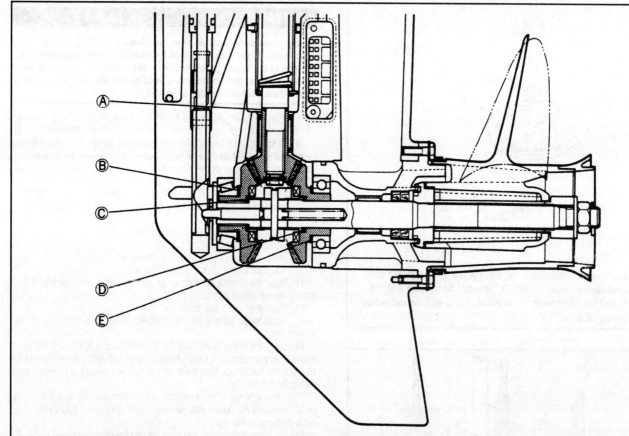

Part Name	Type of Thickness (mm)	Standard Thickness (mm)
Ⓐ Pinion gear back-up shim	1.7, 1.8, 1.9, 2.0, 2.1, 2.2	2.0
Ⓑ Forward gear back-up shim	0.8, 0.9, 1.0, 1.1, 1.2, 1.3, 1.4, 1.5	1.2
Ⓒ Forward gear thrust washer	1.5	1.5
Ⓓ Reverse gear thrust washer	0.8, 0.9, 1.0, 1.1, 1.2, 1.3	2.0
Ⓔ Reverse gear back-up shim	0.2, 0.5, 0.8, 1.0	1.5

Fig. 24 Back-up shim and thrust washer mounting locations and shim thickness chart

LOWER UNIT 8-23

CHECKING AND ADJUSTING TOOTH CONTACT (FORWARD/PINION GEARS)
♦ See accompanying illustrations

Check tooth contact pattern using the following procedure.
 1. Too correctly assess tooth contact, smear a light coating of Prussian Blue compound on the convex surface of the forward gear.
 2. Install the propeller shaft and bearing housing assembly (minus the reverse gear and internal components).
 3. Push the propeller shaft inward and hold it in that position. Rotate the driveshaft clockwise 5–6 times by hand.
 4. Then pull out the propeller shaft and bearing housing and then check the gear tooth contact pattern.
 5. This is the optimum tooth contact. Doing a shim adjustment may be necessary to obtain this contact pattern.
 6. An example of incorrect top side toe contact. To correct this condition, decrease the forward gear shim thickness and slightly increase the pinion gear shim thickness.

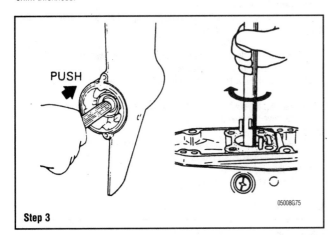
Step 3

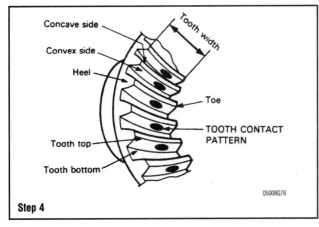
Step 4

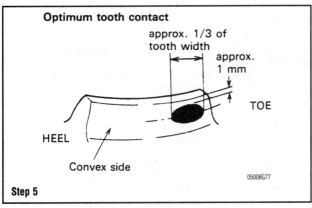
Step 5

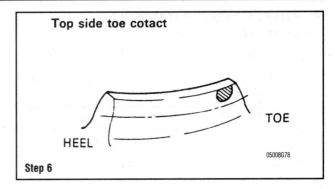
Step 6

➡ **Do not set the tooth contact in this position (top side toe contact). Damage and chipping of the pinion and forward gear may result.**

 7. An example of incorrect bottom side toe contact. To correct this condition, increase the forward gear shim thickness and slightly decrease the pinion gear shim thickness.

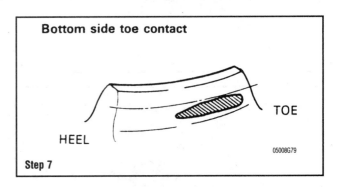
Step 7

➡ **Do not set the tooth contact in this position (bottom side toe contact). Damage and chipping of the pinion gear may result.**

 8. After adjusting the tooth contact pattern, install the propeller shaft, bearing housing assembly, reverse gear and all related shims and washers.
 9. Recheck the amount of backlash by slightly rotating the drive shaft by hand. Backlash should not be less than when checked at the start of this entire procedure.
 10. If backlash is less, reduce the reverse gear back-up shim thickness.

PINION GEAR TO FORWARD GEAR THRUST PLAY

Obtain the following special tools to perform the measurement:
 • Gear adjusting set (09951–09510)

 1. Fit the propeller shaft to the forward gear and press the forward gear forward by pushing with the propeller shaft. (In this case, first remove the push rod from the propeller shaft)
 2. Set the special tool up on the driveshaft.
 3. Push the driveshaft down. Hold this position and place the gauge with its rod pushed in approximately 2 mm.
 4. Still keeping this position, zero the dial gauge. Then pull up on the driveshaft and read the maximum play on the gauge.

If the amount of backlash is larger than specified, increase the thickness of each of the pinion gear back-up shim(s) or forward gear back-up shim(s). If smaller, decrease the thickness of each shim.

 5. When the pinion and forward gear back-up shim are determined to be the correct size, note the amount of thrust play measured. The thrust play measurement will be necessary for the adjustment of the reverse gear.

REVERSE GEAR THRUST PLAY ADJUSTMENT

 1. Perform the same procedure as above with the exception of installing the propeller shaft and propeller shaft bearing housing into the gearcase.
 2. If the measurement of the play is equal to the pinion gear to forward gear measurement, the condition is correct.
 3. If the amount of play is smaller, decrease the thickness of the reverse gear back-up shim.

8-24 LOWER UNIT

PROPELLER SHAFT THRUST PLAY ADJUSTMENT

♦ See accompanying illustrations

1. Install the dial gauge adaptor plate and dial gauge with the long rod on the propeller shaft.
2. Push the propeller shaft inward. Hold the shaft in this position and preload the dial gauge 2 mm. Now zero the gauge.
3. Pull the driveshaft slowly out and read the amount of play on the gauge. Maximum propeller shaft thrust play is 0.008–0.016 in. (0.2–0.4 mm).

If the amount of thrust play is larger than specified, increase the thickness of the reverse gear thrust washer. If smaller, decrease the thickness of the thrust washer.

Step 2

Step 3

DT35 and DT40

DISASSEMBLY

♦ See Figure 26

1. Remove the oil vent screw and oil fill screw and drain the gear oil into a suitable container.
2. Loosen the clutch rod lock nut.
3. Loosen the clutch rod turn buckle, and separate the clutch rod from the shift rod.
4. Separate the gearcase assembly from the drive shaft housing by removing the six attaching bolts.
5. Place the gearcase in suitable holding fixture or padded vice.
6. Remove the trim tab.
7. Remove the propeller.
8. Remove the bearing housing bolts.
9. Use the following special tools to remove the propeller shaft bearing housing:
   - Slide hammer (09930–30102)
   - Propeller shaft bearing housing remover (09930–30161)
10. Remove the bearing housing from the gearcase.
11. Remove the water pump seal tube from the pump case.
12. Remove the water pump assembly.
13. Using the driveshaft holder tool (09921–29610) and a wrench, remove the pinion gear nut.

➡Make sure to pad the gearcase with rag so that the wrench won't damage or distort the gearcase.

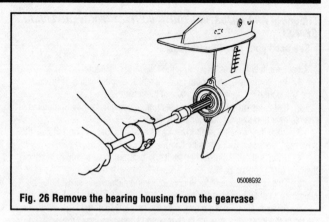
Fig. 26 Remove the bearing housing from the gearcase

14. Take out the forward gear, shim(s) and bearing.
15. Remove the pinion gear.
16. Remove the driveshaft bearing housing and shift rod assembly from the gearcase.
17. Remove the driveshaft.
18. Remove the driveshaft spring housing from the gearcase.
19. Remove the two pinion washers.
20. Pull off the driveshaft spring pin and remove the shaft spring from the driveshaft.
21. Pull out the push rode from the inboard end of the propeller shaft.
22. Remove the reverse gear and bearing housing from the propeller shaft assembly.
23. Remove the clutch dog spring from around the shift dog clutch.
24. Using a drift, remove the pin from the clutch dog shifter, and slide the shifter off the propeller shaft.
25. Remove the push pin and return spring from the end of the propeller shaft.

CLEANING & INSPECTION

♦ See Figures 27, 28 and 29

Wash all parts completely and dry them using compressed air. Inspect each part and service it, as necessary, or replace the part if it does not meet specification. Parts to be inspected and items to be checked are as follows:
- All bearings for wear and damage
- Propeller shaft and driveshaft for wear at oil seal contact points
- All gear teeth for damage or wear
- Dogs on clutch dog shifter for wear and damage
- Dogs on the forward and reverse gears for damage and wear
- Shifting cam and pushrod for wear
- Perform a gearcase pressure test to check the seals. Use the oil leakage tester (09950–69511) and air pump assembly (09821–00004) and pressurize the gearcase to 14.22 psi (1.0 kg/cm). With the gearcase pressurized, spray soapy water onto the seals and check for escaping air bubbles.
- O-rings and oil seals for cracks, tears and wear
- Propeller for nicks, bent blades or other damage and wear
- Cooling circuit for clogging or other obstructions
- Gearcase for rusting, pitting and distortion.

ASSEMBLY

Liberally apply outboard motor gear oil to the forward gear, drive shaft and propeller shaft before assembly.

1. Install the push pin and return spring into the end of the propeller shaft.
2. Slide the shifter onto the propeller shaft. The letter "F" marked on the shifter faces toward the forward gear. Then, using a drift, install the pin through the clutch dog shifter.
3. Install the clutch dog spring around the dog clutch shifter.
4. Install the reverse gear and bearing housing onto the propeller shaft assembly.
5. Install the push rode into the inboard end of the propeller shaft.
6. Install the shaft spring and pin onto the driveshaft.
7. Install the two pinion washers.
8. Install the driveshaft spring housing onto the gearcase.
9. Install the driveshaft.

LOWER UNIT 8-25

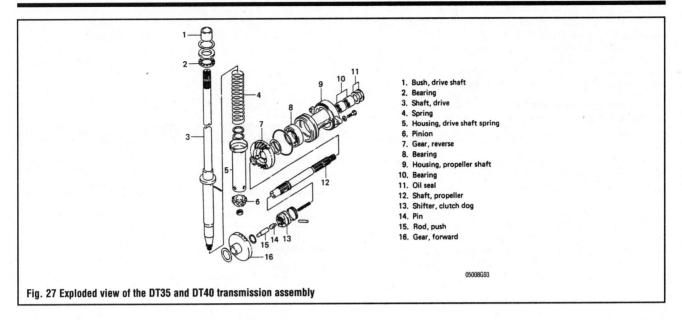

1. Bush, drive shaft
2. Bearing
3. Shaft, drive
4. Spring
5. Housing, drive shaft spring
6. Pinion
7. Gear, reverse
8. Bearing
9. Housing, propeller shaft
10. Bearing
11. Oil seal
12. Shaft, propeller
13. Shifter, clutch dog
14. Pin
15. Rod, push
16. Gear, forward

Fig. 27 Exploded view of the DT35 and DT40 transmission assembly

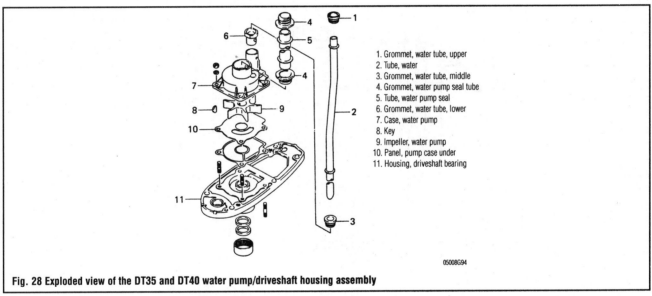

1. Grommet, water tube, upper
2. Tube, water
3. Grommet, water tube, middle
4. Grommet, water pump seal tube
5. Tube, water pump seal
6. Grommet, water tube, lower
7. Case, water pump
8. Key
9. Impeller, water pump
10. Panel, pump case under
11. Housing, driveshaft bearing

Fig. 28 Exploded view of the DT35 and DT40 water pump/driveshaft housing assembly

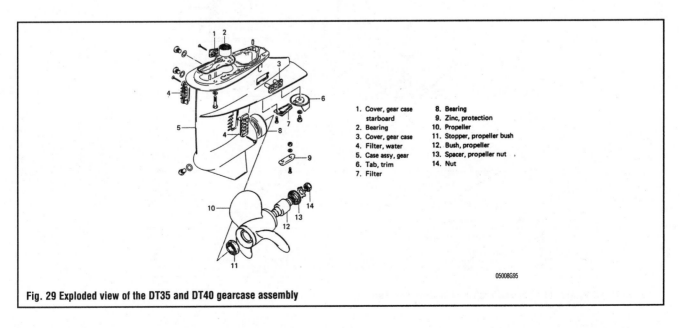

1. Cover, gear case starboard
2. Bearing
3. Cover, gear case
4. Filter, water
5. Case assy, gear
6. Tab, trim
7. Filter
8. Bearing
9. Zinc, protection
10. Propeller
11. Stopper, propeller bush
12. Bush, propeller
13. Spacer, propeller nut
14. Nut

Fig. 29 Exploded view of the DT35 and DT40 gearcase assembly

8-26 LOWER UNIT

10. Apply silicone sealant to both surfaces and install the driveshaft bearing housing and shift rod assembly into the gearcase.
11. Install the forward gear, shim(s) and bearing.
12. Install the pinion gear.
13. Using the driveshaft holder tool (09921–29610) and a wrench, tighten the pinion gear nut to 21.7–28.9 ft. lbs. (30–40 Nm).

→**Make sure to pad the gearcase with rag so that the wrench won't damage or distort the gearcase.**

14. Install the water pump assembly. Make sure to fit the impeller key onto the driveshaft before installing the impeller.
15. After mounting the water pump impeller, put on the pump case while turning the driveshaft clockwise by hand. Force the case onto the impeller being rotated. This will ensure that the vanes are turning in the correct direction. Tighten the mounting nuts to 4.5–7.0 ft. lbs. (6–10 Nm).
16. Install the water pump seal tube into the pump case.
17. Before installing the bearing housing into the gearcase, apply a coat of water proof marine grease onto the push rod and O-ring.
18. Use the following special tools to install the bearing housing:
    - Propeller shaft bearing housing installer (09922–59410)
    - Installer hammer (09922–59420)
19. Install the bearing housing bolt and tighten to 11.5–14.5 ft. lbs. (15–20 Nm).
20. Install the propeller and tighten the nut, while using the drive shaft holder tool, to 36.0–43.5 ft. lbs. (50–60 Nm).
21. Install the trim tab.
22. Apply water proof marine grease to the splines of the driveshaft.
23. Install the gearcase assembly onto the drive shaft housing by the six attaching bolts.
24. Attach the clutch rod and shift to the clutch rod turn buckle.
25. Tighten the clutch rod lock nut.
26. Fill the gearcase through the drain hole with the proper amount of gear oil (20.6 oz.) until oil drips from the vent hole, then install and tighten the screws.

✱✱ CAUTION

Recheck the oil level and top off the gearcase after the initial operation. Usually the level will need to be topped off. Failure to check this could result in damage to the gearcase due to lack of lubrication.

SHIMMING PROCEDURE

Transmission Gear Adjustment

▶ See Figure 30

PINION GEAR TO FORWARD GEAR BACKLASH

To measure the backlash, hold the driveshaft steady and move the forward gear back and forth. Check the backlash at the heel of the forward gear.

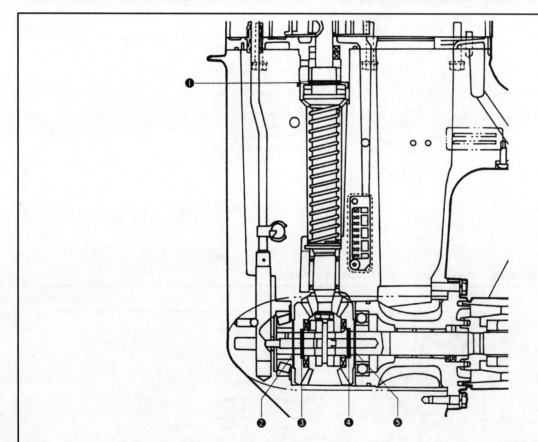

	Parts Name	Type of Thickness	Standard Thickness
❶	Pinion gear back-up shim & thrust washer (mm)	0.5, 0.6, 0.7, 0.8, 0.9, 1.0	2.0
❷	Forward gear back-up shim & thrust washer (mm)	0.5, 0.6, 0.7, 0.8, 0.9, 1.0	1.2
❸	Propeller shaft front thrust washer (mm)	2.0, 2.2	2.0
❹	Propeller shaft rear thrust washer (mm)	1.8, 1.9, 2.0, 2.1, 2.2, 2.3	2.0
❺	Reverse gear back-up shim & thrust washer (mm)	0.5, 0.6, 0.7, 0.8, 0.9, 1.0	1.5

Fig. 30 Thrust washer and shim locations and shim thickness chart

LOWER UNIT 8-27

If the amount of backlash is larger than 0.0039–0.0079 in. (0.1–0.2 mm), increase the thickness of each of the pinion gear back-up shim(s). If the backlash is smaller, decrease the thickness of each shim.

❈❈ CAUTION

In setting the dial gauge, set the gauge rod end to contact the convex side of the forward gear tooth heel end. Use care to not have the rod contact on the neighboring tooth.

CHECKING AND ADJUSTING TOOTH CONTACT (FORWARD/PINION GEARS)

♦ See accompanying illustrations

Check tooth contact pattern using the following procedure.
1. Too correctly assess tooth contact, smear a light coating of Prussian Blue compound on the convex surface of the forward gear.
2. Install the propeller shaft and bearing housing assembly (minus the reverse gear and internal components).
3. Push the propeller shaft inward and hold it in that position. Rotate the driveshaft clockwise 5–6 times by hand.
4. Then pull out the propeller shaft and bearing housing and then check the gear tooth contact pattern.

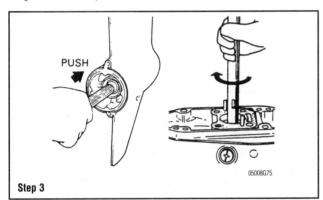
Step 3

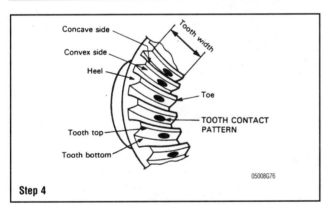
Step 4

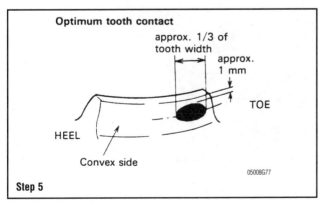
Step 5

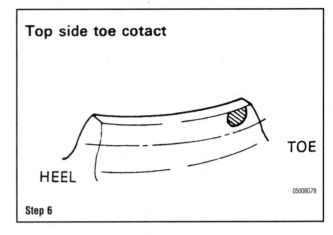
Step 6

5. This is the optimum tooth contact. Doing a shim adjustment may be necessary to obtain this contact pattern.
6. An example of incorrect top side toe contact. To correct this condition, decrease the forward gear shim thickness and slightly increase the pinion gear shim thickness.

➡**Do not set the tooth contact in this position (top side toe contact). Damage and chipping of the pinion and forward gear may result.**

7. An example of incorrect bottom side toe contact. To correct this condition, increase the forward gear shim thickness and slightly decrease the pinion gear shim thickness.

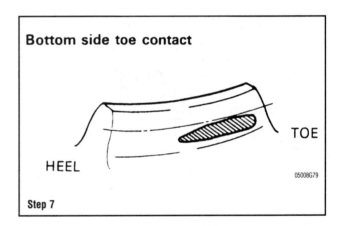
Step 7

➡**Do not set the tooth contact in this position (bottom side toe contact). Damage and chipping of the pinion gear may result.**

8. After adjusting the tooth contact pattern, install the propeller shaft, bearing housing assembly, reverse gear and all related shims and washers.
9. Recheck the amount of backlash by slightly rotating the drive shaft by hand. Backlash should not be less than when checked at the start of this entire procedure.
10. If backlash is less, reduce the reverse gear back-up shim thickness.

PINION GEAR TO FORWARD GEAR THRUST PLAY

Obtain the following special tools to perform the measurement:
• Gear adjusting set (09951–09510)
1. Fit the propeller shaft to the forward gear and press the forward gear forward by pushing with the propeller shaft. (In this case, first remove the push rod from the propeller shaft)
2. Set the special tool up on the driveshaft.
3. Push the driveshaft down. Hold this position and place the gauge with its rod pushed in approximately 2 mm.
4. Still keeping this position, zero the dial gauge. Then pull up on the driveshaft and read the maximum play on the gauge.

If the amount of backlash is larger than specified, increase the thickness of each of the pinion gear back-up shim(s) or forward gear back-up shim(s). If smaller, decrease the thickness of each shim.

8-28 LOWER UNIT

5. When the pinion and forward gear back-up shim are determined to be the correct size, note the amount of thrust play measured. The thrust play measurement will be necessary for the adjustment of the reverse gear.

REVERSE GEAR THRUST PLAY ADJUSTMENT

Perform the same procedure as above with the exception of installing the propeller shaft and propeller shaft bearing housing into the gearcase.

1. If the measurement of the play is equal to the pinion gear to forward gear measurement, the condition is correct. If the amount of play is smaller, decrease the thickness of the reverse gear back-up shim.

PROPELLER SHAFT THRUST PLAY ADJUSTMENT
◆ See accompanying illustrations

1. Install the dial gauge adaptor plate and dial gauge with the long rod on the propeller shaft.
2. Push the propeller shaft inward. Hold the shaft in this position and preload the dial gauge 2 mm. Now zero the gauge.
3. Pull the driveshaft slowly out and read the amount of play on the gauge. Maximum propeller shaft thrust play is 0.008–0.016 in. (0.2–0.4 mm).

If the amount of thrust play is larger than specified, increase the thickness of the reverse gear thrust washer. If smaller, decrease the thickness of the thrust washer.

Step 2

Step 3

DT55 and DT65

DISASSEMBLY

◆ See accompanying illustrations

1. Separate the clutch shaft and clutch rod.
2. Remove the gearcase.
3. Secure the gearcase in a suitable holding fixture or a vise with protective jaws. If protective jaws are not available, position the gearcase upright with the skeg between wooden blocks in a vise.
4. Remove the water pump.

Step 1

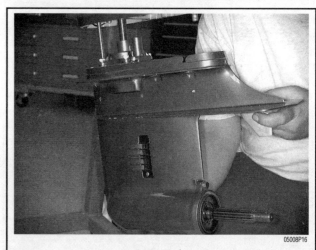
Step 2

Step 4

LOWER UNIT   8-29

5. Remove the 2 bolts holding the bearing housing.
6. Using a soft mallet, tap on the bearing housing to loosen it.
7. Remove the bearing housing from the gear case.
8. Then remove the propeller shaft and reverse gear.
9. Fit an appropriate size box-end wrench over the pinion nut and pad the sides of the gearcase housing bore to prevent distortion or damage from contact with the wrench.
10. Holding the pinion nut with the wrench installed, turn the driveshaft counterclockwise to loosen the pinion nut.
11. Remove the tools. Remove the pinion nut and pinion gear.
12. Remove the forward gear and shim(s).
13. Unbolt and remove the driveshaft bearing housing.

✱✱ CAUTION

The bolts must be tightened evenly otherwise the bearing housing will be distorted and damaged during removal.

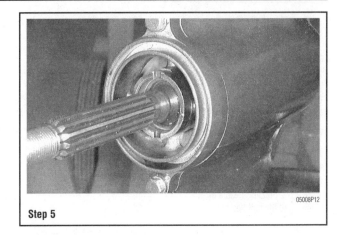
Step 5

Step 6

Step 7

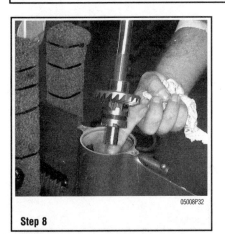
Step 8

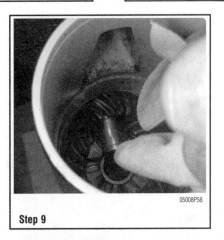
Step 9

Step 10

Step 11

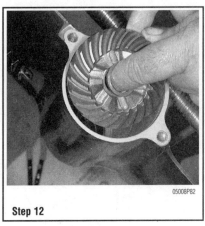
Step 12

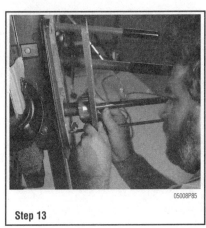
Step 13

8-30 LOWER UNIT

14. To separate the bearing housing, install two 6 mm bolts.
15. Tighten the bolts evenly and alternately and force the bearing housing free.
16. Remove the driveshaft and bearing housing from the gearcase as an assembly.
17. Remove the driveshaft bearing housing snap ring holding the preload spring collar.
18. Remove the preload spring collar or pre load spring and washers.
19. Pull out the push rod from the propeller shaft.
20. Remove the clutch dog spring from the shifter.
21. Use a small drift to remove the pin from the clutch dog shifter.
22. Slide the clutch dog shifter off the propeller shaft and lay out the shifter components.
23. Remove and discard the propeller shaft bearing housing oil seals
24. The housing needle bearings.
25. And the ball bearing.

CLEANING & INSPECTION

▶ See Figure 31

Wash all parts completely and dry them using compressed air. Inspect each part and service it, as necessary, or replace the part if it does not meet specification. Parts to be inspected and items to be checked are as follows:
- All bearings for wear and damage
- Propeller shaft and driveshaft for wear at oil seal contact points
- All gear teeth for damage or wear
- Dogs on clutch dog shifter for wear and damage
- Dogs on the forward and reverse gears for damage and wear
- Shifting cam and pushrod for wear
- Check the magnet for sign of debris in the gear oil
- Perform a gearcase pressure test to check the seals. Use the oil leakage tester (09950–69511) and air pump assembly (09821–00004) and pressurize

Step 16

Step 20

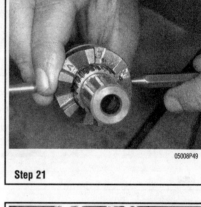
Step 21

Step 22

Step 23

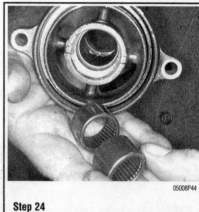
Step 24

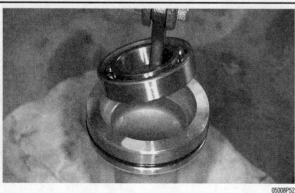
Step 25

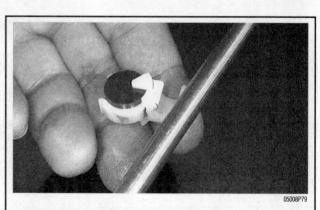
Fig. 31 Check the magnet for sign of debris in the gear oil

LOWER UNIT 8-31

the gearcase to 14.22 psi (1.0 kg/cm). With the gearcase pressurized, spray soapy water onto the seals and check for escaping air bubbles.
- O-rings and oil seals for cracks, tears and wear
- Propeller for nicks, bent blades or other damage and wear
- Cooling circuit for clogging or other obstructions
- Gearcase for rusting, pitting and distortion.

ASSEMBLY

♦ See accompanying illustrations

1. Replace the bearings and seals in the propeller bearing housing.
2. When replacing the oil seal, make sure to position the lip and spring to the outside and apply waterproof marine grease to the seal lip and install the retaining ring into the groove.
3. The clutch dog shifter is marked for easy assembly. The end with the letters "REV" is meant to face the Reverse gear. Be sure to mount the shifter correctly on the propeller shaft.
4. Install the clutch return spring, push pin and push rod into the propeller shaft.
5. Bring the push pin to the slot, making the hole of the push pin visible in the slot.
6. Slide the clutch dog shifter over the slot, aligning the pin hole to the hole in the push pin and drive the spring pin into the dog shifter.
7. Install the clutch dog spring into the groove on the dog shifter so that the spring pin does not come out.

→After connecting the dog shifter to the push pin, check to be sure that the spring pin is all the way in, with its driven end flush with the surface of the shifter.

8. After sliding the spring onto the shaft, make sure the end of the spring fits into the notch provided in the inner race of the drive shaft bearing.
9. Be careful not to miss the driveshaft lock washer when putting the collar on the preload spring. Fit the washer into the collar, making the tab on the washer show out of the slip provided in the top end of the collar, and then put on the collar. After this, insert the thrust washer at the near side.
10. After putting on the preload spring collar, push down the collar all the way to compress the preload spring inside, and then install the circlip to retain the collar.
11. When installing the driveshaft bearing housing, apply silicone sealant to the gearcase mating surface
12. Apply water proof marine grease to the gearcase O-ring, shift rod guide O-ring and inside the shift rod boot.

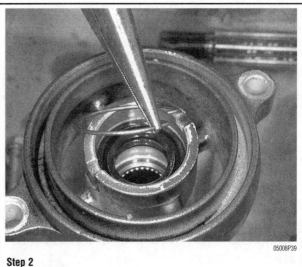
Step 2

Step 6

Step 7

Step 8

Step 9

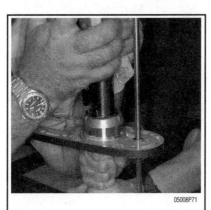
Step 10

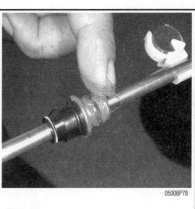
Step 12

8-32 LOWER UNIT

13. Install the driveshaft, shift rod and driveshaft bearing assembly onto the gearcase.
14. Tighten the bolts evenly and alternately.
15. When installing the pinion nut, apply a thread locking compound to the threads, pad the gearcase to prevent damage from the wrench and tighten the nut to 21.5–29.ft. lbs. (30–40 Nm).
16. Install the forward gear and shim(s). Be sure to have the thrust washer and shim fitted between the forward gear and the tapered roller bearing in order to secure a proper backlash between this gear and the pinion gear. Make absolutely sure that the thrust washer and shim are those that are found to be correct after measuring or adjusting the backlash.
17. Install the reverse gear and thrust washer onto the propeller.

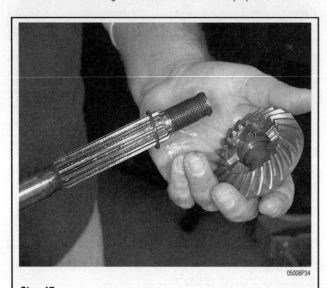
Step 17

18. Before installing the propeller shaft assembly with its bearing housing, apply water proof marine grease to the push rod O-ring.
19. Before inserting the propeller shaft assembly into the gearcase, check to be sure that the cam portion is facing toward the propeller shaft.

20. Use the following special tools to install the housing into the gearcase:
    - Propeller shaft housing installer (09922–59410)
    - Bearing installer (09922–59420)
21. When installing the housing, tighten the attaching bolts to 11.0–14.5 ft. lbs. (15–20 Nm).
22. Install the water pump assembly.
23. When assembling the gearcase assembly to the driveshaft housing, apply silicone sealant to both joining faces, and assemble after aligning the dowel pins, water tube wand water pump seal tube. Also apply silicone sealant to the tightening bolt bodies and tighten the bolts to: 8 mm: 11.0–14.5 ft. lbs. (15–20 Nm); 10 mm: 24.5–29.5 ft. lbs. (34–41 Nm).

➥When installing the gearcase assembly to the engine, apply some waterproof marine grease to the splines of the driveshaft.

24. Refill the gearcase with new gear oil. After the initial running of the engine, recheck the oil level and refill if necessary. If this is procedure is not followed, it may lead to damage due to lack of lubrication.
25. Connect the clutch shaft and clutch rod.

SHIMMING PROCEDURE

Transmission Gear Adjustment

♦ See Figure 32

PINION GEAR TO FORWARD GEAR BACKLASH

♦ See Figure 33

To measure the backlash, hold the driveshaft steady and move the forward gear back and forth. Check the backlash at the heel of the forward gear.

If the amount of backlash is larger than 0.004–0.008 in. (0.1–0.2 mm), increase the thickness of each of the pinion gear back-up shim(s). If the backlash is smaller, decrease the thickness of each shim.

✴✴ CAUTION

In setting the dial gauge, set the gauge rod end to contact the convex side of the forward gear tooth heel end. Use care to not have the rod contact on the neighboring tooth.

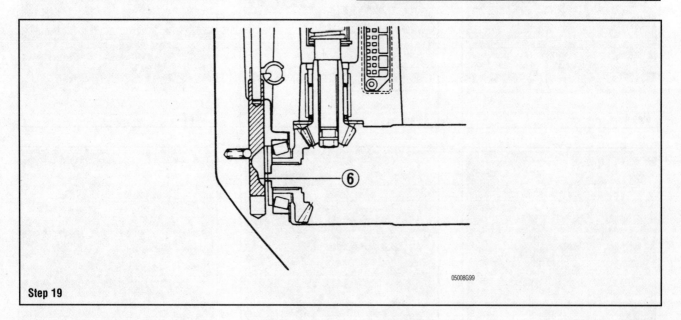
Step 19

LOWER UNIT 8-33

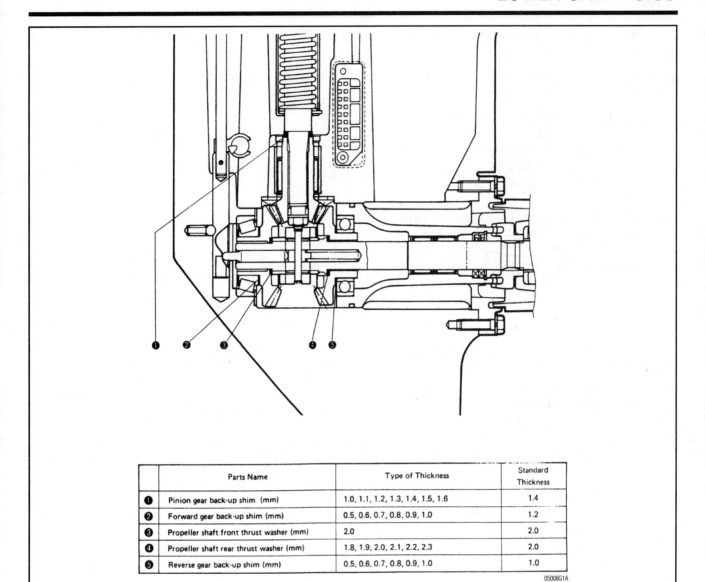

	Parts Name	Type of Thickness	Standard Thickness
❶	Pinion gear back-up shim (mm)	1.0, 1.1, 1.2, 1.3, 1.4, 1.5, 1.6	1.4
❷	Forward gear back-up shim (mm)	0.5, 0.6, 0.7, 0.8, 0.9, 1.0	1.2
❸	Propeller shaft front thrust washer (mm)	2.0	2.0
❹	Propeller shaft rear thrust washer (mm)	1.8, 1.9, 2.0, 2.1, 2.2, 2.3	2.0
❺	Reverse gear back-up shim (mm)	0.5, 0.6, 0.7, 0.8, 0.9, 1.0	1.0

Fig. 32 Thrust washer and shim locations and shim thickness chart

2. Install the propeller shaft and bearing housing assembly (minus the reverse gear and internal components).
3. Push the propeller shaft inward and hold it in that position. Rotate the driveshaft clockwise 5–6 times by hand.

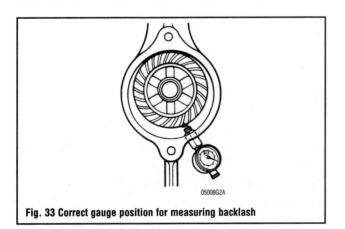

Fig. 33 Correct gauge position for measuring backlash

CHECKING AND ADJUSTING TOOTH CONTACT (FORWARD/PINION GEARS)

▶ See accompanying illustrations

Check tooth contact pattern using the following procedure.
1. Too correctly assess tooth contact, smear a light coating of Prussian Blue compound on the convex surface of the forward gear.

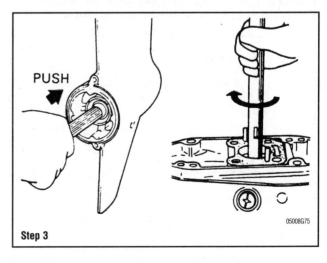

Step 3

8-34 Lower Unit

4. Then pull out the propeller shaft and bearing housing and then check the gear tooth contact pattern.

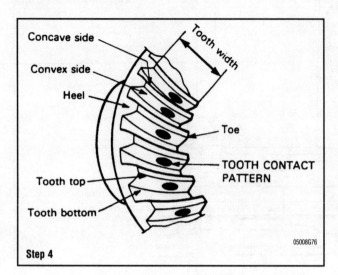

Step 4

5. This is the optimum tooth contact. Doing a shim adjustment may be necessary to obtain this contact pattern.

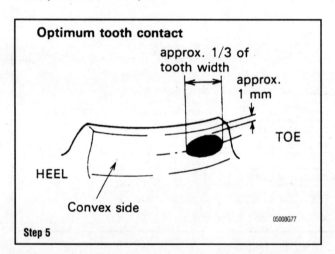

Step 5

6. An example of incorrect top side toe contact. To correct this condition, decrease the forward gear shim thickness and slightly increase the pinion gear shim thickness.

➡ Do not set the tooth contact in this position (top side toe contact). Damage and chipping of the pinion and forward gear may result.

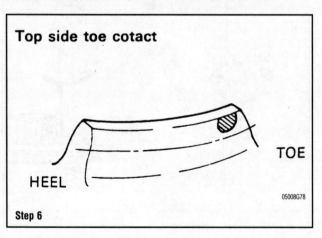

Step 6

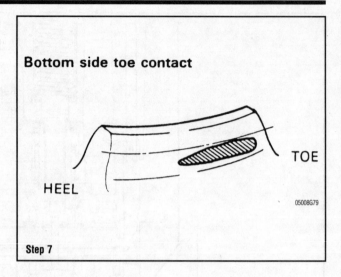

Step 7

7. An example of incorrect bottom side toe contact. To correct this condition, increase the forward gear shim thickness and slightly decrease the pinion gear shim thickness.

➡ Do not set the tooth contact in this position (bottom side toe contact). Damage and chipping of the pinion gear may result.

8. After adjusting the tooth contact pattern, install the propeller shaft, bearing housing assembly, reverse gear and all related shims and washers.
9. Recheck the amount of backlash by slightly rotating the drive shaft by hand. Backlash should not be less than when checked at the start of this entire procedure.
10. If backlash is less, reduce the reverse gear back-up shim thickness.

PINION GEAR TO FORWARD GEAR THRUST PLAY

Obtain the following special tools to perform the measurement:
- Gear adjusting set (09951–09510)

1. Fit the propeller shaft to the forward gear and press the forward gear forward by pushing with the propeller shaft. (In this case, first remove the push rod from the propeller shaft)
2. Set the special tool up on the driveshaft.
3. Push the driveshaft down. Hold this position and place the gauge with its rod pushed in approximately 2 mm.
4. Still keeping this position, zero the dial gauge. Then pull up on the driveshaft and read the maximum play on the gauge.

If the amount of backlash is larger than specified, increase the thickness of each of the pinion gear back-up shim(s) or forward gear back-up shim(s). If smaller, decrease the thickness of each shim.

5. When the pinion and forward gear back-up shim are determined to be the correct size, note the amount of thrust play measured. The thrust play measurement will be necessary for the adjustment of the reverse gear.

REVERSE GEAR THRUST PLAY ADJUSTMENT

Perform the same procedure as above with the exception of installing the propeller shaft and propeller shaft bearing housing into the gearcase.

1. If the measurement of the play is equal to the pinion gear to forward gear measurement, the condition is correct. If the amount of play is smaller, decrease the thickness of the reverse gear back-up shim.

PROPELLER SHAFT THRUST PLAY ADJUSTMENT

♦ See accompanying illustrations

1. Install the dial gauge adaptor plate and dial gauge with the long rod on the propeller shaft.
2. Push the propeller shaft inward. Hold the shaft in this position and preload the dial gauge 2 mm. Now zero the gauge.
3. Pull the driveshaft slowly out and read the amount of play on the gauge. Maximum propeller shaft thrust play is 0.008–0.016 in. (0.2–0.4 mm).

If the amount of thrust play is larger than specified, increase the thickness of the reverse gear thrust washer. If smaller, decrease the thickness of the thrust washer.

LOWER UNIT 8-35

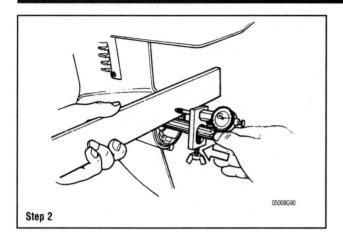
Step 2

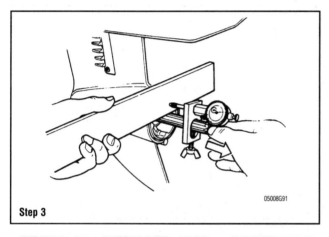
Step 3

DT75 and DT85

DISASSEMBLY

♦ **See accompanying illustrations**

1. Remove the gearcase.
2. Secure the gearcase in a suitable holding fixture or a vise with protective jaws. If protective jaws are not available, position the gearcase upright with the gearcase skeg held between wooden blocks in a vise.
3. Remove the trim tab.
4. Remove the propeller.
5. Remove the water pump.
6. Remove the seal ring and housing washer.

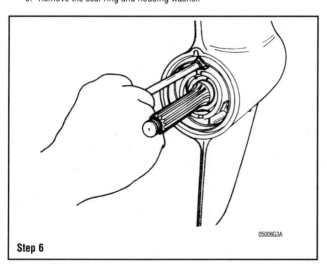
Step 6

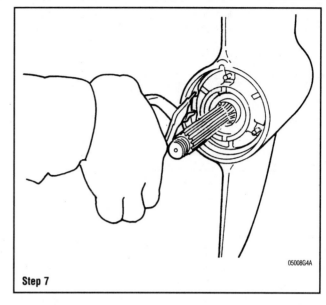
Step 7

7. Remove the snap ring with a pair of snap ring pliers.
8. Install the bearing housing remover (09930-39410) and special bolts (09950-59520)

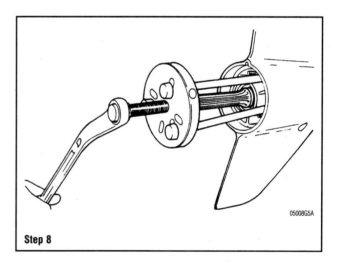
Step 8

9. Remove the propeller shaft bearing housing assembly.
10. Remove and discard the propeller shaft bearing housing oil seals and bearing.
11. Attach driveshaft holder tool (09950–79510) to the top of the driveshaft.

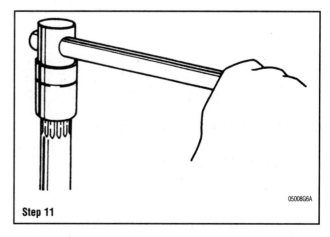
Step 11

8-36 LOWER UNIT

12. Fit an appropriate size box-end wrench over the pinion nut and pad the sides of the gearcase prop shaft bore to prevent the wrench from damaging or distorting the gearcase prop shaft bore.
13. Holding pinion nut with the wrench, turn the driveshaft counterclockwise to loosen the pinion nut.
14. Remove the tools. Remove the pinion nut and pinion gear.
15. Remove the driveshaft bearing housing and the shift rod.
16. Remove the driveshaft, preload spring and thrust bearing as an assembly.
17. Remove the driveshaft spring collar from the gearcase bore.
18. Remove the driveshaft thrust washers from the gearcase bore.
19. Remove the forward gear and thrust washer.
20. Slide all components except the clutch dog from the propeller shaft. Disassemble the propeller shaft clutch assembly.
21. Use a drift to drive the pin out of the clutch dog shifter.
22. From the end of the propeller shaft, remove the push pin and return spring.

CLEANING & INSPECTION

▶ See Figures 34, 35 and 36

Wash all parts completely and dry them using compressed air. Inspect each part and service it, as necessary, or replace the part if it does not meet specification. Parts to be inspected and items to be checked are as follows:
- All bearings for wear and damage
- Propeller shaft and driveshaft for wear at oil seal contact points
- All gear teeth for damage or wear
- Dogs on clutch dog shifter for wear and damage
- Dogs on the forward and reverse gears for damage and wear
- Shifting cam and pushrod for wear
- Check the magnet for sign of debris in the gear oil
- Perform a gearcase pressure test to check the seals. Use the oil leakage tester (09950–69511) and air pump assembly (09821–00004) and pressurize

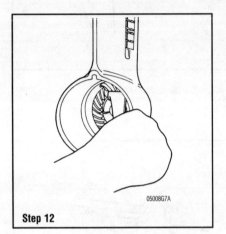
Step 12

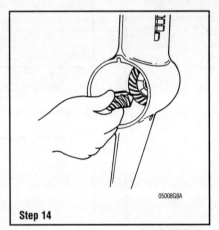
Step 14

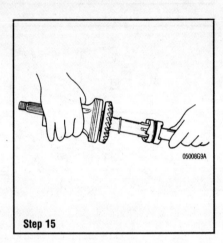
Step 15

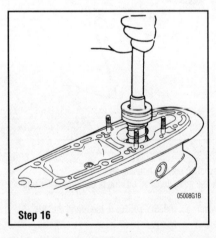
Step 16

Step 17

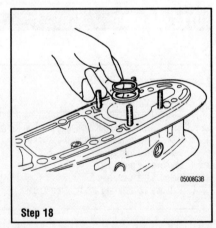
Step 18

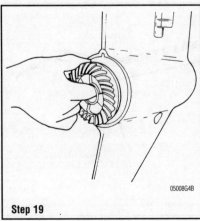
Step 19

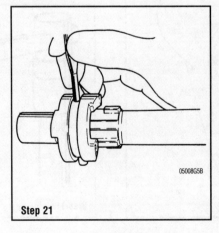
Step 21

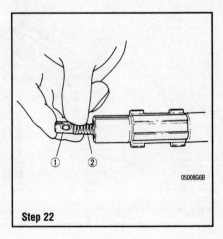
Step 22

LOWER UNIT 8-37

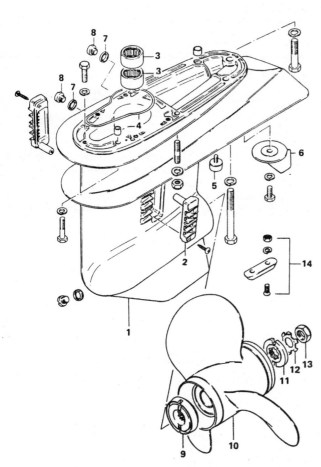

1. Gear case set
2. Water filter
3. Pinion bearing
4. Gear case seal ring
5. Protection zinc
6. Trim tab
7. Gasket
8. Drain plug
9. Propeller bushing stopper
10. Propeller
11. Propeller nut spacer
12. Lock nut washer
13. Nut
14. Protection zinc assy

1. Drive shaft
2. Shim
3. Thrust washer
4. Thrust bearing
5. Spring
6. Thrust washer
7. Pinion nut
8. Spring collar
9. Pinion/Gear set
10. Forward gear radial bearing
11. Thrust bearing
12. Thrust washer
13. Washer
14. Thrust washer
15. Push rod
16. Push pin
17. Clutch dog shifter
18. Pin
19. Clutch dog shifter spring
20. Return spring
21. Propeller shaft
22. Shim
23. Thrust washer
24. Shim
25. Reverse gear bearing
26. O ring
27. Propeller shaft housing
28. Propeller shaft bearing
29. Propeller shaft oil seal
30. Ring
31. Housing ring
32. Housing washer
33. Seal ring

Fig. 34 Exploded view of the DT75 and DT85 lower unit component locations

8-38 Lower Unit

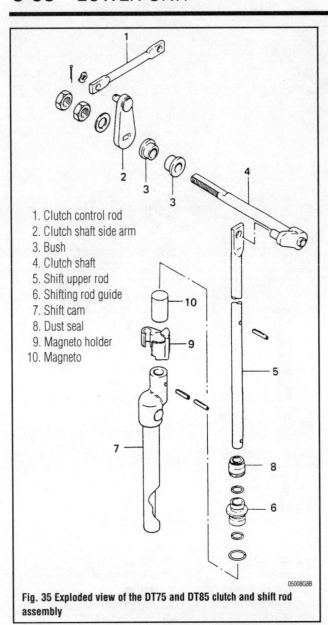

1. Clutch control rod
2. Clutch shaft side arm
3. Bush
4. Clutch shaft
5. Shift upper rod
6. Shifting rod guide
7. Shift cam
8. Dust seal
9. Magneto holder
10. Magneto

Fig. 35 Exploded view of the DT75 and DT85 clutch and shift rod assembly

Fig. 36 Check the magnet for sign of debris in the gear oil

the gearcase to 14.22 psi (1.0 kg/cm). With the gearcase pressurized, spray soapy water onto the seals and check for escaping air bubbles.
- O-rings and oil seals for cracks, tears and wear
- Propeller for nicks, bent blades or other damage and wear
- Cooling circuit for clogging or other obstructions
- Gearcase for rusting, pitting and distortion.

ASSEMBLY

♦ See accompanying illustrations

1. Install a new propeller shaft housing bearing with an appropriate installer.
2. Install new propeller shaft housing oil seals and a new housing O-ring. Coat the seal lips and O-ring with an appropriate waterproof marine grease.
3. Remove and discard the driveshaft bearing housing oil seal. Install a new seal with an appropriate installer. Coat seal lips with waterproof marine grease.
4. Lubricate all parts with Suzuki Outboard Motor Gear Oil.
5. Install the driveshaft spring thrust washer in the sequence shown.

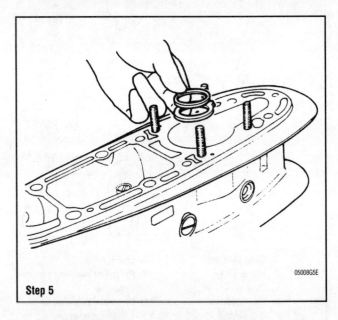

Step 5

6. Install the driveshaft spring collar into the gearcase bore.
7. Install the driveshaft, preload spring and thrust bearing as an assembly.
8. Install the shift rod assembly.
9. Install the forward gear in this order: thrust bearing, thrust washer, forward gear back-up shim and the forward gear.
10. Install the pinion gear into the propeller shaft bore and mesh with the forward gear.
11. Holding the pinion gear in place, install the driveshaft assembly into the gearcase. Rotate shaft to align its splines with those of the pinion gear and seat the gear on the shaft.
12. Coat the pinion nut threads with a small amount of thread locking compound.
13. Position nut over driveshaft threads in prop shaft bore. Start nut by hand.
14. Hold pinion gear nut with an appropriate size box-end wrench and pad the sides of the gearcase prop shaft bore to prevent the wrench from damaging or distorting the prop shaft bore.
15. Attach the driveshaft holder tool (09950–79510) to the top of the driveshaft. Rotate driveshaft and tighten the pinion gear nut to 43.5–50.5 ft. lbs. (60–70 Nm).
16. Check the forward gear backlash. If not within specification, refer to the shimming section.
17. Temporarily install the propeller shaft/bearing housing assembly. Check

Lower Unit 8-39

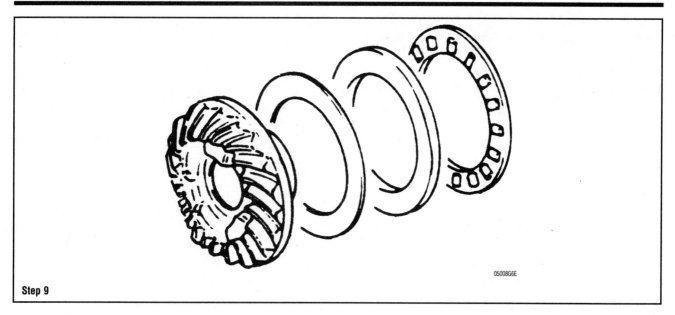

Step 9

pinion gear depth, reverse gear backlash and gear tooth contact pattern. If not within specification, refer to the shimming section.

18. Coat gearcase and driveshaft bearing housing mating surfaces with a silicone sealant and install the housing. Tighten fasteners securely.

19. Lubricate the propeller shaft bearing housing O-ring and the shift mechanism pushrod with waterproof marine grease.

20. Reassemble the propeller shaft and clutch dog shifter assembly. The end of the shifter with the letter "F" is meant to face the forward gear.

21. First insert the clutch push pin and return spring into the propeller shaft.

22. Next bring the push pin to the slot, making the hole of the push pin visible in the slot.

23. Then slide the dog shifter over to the slot, aligning its pin hole to the hole.

24. Drive the spring pin into the dog shifter and push pin.

➟**After connecting the dog shifter to the push pin, check to be sure that the spring pin is all the way in, with its driven end flush with the surface of the shifter.**

25. When installing the clutch push rod, liberally coat the push rod with waterproof marine grease. Then install the clutch push rod into the propeller shaft.

26. Install the propeller shaft/bearing housing assembly into the propeller shaft bore. Use shaft housing installer (09922-59410) and installer handle (09922-59420).

27. Coat the bearing housing bolts with a small amount of thread locking compound and tighten securely.

28. Install the water pump.

29. Install the gearcase. Fill with recommended type and quantity of lubricant.

30. Check gearcase lubricant level after engine has been run. Change the lubricant after the first 10 hours of operation.

SHIMMING PROCEDURE

The DT75 and DT85 transmission gears (forward, reverse and pinion) are designed so that the tooth bearing can be adjusted without using special tools.

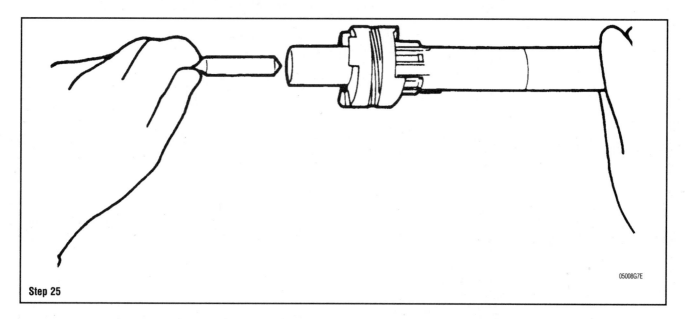

Step 25

8-40 Lower Unit

On these models, all gears both spare and mounted, are subjected to tooth bearing adjustment at the factory and are marked with the difference in millimeters between the designed dimensions and the actual dimensions. At the same time, the difference is stamped on the anti-ventilation plate (where the trim tab is mounted).

When gears or the gearcase are replaced by a new component, it is necessary to compare the values indicated on both the old parts and the new parts to determine the thickness of the shim to be used.

Transmission Gear Adjustment

♦ See Figures 37 and 38

SHIM ADJUSTMENT PROCEDURE

♦ See accompanying illustrations

When only gears are replaced, shim adjustment should be made by comparing the values that are stamped on the gears.

1. Pinion gear shim adjustment. When the actual dimension is larger than the standard dimension as shown in the illustration, the gear has a "+ mm" mark on it. On the contrary, if smaller it has a "- mm" mark. For example, when the actual dimension is 0.2 mm longer, the pinion gear is marked "+0.2 mm". Accordingly, the thickness of the shim mounted is 1.0 mm which is 0.2 mm thinner than the standard thickness of 1.2 mm.

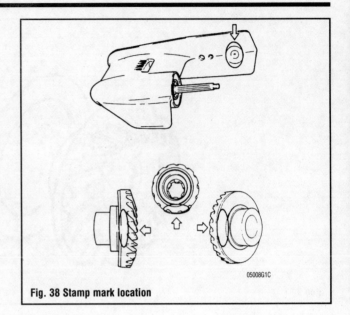

Fig. 38 Stamp mark location

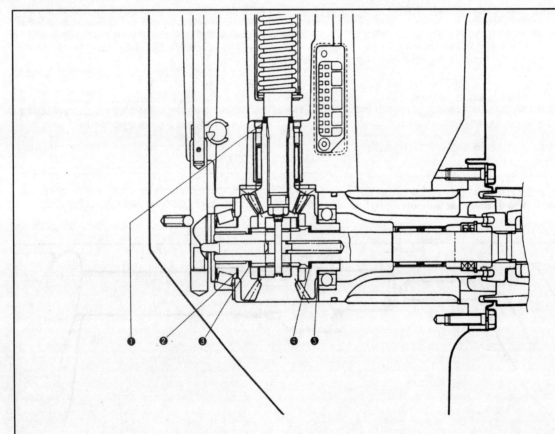

	Parts Name	Type of Thickness	Standard Thickness
❶	Pinion gear back-up shim (mm)	1.0, 1.1, 1.2, 1.3, 1.4, 1.5, 1.6	1.4
❷	Forward gear back-up shim (mm)	0.5, 0.6, 0.7, 0.8, 0.9, 1.0	1.2
❸	Propeller shaft front thrust washer (mm)	2.0	2.0
❹	Propeller shaft rear thrust washer (mm)	1.8, 1.9, 2.0, 2.1, 2.2, 2.3	2.0
❺	Reverse gear back-up shim (mm)	0.5, 0.6, 0.7, 0.8, 0.9, 1.0	1.0

Fig. 37 DT75 and DT85 shim locations

LOWER UNIT 8-41

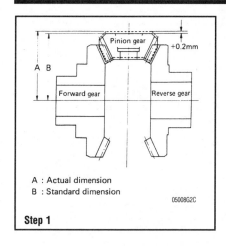

Step 1
A : Actual dimension
B : Standard dimension

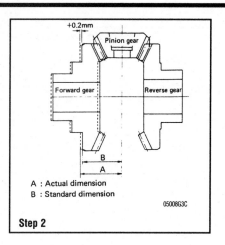

Step 2
A : Actual dimension
B : Standard dimension

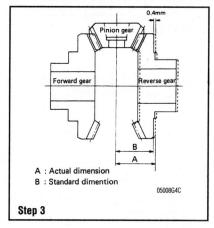

Step 3
A : Actual dimension
B : Standard dimention

	Mark on new gear								
	− 0.4	− 0.3	− 0.2	− 0.1	0	+ 0.1	+ 0.2	+ 0.3	+ 0.4
− 0.4		− 0.1	− 0.2	− 0.3	− 0.4	− 0.5	− 0.6	− 0.7	− 0.8
− 0.3	+ 0.1		− 0.1	− 0.2	− 0.3	− 0.4	− 0.5	− 0.6	− 0.7
− 0.2	+ 0.2	+ 0.1		− 0.1	− 0.2	− 0.3	− 0.4	− 0.5	− 0.6
− 0.1	+ 0.3	+ 0.2	+ 0.1		− 0.1	− 0.2	− 0.3	− 0.4	− 0.5
0	+ 0.4	+ 0.3	+ 0.2	+ 0.1		− 0.1	− 0.2	− 0.3	− 0.4
+ 0.1	+ 0.5	+ 0.4	+ 0.3	+ 0.2	+ 0.1		− 0.1	− 0.2	− 0.3
+ 0.2	+ 0.6	+ 0.5	+ 0.4	+ 0.3	+ 0.2	+ 0.1		− 0.1	− 0.2
+ 0.3	+ 0.7	+ 0.6	+ 0.5	+ 0.4	+ 0.3	+ 0.2	+ 0.1		− 0.1
+ 0.4	+ 0.8	+ 0.7	+ 0.6	+ 0.5	+ 0.4	+ 0.3	+ 0.2	+ 0.1	

(Left axis label: Mark on old gear)

Step 4

2. Forward gear shim adjustment. The forward gear is also marked as to the differences between the actual and standard dimensions. When the actual dimension is larger than the standard, the gear has a "+ mm" mark. When the dimension is smaller, it has a "- mm" mark. For example, when the actual dimension is 0.2mm larger than the standard, the gear is marked "+0.2mm". Accordingly, the thickness of the shim for the forward gear is 1.0 mm, which is 0.2 mm thinner than the standard thickness of 1.2 mm. When the forward gear is replaced, the shim adjustment can be done in the same manner as the pinion gear.

3. Reverse gear shim adjustment. The reverse gear is also marked the same as the forward gear. On the reverse gear, the adjustment shim is available in only two types: 0.2 mm and 0.5 mm. Therefore, a desired shim thickness may not be available in some instances. In this case, a thinner shim should be used so that the sum of the actual gear and shim thickness. It should be noted that when the desired shim thickness is not obtainable, it will be 0.1 mm or 0.3 mm. In other cases, either one of the 0.2 mm and 0.5 mm shims will be suitable.

4. This shim thickness chart shows the increases and decreases in shim thickness according to the thickness of the old and new gears.

➡ **The chart shows the thickness of new gears on the top line and old gears on the left. Each box in the chart shows an increase or decrease according to the old and new gears.**

5. Forward and pinion gear adjustment when the gearcase is replaced. As shown, the difference between the actual and standard dimensions is marked on where the trim tab is located. When the gearcase is replaced, compare the values on both old and new gearcases, and make an adjustment on shim thickness. The stamped marks on the cases are: F mm (on the forward gear side); P mm (on the pinion gear side).

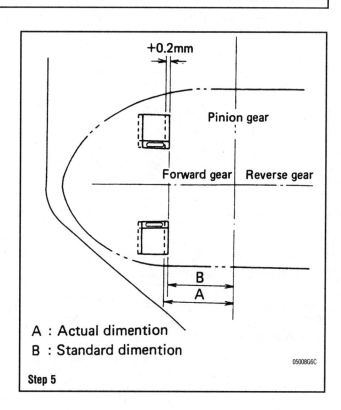

Step 5
A : Actual dimention
B : Standard dimention

8-42 LOWER UNIT

	Mark on new gear case								
Mark on old gear case	−0.4	−0.3	−0.2	−0.1	0	+0.1	+0.2	+0.3	+0.4
−0.4		+0.1	+0.2	+0.3	+0.4	+0.5	+0.6	+0.7	+0.8
−0.3	−0.1		+0.1	+0.2	+0.3	+0.4	+0.5	+0.6	+0.7
−0.2	−0.2	−0.1		+0.1	+0.2	+0.3	+0.4	+0.5	+0.6
−0.1	−0.3	−0.2	−0.1		+0.1	+0.2	+0.3	+0.4	+0.5
0	−0.4	−0.3	−0.2	−0.1		+0.1	+0.2	+0.3	+0.4
+0.1	−0.5	−0.4	−0.3	−0.2	−0.1		+0.1	+0.2	+0.3
+0.2	−0.6	−0.5	−0.4	−0.3	−0.2	−0.1		+0.1	+0.2
+0.3	−0.7	−0.6	−0.5	−0.4	−0.3	−0.2	−0.1		+0.1
+0.4	−0.8	−0.7	−0.6	−0.5	−0.4	−0.3	−0.2	−0.1	

Step 6

6. When replacing the gearcase, choose forward gear and pinion gear adjustment values according to the chart. The chart shows the marks on new gearcases on the top and those on old gearcases on the right. Each box indicates an increase or decrease in shim thickness on the forward gear side. Example: When the old gearcase has a F −0.2 mark, and the new gearcase has a F +0.1 mark, a +0.3 mm thick shim should be added to the value measured.

PINION GEAR TO FORWARD GEAR BACKLASH

▶ See Figure 39

To measure the backlash, hold the driveshaft steady and move the forward gear back and forth. Check the backlash at the heel of the forward gear.

If the amount of backlash is larger than 0.002–0.012 in. (0.05–0.3 mm), increase the thickness of each of the pinion gear back-up shim(s). If the backlash is smaller, decrease the thickness of each shim.

✲✲ CAUTION

In setting the dial gauge, set the gauge rod end to contact the convex side of the forward gear tooth heel end. Use care to not have the rod contact on the neighboring tooth.

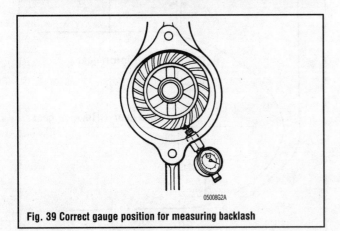

Fig. 39 Correct gauge position for measuring backlash

CHECKING AND ADJUSTING TOOTH CONTACT (FORWARD/PINION GEARS)

▶ See accompanying illustrations

1. Coat the entire surface of the forward gear with red lead paste and install the propeller shaft and related parts into the gearcase. It is unnecessary to install the propeller shaft housing nuts, housing ring and housing washer.

2. While pushing the propeller shaft, hold it firmly by hand so that it dies not turn. Using the driveshaft holder (09950–79510), slowly rotate the driveshaft about 5 turns and then remove the forward gear from the propeller shaft. Check the red lead paste to see whether the gear tooth contact is correct or not.

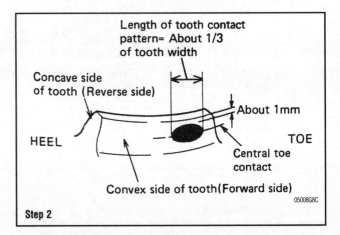

Step 2

3. This illustration shows a top side toe contact. In this case, decrease the forward gear shim thickness and increase the pinion gear shim thickness.

➥ The top side toe contact will result in a chipped forward gear tooth or damage to a tooth bottom of the pinion gear.

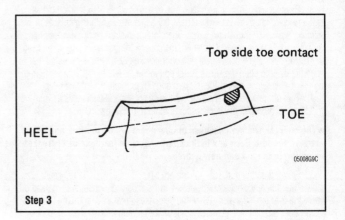

Step 3

4. This illustration shows a bottom side toe contact. In this case, increase the forward gear shim thickness and decrease the pinion gear shim thickness.

LOWER UNIT 8-43

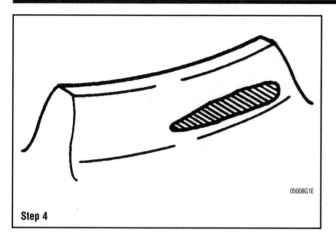
Step 4

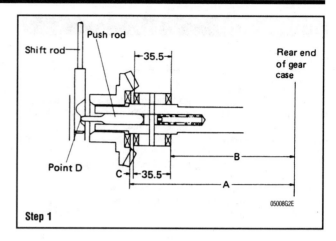
Step 1

➡ **If this tooth bearing position is not correct, the pinion gear tooth may be chipped. This problem must be corrected.**

PINION GEAR TO FORWARD GEAR THRUST PLAY

Obtain the following special tools to perform the measurement:
- Gear adjusting set (09951–09510)

1. Fit the propeller shaft to the forward gear and press the forward gear forward by pushing with the propeller shaft. (In this case, first remove the push rod from the propeller shaft)
2. Set the special tool up on the driveshaft.
3. Push the driveshaft down. Hold this position and place the gauge with its rod pushed in approximately 2 mm.
4. Still keeping this position, zero the dial gauge. Then pull up on the driveshaft and read the maximum play on the gauge.

If the amount of backlash is larger than specified, increase the thickness of each of the pinion gear back-up shim(s) or forward gear back-up shim(s). If smaller, decrease the thickness of each shim.

5. When the pinion and forward gear back-up shim are determined to be the correct size, note the amount of thrust play measured. The thrust play measurement will be necessary for the adjustment of the reverse gear.

REVERSE GEAR THRUST PLAY ADJUSTMENT

1. Perform the same procedure as above with the exception of installing the propeller shaft and propeller shaft bearing housing into the gearcase.
2. If the measurement of the play is equal to the pinion gear to forward gear measurement, the condition is correct. If the amount of play is smaller, decrease the thickness of the reverse gear back-up shim.

PROPELLER SHAFT THRUST PLAY ADJUSTMENT

1. Install the dial gauge adaptor plate and dial gauge with the long rod on the propeller shaft.
2. Push the propeller shaft inward. Hold the shaft in this position and preload the dial gauge 2 mm. Now zero the gauge.
3. Pull the driveshaft slowly out and read the amount of play on the gauge. Maximum propeller shaft thrust play is 0.002–0.012 in. (0.05–0.3 mm).

If the amount of thrust play is larger than specified, increase the thickness of the reverse gear thrust washer. If smaller, decrease the thickness of the thrust washer.

ADJUSTING THE CLUTCH ON GEARCASE SIDE
▶ See accompanying illustrations

✱✱ CAUTION

The following adjustments should be carried out only after the tooth contact adjustments have been completed.

1. Set the detent shift position at the neutral position.
2. With the propeller shaft removed, measure the dimension from the rear end of the gearcase to the dog section surface of the forward gear (dimension "A") with vernier calipers.
3. Gently insert the propeller shaft assembly into the gearcase until you can confirm that the forward end of the pushrod attached at the front end of the propeller shaft assembly lightly contacts the detent shift side (point "D"). Stop the

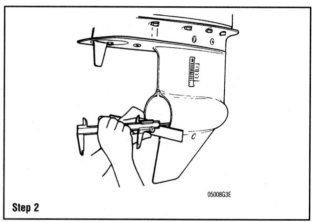
Step 2

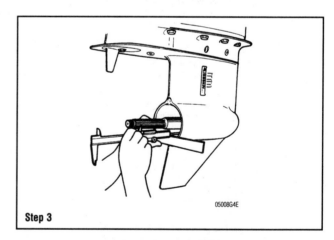
Step 3

insertion of the propeller shaft assembly when the pushrod has reached the point D. Measure the dimension B.

4. After dimensions "A" and "B" have been measured, calculate the clearance dimension at the point C, using the following formula: C = A—B—35.5 = 0.9 to 1.4 mm.

✱✱ CAUTION

If the clearance dimension at the point C is within a range of 0.9 mm to 1.4 mm, the adjustment is satisfactory. However, if the clearance does not conform to the specification, select a pushrod from among the following pushrods available to bring the clearance within specification.

- 2.44 in. (62 mm): No identification mark
- 2.42 in. (61.5 mm): One line
- 2.46 in. (62.5 mm): Two lines

8-44 LOWER UNIT

DT115 and DT140

DISASSEMBLY

1. Remove the gearcase.
2. Secure the gearcase in a suitable holding fixture or a vise with protective jaws. If protective jaws are not available, position the gearcase upright with the gearcase skeg held between wooden blocks in a vise.
3. Remove the water pump.
4. Remove the 2 bolts holding the bearing housing.
5. Install the bearing housing remover (09930-39411) and special bolts (09930-39430).
6. Remove the propeller shaft/bearing housing assembly.
7. Remove the driveshaft bearing housing fasteners.
8. Pry the driveshaft bearing housing free and slide it up and off of the driveshaft.
9. Carefully remove the driveshaft thrust bearing and shim from the bearing housing.
10. Attach driveshaft holder tool (09921-29410) to the top of the driveshaft.
11. Fit an appropriate size box-end wrench over the pinion nut and pad the sides of the gearcase prop shaft bore to prevent the wrench from damaging or distorting the gearcase prop shaft bore.
12. Holding pinion nut with the wrench, turn the driveshaft counterclockwise to loosen the pinion nut.
13. Remove the tools. Remove the pinion nut and pinion gear.
14. Remove the forward gear, thrust washer and shim(s).
15. Remove the driveshaft, preload spring and thrust bearing as an assembly.
16. Remove the driveshaft spring collar from the gearcase bore.
17. Remove the driveshaft thrust washers from the gearcase bore.
18. Remove the clutch rod and shift rod guide housing from the gearcase.
19. Slide all components except the clutch dog from the propeller shaft.
20. Disassemble the propeller shaft clutch assembly.
21. Clean and inspect all parts.
22. If the inspection of the forward gear bearing indicates replacement is necessary, perform the following procedure:
 • Install bearing housing remover (09930-39411) using the 2 long bolts from tool (09930-39430).
 • Remove forward gear bearing housing with a slide hammer (09930-30102).
 • Use an appropriate installer to install a new bearing
23. Remove and discard the propeller shaft bearing housing oil seals and bearing.
24. Install a new propeller shaft housing bearing with an appropriate installer.
25. Install new propeller shaft housing oil seals and a new housing O-ring. Coat the seal lips and O-ring with an appropriate waterproof marine grease.
26. Remove and discard the driveshaft bearing housing oil seal. Install a new seal with an appropriate installer. Coat seal lips with waterproof marine grease.

CLEANING & INSPECTION

♦ See Figures 40 and 41

Wash all parts completely and dry them using compressed air. Inspect each part and service it, as necessary, or replace the part if it does not meet specification. Parts to be inspected and items to be checked are as follows:

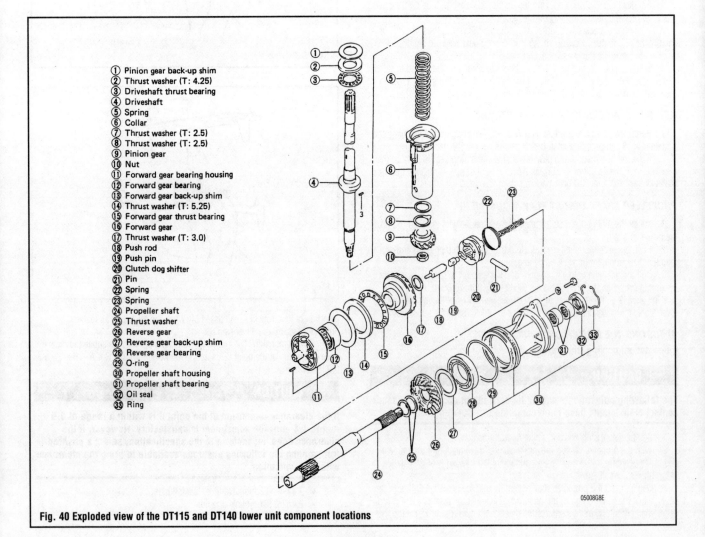

Fig. 40 Exploded view of the DT115 and DT140 lower unit component locations

Lower Unit 8-45

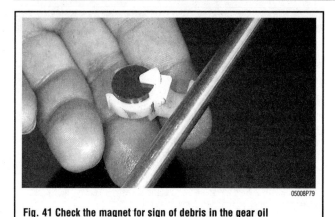

Fig. 41 Check the magnet for sign of debris in the gear oil

- All bearings for wear and damage
- Propeller shaft and driveshaft for wear at oil seal contact points
- All gear teeth for damage or wear
- Dogs on clutch dog shifter for wear and damage
- Dogs on the forward and reverse gears for damage and wear
- Shifting cam and pushrod for wear
- Check the magnet for sign of debris in the gear oil
- Perform a gearcase pressure test to check the seals. Use the oil leakage tester (09950–69511) and air pump assembly (09821–00004) and pressurize the gearcase to 14.22 psi (1.0 kg/cm). With the gearcase pressurized, spray soapy water onto the seals and check for escaping air bubbles.
- O-rings and oil seals for cracks, tears and wear
- Propeller for nicks, bent blades or other damage and wear
- Cooling circuit for clogging or other obstructions
- Gearcase for rusting, pitting and distortion.

ASSEMBLY

♦ See Figure 42

1. Install a new propeller shaft housing bearing with an appropriate installer.
2. Install new propeller shaft housing oil seals and a new housing O-ring. Coat seal lips and O-ring with waterproof marine grease.
3. Remove and discard the driveshaft bearing housing oil seal. Install a new seal with an appropriate installer. Coat seal lips with waterproof marine grease.
4. Assemble the propeller shaft components according to the illustrations.
5. Lubricate all parts with Suzuki Outboard Motor Gear Oil.
6. Install the clutch rod and shift rod guide into the gearcase.
7. Before inserting the shift rod guide housing assembly, check to be sure that the cam portion is facing toward the propeller shaft.
8. Make sure that the dowel pins are in position. Tighten the bolts securely and make sure the unit shifts smoothly.
9. Install the two-driveshaft thrust washers into the gearcase bore.
10. Install the driveshaft spring collar into the gearcase bore.
11. Install the driveshaft, preload spring and thrust bearing as an assembly.
12. Install the shift rod assembly.
13. Install the shim, thrust washer and thrust bearing on the forward gear in that order. Install the forward gear assembly into the gearcase.

➥When installing the forward gear bearing housing, align the groove in the gearcase with the pin on the housing.

14. Install the pinion gear into the propeller shaft bore and mesh with the forward gear.
15. Holding the pinion gear in place, install the driveshaft assembly into the gearcase. Rotate shaft to align its splines with those of the pinion gear and seat gear on shaft.
16. Coat the pinion nut threads with a small amount of thread locking compound.
17. Position nut over driveshaft threads in prop shaft bore. Start nut by hand.
18. Hold pinion gear nut with an appropriate size box-end wrench and pad side of prop shaft bore to prevent damage from contact with the wrench.

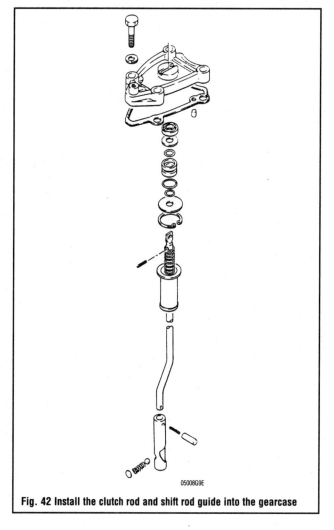

Fig. 42 Install the clutch rod and shift rod guide into the gearcase

19. Attach driveshaft holder tool (09921–29410) to the top of the driveshaft. Rotate driveshaft and tighten the pinion gear nut to specification.
20. Check forward gear backlash.
21. Temporarily install the propeller shaft/bearing housing assembly. Check pinion gear depth, reverse gear backlash and gear tooth contact pattern.
22. Coat gearcase and driveshaft bearing housing mating surfaces with a silicone sealer and install the housing.
23. Lubricate the propeller shaft bearing housing O-ring and the shift mechanism pushrod with waterproof marine grease.
24. Install the propeller shaft/bearing housing assembly into the propeller shaft bore. Use shaft housing installer (09922–59410) and installer handle (09922–59420).
25. Coat the bearing housing bolts with a small amount of thread locking compound and tighten securely.
26. Install the water pump.
27. Install the gearcase. Fill with recommended type and quantity of lubricant.
28. Check gearcase lubricant level after engine has been run. Change the lubricant after 10 hours of operation (break-in period).

SHIMMING PROCEDURE

Pinion gear adjustment

♦ See Figure 43

Before installing the driveshaft assembly into the gearcase, perform the pinion gear adjustment using the following method.

1. Assemble the bearing "3", thrust washer "2", shim "1" and driveshaft bearing housing to the driveshaft.

8-46 LOWER UNIT

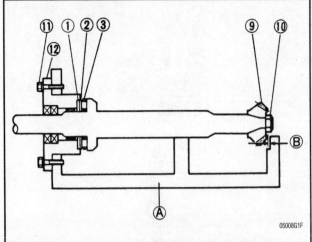

Fig. 43 Assemble the bearing "3", thrust washer "2", shim "1" and driveshaft bearing housing to the driveshaft

✱✱ CAUTION

Use a thinner shim "1" than the standard shim so that the clearance "B" will exist.

2. Position the shimming gauge "A" (09951–09420) horizontally in a vise and tighten securely.
3. Insert the driveshaft through the shimming gauge "A" opening and assemble the pinion gear "9" and nut "10" to the driveshaft. Tighten the pinion nut "10" and driveshaft bearing housing bolt "11" to the shimming gauge "A".
4. Hold the driveshaft against the bearing housing "12" while measuring the clearance between the gauge and flat edge of the pinion gear "9" with a feeler gauge. Measured clearance "B" plus thickness of the shim "1" is the total shim thickness to be used in the gear housing reassembly to obtain the correct pinion gear position.

➡ Always use ONE shim.

Forward Gear Adjustment

After installing the forward gear and driveshaft assembly in the gearcase, using the forward gear back-up shim that was removed during disassembly, perform the forward gear adjustment using the following method.

1. Remove the gearcase drain plug and install the gear adjusting gauge (09951–09510).

✱✱ CAUTION

When adjusting the dial gauge, align the gauge rod end so that it contacts the convex side of the forward gear tooth at the heel end. Do not allow the rod to contact the neighboring tooth.

2. Read the backlash by moving the forward gear slightly back and forth by hand.
If the amount of backlash is larger than specified: 0.006–0.012 in. (0.15–0.3 mm), increase the thickness of the forward gear back-up shim. If smaller, decrease the thickness of the forward gear back-up shim.

Propeller Shaft Thrust Play Adjustment

1. Install the dial gauge adaptor plate and dial gauge with long rod on the propeller shaft.
2. Push the propeller shaft inward. Hold the shaft in this position and preload the dial indicator 2 mm. Then zero the dial.
3. With the shaft pushed in, slowly pull out the shaft and read the maximum play indicated on the gauge. Play should read 0.008–0.016 in. (0.2–0.4 mm).
If the amount of thrust play is larger than specified, increase the thickness of the reverse gear thrust washer. If smaller, decrease the thickness.
Thrust washers are available in the following thicknesses: 1.0, 1.2, 1.3, 1.4, 1.6 mm. Standard thickness is 2.4 mm.

V4 and V6

DISASSEMBLY

▶ See accompanying illustrations

1. Remove the gearcase.
2. Secure the gearcase in a suitable holding fixture or a vise with protective jaws. If protective jaws are not available, position the gearcase upright with the gearcase skeg placed between wooden blocks in a vise.
3. Push down the clutch rod and place the transmission in forward gear.
4. Pull off the cotter pin on the propeller.
5. Remove the propeller nut while using the driveshaft holder tool (09921–28710) to hold the splined end of the driveshaft. Remove the spacer, washer propeller stopper and then the propeller.
6. Remove the bolts and lift off the water pump case and impeller.

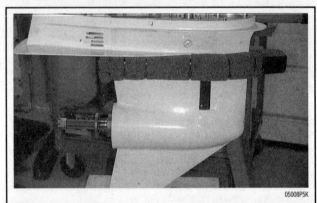

Step 2

Step 3

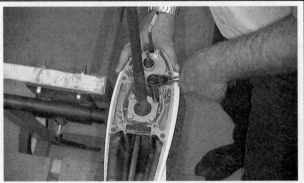

Step 6

LOWER UNIT 8-47

Step 7

Step 8

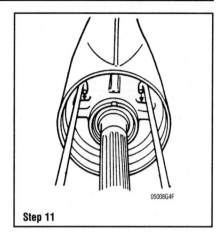
Step 11

7. Remove the key. Remove the pump case lower plate and the gasket.
8. Use the shift rod and shift the transmission back to neutral. Remove the bolts and lift off the shift unit.
9. Straighten the tabs on the lockwasher. Remove the key from the groove in the gearcase and lockwasher.
10. Install the propeller shaft stopper remover/installer tool (09951–18710) and remove the stopper and lockwasher.
11. Install the flywheel rotor remover tool (09930–39411) and propeller shaft housing remover arms (09950–58710) onto the housing.
12. Remove the propeller shaft/bearing housing assembly.
13. Attach the driveshaft holder tool (09921–29410) to the top of the driveshaft.

➡**Fit an appropriate size box-end wrench over the pinion nut and pad the sides of the gearcase prop shaft bore to prevent the wrench from damaging or distorting the prop shaft bore.**

14. Holding pinion nut with the wrench installed, loosen the pinion nut by turning the driveshaft.
15. Remove the tools. Remove the pinion nut and pinion gear.
16. Remove the forward gear, thrust washer and shim(s).
17. Remove the bolts holding the driveshaft bearing housing.
18. Carefully remove the driveshaft bearing housing and the shift rod.
19. Remove the driveshaft thrust washers, protector and the driveshaft spring collar from the gearcase bore.
20. Slide all components except the clutch dog from the propeller shaft.
21. Disassemble the propeller shaft clutch assembly.

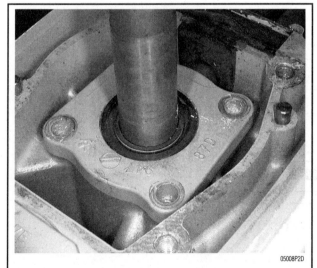
Step 17

CLEANING & INSPECTING

◆ See Figures 44, 45, 46 and 47

Wash all parts completely and dry them using compressed air. Inspect each part and service it, as necessary, or replace the part if it does not meet specification. Parts to be inspected and items to be checked are as follows:
- All bearings for wear and damage
- Propeller shaft and driveshaft for wear at oil seal contact points
- All gear teeth for damage or wear
- Dogs on clutch dog shifter for wear and damage
- Dogs on the forward and reverse gears for damage and wear
- Shifting cam and pushrod for wear
- Check the magnet for sign of debris in the gear oil
- Perform a gearcase pressure test to check the seals. Use the oil leakage tester (09950–69511) and air pump assembly (09821–00004) and pressurize the gearcase to 14.22 psi (1.0 kg/cm). With the gearcase pressurized, spray soapy water onto the seals and check for escaping air bubbles.
- O-rings and oil seals for cracks, tears and wear
- Propeller for nicks, bent blades or other damage and wear
- Cooling circuit for clogging or other obstructions
- Gearcase for rusting, pitting and distortion.

ASSEMBLY

1. Clean and inspect all parts.
2. Inspect the pinion gear adjustment of the driveshaft.
3. Remove and discard the propeller shaft/bearing housing oil seal and O-ring.
4. Inspect the propeller shaft/bearing housing bearings. Replace if necessary.
5. Install new propeller shaft/bearing housing oil seal and a new housing O-ring. Coat seal lips and O-ring with waterproof marine grease.
6. Assemble the propeller shaft components according to the illustrations, noting the following.
7. Lubricate all parts with Suzuki Outboard Motor Gear Oil.
8. Align the groove in the driveshaft spring protector with the tongue in the gearcase and install driveshaft spring protector into the gearcase.
9. Install the protector and the driveshaft thrust washers.
10. Carefully install the driveshaft and the driveshaft bearing housing.
11. Install the bolts holding the driveshaft bearing housing. Tighten the bolts securely.
12. Install the forward gear, thrust washer and shim(s).
13. Install the pinion gear into the propeller shaft bore and mesh with the forward gear.
14. Coat the pinion nut threads with thread locking compound.
15. Position nut over driveshaft threads in prop shaft bore. Start nut by hand.
16. Attach driveshaft holder tool (09921–29410) to the top of the driveshaft. Rotate driveshaft and tighten the pinion gear nut to specification.

8-48 LOWER UNIT

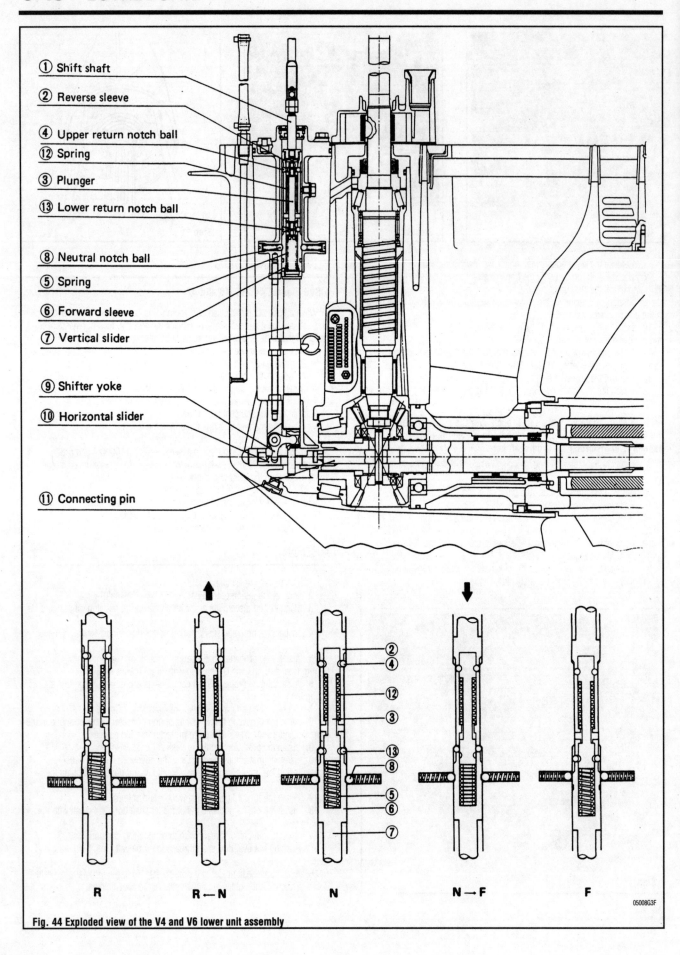

Fig. 44 Exploded view of the V4 and V6 lower unit assembly

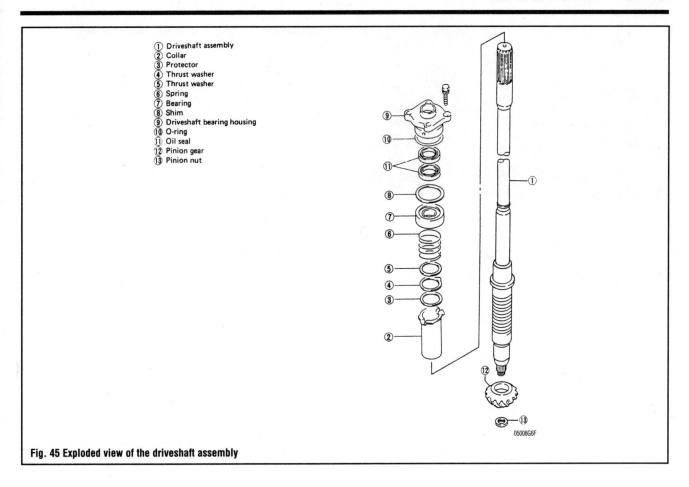

Fig. 45 Exploded view of the driveshaft assembly

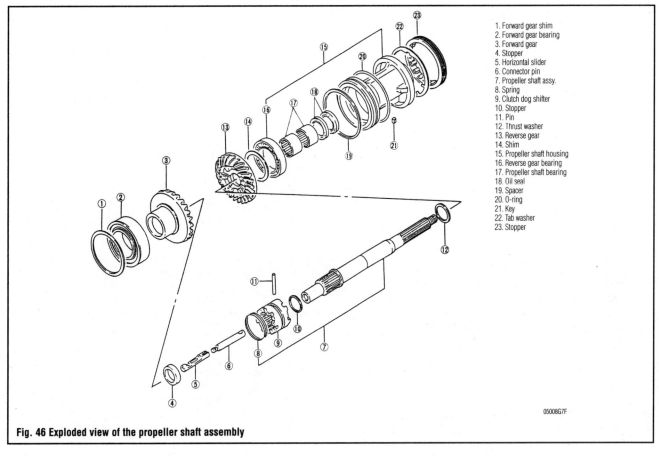

Fig. 46 Exploded view of the propeller shaft assembly

8-50 LOWER UNIT

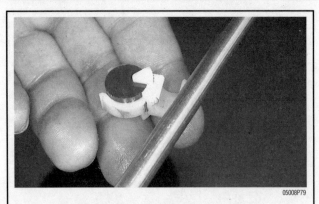

Fig. 47 Check the magnet for sign of debris in the gear oil

17. Prior to installing the propeller shaft/bearing housing assembly into the gearcase, rotate the propeller shaft so that the flat surface on the end of the horizontal slider is facing UP.
18. Install the propeller shaft/bearing housing into the gearcase. Use shaft housing installer tool (09922–59410) and installer handle (09922–59420).
19. Make sure the groove in the gearcase and the bearing housing are aligned. Install the key into the groove in the gearcase and bearing housing.
20. Install the lockwasher and align the groove in the lockwasher with the tongue in the gearcase.
21. Check forward and reverse gear backlash.
22. Apply a small amount of thread locking compound, to the threads of the stopper.
23. Position the stopper with the "OFF" mark facing toward the outside and install the stopper.
24. Use the same special tool used to remove the stopper and tighten the stopper to specification. If necessary, slightly tighten the stopper until it is aligned with the lockwasher.
25. Bend down the lockwasher tab to lock the stopper in place.
26. Apply waterproof marine grease to the O-ring seal on the shift unit. Make sure the shift rod is in **NEUTRAL**.
27. Make sure the locating dowel is in place in the gearcase and install the shift unit and bolts. Tighten the bolts securely.
28. Install the water pump.
29. Install the gearcase. Fill with recommended type and quantity of lubricant.
30. Check gearcase lubricant level after engine has been run. Change the lubricant after 10 hours of operation (break-in period).

SHIMMING PROCEDURE

Pinion Gear Adjustment

♦ See Figure 48

1. Assemble the bearing "7" and driveshaft bearing housing "9" onto the driveshaft.

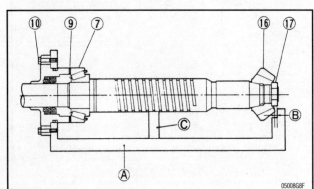

Fig. 48 Position the shimming gauge (09951–08720) "A" horizontally in a vice

➡ Use a thinner pinion shim "9" than the standard shim so that a clearance "B" exists.

2. Position the shimming gauge (09951–08720) "A" horizontally in a vice.
3. Insert the driveshaft through the shimming gauge "A" opening and assemble the pinion gear "16" and nut "17" and driveshaft bearing housing bolt "10" to the shimming gauge "A". Adjust the support "C" so the driveshaft is parallel to the tool "A".
4. Hold the driveshaft against the bearing housing "10" while measuring the clearance "B" between the gauge and the flat edge of the pinion gear "16" with a feeler gauge. Measured clearance plus shim "9" is the total shim thickness to be used in the gear housing reassembly in order to achieve the correct pinion gear position.

➡ Always use ONE shim.

5. Remove the driveshaft from the tool, and change the shim if needed, and reinstall into the gearcase.

Forward and Reverse Gear Adjustment

FORWARD

♦ See Figure 49

1. Set the gear holder (09951–98720) onto the propeller shaft.
2. Turning the bolt clockwise, tighten securely.
3. Set the dial gauge (09900–20606) and backlash indicator tool (09952–08710) on the driveshaft.
4. Read the backlash on the gauge by lightly moving the driveshaft back and forth slightly by hand.
5. If the amount of backlash is larger than specified: 0.020–0.026 in. (0.5–0.65 mm), increase the thickness of the forward gear back-up shim. If smaller, decrease the thickness.

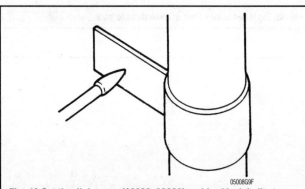

Fig. 49 Set the dial gauge (09900–20606) and backlash indicator tool (09952–08710) on the driveshaft

REVERSE

1. Turn the bolt counterclockwise, tighten securely.
2. Read the backlash on the gauge by moving the driveshaft slightly back and forth.
3. If the amount of backlash is larger than specified: 0.028–0.033 in. (0.7–0.85 mm), increase the thickness of the reverse gear back-up shim. If smaller, decrease the thickness of the shim.

➡ Always use ONE shim.

Reverse Gear Thrust

1. Install the dial indicator onto the propeller shaft.
2. Push the propeller shaft inwards and hold this position. Preload the gauge approximately 2 mm and zero the gauge.
3. Slowly pull out the propeller shaft and read the maximum play: 0.004–0.008 in. (0.10–0.20 mm) on the gauge.
4. If the amount of thrust play is larger than specified, increase the thickness of the reverse gear thrust washer. If smaller, decrease the thickness of the thrust washer.

LOWER UNIT 8-51

JET DRIVE

Description and Operation

The jet drive unit is designed to permit boating in areas prohibited to a boat equipped with a conventional propeller drive system. The housing of the jet drive barely extends below the hull of the boat allowing passage in ankle deep water, white water rapids and over sand bars or in shoal water which would foul a propeller drive.

The jet drive provides reliable propulsion with a minimum of moving parts. Simply stated, water is drawn into the unit through an intake grille by an impeller driven by a driveshaft off the crankshaft of the powerhead. The water is immediately expelled under pressure through an outlet nozzle directed away from the stern of the boat.

As the speed of the boat increases and reaches planing speed, the jet drive discharges water freely into the air and only the intake grille makes contact with the water.

The jet drive is provided with a gate arrangement and linkage to permit the boat to be operated in reverse. When the gate is moved downward over the exhaust nozzle, the pressure stream is reversed by the gate and the boat moves sternward.

Conventional controls are used for powerhead speed, movement of the boat, shifting and power trim and tilt.

Model Identification and Serial Numbers

♦ See Figure 50

A model letter identification is stamped on the rear, port side of the jet drive housing. A serial number for the unit is stamped on the starboard side of the jet drive housing, as indicated in the accompanying illustration.

The jet drives that are used with the outboard units covered in this manual are: PU40, PU55, PU85 and PU140. These letters are embossed on the port side of the jet drive housing.

These numbers reflect the specific size of engine which they are attached to.

For the most part, jet drive units are identical in design, function and operation. Differences lie in size and securing hardware.

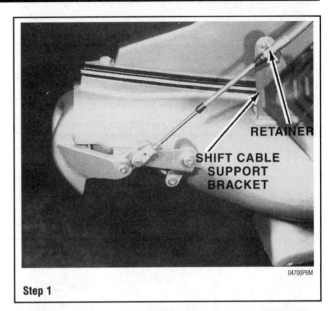
Step 1

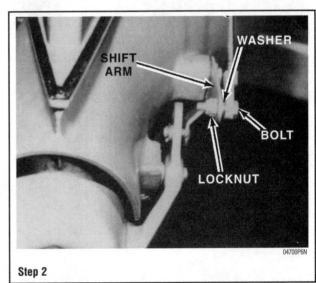
Step 2

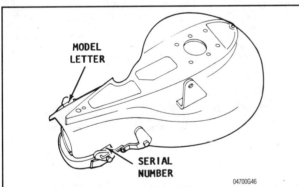
Fig. 50 The model letter designation and the serial numbers are embossed on the jet drive housing

Jet Drive Assembly

REMOVAL & INSTALLATION

♦ See accompanying illustrations

1. Remove the two bolts and retainer securing the shift cable to the shift cable support bracket.
2. Remove the locknut, bolt and washer securing the shift cable to the shift arm. Try not to disturb the length of the cable.
3. Remove the six bolts securing the intake grille to the jet casing.
4. Ease the intake grille from the jet drive housing.

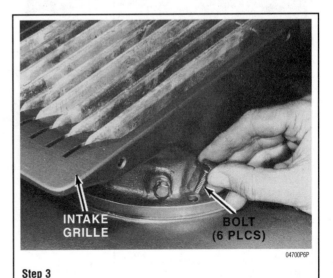
Step 3

8-52 LOWER UNIT

5. Pry the tab or tabs of the tabbed washer away from the nut to allow the nut to be removed.
6. Loosen and then remove the nut.
7. Remove the tabbed washer and spacers. Make a careful count of the spacers behind the washer. If the unit is relatively new, there could be as many as eight spacers stacked together. If less than eight spacers are removed from behind the washer, the others will be found behind the jet impeller, which is removed in the following step. A total of eight spacers will be found.
8. Remove the jet impeller from the shaft. If the impeller is frozen to the shaft, obtain a block of wood and a hammer. Tap the impeller in a clockwise direction to release the shear key.
9. Slide the nylon sleeve and shear key free of the driveshaft and any spacers found behind the impeller. Make a note of the number of spacers at both locations—behind the impeller and on top of the impeller, under the nut and tabbed washer.
10. One external bolt and four internal bolts are used to secure the jet drive to the intermediate housing. The external bolt is located at the aft end of the anti-cavitation plate.
11. The four internal bolts are located inside the jet drive housing, as indicated in the accompanying illustration. Remove the five attaching bolts.
12. Lower the jet drive from the intermediate housing. Remove the locating pin from the forward starboard side (or center forward, depending on the model being serviced) of the upper jet housing.
13. Remove the locating pin from the aft end of the housing. This pin and the one removed in the previous step should be of identical size.

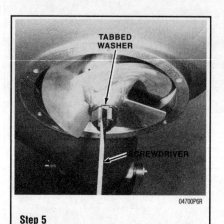
Step 5

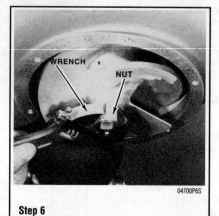
Step 6

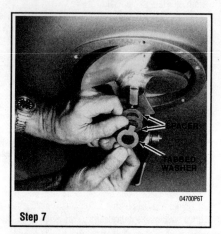
Step 7

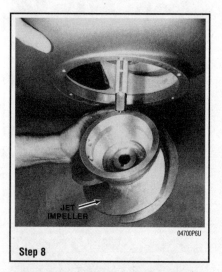
Step 8

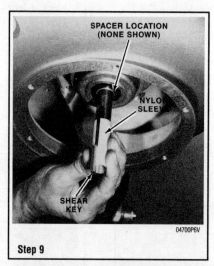
Step 9

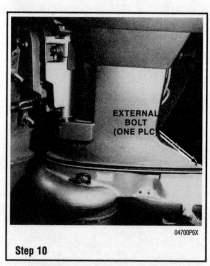
Step 10

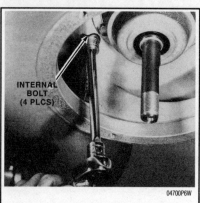
Step 11

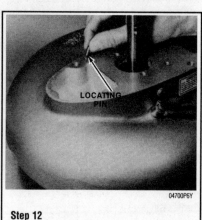
Step 12

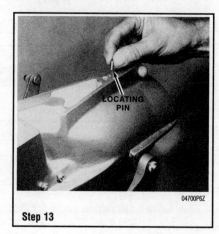
Step 13

LOWER UNIT 8-53

14. Remove the four bolts and washers from the water pump housing. Pull the water pump housing, the inner cartridge and the water pump impeller, up and free of the driveshaft. Next, remove the outer gasket, the steel plate and the inner gasket.

15. Remove the two small locating pins and lift the aluminum spacer up and free of the drive shaft.

Remove the driveshaft and bearing assembly from the housing.

Remove the large thick adaptor plate from the intermediate housing. This plate is secured with seven bolts and lock washers. Lower the adaptor plate from the intermediate housing and remove the two small locating pins, one on the forward port side and another from the last aft hole in the adaptor plate. Both pins are identical in size.

To install:

16. Install the other small locating pin into the forward starboard side (or center forward end, depending on the model being serviced).

17. Raise the jet drive unit up and align it with the intermediate housing, with the small pins indexed into matching holes in the adapter plate. Install the four internal bolts.

18. Install the one external bolt at the aft end of the anti-cavitation plate. Tighten all bolts to a torque value of 11 ft. lbs. (15Nm).

19. Place the required number of spacers up against the bearing housing. Slide the nylon sleeve over the driveshaft and insert the shear key into the slot of the nylon sleeve with the key resting against the flattened portion of the driveshaft.

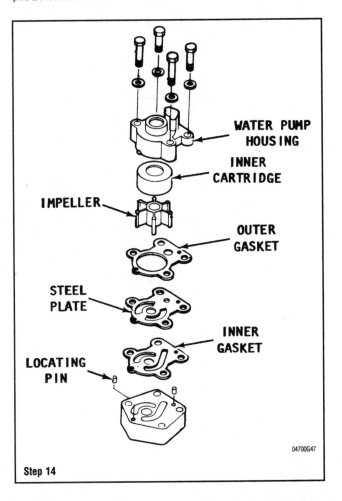
Step 14

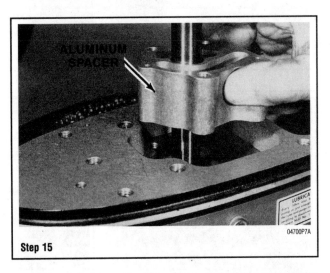
Step 15

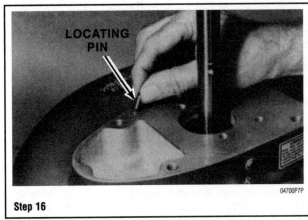
Step 16

Step 17

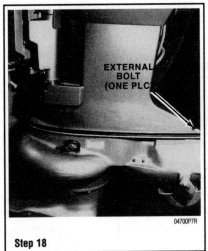
Step 18

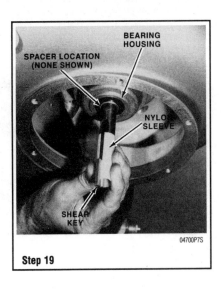
Step 19

8-54 LOWER UNIT

Step 20

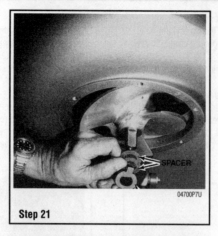
Step 21

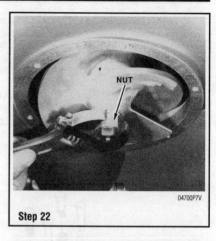
Step 22

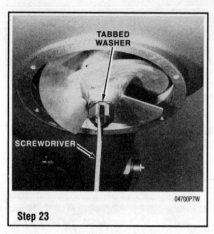
Step 23

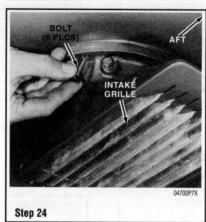
Step 24

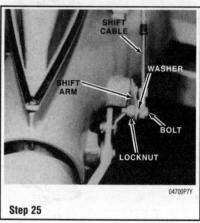
Step 25

20. Slide the jet impeller up onto the driveshaft, with the groove in the impeller collar indexing over the shear key.

21. Place the remaining spacers over the driveshaft.

22. Tighten the nut to a torque value of 17 ft. lbs. (23Nm). If neither of the two tabs on the tabbed washer aligns with the sides of the nut, remove the nut and washer. Invert the tabbed washer. Turning the washer over will change the tabs by approximately 15°. Install and tighten the nut to the required torque value. The tabbed washer is designed to align with the nut in one of the two positions described.

23. Bend the tabs up against the nut to prevent the nut from backing off and becoming loose.

24. Install the intake grille onto the jet drive housing with the slots facing aft. Install and tighten the six securing bolts. Tighten ¼ in. bolts to a torque value of 5 ft. lbs. (7Nm). Tighten 5/16 in. bolts to 11 ft. lbs. (15Nm).

25. Slide the bolt through the end of the shift cable, washer and into the shift arm. Install the locknut onto the bolt and tighten the bolt securely.

26. Install the shift cable against the shift cable support bracket and secure it in place with the two bolts.

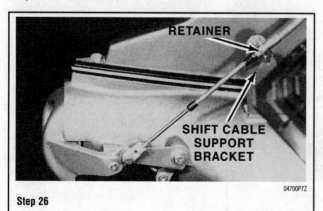
Step 26

ADJUSTMENT

Cable Alignment And Free Play

♦ See Figs. 51 and 52

1. Move the shift lever downward into the forward position. The leaf spring should snap over on top of the lever to lock it in position.

2. Remove the locknut, washer and bolt from the threaded end of the shift cable. Push the reverse gate firmly against the rubber pad on the underside of the jet drive housing.

Check to be sure the link between the reverse gate and the shift arm is hooked into the LOWER hole on the gate.

Hold the shift arm up until the link rod and shift arm axis form an imaginary straight line, as indicated in the accompanying illustration. Adjust the length of the shift cable by rotating the threaded end, until the cable can be installed back onto the shift arm without disturbing the imaginary line. Pass the nut through the cable end, washer and shift arm. Install and tighten the locknut.

Neutral Stop Adjustment

See Figures 53, 54, and 55

In the forward position, the reverse gate is neatly tucked underneath and clear of the exhaust jet stream.

In the reverse position, the gate swings up and blocks the jet stream deflecting the water in a forward direction under the jet housing to move the boat sternward.

In the neutral position, the gate assumes a happy medium—a balance between forward and reverse when the powerhead is operating at IDLE speed. Actually, the gate is deflecting some water to prevent the boat from moving forward, but not enough volume to move the boat sternward.

✴✴ WARNING

The gate must be properly adjusted for safety of boat and passengers. Improper adjustment could cause the gate to swing up to the

LOWER UNIT 8-55

Fig. 51 Move the shift lever downward into the forward position. The leaf spring should snap over on top of the lever to lock it in position

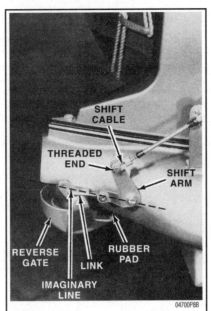

Fig. 52 Remove the locknut, washer and bolt from the threaded end of the shift cable

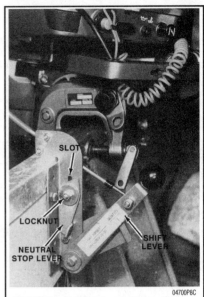

Fig. 53 Loosen, but do not remove the locknut on the neutral stop lever. Check to be sure the lever will slide up and down along the slot in the shift lever bracket

Fig. 54 Start the powerhead and allow it to operate only at IDLE speed. With the neutral stop lever in the down position, move the shift lever until the jet stream forces on the gate are balanced. Balanced means the water discharged is divided in both directions and the boat moves neither forward nor sternward. The gate is then in the neutral position with the powerhead at idle speed

reverse position while the boat is moving forward causing serious injury to boat or passengers.

1. Loosen, but do not remove the locknut on the neutral stop lever. Check to be sure the lever will slide up and down along the slot in the shift lever bracket.

➡ The following procedure must be performed with the boat and jet drive in a body of water. Only with the boat in the water can a proper jet stream be applied against the gate for adjustment purposes.

✱✱ CAUTION

Water must circulate through the lower unit to the powerhead anytime the powerhead is operating to prevent damage to the water pump in the lower unit. Just five seconds without water will damage the water pump impeller.

2. Start the powerhead and allow it to operate only at IDLE speed. With the neutral stop lever in the down position, move the shift lever until the jet stream forces on the gate are balanced. Balanced means the water discharged is divided in both directions and the boat moves neither forward nor sternward. The gate is then in the neutral position with the powerhead at idle speed.

3. Move the neutral stop lever up against the shift lever until the stop lever barely makes contact with the shift lever. Tighten the locknut to maintain this new adjusted position. Shut down the powerhead.

➡ The reverse gate may not swing to the full up position in reverse gear after the previous steps have been performed. Do not be concerned. This condition is acceptable, because water pressure in reverse will close the gate fully under normal operation.

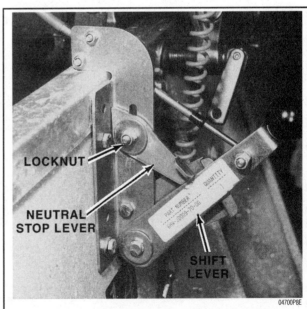

Fig. 55 Move the neutral stop lever up against the shift lever until the stop lever barely makes contact with the shift lever. Tighten the locknut to maintain this new adjusted position. Shut down the powerhead

8-56 Lower Unit

Trim Adjustment

▶ See Figure 56

1. During operation, if the boat tends to pull to port or starboard, the flow fins may be adjusted to correct the condition. These fins are located at the top and bottom of the exhaust tube.
2. If the boat tends to pull to starboard, bend the trailing edge of each fin approximately 1/16 in. (1.5mm) toward the starboard side of the jet drive. Naturally, if the boat tends to pull to port, bend the fins toward the port side.

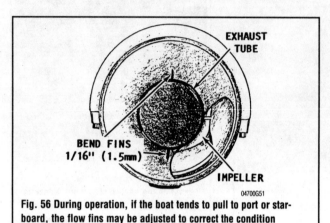

Fig. 56 During operation, if the boat tends to pull to port or starboard, the flow fins may be adjusted to correct the condition

DISASSEMBLY

1. Remove the locating pin from the forward starboard side (or center forward, depending on the model being serviced) of the upper jet housing.

→There will be a total of six locating pins to be removed in the following steps. Make careful note of the size and location of each when they are removed, as an assist during assembling.

2. Remove the locating pin from the aft end of the housing. This pin and the one removed in the previous step should be of identical size.

3. Remove the four bolts and washers from the water pump housing.
Pull the water pump housing, the inner cartridge and the water pump impeller, up and free of the driveshaft. Remove the Woodruff key from its recess in the driveshaft. Next, remove the outer gasket, the steel plate and the inner gasket.
4. Remove the two small locating pins and lift the aluminum spacer up and free of the driveshaft.
5. Remove the driveshaft and bearing assembly from the housing.
Remove the large thick adapter plate from the intermediate housing. This plate is secured with seven bolts and lock-washers. Lower the adapter plate from the intermediate housing and remove the two small locating pins, one on the forward port side and another from the last aft hole in the adapter plate. Both pins are identical size.

CLEANING & INSPECTING

▶ See Figures 57 and 58

Wash all parts, except the driveshaft assembly, in solvent and blow them dry with compressed air. Rotate the bearing assembly on the driveshaft to inspect the bearings for rough spots, binding and signs of corrosion or damage.
Saturate a shop towel with solvent and wipe both extensions of the driveshaft.

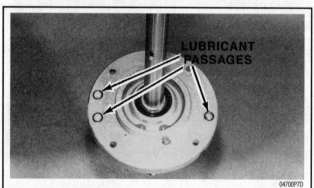

Fig. 57 Take extra precautions to prevent solvent from entering the lubrication passages

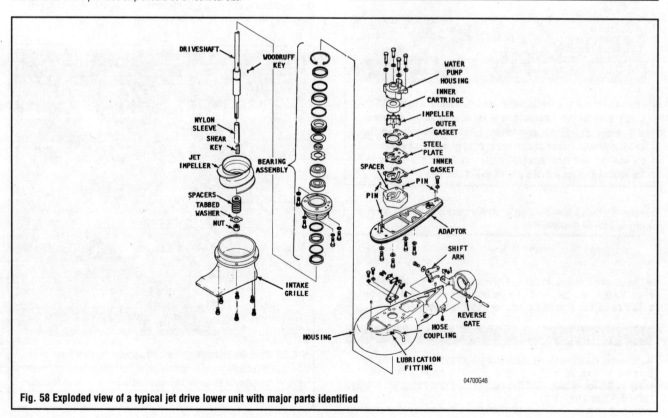

Fig. 58 Exploded view of a typical jet drive lower unit with major parts identified

LOWER UNIT 8-57

Bearing Assembly

♦ See Figure 59

Lightly wipe the exterior of the bearing assembly with the same shop towel. Do not allow solvent to enter the three lubricant passages of the bearing assembly. The best way to clean these passages is not with solvent—because any solvent remaining in the assembly after installation will continue to dissolve good useful lubricant and leave bearings and seals dry. This condition will cause bearings to fail through friction and seals to dry up and shrink—losing their sealing qualities.

The only way to clean and lubricate the bearing assembly is after installation to the jet drive—via the exterior lubrication fitting.

If the old lubricant emerging from the hose coupling is a dark, dirty, gray color, the seals have already broken down and water is attacking the bearings. If such is the case, it is recommended the entire driveshaft bearing assembly be taken to the dealer for service of the bearings and seals.

Dismantling Bearing Assembly

A complicated procedure must be followed to dismantle the bearing assembly including torching off the bearing housing. Naturally, excessive heat might ruin the seals and bearings. Therefore, the best recommendation is to leave this part of the service work to the experts at your local Honda dealership.

Driveshaft and Associated Parts

Inspect the threads and splines on the driveshaft for wear, rounded edges, corrosion and damage.

Carefully check the driveshaft to verify the shaft is straight and true without any sign of damage.

Inspect the jet drive housing for nicks, dents, corrosion, or other signs of damage. Nicks may be removed with No. 120 and No. 180 emery cloth.

Reverse Gate

Inspect the gate and its pivot points. Check the swinging action to be sure it moves freely the entire distance of travel without binding.

Inspect the slats of the water intake grille for straightness. Straighten any bent slats, if possible. Use the utmost care when prying on any slat, as they tend to break if excessive force is applied. Replace the intake grille if a slat is lost, broken, or bent and cannot be repaired. The slats are spaced evenly and the distance between them is critical, to prevent large objects from passing through and becoming lodged between the jet impeller and the inside wall of the housing.

Jet Impeller

♦ See Figure 60

The jet impeller is a precisely machined and dynamically balanced aluminum spiral. Observe the drilled recesses at exact locations to achieve this delicate balancing. Some of these drilled recesses are clearly shown in the accompanying illustration.

Excessive vibration of the jet drive may be attributed to an out-of-balance condition caused by the jet impeller being struck excessively by rocks, gravel or cavitation burn.

The term cavitation burn is a common expression used throughout the world among people working with pumps, impeller blades and forceful water movement. Burns on the jet impeller blades are caused by cavitation air bubbles exploding with considerable force against the impeller blades. The edges of the blades may develop small dime size areas resembling a porous sponge, as the aluminum is actually eaten by the condition just described.

Excessive rounding of the jet impeller edges will reduce efficiency and performance. Therefore, the impeller should be inspected at regular intervals.

If rounding is detected, the impeller should be placed on a work bench and the edges restored to as sharp a condition as possible, using a file. Draw the file in only one direction. A back-and-forth motion will not produce a smooth edge. Take care not to nick the smooth surface of the jet impeller. Excessive nicking or pitting will create water turbulence and slow the flow of water through the pump.

Inspect the shear key. A slightly distorted key may be reused although some difficulty may be encountered in assembling the jet drive. A cracked shear key should be discarded and replaced with a new key.

Water Pump

♦ See Figure 61

Clean all water pump parts with solvent and then blow them dry with compressed air. Inspect the water pump housing for cracks and distortion, possibly caused from overheating. Inspect the steel plate, the thick aluminum spacer and the water pump cartridge for grooves and/or rough spots. If possible always install a new water pump impeller while the jet drive is disassembled. A new water pump impeller will ensure extended satisfactory service and give peace of mind to the owner. If the old water pump impeller must be returned to service, never install it in reverse of the original direction of rotation. Installation in reverse will cause premature impeller failure.

If installation of a new water pump impeller is not possible, check the sealing surfaces and be satisfied they are in good condition. Check the upper, lower and ends of the impeller vanes for grooves, cracking and wear. Check to be sure the indexing notch of the impeller hub is intact and will not allow the impeller to slip.

ASSEMBLING

♦ See Figure 62

Identify the two small locating pins used to index the large thick adapter plate to the intermediate housing. Insert one pin into the last hole aft on the topside of the plate. Insert the other pin into the hole forward toward the port side, as shown.

Lift the plate into place against the intermediate housing with the locating pins indexing with the holes in the intermediate housing. Secure the plate with the five (or seven) bolts.

➡ On the five bolt model, one of the five bolts is shorter than the other four. Install the short bolt in the most aft location.

Tighten the long bolts to a torque value of 22 ft. lbs. (30Nm). Tighten the short bolt to a torque value of 11 ft. lbs. (15Nm).

1. Place the driveshaft bearing assembly into the jet drive housing. Rotate the bearing assembly until all bolt holes align. There is only one correct position.

➡ If installing a new jet impeller, place all eight spacers at the lower or nut end of the impeller and skip the following step.

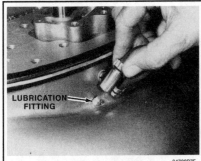

Fig. 59 Cleaning and lubricating the bearing assembly is best accomplished by completely replacing the old lubricant

Fig. 60 The slats of the grille must be carefully inspected and any bent slats straightened for maximum performance of the jet drive

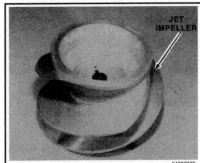

Fig. 61 The edges of the jet impeller should be kept as sharp as possible for maximum jet drive efficiency

8-58 LOWER UNIT

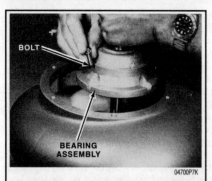

Fig. 62 Place the driveshaft bearing assembly into the jet drive housing. Rotate the bearing assembly until all bolt holes align

Fig. 63 The clearance between the outer edge of the jet drive impeller and the water intake housing cone wall should be maintained at approximately 1/32 in. (0.8mm)

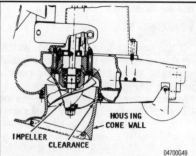

Fig. 64 Spacers are used depending on the model being serviced. When new, all spacers are located at the tapered (or nut) end of the impeller

Shimming Jet Impeller

♦ See Figures 63 and 64

1. The clearance between the outer edge of the jet drive impeller and the water intake housing cone wall should be maintained at approximately 1/32 in. (0.8mm). This distance can be visually checked by shining a flashlight up through the intake grille and estimating the distance between the impeller and the casing cone, as indicated in the accompanying illustrations. It is not humanly possible to accurately measure this clearance, but by observing closely and estimating the clearance, the results should be fairly accurate.

After continued use, the clearance will increase. The spacers previously removed are used to position the impeller along the driveshaft with a desired clearance of 1/32 in. (0.8mm) between the jet impeller and the housing wall.

2. Spacers are used depending on the model being serviced. When new, all spacers are located at the tapered (or nut) end of the impeller. As the clearance increases, the spacers are transferred from the tapered (nut) end and placed at the wide (intermediate housing) end of the jet impeller.

This procedure is best accomplished while the jet drive is removed from the intermediate housing.

Secure the driveshaft with the attaching hardware. Installation of the shear key and nylon sleeve is not vital to this procedure. Place the unit on a convenient work bench. Shine a flashlight through the intake grille into the housing cone and eyeball the clearance between the jet impeller and the cone wall, as indicated in the accompanying line drawing. Move spacers one-at-a-time from the tapered end to the wide end to obtain a satisfactory clearance. Dismantle the driveshaft and note the exact count of spacers at both ends of the bearing assembly. This count will be recalled later during assembly to properly install the jet impeller.

Water Pump Assembling

♦ See Figures 65 and 66

1. Place the aluminum spacer over the driveshaft with the two holes for the indexing pins facing upward. Fit the two locating pins into the holes of the spacer.

➡ The manufacturer recommends no sealant be used on either side of the water pump gaskets.

2. Slide the inner water pump gasket (the gasket with two curved openings) over the driveshaft. Position the gasket over the two locating pins. Slide the steel plate down over the driveshaft with the tangs on the plate facing downward and with the holes in the plate indexed over the two locating pins.

Check to be sure the tangs on the plate fit into the two curved openings of the gasket beneath the plate. Now, slide the outer gasket (the gasket with the large center hole) over the driveshaft. Position the gasket over the two locating pins.

Fit the Woodruff key into the driveshaft. Just a dab of grease on the key will help to hold the key in place. Slide the water pump impeller over the driveshaft with the rubber membrane on the top side and the keyway in the impeller indexed over the Woodruff key. Take care not to damage the membrane. Coat the impeller blades with Hondaline Grease or equivalent water resistant lubricant.

Install the insert cartridge, the inner plate and finally the water pump housing over the driveshaft. Rotate the insert cartridge counterclockwise over the impeller to tuck in the impeller vanes. Seat all parts over the two locating pins.

➡ On some models, two different length bolts are used at this location.

Tighten the four bolts to a torque value of 11 ft. lbs. (15Nm).

3. Install one of the small locating pins into the aft end of the jet drive housing.

Fig. 65 Place the aluminum spacer over the driveshaft with the two holes for the indexing pins facing upward

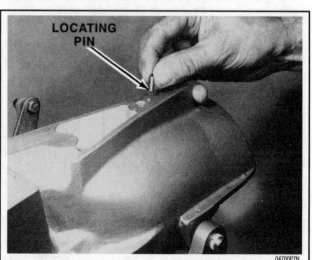

Fig. 66 Install one of the small locating pins into the aft end of the jet drive housing

MANUAL TILT 9-2
DESCRIPTION AND OPERATION 9-2
   SERVICING 9-2
GAS ASSISTED TILT 9-2
DESCRIPTION AND OPERATION 9-2
GAS ASSIST DAMPER 9-2
   TESTING 9-2
   REMOVAL & INSTALLATION 9-2
POWER TILT 9-3
DESCRIPTION AND OPERATION 9-3
TROUBLESHOOTING THE POWER TILT SYSTEM 9-4
TILT MOTOR AND PUMP 9-4
   CHECKING FLUID LEVEL 9-4
   BLEEDING THE SYSTEM 9-4
   TESTING 9-5
   REMOVAL & INSTALLATION 9-5
   OVERHAUL 9-5
TILT CYLINDER 9-5
   REMOVAL & INSTALLATION 9-5
   OVERHAUL 9-5
TILT SWITCH 9-6
POWER TRIM/TILT 9-6
DESCRIPTION AND OPERATION 9-6
TROUBLESHOOTING THE POWER TRIM/TILT SYSTEM 9-7
TRIM/TILT PUMP 9-8
   TESTING 9-8
   REMOVAL & INSTALLATION 9-9
TRIM/TILT MOTOR 9-9
   TESTING 9-9
   REMOVAL & INSTALLATION 9-10
   DISASSEMBLY 9-11
   CLEANING & INSPECTION 9-12
   ASSEMBLY 9-13
TRIM/TILT CYLINDER 9-14
   TESTING 9-14
   REMOVAL & INSTALLATION 9-14
   DISASSEMBLY 9-15
   CLEANING & INSPECTION 9-15
   ASSEMBLY 9-15
TRIM/TILT SWITCH 9-15
TRIM/TILT RELAY 9-15
   TESTING 9-15
   REMOVAL & INSTALLATION 9-15

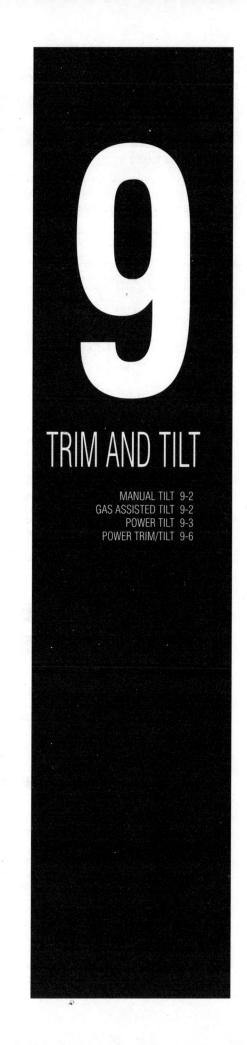

TRIM AND TILT

MANUAL TILT 9-2
GAS ASSISTED TILT 9-2
POWER TILT 9-3
POWER TRIM/TILT 9-6

9-2 TRIM AND TILT

MANUAL TILT

Description and Operation

▶ See Figures 1 and 2

All outboard installations are equipped with some means of raising or lowering (pivoting) the outboard for efficient operation under various load, boat design, water conditions, and for trailering to and from the water. By pivoting

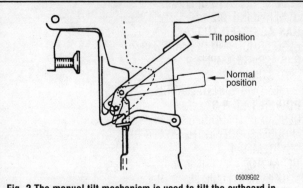

Fig. 2 The manual tilt mechanism is used to tilt the outboard in relation to the stern bracket

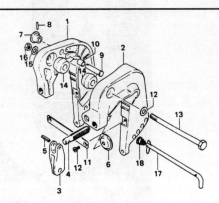

1. Clamp bracket (RH)
2. Clamp bracket (LH)
3. Clamp handle
4. Clamp screw
5. Clamp handle pin
6. Clamp plate
7. Tilt stopper knob
8. Tilt stopper knob pin
9. Tilt stopper pin
10. Bushing
11. Clamp bracket plate
12. Screw
13. Bolt
14. Swivel bracket bushing
15. Swivel bracket washer
16. Swivel bracket nut
17. Tilt lock pin
18. Tilt lock pin spring

Fig. 1 Typical manual tilt bracket assembly used on low horsepower outboards

the outboard, the correct trim angle can be achieved to ensure maximum performance and fuel economy as well as a more comfortable ride for the crew and passengers.

The manual tilt mechanism is used to tilt the outboard in relation to the stern bracket. To adjust the outboard angle, the tilt lever is moved to the tilt position. This disengages the release rod and causes the reverse lock to disengage from the adjusting pin. The unit can be set at varying angles by raising or lowering the outboard. To release the tilt mechanism, return the tilt lever to the normal position, then raise and lower the extension case slightly.

➡ The tilt lever should be kept in the normal position whenever the outboard is operating.

SERVICING

Service procedures for the manual tilt system are confined to general lubrication and inspection. If individual components should wear or break, replacement of the defective components is necessary.

GAS ASSISTED TILT

Description and Operation

The gas assisted tilt system consists of a single shock absorber. The shock absorber contains a high pressure gas chamber located in the upper portion of the cylinder bore above the piston assembly. The piston contains a down relief valve and an absorber relief valve. Below the piston assembly, the lower cylinder bore contains an oil chamber. This lower chamber is connected to the upper chamber above the piston by internal passages.

When the outboard is tilted upward, pressurized gas, sealed in the cylinder, expands to extend the piston rod in the cylinder and assist the tilting action.

Gas Assist Damper

TESTING

Testing of the gas assisted damper is a fairly simple operation. Since the operation of the cylinder is solely reliant on the action of the gas pressure on the piston, if the damper fails to provide the appropriate amount of lifting assistance and the steering tube adjustment is not set too tight, then it can be assumed that the damper is faulty.

REMOVAL & INSTALLATION

1. Place the tilt lever in the tilt position.
2. Position and support the outboard in its fully tilted position.
3. Lower the manual tilt lock levers to keep the outboard tilted during this procedure.

✴ CAUTION

To prevent outboard damage or serious personal injury, it is a good idea to provide some additional means of keeping the outboard tilted in the event the tilt lock levers fail.

4. Remove the snapring and slide the upper cylinder pin from the mounting frame.
5. Pivot the damper away from the mounting frame.
6. Remove the securing bolt from the lower damper pin and then slide the pin from its bore.
7. Remove the damper assembly from the stern bracket.

➡ Store the gas assisted damper assembly vertically with the upper cylinder bushing upward.

To install:
8. Position the damper assembly on the stern bracket.
9. Lubricate and install the lower damper pin into its bore. Install the securing bolt and tighten it securely.

➡ The groove in the lower damper pin must align with the securing bolt hole.

10. Pivot the damper toward the stern bracket.
11. Install the upper cylinder pin on the stern bracket and install the snapring to secure it.
12. Remove the supports from the outboard.
13. Raise the manual tilt lock levers and test the gas assisted tilt assembly for proper operation.

TRIM AND TILT

POWER TILT

Description and Operation

♦ See Figure 3

➡ The power tilt system is used primarily on the DT35 and DT40 models.

The power tilt system consists of a housing with an electric motor, gear driven hydraulic pump, hydraulic reservoir and a single tilt cylinder. The cylinder performs a double function as a tilt cylinder and also as a shock absorber, should the lower unit strike an underwater object while the boat is underway.

✳✳ WARNING

The power tilt system is not designed for use as a power trim system. Damage to the system could result if misused.

The necessary valves, check valves, relief valves, and hydraulic passageways are incorporated internally and externally for efficient operation. A manual release valve is provided to permit the outboard unit to be raised or lowered should the battery fail to provide the necessary current to the electric motor or if a malfunction should occur in the hydraulic system.

The gear driven pump operates in much the same manner as an oil circulation pump installed on motor vehicles. The gears rotate in either direction, depending on the desired cylinder movement. One side of the pump is considered the suction side, and the other the pressure side, when the gears rotate in a given direction. These sides are reversed, the suction side becomes the pressure side and the pressure side becomes the suction side when gear movement is changed to the opposite direction.

➡ As a convenience, on some models an auxiliary tilt switch is installed on the exterior of the engine case.

When the up portion of the tilt switch is depressed, the tilt motor works in a clockwise direction. Pump hydraulic pressure then opens the up pressure main check valve while moving the spool valve to open the down pressure main check valve. Oil at pump pressure then flows through the up pressure main check valve and outlet tube to the down side of the tilt cylinder. Oil pressure raises the piston in the cylinder forcing oil from the up side of the cylinder to return to the valve body through the outlet tube, the down pressure main check valve and the spool valve chamber. Because of the difference in volume between the up and down side of the cylinder, the down throttling check valve opens to add oil from the reservoir to allow a complete upward stroke. When the tilt cylinder reaches full stroke and the up tilt operation is continued, oil flow from the pump will return to the reservoir through the up blow valve.

When the down portion of the tilt switch is depressed, the tilt motor works in a counterclockwise direction. Pump hydraulic pressure then opens the down pressure main check valve while moving the spool valve to open the down pressure main check valve. Oil at pump pressure then flows through the down pressure main check valve and outlet tube to the up side of the tilt cylinder. Oil pressure lowers the piston in the cylinder forcing oil from the down side of the cylinder to return to the valve body through the inlet tube, the up pressure main check valve and the spool valve chamber. When the tilt cylinder reaches full down position and down tilt operation is continued, oil flow from the pump will return to the reservoir through the down throttling valve.

In the event the outboard lower unit should strike an underwater object while the boat is underway, a sudden high impact pressure is created in the up side of the tilt cylinder. This pressure opens the shock absorber valve in the tilt cylinder piston allowing oil to transfer to the down side of the piston and opens the down check valve allowing oil to return to the reservoir. The opening of these valves enable the tilt rod to extend.

A manual relief valve allows easy manual tilt of the outboard should electric power be lost. The valve opens when the screw is turned counterclockwise, allowing fluid to flow through the manual valve. When the relief valve screw is turn fully clockwise, the manual valve closed and the outboard locked in position.

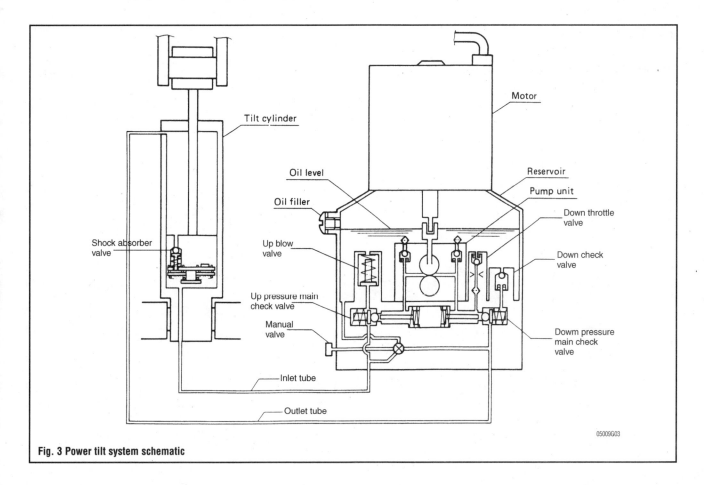

Fig. 3 Power tilt system schematic

9-4 TRIM AND TILT

Troubleshooting the Power Tilt System

Any time a problem develops in the power tilt system the first step is to determine whether it is electrical or hydraulic in nature. After the determination is made, then the appropriate steps can be taken to remedy the problem.

The first step in troubleshooting is to make sure all the connectors are properly plugged in and that all the terminals and wires are free of corrosion. The simple act of disconnecting and connecting a terminal may sometimes loosen corrosion that preventing a proper electrical connection. Inspect each terminal carefully and coat each with dielectric grease to prevent corrosion.

The next step is to make sure the battery is fully charged and in good condition. While checking the battery, perform the same maintenance on the battery cables as you did on the electrical terminals. Disconnect the cables (negative side first), clean and coat them and then reinstall them. If the battery is past its useful life, replace it. If it only requires a charge, charge it.

Check the power tilt fuse (as appropriate). Many systems will have a fuse to prevent large current draws from damaging the system. If this fuse is blown, the system will cease to function. This is a good indicator that you may have problems elsewhere in the electric system. Fuses don't blow without cause.

After inspecting the electrical side of the system, check the hydraulic fluid level and top it off as necessary. Remember to position the motor properly (full tilt up or down) to get an accurate measurement of fluid level. A slight decrease in the level of hydraulic fluid may cause the system to act sporadically.

Finally, make sure the manual release valve is in the proper position. A slightly open manual release valve may prevent the system from working properly and mimic other more serious problems.

Just remember to check the simple things first. If these simple tests do not diagnose the cause of the problem, then it is time to investigate more deeply. Perform the hydraulic pressure tests in this section to determine if the pump is making adequate pressure. Inspect the entire power tilt electrical harness with a multimeter, checking for excessive resistance and proper voltage.

Tilt Motor and Pump

CHECKING FLUID LEVEL

♦ See Figures 4 and 5

➡ On late model tilt units (from serial # 613245), the engine must be in the fully up tilt position prior to checking fluid level. On early model units (up to serial # 613244), the outboard must be in the full tilt down position. If you are unsure as to which model you own, refer to your owners manual for specific instructions.

1. Place the outboard in the correct position.
2. Remove the oil plug on the side of the pump and visually inspect the fluid level. The hydraulic fluid level should be at the lower edge of the oil plug hole. If the level is below specification, add Automatic Transmission Fluid (ATF) until the level is as specified.
3. After adding hydraulic fluid, bleed the system to remove air from the reservoir and fluid passages.

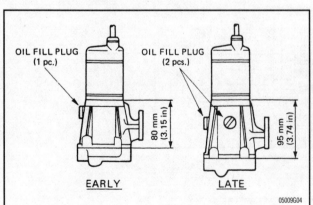

Fig. 4 Oil plug location and identification of early and late model pumps

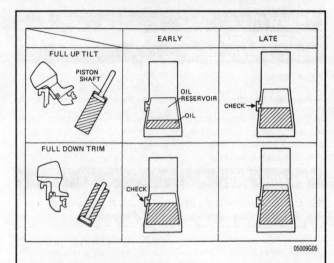

Fig. 5 Insure the powerhead is in the proper position prior to checking the hydraulic fluid

BLEEDING THE SYSTEM

♦ See Figure 6

➡ The hydraulic bleeding sequence should be performed whenever the tilt assembly has been removed for service.

1. Place the outboard in the fully down position.
2. Remove the filler plug and, as necessary, fill the reservoir to the bottom of the filler plug hole with automatic transmission fluid.
3. Wipe the area around the oil filler plug to remove any dirt that may contaminate the power tilt unit.
4. Remove the oil filler plug and install an oil filler tank adapter (55181-95200) onto the reservoir. This adapter attaches to an external reservoir to keep the hydraulic unit filled with fluid during the bleeding process.

➡ If an oil filler tank adapter is not available, you can bleed the system by keeping the reservoir constantly fill after each time the outboard is raised and lowered.

5. Raise the motor up to the full tilt position by operating the tilt switch and support the motor with the tilt lock arm.
6. Turn the manual release valve two full turns counterclockwise. Do not over rotate the manual release valve.

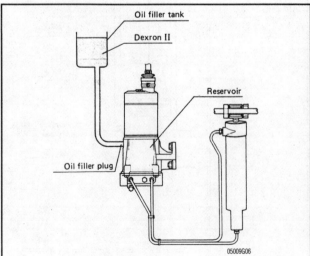

Fig. 6 The best way to bleed the system is to install an oil filler tank and adapter to the pump

TRIM AND TILT 9-5

7. Release the tilt lock arm and allow the motor to fall to the full down position slowly.
8. Repeat the last three steps until no more air bubbles are seen in the line between the oil tank and the external reservoir.
9. Place the motor in the full tilt position (from serial # 613245) or the full down position (up to serial # 613244) and remove the adapter.
10. Add fluid until the level reaches the bottom of the oil filler hole.
11. Tighten the filler plug to 2–3 ft. lbs. (3–5 Nm).
12. Raise and lower the motor 5–6 times using the power tilt switch.

TESTING

Electric Motor

1. Ensure the manual release valve is in the manual tilt position.
2. Disconnect the tilt motor wiring harness at the quick connect fittings.
3. Using jumper cables, momentarily make contact between the disconnected leads and a fully charged battery.

➡ Make the contact only as long as necessary to hear the electric motor rotating.

4. Reverse the leads on the battery posts and again listen for the sound of the motor rotating. The motor should rotate with the leads making contact with the battery in either direction.
5. If the motor operates properly, the problem may be in the tilt switch or associated wiring.
6. If the motor does not operate as specified, it may be faulty.

Hydraulic Pump

♦ See Figure 7

TILT DOWN PRESSURE TEST

1. Remove the outlet tube from ports "A" and "B" as shown in the illustration.
2. Connect a minimum 3000 psi test gauge between the ports.
3. Operate the power tilt switch down until the tilt rod reaches the bottom of its stroke.
4. Tilt down pressure should be 430–1000 psi (3000–7000 kPa).
5. If the pressure reading does not remain steady or does not reach specified pressure, the pump and valve body may be faulty.

TILT UP PRESSURE TEST

1. Remove the outlet tube from ports "C" and "D" as shown in the illustration.
2. Connect a minimum 3000 psi test gauge between the ports.
3. Operate the power tilt switch up until the tilt rod reaches the full stroke.

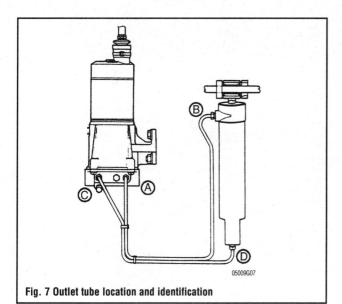

Fig. 7 Outlet tube location and identification

4. Tilt down pressure should be 1850–2420 psi (13000–17000 kPa).
5. If the pressure reading does not remain steady or does not reach specified pressure, the pump and valve body may be faulty.

➡ After the following test have been performed, check the tube connections for leaks, check the oil level in the reservoir and perform a system bleeding.

REMOVAL & INSTALLATION

The tilt motor and hydraulic pump assembly is mounted on the side of the stern bracket.
1. Label and disconnect the motor electrical harness.
2. Place a drain pan under the hydraulic lines to catch any spilled fluid.
3. Label the hydraulic lines for proper positioning.
4. Use a flare nut wrench to loosen the fittings. Disconnect the lines from the hydraulic unit and cap them to prevent the entry of dirt.
5. Remove the bolts/nuts mounting the pump to the stern bracket, then lift the pump from the bracket.

To install:
6. Position the pump on the stern bracket and install the mounting bolts/nuts.
7. Connect the lines to the hydraulic unit and using a flare nut wrench, tighten the fittings securely.
8. Connect the motor electrical harness.
9. Fill the fluid reservoir to capacity and bleed the air from the system.
10. Check the power tilt system for proper operation.

OVERHAUL

Overhaul procedures for the power tilt motor are identical to those for the externally mounted power trim/tilt motor. Refer to the procedure for externally mounted power trim/tilt motor for further information

Tilt Cylinder

REMOVAL & INSTALLATION

1. Turn the manual valve screw to allow the outboard to be manually tilted. Position and support the outboard in its fully tilted position.
2. Label and disconnect the wiring harnesses and feed them through the hole in the stern bracket.
3. Remove the snapring and slide the upper cylinder pin from the mounting frame.
4. Remove the assembly from the stern bracket.

➡ Store the assembly vertically with the upper cylinder bushing upward. Never store the damper horizontally with the distance collar facing upward.

To install:
5. Lubricate and install the cylinder bushings.
6. Position the assembly on the stern bracket.
7. Install the distance collar and through bolt on the stern bracket. Tighten the through bolt to 25 ft. lbs. (35Nm).
8. Pivot the assembly toward the mounting frame.
9. After positioning the wave washers, lubricate and install the upper cylinder pin on the mounting frame. Install the E-ring to secure the pin in place.
10. Feed the assembly wiring harnesses through the hole in the stern bracket and connect them.
11. Install the self locking nut and tighten to 25 ft. lbs. (35Nm). Install the steering tube cap.
12. Install the outboard on the boat and test the power tilt assembly for proper operation.

OVERHAUL

Overhaul procedures for the power tilt cylinder are identical to those for the power trim/tilt cylinders. Refer to the procedure for power trim/tilt cylinders for further information.

9-6 TRIM AND TILT

➡ Overhaul procedures for the tilt cylinder is confined to removal of the end cap, removing the piston and replacing the O-rings. A pin wrench or spanner wrench is required to remove the end cap. Even with the tool, removal of the end cap is not a simple task. The elements, especially if the unit has been used in a salt water atmosphere, will have their corrosive affect on the threads. Any attempt to break the end cap loose may very likely elongate the two holes provided for the wrench. Once the holes are damaged, all hope of removing the end are lost. The only solution in such a case is to replace the cylinder as a unit.

Tilt Switch

Complete diagnosis, testing and servicing procedures for the tilt switch are located in the "Remote Control" section of this manual.

POWER TRIM/TILT

Description and Operation

♦ See Figures 8, 9 and 10

The power trim/tilt systems consist of a housing with an electric motor, gear driven hydraulic pump, hydraulic reservoir and at least two trim/tilt cylinders. The cylinders perform a double function as trim/tilt cylinders and also as a shock absorbers, should the lower unit strike an underwater object while the boat is underway.

The necessary valves, check valves, relief valves, and hydraulic passageways are incorporated internally and externally for efficient operation. A manual release valve is provided to permit the outboard unit to be raised or lowered should the battery fail to provide the necessary current to the electric motor or if a malfunction should occur in the hydraulic system.

The gear driven pump operates in much the same manner as an oil circulation pump installed on motor vehicles. The gears rotate in either direction, depending on the desired cylinder movement. One side of the pump is considered the suction side, and the other the pressure side, when the gears rotate in a given direction. These sides are reversed, the suction side becomes the pressure side and the pressure side becomes the suction side when gear movement is changed to the opposite direction.

Depending on the model, up to two relays may be used for the electric motor. The relays are usually located at the bottom cowling pan, where they are fairly well protected from moisture.

Fig. 8 Typical external pump power trim/tilt system. The pump (right) is mounted to the stern bracket and the cylinders (left) are positioned under the outboard

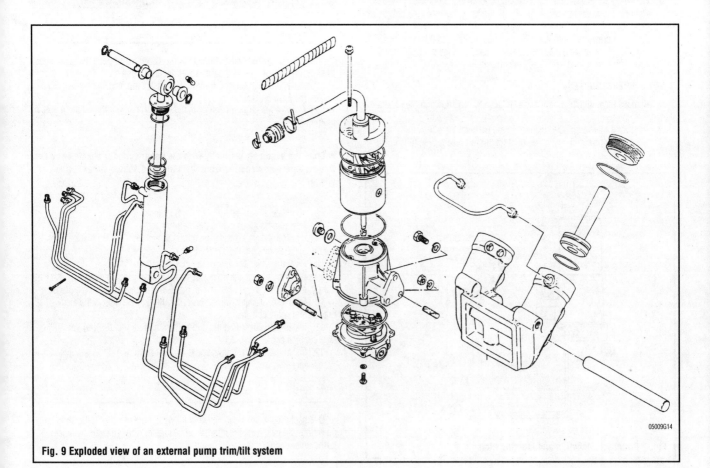

Fig. 9 Exploded view of an external pump trim/tilt system

TRIM AND TILT

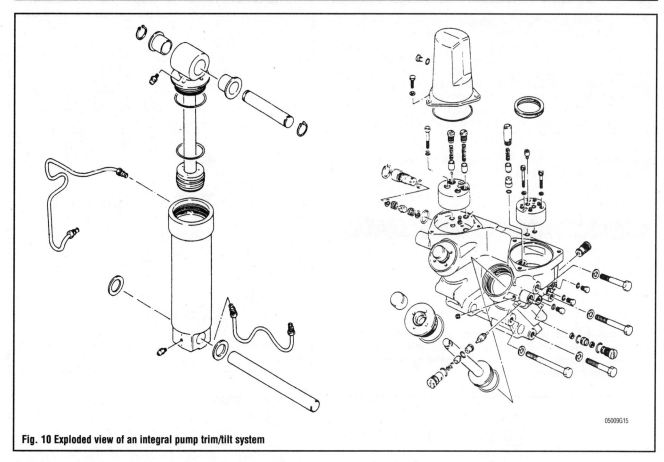

Fig. 10 Exploded view of an integral pump trim/tilt system

➥As a convenience, on some models an auxiliary trim/tilt switch is installed on the exterior cowling.

When the up portion of the trim/tilt switch is depressed, the up circuit, through the relay, is closed and the electric motor rotates in a clockwise direction. Pressurized oil from the pump passes through a series of valves to the lower chamber of the trim cylinders, the pistons are extended and the outboard unit is raised. The fluid in the upper chamber of the pistons is routed back to the reservoir as the piston is extended. When the desired position for trim is obtained, the switch on the control handle is released and the outboard is held stationary.

If the trim cylinder pistons should become fully extended, such as in a tilt up situation, fluid pressure in the lower chamber of the trim cylinders increases. This increase in pressure opens an up relief valve and the fluid is routed to the reservoir. The sound of the electric motor and the pump will have a noticeable change.

When the down portion of the trim/tilt switch is depressed, the down circuit, through the relay, is closed and the electric motor rotates in a counterclockwise direction. The pressure side of the pump now becomes the suction side and the original suction side becomes the pressure side. Pressurized oil from the pump passes through a series of valves to the upper chamber of the trim cylinders, the pistons are retracted and the outboard unit is lowered. The fluid in the lower chamber of the pistons is routed back to the reservoir as the retracted is extended. When the desired position for trim is obtained, the switch on the control handle is released and the outboard is held stationary.

If the trim cylinder pistons should become fully retracted, such as in a tilt down situation, fluid pressure in the upper chamber of the trim cylinders increases. This increase in pressure opens an up relief valve and the fluid is routed to the reservoir. The sound of the electric motor and the pump will have a noticeable change.

In the event the outboard lower unit should strike an underwater object while the boat is underway, the tilt piston would be suddenly and forcibly extended, moved upward. For this reason, the lower end of the tilt piston is capped with a free piston. This free piston normally moves up and down with the tilt piston.

The free piston also moves upward but at a much slower rate than the tilt piston. The action of the tilt piston separating from the free piston causes two actions. First, the hydraulic fluid in the upper chamber above the piston is com-

pressed and pressure builds in this area. Second, a vacuum is formed in the area between the tilt piston and the free piston.

This vacuum in the area between the two pistons sucks fluid from the upper chamber. The fluid fills the area slowly and the shock of the lower unit striking the object is absorbed. After the object has been passed the weight of the outboard unit tends to retract the piston. The fluid between the tilt piston and the free piston is compressed and forced through check valves to the reservoir until the free piston reaches its original neutral position.

A manual relief valve, located on the stern bracket, allows easy manual tilt of the outboard should electric power be lost. The valve opens when the screw is turned counterclockwise, allowing fluid to flow through the manual passage. When the relief valve screw is turn fully clockwise, the manual passage is closed and the outboard lock in position.

A thermal valve is used to protect the trim/tilt motor and allow it to maintain a designated trim angle. Oil in the upper chamber is pressurized when force is applied to the outboard from the rear while cruising. Oil is directed through the right side check valve and activates the thermal valve to release oil pressure and lessen the strain on the motor and pump.

Troubleshooting the Power Trim/Tilt System

Any time a problem develops in the power trim/tilt system the first step is to determine whether it is electrical or hydraulic in nature. After the determination is made, then the appropriate steps can be taken to remedy the problem.

The first step in troubleshooting is to make sure all the connectors are properly plugged in and that all the terminals and wires are free of corrosion. The simple act of disconnecting and connecting a terminal may sometimes loosen corrosion that preventing a proper electrical connection. Inspect each terminal carefully and coat each with dielectric grease to prevent corrosion.

The next step is to make sure the battery is fully charged and in good condition. While checking the battery, perform the same maintenance on the battery cables as you did on the electrical terminals. Disconnect the cables (negative side first), clean and coat them and then reinstall them. If the battery is past its useful life, replace it. If it only requires a charge, charge it.

Check the power trim/tilt fuse (as appropriate). Many systems will have a

9-8 TRIM AND TILT

fuse to prevent large current draws from damaging the system. If this fuse is blown, the system will cease to function. This is a good indicator that you may have problems elsewhere in the electric system. Fuses don't blow without cause.

After inspecting the electrical side of the system, check the hydraulic fluid level and top it off as necessary. Remember to position the motor properly (full tilt up or down) to get an accurate measurement of fluid level. A slight decrease in the level of hydraulic fluid may cause the system to act sporadically.

Finally, make sure the manual release valve is in the proper position. A slightly open manual release valve may prevent the system from working properly and mimic other more serious problems.

Just remember to check the simple things first. If these simple tests do not diagnose the cause of the problem, then it is time to investigate more deeply. Perform the hydraulic pressure tests in this section to determine if the pump is making adequate pressure. Inspect the entire power trim/tilt electrical harness with a multimeter, checking for excessive resistance and proper voltage.

Trim/Tilt Pump

TESTING

External Pump

♦ See Figures 11, 12 and 13

➥Early tilt/trim system tests used a pressure gauge connected directly to the pump. While this method was adequate, later tests use a more system oriented approach and connect a pressure gauge inline between the pump and the cylinders.

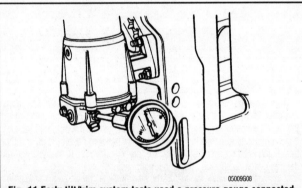

Fig. 11 Early tilt/trim system tests used a pressure gauge connected directly to the pump

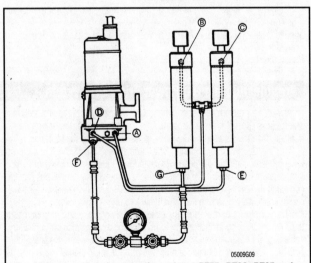

Fig. 12 Pressure gauge connection points—DT55, DT60, DT65 and 1992 and prior DT75 and DT85

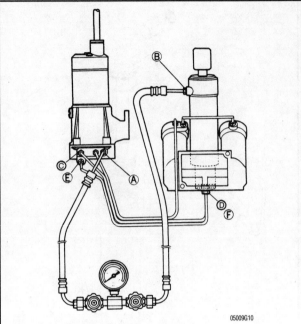

Fig. 13 Pressure gauge connection points—DT115 and DT140 and 1993 and later DT75 and DT85

TILT DOWN PRESSURE TEST

1. Remove the outlet tube from ports (A) and (B) as shown in the illustration.
2. Connect a minimum 3000 psi test gauge between ports (A) and (B).
3. Operate the power tilt switch down until the tilt rod reaches the bottom of its stroke.
4. Tilt down pressure should be 400–800 psi (3000–7000 kPa).
5. If the pressure reading does not remain steady or does not reach specified pressure, the pump may be faulty.
6. Remove the test gauge and connect the outlet tube securely.

TILT UP PRESSURE TEST

1. Remove the outlet tube from ports (C) and (D) as shown in the illustration.
2. Connect a minimum 3000 psi test gauge between the ports.
3. Operate the power tilt switch up until the tilt rod reaches the full stroke.
4. Tilt down pressure should be 1850–2400 psi (13000–17000 kPa).
5. If the pressure reading does not remain steady or does not reach specified pressure, the pump may be faulty.
6. Remove the test gauge and connect the outlet tube securely.
7. Perform the same test with ports (F) and (G).
8. After the following tests have been performed, check the tube connections for leaks, check the oil level in the reservoir and perform a system bleeding.

Integral Pump

♦ See Figure 14

TILT DOWN PRESSURE TEST

1. Raise the engine to the full tilt position and lower the tilt lock lever.
2. Remove the outlet tube from ports (A) and (B) as shown in the illustration.
3. Connect a minimum 3000 psi test gauge between the ports.
4. Close test gauge valve (B) and open test gauge valve (A.)
5. Operate the power tilt switch up until the tilt rod reaches the full stroke.
6. The specified oil pressure is 1280-1700 psi (8826-11720 kPa).
7. If the pressure reading does not remain steady or does not reach specified pressure, the pump may be faulty.

TILT UP PRESSURE TEST

1. Close test gauge valve (A) and open test gauge valve (B.)

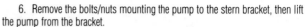
TRIM AND TILT 9-9

6. Remove the bolts/nuts mounting the pump to the stern bracket, then lift the pump from the bracket.

To install:

7. Position the pump on the stern bracket and install the mounting bolts/nuts.
8. Connect the lines to the hydraulic unit and using a flare nut wrench, tighten the fittings securely.
9. Connect the motor electrical harness.
10. Fill the fluid reservoir to capacity and bleed the air from the system.
11. Check the power tilt system for proper operation.

Integral Pump

♦ See Figure 15

The tilt/trim pump on these units is integral with the trim cylinder assembly and cannot be removed separately. If pump service is required, the entire trim/tilt assembly must be removed and disassembled to access the proper parts of the housing where the pump is located.

Trim/Tilt Motor

TESTING

1. Ensure the manual release valve is in the manual tilt position.
2. Disconnect the trim/tilt motor wiring harness at the quick connect fittings.

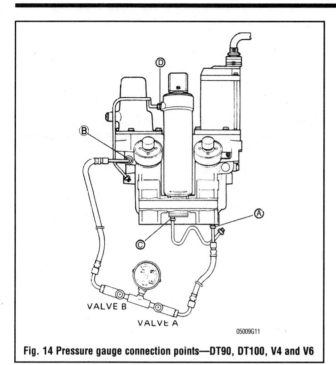

Fig. 14 Pressure gauge connection points—DT90, DT100, V4 and V6

Step 1

2. Operate the power tilt switch down until the tilt rod reaches the bottom of its stroke.
3. The specified oil pressure is 400-800 psi (2,800-5,600 kPa).
4. If the pressure reading does not remain steady or does not reach specified pressure, the pump may be faulty.
5. Check the oil level, refill and bleed the system as described in this chapter.
6. Remove the test gauge and connect the outlet tube securely.
7. After the following tests have been performed, check the tube connections for leaks, check the oil level in the reservoir and perform a system bleeding.

REMOVAL & INSTALLATION

External Pump

♦ See accompanying illustrations

1. External pumps are mounted on the stern bracket and easily serviced.
2. Label and disconnect the motor electrical harness.
3. Label the hydraulic lines for proper positioning.
4. Use a flare nut wrench to loosen the fittings.
5. Disconnect the lines from the hydraulic unit and cap them to prevent the entry of dirt.

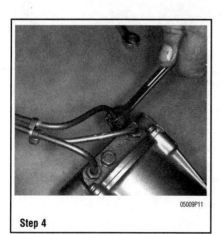

Step 4

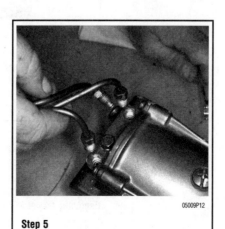

Step 5

Step 7

9-10 TRIM AND TILT

Fig. 15 Typical integral pump trim/tilt system used on a V4 or V6 outboard

3. Using jumper cables, momentarily make contact between the disconnected leads and a fully charged battery.

→Make the contact only as long as necessary to hear the electric motor rotating.

4. Reverse the leads on the battery posts and again listen for the sound of the motor rotating. The motor should rotate with the leads making contact with the battery in either direction.
5. If the motor operates properly, the problem may be in the trim/tilt switch or associated wiring.
6. If the motor does not operate as specified, it may be faulty.

REMOVAL & INSTALLATION

Integral Pump

♦ See Figure 16

→On some models it may be necessary to remove the trim/tilt assembly prior to removing the trim/tilt motor.

1. Raise the outboard to the full tilt position and safely support it in this position.
2. Remove the trim/tilt assembly as necessary to gain access to the trim/tilt motor.
3. Label and disconnect the trim/tilt motor wiring harness.
4. Remove motor-to-pump attaching screws.
5. Lift the motor from the manifold, noting the position of the drive joint and the O-ring prior to removing them.
6. Discard the O-ring.

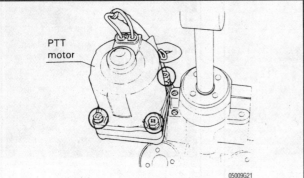

Fig. 16 Integral pump motors are usually attached to the hydraulic unit by four screws

To install:
7. Insure the mating surfaces of the motor and manifold are clean.
8. Install a new O-ring and position the drive joint in place.
9. Check the level of fluid in the gear pump area. Add fluid until the filter is covered.
10. Position the motor in place, insuring the drive joint is fully inserted into the gear pump assembly.

→When installing the motor, insure the tip of the armature shaft is firmly snugged into the drive joint. Double check this alignment prior to tightening the attaching screws.

11. Install and securely tighten the motor-to-pump attaching screws.
12. Connect the trim/tilt motor wiring harness.
13. Install the trim/tilt assembly if removed.
14. Lower the outboard.
15. Bleed the trim/tilt hydraulic system.
16. Check the trim/tilt system for proper operation.

External Pump

♦ See accompanying illustrations

→On some models it may be necessary to remove the trim/tilt assembly prior to removing the trim/tilt motor.

1. Raise the outboard to the full tilt position and safely support it in this position.
2. Remove the trim/tilt assembly as necessary to gain access to the trim/tilt motor.
3. Label and disconnect the trim/tilt motor wiring harness.
4. Remove motor-to-pump attaching screws.
5. Lift the motor from the manifold, noting the position of the drive joint and the O-ring prior to removing them.
6. Discard the O-ring.

To install:
7. Insure the mating surfaces of the motor and manifold are clean.
8. Install a new O-ring and position the drive joint in place.
9. Check the level of fluid in the gear pump area. Add fluid until the filter is covered.

Step 9

10. Position the motor in place, insuring the drive joint is fully inserted into the gear pump assembly.

→When installing the motor, insure the tip of the armature shaft is firmly snugged into the drive joint. Double check this alignment prior to tightening the attaching screws.

11. Install and securely tighten the motor-to-pump attaching screws.
12. Connect the trim/tilt motor wiring harness.

TRIM AND TILT 9-11

Step 10

13. Install the trim/tilt assembly if removed.
14. Lower the outboard.
15. Bleed the trim/tilt hydraulic system.
16. Check the trim/tilt system for proper operation.

DISASSEMBLY

Integral Pump

♦ See accompanying illustrations

1. Remove the screw attaching the wire holder and carefully pull the wires from the motor.
2. Remove the wire grommets from their position in the motor carefully.

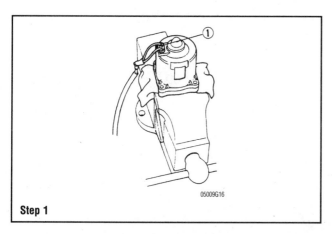
Step 1

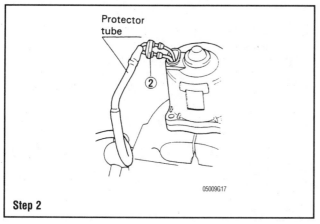
Step 2

3. Remove the screws attaching the field case to the rear brush cover.
4. Remove the field case with the armature installed from the rear brush cover.

➟ Push the motor wiring harness toward the case removing the armature assembly.

5. Remove the armature from the field case.
6. Once removed take care to keep the components clean by placing them on a clean sheet of paper. The smallest amount of dirt may hinder proper motor operation.

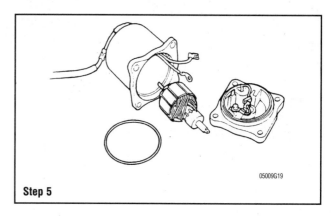
Step 5

External Pump

♦ See accompanying illustrations

1. Loosen the water tight connection that surrounds the wiring harness at the at the top of the motor and slide the nut and boot up the wire.
2. Remove the screws attaching the upper cover to the yoke assembly.

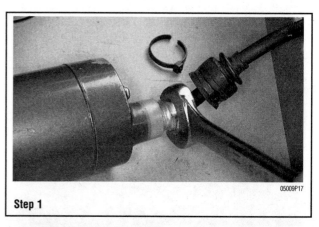
Step 1

Step 2

9-12 TRIM AND TILT

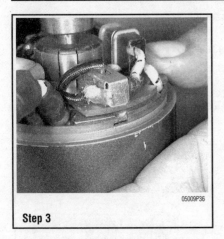
Step 3

Step 4

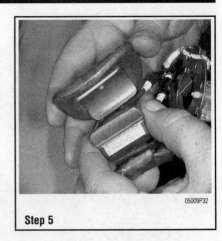
Step 5

3. Carefully lift the brush holder from the yoke.
4. Lift the armature from the yoke making sure to keep the commutator clean.

➡ On some models, the field windings are an integral part of the yoke and cannot be removed.

5. Carefully remove the brush holder and field windings from the yoke.
6. Remove the yoke from the pump housing and discard the O-ring.

CLEANING & INSPECTION

▶ See accompanying illustrations

1. Clean all components with a electrical contact cleaner and dry using compressed air.
2. Measure brush length using calipers. Standard brush length should be 0.51 in. (13mm) and the service limit is .035 in. (9mm). If brush length is beyond the service limit, replace the brushes.
3. Check for continuity between the brush and the terminal on the breaker assembly. Replace the breaker assembly if there is no continuity.

4. Inspect the mica depth on the commutator. If the mica depth is less than 0.020 in. (0.5mm) or the grooves are clogged, use a hacksaw blade or small file to deepen the grooves.
5. Check for continuity between each section of the armature. If an open circuit exists between any two segments of the armature it is faulty and should be replaced.
6. Check for continuity between the commutator and armature coil core. If continuity exists the armature is faulty and should be replaced.
7. Check for continuity between the commutator and armature shaft. If continuity exists the armature is faulty and should be replaced.

Step 2

Step 4

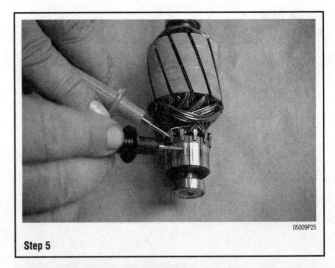
Step 5

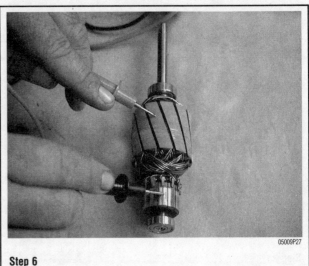
Step 6

TRIM AND TILT

ASSEMBLY

Integral Pump

▶ See accompanying illustrations

1. Install a new O-ring on the pump housing and fit the yoke in place.
2. Install the armature in the field case.

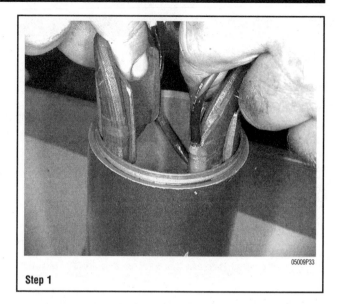
Step 1

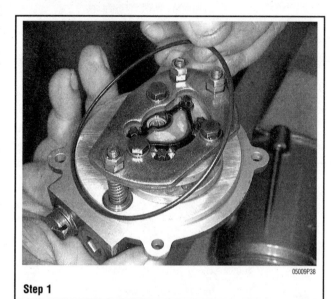
Step 1

➡ Pull the motor wires through the field case while installing the armature.

3. Check the armature for smooth rotation.
4. Install a new O-ring.
5. Install the rear brush cover on the field case.
6. Align any matchmarks on the field case and install the screws attaching the field case to the rear brush cover.
7. Install the wire grommets on the motor carefully.
8. Install the screw attaching the wire holder.

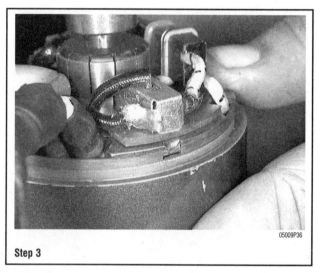
Step 3

4. Place the upper cover on the assembly and securely tighten the attaching screws.
5. Secure the water tight connection that surrounds the wiring harness at the at the top of the motor.

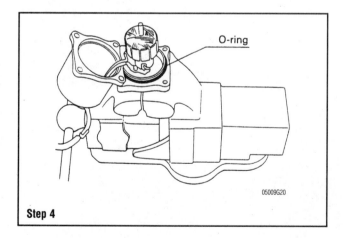
Step 4

External Pump

▶ See accompanying illustrations

1. Slide the field windings into the yoke, as applicable.
2. Slide the armature carefully into the yoke.
3. Place the brush holder over the armature shaft and using a pick to hold the brushes back.

Step 4

TRIM AND TILT

Trim/Tilt Cylinder

TESTING

▶ See Figure 14

➡ This test can only be performed on the integral pump models.

1. Raise the engine to the full tilt position and lower the tilt lock lever.
2. Remove the outlet tube from ports "A" and "C" as shown in the illustration.
3. Connect a minimum 3000 psi test gauge between the ports.
4. Operate the power tilt switch up until the tilt rod reaches the full stroke.
5. Close test gauge valve "B" before releasing the switch.
6. If the pressure reading does not remain steady, the tilt cylinder must be disassembled an inspected for a damaged seal ring, scored cylinder or shock valve problem. Replace if necessary.
7. Remove the test gauge and connect the tubes securely.
8. Remove the outlet tube from ports "B" and "D" as shown in the illustration.
9. Repeat the tests as previously performed.
10. After the following tests have been performed, check the tube connections for leaks, check the oil level in the reservoir and perform a system bleeding.

REMOVAL & INSTALLATION

External Pump

▶ See accompanying illustrations

1. Tilt the outboard to the full tilt position and secure it with the tilt lock levers.

➡ If the trim/tilt system is inoperable, open the manual release valve and manually raise the outboard.

2. Label and disconnect the tilt/trim electrical harness.
3. Disconnect and cap the hydraulic lines.
4. Remove the cotter pins and/or snap rings that hold the tilt rod pins in place.
5. Remove the tilt rod pins. It may be necessary to use a drift to push the pins through the bore.
6. Some tilt rods are held in by a single pin, while others use two separate pins.
7. Remove the fasteners holding the cylinder support bracket to the clamp bracket assembly.
8. Remove the cylinder support bracket with the cylinders attached.
9. Remove the lower cylinder shaft fasteners.
10. Slide the shaft from the support bracket and cylinders.
11. Remove the cylinders.

To install:

12. Install the cylinders.
13. Lubricate and install the lower cylinder shaft.
14. Install the lower cylinder shaft fasteners.
15. Install the cylinder support bracket with the cylinders attached.
16. Install and securely tighten the fasteners holding the cylinder support bracket to the clamp bracket assembly.
17. Lubricate and install the tilt rod pins.
18. Install the tilt rod pin fasteners.
19. Connect the hydraulic lines taking special care not to crossthread the fittings.

Step 4

Step 5

Step 6

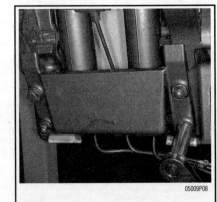
Step 7

Step 8

Step 10

Step 18

TRIM AND TILT 9-15

20. Connect the tilt/trim electrical harness.
21. Release the tilt lock levers and lower the outboard.
22. Adjust the fluid level in the pump and bleed the hydraulic system.
23. Check tilt/trim system for proper operation.

Integral Pump

♦ See accompanying illustration

1. Tilt the outboard to the full tilt position and secure it with the tilt lock levers.
2. It is also wise to provide some sort of auxiliary support to hold the outboard in place during servicing.

➡ If the trim/tilt system is inoperable, open the manual release valve and manually raise the outboard.

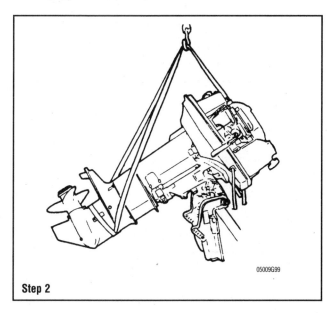
Step 2

3. Label and disconnect the trim/tilt assembly wiring harnesses.
4. Disconnect the negative, then the positive battery cables.
5. Remove the wiring harness stops and holders and feed the harness through the engine.
6. Disconnect the trim sensor.
7. Remove the snap ring at each end of the tilt cylinder upper pin.
8. Carefully drive the upper pin out.
9. Remove the clamp bracket pin nut and the tilt lock pin nut.
10. Remove the tilt/trim assembly.

To install:
11. Position the tilt/trim assembly in place and secure with the proper fasteners.
12. Lubricate and install the clamp bracket pin nut and the tilt lock pin nut.
13. Lubricate and install the upper pin and secure with a snap ring.
14. Feed the harness through the engine case and install the wiring harness stops and holders.
15. Connect the trim/tilt assembly wiring harnesses. Connect the trim sensor.
16. Connect the negative, then the positive battery cables.
17. Tilt the outboard to the full tilt position and secure it with the tilt lock levers.
18. Adjust the level of fluid in the hydraulic pump and bleed hydraulic system.

DISASSEMBLY

➡ Overhaul procedures for the trim/tilt cylinders are confined to removal of the end cap, removing the piston and replacing the O-rings. A pin wrench or spanner wrench is required to remove the end cap. Even with the tool, removal of the end cap is not a simple task. The elements, especially if the unit has been used in a salt water atmosphere, will have their corrosive affect on the threads. Any attempt to break the end cap loose may very likely elongate the two holes provided for the wrench. Once the holes are damaged, all hope of removing the end are lost. The only solution in such a case is to replace the cylinder as a unit.

1. Using a pin wrench or equivalent, loosen the cylinder cap.
2. Unscrew the cylinder cap on each rod and carefully remove the rods from the cylinders.
3. Remove and discard the O-rings on the piston rod.
4. Remove and discard the O-ring and oil seal on the cylinder cap.

CLEANING & INSPECTION

1. Inspect the piston for damage or wear and replace as necessary.
2. Inspect the manual valve for damage or wear and replace as necessary.
3. Inspect and clean the pump filter for damage and replace as necessary. The filter may be cleaned using compressed air.
4. Inspect the bore for grooving or damage and replace as necessary.

ASSEMBLY

1. Lubricate the bore and piston with trim/tilt fluid.
2. Install a new O-ring on the cylinder cap.
3. Install a new O-ring on the piston rod.
4. Place the cylinder cap on the rod and insert the rod into the bore.
5. Using a pin wrench or equivalent, tighten the cylinder cap securely.

Trim/Tilt Switch

Complete diagnosis, testing and servicing procedures for the trim/tilt switch are located in the "Remote Control" section of this manual.

Trim/Tilt Relay

TESTING

Two Terminal Relay

♦ See Figure 17

1. Remove the relay from the outboard.
2. Connect the relay to a 12 volt source and ground.
3. Check for continuity between the relay terminals.
4. Continuity should exist with 12 volts connected to the relay.
5. Continuity should not exist when 12 volts is removed from the relay.
6. If the relay does not function as stated, it may be faulty.

➡ Never assume a relay is bad, always test the relay prior to replacing it. If the relay tests good, check for opens or shorts in the wires connected to the relay.

Four Terminal Relay

♦ See Figure 18

1. Remove the relay from the outboard.
2. Check for continuity between the relay terminals.
3. Continuity should exist between terminals 1, 2 and 3. Continuity should not exist between terminals 3 and 4
4. Connect the relay to a 12 volt source and ground.
5. Continuity should now exist between terminals 2, 3 and 4.
6. If the relay does not function as stated, it may be faulty.

➡ Never assume a relay is bad, always test the relay prior to replacing it. If the relay tests good, check for opens or shorts in the wires connected to the relay.

REMOVAL & INSTALLATION

♦ See Figures 19 and 20

1. Most trim/tilt relays are located behind a plastic electronic cover under and mounted on the power head. Several types of fasteners are used including screws, tabs and grommets. When removing the relays, first disconnect the negative battery cables. Then label and disconnect the wires connected to the relay.

9-16 TRIM AND TILT

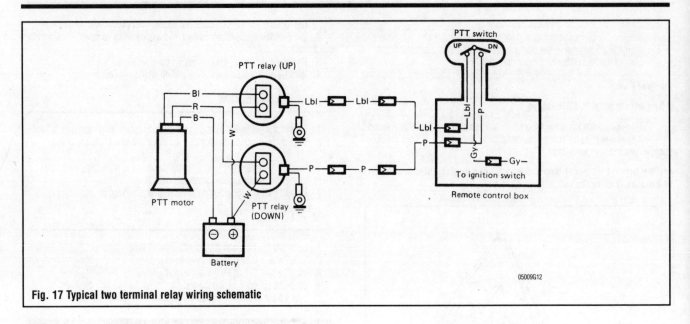

Fig. 17 Typical two terminal relay wiring schematic

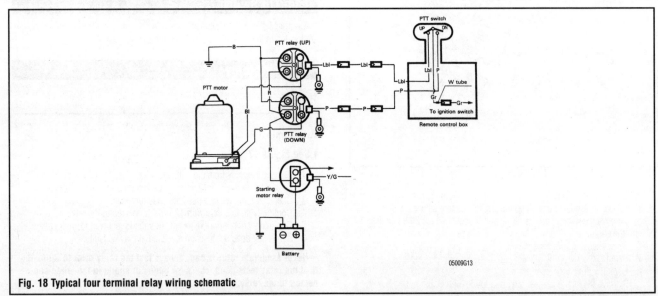

Fig. 18 Typical four terminal relay wiring schematic

Fig. 19 Two terminal relays mounted on the powerhead of a DT55

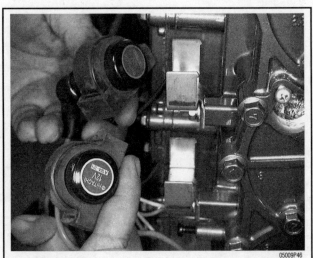

Fig. 20 Some relays are mounted in grommets which slip onto tabs secured to the powerhead

REMOTE CONTROL BOX 10-2
DESCRIPTION AND OPERATION 10-2
TROUBLESHOOTING THE REMOTE
 CONTROLS 10-2
REMOTE CONTROL BOX 10-2
 REMOVAL & INSTALLATION 10-2
REMOTE CONTROL CABLES 10-2
 REMOVAL & INSTALLATION 10-2
 ADJUSTMENT 10-2
NEUTRAL START SWITCH 10-2
 TESTING 10-2
 REMOVAL & INSTALLATION 10-2
ENGINE STOP SWITCH 10-5
 TESTING 10-5
 REMOVAL & INSTALLATION 10-5
EMERGENCY STOP SWITCH 10-5
 TESTING 10-5
 REMOVAL & INSTALLATION 10-5
IGNITION SWITCH 10-6
 TESTING 10-6
 REMOVAL & INSTALLATION 10-6
TRIM/TILT SWITCH 10-6
 TESTING 10-6
 REMOVAL & INSTALLATION 10-6
WARNING BUZZER 10-7
 TESTING 10-7
 REMOVAL & INSTALLATION 10-7
TILLER HANDLE 10-7
DESCRIPTION AND OPERATION 10-7
TROUBLESHOOTING THE TILLER
 HANDLE 10-8
TILLER HANDLE 10-8
 REMOVAL & INSTALLATION 10-8
THROTTLE CABLE 10-10
 REMOVAL & INSTALLATION 10-10
 ADJUSTMENT 10-10
ENGINE STOP SWITCH 10-10
 TESTING 10-10
 REMOVAL & INSTALLATION 10-10
EMERGENCY STOP SWITCH 10-10
 TESTING 10-10
 REMOVAL & INSTALLATION 10-10
ENGINE START SWITCH 10-10
 TESTING 10-10
 REMOVAL & INSTALLATION 10-10

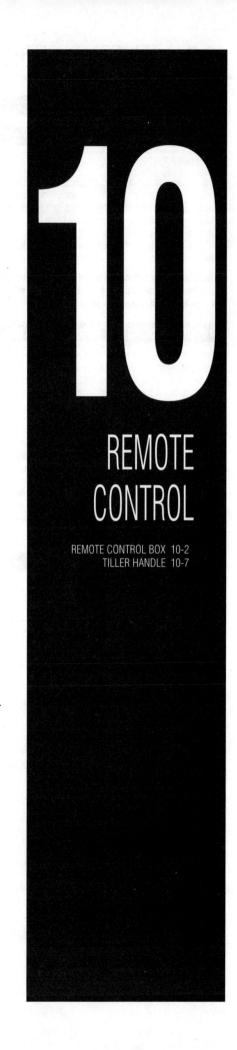

10
REMOTE CONTROL

REMOTE CONTROL BOX 10-2
TILLER HANDLE 10-7

10-2 REMOTE CONTROL

REMOTE CONTROL BOX

Description and Operation

♦ See Figure 1

The remote control box allows the person at the helm to control the throttle operation and shift movements from a location other than where the outboard is mounted. In most cases, the remote control box is mounted approximately halfway forward (midship) on the starboard side of the boat.

The control box usually houses a key switch, engine stop switch, choke switch, neutral safety switch, warning buzzer and the necessary wiring and cable hardware to connect the control box to the outboard unit.

➡ There are many different types of remote control assemblies that can be installed on your boat. The remote control assemblies covered in this manual are the ones most commonly used by the manufacturer.

Troubleshooting the Remote Controls

One of the things taken for granted on most boats is the engine controls and cables. Depending on how they are originally routed, they will either last the life of the vessel or can be easily damaged. These cables should be routinely maintained by careful inspection for kinks or other damage and lubrication with a marine grade grease.

If the cables do not operate properly, have a helper operate the controls at the helm while you observe the cable and linkage operation at the powerhead. Make sure nothing is binding, bent or kinked. Check the hardware that secures the cables to the boat and powerhead to make sure they are tight. Inspect the clevis and cotter pins in the ends of the cables and also give some attention to the cable release hardware.

Another area of concern is the neutral start switch that is sometimes located in the control box and sometimes on the powerhead. This switch may fall out of adjustment and prevent the engine from being started. It is easily inspected using a multimeter by performing a continuity check.

Remote Control Box

♦ See Figures 2, 3 and 4

REMOVAL & INSTALLATION

Side/Panel Mount

1. Remove the bolts/screws securing the control box to the boat.
2. Remove the rear cover of the control box.
3. Label and disconnect the control cables.
4. Label and disconnect the control box electrical harness.
5. Remove the control box from the boat.

To install:

6. Connect the control box electrical harness.
7. Connect the control cables and properly adjust them, as needed.
8. Install the rear cover of the control box, tightening the screws securely.
9. Position the control box in the boat.
10. Install and securely tighten the bolts/screws securing the control box to the boat.

Top Mount

1. Remove the remote control box housing.
2. Remove the bolts/screws securing the control box to the boat.
3. Label and disconnect the control cables.
4. Label and disconnect the control box electrical harness.
5. Remove the control box from the boat.

To install:

6. Connect the control box electrical harness.
7. Connect the control cables and properly adjust them, as needed.
8. Position the control box in the boat.
9. Install and securely tighten the bolts/screws securing the control box to the boat.
10. Install the remote control box housing.

Remote Control Cables

REMOVAL & INSTALLATION

♦ See Figure 5

1. Remove the engine cover.
2. Loosen the locknut and slacken the remote control cable adjustments at the throttle lever.
3. Disconnect the end of the remote control cables from the throttle lever.
4. Remove the screw attaching the throttle reel rod to the throttle grip.
5. Remove the remote control cables from the powerhead.
6. Remove the remote control cables attachments along the side of the boat.
7. Remove the remote control unit from the boat.
8. Remove the remote control back cover(s) and disconnect remote control cables.
9. Remove the remote control cables from the boat.

To install:

10. Position the remote control cables in the boat. Insure all bends are smooth and the cable is not kinked.
11. Install the remote control cable attachments along the side of the boat.
12. If the remote control cables are new, screw the cable ends on the cables approximately 0.43 in. (11mm)
13. Lubricate the cable end with a waterproof marine grease.
14. Connect the remote control cables to the remote control unit.
15. Lubricate the cable end with a waterproof marine grease.
16. Install the remote control cables into position on the engine case.
17. Connect the end of the remote control cables to the throttle lever.
18. Adjust the remote control cables and tighten the locknut securely.
19. Install the remote control back cover(s).
20. Install the remote control unit on the boat.
21. Check for proper throttle operation.
22. Install the engine cover.

ADJUSTMENT

♦ See Figure 6

There are several adjustments that can be made depending on the type of remote control box and engine size. General adjustments are aimed at making sure the remote control can be shifted into reverse and forward positions, full throttle can be achieved in the forward position and reverse throttle is restricted to 3000 rpm to prevent engine run away.

Since the number and type of adjustments vary with each remote control and engine combination, it is recommended that you refer to the rigging instructions for your individual combination when adjusting the remote control cables.

Neutral Start Switch

TESTING

♦ See Figure 7

1. Disconnect the neutral safety switch wiring harness.
2. Connect a multimeter between the switch harness leads.
3. With the remote control lever in the neutral position, continuity should exist. With the remote control lever in the forward or reverse position, continuity should not exist.
4. If the switch does not function as specified, there is a short in either the switch or harness and the switch should be replaced.
5. If the switch functions properly, there may be a problem in the powerhead wiring harness.

REMOVAL & INSTALLATION

1. Remove the control box from the side of the boat and open the side covers to allow access to the internal components.

REMOTE CONTROL 10-3

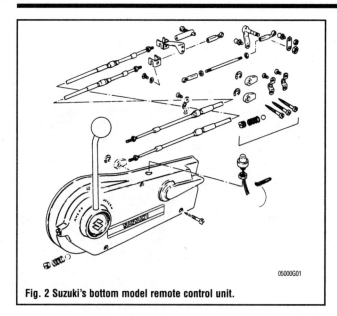

Fig. 2 Suzuki's bottom model remote control unit.

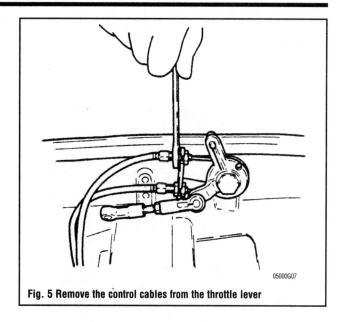

Fig. 5 Remove the control cables from the throttle lever

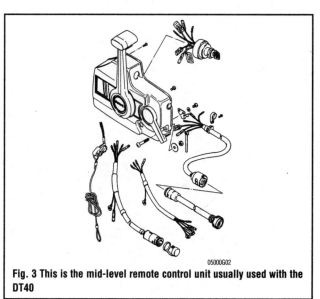

Fig. 3 This is the mid-level remote control unit usually used with the DT40

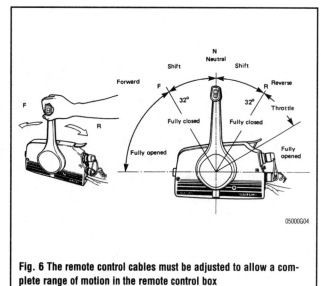

Fig. 6 The remote control cables must be adjusted to allow a complete range of motion in the remote control box

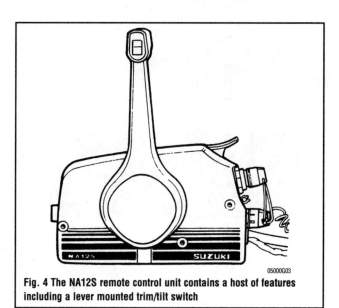

Fig. 4 The NA12S remote control unit contains a host of features including a lever mounted trim/tilt switch

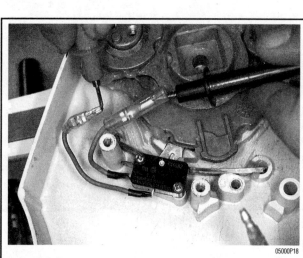

Fig. 7 With the remote control lever in the neutral position, continuity should exist. With the remote control lever in the forward or reverse position, continuity should not exist

10-4 REMOTE CONTROL

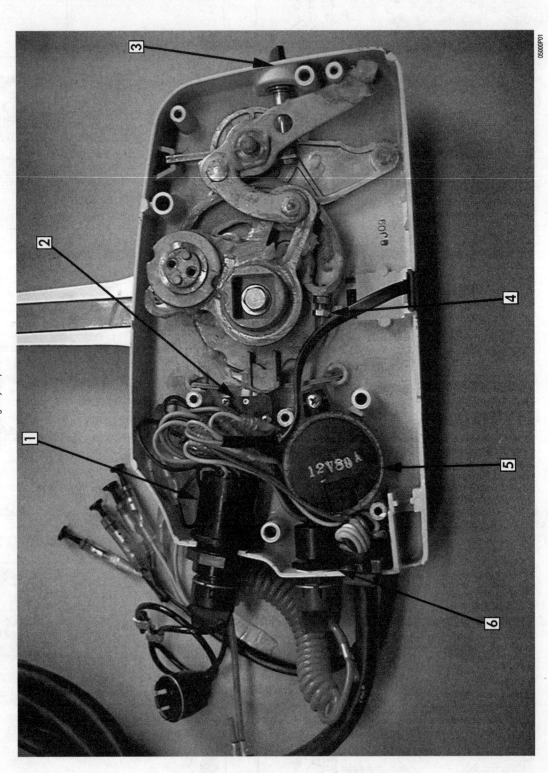

1. Ignition switch
2. Neutral start switch
3. Throttle tensioner
4. Reverse idle speed screw
5. Warning buzzer
6. Emergency stop switch

Fig. 1 The control box typically houses a key switch, engine stop switch, choke switch, neutral safety switch, warning buzzer and the necessary wiring and cable hardware

REMOTE CONTROL

2. Disconnect the neutral safety switch wiring harness.
3. Remove any wire straps that connect the switch to the control box.
4. Remove any retaining nuts/screws that secure the switch to the control box.
5. Remove the switch from the control box.

To install:
6. Install the switch on the control box.
7. As required, install the switch retaining nut/screw and tighten securely.
8. Install any wire straps that connect the switch to the control box or bracket.
9. Connect the neutral safety switch wiring harness.
10. Test the switch for proper operation.
11. Install the control box side covers and mount the control box in the boat.

Engine Stop Switch

TESTING

1. Disconnect the engine stop switch wiring harness.
2. Connect a multimeter between the switch harness leads.
3. With the switch engaged (button pushed), continuity should exist. With the switch released (button not pushed), continuity should not exist.
4. If the switch does not function as specified there is a short in either the switch or harness and the switch should be replaced.
5. If the switch functions properly, there may be a problem in the powerhead wiring harness.

REMOVAL & INSTALLATION

1. Remove the control box from the side of the boat and open the side covers to allow access to the internal components.
2. Disconnect the engine stop switch wiring harness.
3. Remove any wire straps that connect the switch to the control box.
4. Remove any retaining nuts/screws that secure the switch to the control box.

➡ Some switches are screwed into the end of the control box and simply unscrew from the handle.

5. Remove the switch from the control box.

To install:
6. Install the switch on the control box.
7. As required, install the switch retaining nut/screw and tighten securely.
8. Install any wire straps that connect the switch to the control box or bracket.
9. Connect the engine stop switch wiring harness.
10. Test the switch for proper operation.
11. Install the control box side covers and mount the control box in the boat.

Emergency Stop Switch

TESTING

♦ See Figure 8

1. Disconnect the engine stop switch wiring harness.
2. Connect a multimeter between the switch harness leads.
3. With the switch engaged (stop switch lanyard pulled), continuity should exist. With the switch released (stop switch lanyard in position), continuity should not exist.
4. If the switch does not function as specified there is a short in either the switch or harness and the switch should be replaced.
5. If the switch functions properly, there may be a problem in the powerhead wiring harness.

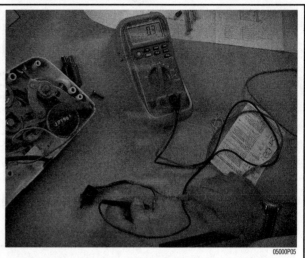

Fig. 8 Connect a multimeter between the switch harness leads. With the switch released (as shown), continuity should not exist

REMOVAL & INSTALLATION

♦ See Figures 9 and 10

1. Remove the control box from the side of the boat and open the side covers to allow access to the internal components.
2. Disconnect the emergency stop switch wiring harness.
3. Remove any wire straps that connect the switch to the control box.
4. Remove any retaining nuts/screws that secure the switch to the control box.

➡ Some switches are screwed into the end of the control box and simply unscrew from the handle.

5. Remove the switch from the control box.

To install:
6. Install the switch on the control box.
7. As required, install the switch retaining nut/screw and tighten securely.
8. Install any wire straps that connect the switch to the control box or bracket.
9. Connect the emergency stop switch wiring harness.

Fig. 9 The emergency switch is held in place on the remote control box by a locknut

10-6 REMOTE CONTROL

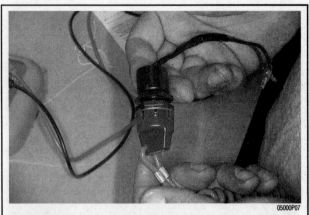

Fig. 10 Here is the complete switch assembly (switch and lanyard) after being removed from the control box

2. Disconnect the engine stop switch wiring harness.
3. Remove any wire straps that connect the switch to the box.
4. Remove the switch attaching nut.
5. Remove the switch from the control box.

To install:
6. Install the switch in the control box.
7. Install the switch attaching nut and tighten securely.
8. Install any wire straps that connect the switch to the control box.
9. Connect the engine stop switch wiring harness.
10. Test the switch for proper operation.
11. Install the control box side covers and mount the control box in the boat.

Trim/Tilt Switch

TESTING

▶ See Figure 12

1. Disconnect the trim/tilt switch wiring harness.
2. Connect a multimeter between the switch harness terminals as illustrated.
3. With the switch in the stated positions, check for continuity between the various terminals.
4. If the switch functions properly, there may be a problem in the powerhead wiring harness.
5. If the switch does not function as specified, the switch may be faulty.

10. Test the switch for proper operation.
11. Install the control box side covers and mount the control box in the boat.

Ignition Switch

TESTING

▶ See Figure 11

1. Disconnect the ignition switch wiring harness.
2. Connect a multimeter between the switch harness leads.
3. Consult the wiring diagrams for proper test positions and wiring colors.

➡**Continuity in the wiring diagram is indicated by two dots with a line drawn connecting them.**

4. With the switch in the stated positions, check for continuity between the various terminals.
5. If the switch functions properly, there may be a problem in the powerhead wiring harness.
6. If the switch does not function as specified, the switch may be faulty.

REMOVAL & INSTALLATION

1. Remove the control box from the side of the boat and open the side covers to allow access to the internal components.

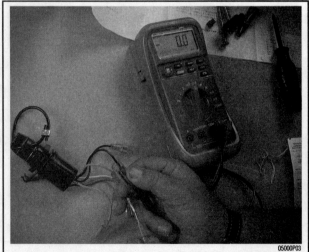

Fig. 11 Connect a multimeter between the switch harness leads and test for continuity

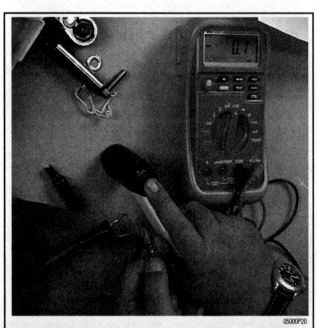

Fig. 12 Connect a multimeter between the switch harness terminals as illustrated and check for continuity between the terminals with the switch in the up and down position

REMOVAL & INSTALLATION

▶ See Figure 13

1. Remove the control box from the side of the boat and open the side covers to allow access to the internal components.
2. Remove the bolt and washer that retain the remote control lever to the control box.
3. Remove the remote control lever taking care to not pull on the tilt/trim wiring harness.
4. Label and disconnect the tilt/trim wiring harness.
5. Remove the screw securing the neutral release lever.
6. Remove the neutral release lever and lever spring.
7. Remove the screw securing the control lever grip.

REMOTE CONTROL 10-7

Fig. 13 To release the trim/tilt harness from the remote control box, pinch this connector with needle nose pliers

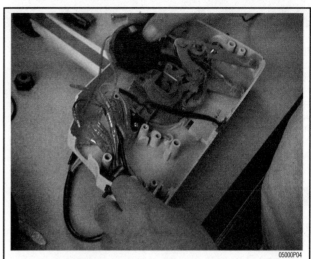

Fig. 14 The warning buzzer must be removed from the remote control box prior to testing

8. Remove the control lever grip from the control lever taking care to not pull on the tilt/trim wiring harness as you thread it though the holes in the control lever.
9. Remove the tilt/trim switch from the lever grip.

To install:
10. Install the tilt/trim switch in the lever grip.
11. Thread the tilt/trim wiring harness through the control lever and install the control lever grip. Secure the grip with the screw.
12. Install the neutral release lever and lever spring, securing them with the setting plate and screw.
13. Connect the tilt/trim wiring harness.
14. Install the remote control lever securing them with the bolt and washer to the remote control box.
15. Install the control box from the side of the boat and open the side covers to allow access to the internal components.

Warning Buzzer

TESTING

▶ See Figures 14 and 15

1. The warning buzzer is tested by simply connecting it to a 12 volt power source.
2. The buzzer should sound when properly connected to power.
3. If the buzzer functions properly, there may be a problem in the wiring harness.
4. If the buzzer does not perform as stated, it may be faulty.

REMOVAL & INSTALLATION

1. Remove the control box from the side of the boat and open the side covers to allow access to the internal components.

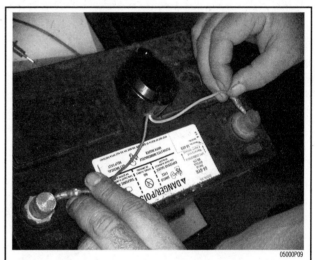

Fig. 15 The warning buzzer is tested by simply connecting it to a 12 volt power source

2. Disconnect the buzzer wiring harness.
3. Remove any wire straps that connect the buzzer wiring harness to the box.
4. Remove the buzzer from the control box.

To install:
5. Install the buzzer in the control box.
6. Install any wire straps that connect the buzzer harness to the control box.
7. Connect the buzzer wiring harness.
8. Test the buzzer for proper operation.
9. Install the control box side covers and mount the control box in the boat.

TILLER HANDLE

Description and Operation

Steering control for most outboards begins at the tiller handle and ends at the propeller. Tiller steering is the most simple form of small outboard control. All components are mounted directly to the engine and are easily serviceable.

Throttle control is performed via a throttle grip mounted to the tiller arm. As the grip is rotated a cable opens and closes the throttle lever on the engine. An adjustment thumbscrew is usually located near the throttle grip to allow adjustment of the turning resistance. In this way, the operator does not have to keep constant pressure on the grip to maintain engine speed.

An emergency engine stop switch is used on most outboards to prevent the engine from continuing to run without the operator in control. This switch is controlled by a small clip which keeps the switch open during normal engine operation. When the clip is removed, a spring inside the switch closes it and completes a ground connection to stop the engine. The clip is connected to a lanyard that is worn around the helmsman's wrist.

Some tiller systems utilize a throttle stopper system which limits throttle opening when the shift lever is in neutral and reverse. This prevents overrevving the engine under no-load conditions and also limits the engine speed when in reverse.

10-8 REMOTE CONTROL

Troubleshooting the Tiller Handle

If the tiller steering system seems loose, first check the engine for proper mounting. Ensure the engine is fastened to the transom securely. Next, check the tiller hinge point where it attaches to the engine and tighten the hinge pivot bolt as necessary.

Excessively tight steering that cannot be adjusted using the tension adjustment is usually due to a lack of lubrication. Once the swivel case bushings run dry, the steering shaft will get progressively tighter and eventually seize. This condition is generally caused by a lack of periodic maintenance.

Correct this condition by lubricating the swivel case bushings and working the outboard back and forth to spread the lubricant. However, this may only be a temporary fix. In severe cases, the swivel case bushings may need to be replaced.

Tiller Handle

REMOVAL & INSTALLATION

Without Cable

1. Remove the engine cover.
2. Remove the tiller handle pivot bolt and carefully remove the tiller handle.
3. Remove the mounting collar, mounting rubber, distance collar and special washer from the tiller handle.

To install:

4. Install the mounting collar, mounting rubber, distance collar and special washer on the tiller handle.
5. Position the tiller handle on the outboard and install the tiller handle pivot bolt. Tighten the pivot bolt securely.
6. Check the handle for smooth operation.
7. Install the engine cover.

Single Cable

◆ See Figure 16

1. Remove the engine cover.
2. Place the throttle grip in the idle position on the tiller arm.
3. Loosen the locknut, slacken the adjustment on the throttle cable and remove the cable at the throttle lever.
4. Label and disconnect the engine stop switch wiring harness.
5. Remove the tiller handle through bolts.
6. Remove the rubber grommets from the stopper, then remove the stopper from the tiller handle.
7. Remove the tiller handle inner bushing and mount.
8. Slide the tiller handle from the engine case.

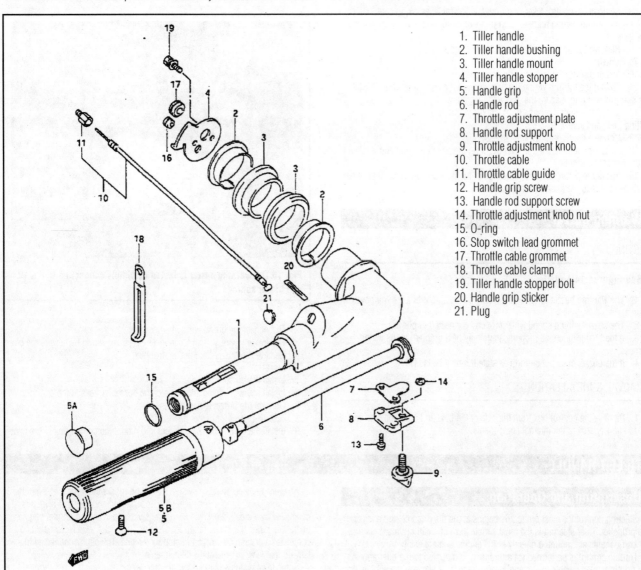

1. Tiller handle
2. Tiller handle bushing
3. Tiller handle mount
4. Tiller handle stopper
5. Handle grip
6. Handle rod
7. Throttle adjustment plate
8. Handle rod support
9. Throttle adjustment knob
10. Throttle cable
11. Throttle cable guide
12. Handle grip screw
13. Handle rod support screw
14. Throttle adjustment knob nut
15. O-ring
16. Stop switch lead grommet
17. Throttle cable grommet
18. Throttle cable clamp
19. Tiller handle stopper bolt
20. Handle grip sticker
21. Plug

Fig. 16 Exploded view of an engine case with a single cable tiller handle

REMOTE CONTROL   10-9

9. Carefully remove the tiller handle from the bracket.

To install:

10. Lubricate the throttle reel where the cable makes contact and the tiller handle where it goes through the engine case.
11. Carefully guide the throttle cable and wiring harness through the hole in the case and install tiller handle.
12. Install the tiller handle inner bushing and mount.
13. Install the stopper on the tiller handle.
14. Install the rubber grommets on the stopper.
15. Remove the tiller handle through bolts and tighten securely.
16. Connect the engine stop switch wiring harness.
17. Install the throttle cable and adjust the throttle cable tension. Tighten the locknut securely.
18. Operate the throttle grip and check for proper operation at full throttle and idle.
19. Install the engine cover.

Dual Cables

♦ See Figure 17

1. Remove the engine cover.
2. Place the throttle grip in the idle position on the tiller arm.
3. Loosen the locknut, slacken the adjustment on the throttle cables and remove the cables at the throttle lever.
4. Label and disconnect the engine stop switch wiring harness.
5. Remove the bolts/nuts attaching the tiller handle bracket bolts.
6. Remove the rubber grommets from the stopper, then remove the stopper from the tiller handle.
7. Remove the tiller handle inner bushing and mount.
8. Slide the tiller handle from the engine case.
9. Carefully remove the tiller handle from the bracket.

To install:

10. Lubricate the throttle reel where the cable makes contact and the tiller handle where it goes through the engine case.
11. Carefully guide the throttle cable and wiring harness through the hole in the case and install tiller handle.
12. Install the tiller handle inner bushing and mount.
13. Install the stopper on the tiller handle.
14. Install the rubber grommets on the stopper.
15. Remove the tiller handle through bolts and tighten securely.
16. Connect the engine stop switch wiring harness.
17. Install the throttle cable and adjust the throttle cable tension. Tighten the locknut securely.
18. Operate the throttle grip and check for proper operation at full throttle and idle.
19. Install the engine cover.

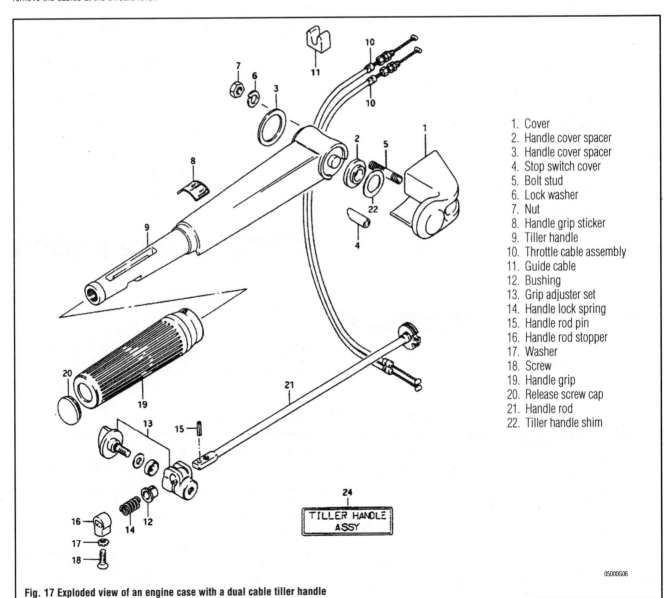

1. Cover
2. Handle cover spacer
3. Handle cover spacer
4. Stop switch cover
5. Bolt stud
6. Lock washer
7. Nut
8. Handle grip sticker
9. Tiller handle
10. Throttle cable assembly
11. Guide cable
12. Bushing
13. Grip adjuster set
14. Handle lock spring
15. Handle rod pin
16. Handle rod stopper
17. Washer
18. Screw
19. Handle grip
20. Release screw cap
21. Handle rod
22. Tiller handle shim

Fig. 17 Exploded view of an engine case with a dual cable tiller handle

10-10 REMOTE CONTROL

Throttle Cable

REMOVAL & INSTALLATION

1. Remove the engine cover.
2. Place the tiller handle in the up position with the throttle grip in the idle position.
3. Loosen the locknut and slacken the throttle cable adjustment at the throttle lever.
4. Disconnect the end of the throttle cable from the throttle lever.
5. Remove the screw attaching the throttle reel rod to the throttle grip.
6. Loosen the friction bolt on the throttle friction block.
7. Lower the throttle reel assembly from the tiller handle and disconnect the throttle cable.
8. Remove the throttle cable from the powerhead.

To install:

9. Install the throttle cable into position on the engine case.
10. Lubricate the throttle cable and throttle reel prior to installation.
11. Connect the throttle cable end to the throttle reel and route the cable around the reel.
12. Install the throttle reel assembly into the tiller arm and tighten the friction bolt to hold it in place.
13. Install the screw attaching the throttle reel rod to the throttle grip.
14. Connect the end of the throttle cable to the throttle lever.
15. Adjust the throttle cable tension and tighten the locknut securely.
16. Check for proper throttle operation.
17. Install the engine cover.

ADJUSTMENT

1. Remove the engine cover.
2. Loosen the locknut holding the cable adjusting nut in place.
3. Adjust cable(s) to zero slack with the throttle in the idle position.
4. Disable the ignition by disconnecting the negative battery cable and removing the spark plugs from the powerhead.
5. Place lower unit in forward and manually spin prop insure full engagement of the gears.
6. Remove the carburetor air box or cover.
7. Turn the throttle grip to the full throttle position.
8. Visually inspect the throttle plate(s) to insure they are centered. If the throttle plate(s) are not perfectly centered in the throttle bore, adjust the cable(s) to obtain perfect centering.
9. Tighten the locknut holding the cable adjusting nut in place.
10. Install the spark plugs and connect the negative battery cable.
11. Install the engine cover.
12. Test the engine at full throttle under load.

Engine Stop Switch

TESTING

1. Disconnect the engine stop switch wiring harness.
2. Connect a multimeter between the switch harness leads.
3. With the switch engaged (button pushed), continuity should exist. With the switch released (button not pushed), continuity should not exist.
4. If the switch does not function as specified there is a short in either the switch or harness and the switch should be replaced.
5. If the switch functions properly, there may be a problem in the powerhead wiring harness.

REMOVAL & INSTALLATION

1. Disconnect the engine stop switch wiring harness.
2. Remove any wire straps that connect the switch to the tiller handle or bracket.
3. Remove any retaining nuts/screws that secure the switch to the tiller handle.

➡ **Some switches are screwed into the end of the tiller handle and simply unscrew from the handle.**

4. Remove the switch from the tiller handle.

To install:

5. Install the switch on the tiller handle.
6. As required, install the switch retaining nut/screw and tighten securely.
7. Install any wire straps that connect the switch to the tiller handle or bracket.
8. Connect the engine stop switch wiring harness.
9. Test the switch for proper operation.

Emergency Stop Switch

TESTING

1. Disconnect the engine stop switch wiring harness.
2. Connect a multimeter between the switch harness leads.
3. With the switch engaged (stop switch lanyard pulled), continuity should exist. With the switch released (stop switch lanyard in position), continuity should not exist.
4. If the switch does not function as specified there is a short in either the switch or harness and the switch should be replaced.
5. If the switch functions properly, there may be a problem in the powerhead wiring harness.

REMOVAL & INSTALLATION

1. Disconnect the emergency stop switch wiring harness.
2. Remove any wire straps that connect the switch to the tiller handle or bracket.
3. Remove any retaining nuts/screws that secure the switch to the tiller handle.

➡ **Some switches are screwed into the end of the tiller handle and simply unscrew from the handle.**

4. Remove the switch from the tiller handle.

To install:

5. Install the switch on the tiller handle.
6. As required, install the switch retaining nut/screw and tighten securely.
7. Install any wire straps that connect the switch to the tiller handle or bracket.
8. Connect the emergency stop switch wiring harness.
9. Test the switch for proper operation.

Engine Start Switch

TESTING

1. Disconnect the engine stop switch wiring harness.
2. Connect a multimeter between the switch harness leads.
3. With the switch engaged (button pressed), continuity should exist. With the switch released (button not depressed), continuity should not exist.
4. If the switch does not function as specified there is a short in either the switch or harness and the switch should be replaced.
5. If the switch functions properly, there may be a problem in the powerhead wiring harness.

REMOVAL & INSTALLATION

1. Disconnect the engine start switch wiring harness.
2. Remove any wire straps that connect the switch to the tiller handle or bracket.
3. Remove any retaining nuts/screws that secure the switch to the engine case.
4. Remove the switch from the engine case.

To install:

5. Install the switch on the engine case.
6. As required, install the switch retaining nut/screw and tighten securely.
7. Install any wire straps that connect the switch wiring harness.
8. Connect the engine start switch wiring harness.
9. Test the switch for proper operation.

HAND REWIND STARTER 11-2
DESCRIPTION AND OPERATION 11-2
TROUBLESHOOTING THE HAND REWIND
    STARTER 11-2
OVERHEAD TYPE STARTER 11-2
DT2, DT2.2 AND DT4 11-2
    REMOVAL & INSTALLATION 11-2
    DISASSEMBLY 11-2
    CLEANING & INSPECTION 11-3
    ASSEMBLY 11-3
DT6 AND DT8 11-4
    REMOVAL & INSTALLATION 11-4
    DISASSEMBLY 11-4
    CLEANING & INSPECTION 11-4
    ASSEMBLY 11-5
DT9.9 AND DT15 11-6
    REMOVAL & INSTALLATION 11-6
    ADJUSTMENT 11-6
    DISASSEMBLY 11-7
    CLEANING & INSPECTION 11-7
    ASSEMBLY 11-7
DT20 AND DT25 11-8
    REMOVAL & INSTALLATION 11-8
    DISASSEMBLY 11-8
    CLEANING & INSPECTION 11-8
    ASSEMBLY 11-9
DT25C, DT30C, DT35C AND
  DT40C 11-9
    REMOVAL & INSTALLATION 11-9
    ADJUSTMENT 11-9
    DISASSEMBLY 11-9
    CLEANING & INSPECTION 11-10
    ASSEMBLY 11-10
BENDIX TYPE STARTER 11-10
DT6 AND DT8 11-10
    REMOVAL & INSTALLATION 11-10
    DISASSEMBLY 11-10
    CLEANING & INSPECTION 11-11
    ASSEMBLY 11-11
DT20 AND DT25 11-11
    REMOVAL & INSTALLATION 11-11
    DISASSEMBLY 11-11
    CLEANING & INSPECTION 11-12
    ASSEMBLY 11-12

11
HAND REWIND STARTER

HAND REWIND STARTER 11-2
OVERHEAD TYPE STARTER 11-2
BENDIX TYPE STARTER 11-10

11-2 HAND REWIND STARTER

HAND REWIND STARTER

Description and Operation

♦ See Figures 1, 2 and 3

The main components of a hand rewind starter (recoil starter) are the cover, rewind spring and pawl arrangement. Pulling the rope rotates the pulley, winds the spring and activates the pawl into engagement with the starter hub at the top of the flywheel. Once the pawl engages the hub, the powerhead is spun as the rope unwinds from the pulley.

Releasing the rope on rewind starter moves the pawl out of mesh with the hub. The powerful clock-type spring recoils the pulley in the reverse direction to rewind the rope to the original position.

Some starters may use a Neutral Start Interlock (NSI) system. The hand rewind starter should not function when the shift handle is in any other position than **NEUTRAL**. This prevents starting in gear and possibly throwing the occupants overboard. Always check for the proper function of this system after any repairs.

Troubleshooting The Hand Rewind Starter

Repair on hand rewind starter units is generally confined to rope, pawl and occasionally spring replacement.

※※ **CAUTION**

When replacing the recoil starter spring extreme caution must be used. The spring is under tension and can be dangerous if not released properly.

Starters which use friction springs to assist pawl action may suffer from bent springs. This will cause the amount of friction exerted to not be correct and the pawl will not be moved into engagement.

Models equipped with a neutral start interlock system may experience a no-start condition due to a misadjusted interlock cable. The hand rewind starter should only function when the shift handle is in the **NEUTRAL** position.

Fig. 1 A typical hand rewind starter assembly sits atop the outboard's flywheel

Fig. 2 On larger outboards with electric starters, an emergency hand-pull rope . . .

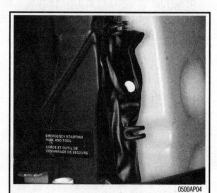

Fig. 3 . . . is usually mounted under the cowl

OVERHEAD TYPE STARTER

DT2, DT2.2 and DT4

REMOVAL & INSTALLATION

1. Remove the engine cover fasteners and lift off the engine covers.
2. Remove the recoil starter assembly attaching bolts and lift the assembly from the powerhead.

To install:

3. Place the recoil starter assembly on the powerhead and tighten the attaching nuts securely.
4. Install the engine covers and secure them with the cover fasteners.
5. Pull the starter knob several times and check for the proper operation of the ratcheting mechanism.

DISASSEMBLY

♦ See Figures 4, 5 and 6

1. Pull the starter rope out as far as it will come and hold the drum with your finger to prevent the rope from rewinding.
2. Hook the rope on the notch of the drum and gently rotate the drum clockwise to release spring tension.

➡ The coil spring will be will attempt to turn the drum quickly. Using your finger, prevent the drum from spinning.

3. With the rope and spring fully stretched and the drum turned all the way, remove the drum securing bolt.
4. Remove the plate that covers the drum and the ratcheting pawl.

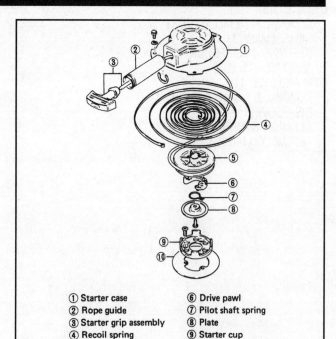

① Starter case
② Rope guide
③ Starter grip assembly
④ Recoil spring
⑤ Sheave drum
⑥ Drive pawl
⑦ Pilot shaft spring
⑧ Plate
⑨ Starter cup
⑩ Magneto insulator

Fig. 4 Exploded view of the starter assembly—DT2 and DT2.2

HAND REWIND STARTER 11-3

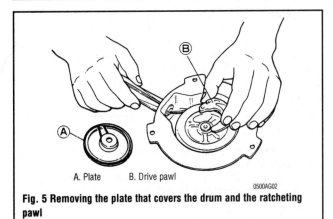

Fig. 5 Removing the plate that covers the drum and the ratcheting pawl

A. Plate   B. Drive pawl

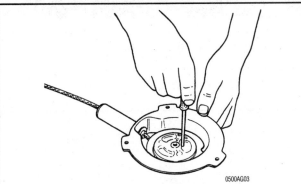

Fig. 6 If the drum will not lift out smoothly, disengage the spring from the drum using screwdriver inserted into the hole on drum

5. Remove the drum by lifting it gently from the housing. If the drum will not lift out smoothly, disengage the spring from the drum using screwdriver inserted into the hole on drum.

➡ It is advisable to wear heavy gloves while removing the spring to prevent your hands from being cut by the sharp spring steel.

✽✽✽ CAUTION

The starter drum spring is under high tension. If the spring should come loose, it may cause serious damage or personal injury. Take all applicable cautions when working with this spring.

6. Carefully remove the coil spring from the starter case.

CLEANING & INSPECTION

♦ See Figure 7

Clean all components and then blow them dry using compressed air. Remove any trace of corrosion and wipe all metal parts with an oil dampened cloth to prevent future corrosion.

Inspect the rope. Replace the rope if it appears to be weak or frayed. If the rope is frayed, check the holes through which the rope passes for rough edges or burrs. Remove the rough edges or burrs with a file and polish the surface until it is smooth.

Inspect the starter return spring end hooks. Replace the spring if it is weak, corroded or cracked. Inspect the inside surface of the starter case and drum for grooves or roughness. Grooves may cause erratic rewinding of the starter rope.

Inspect and lubricate the ratchet mechanism with waterproof grease. Check the mechanism for freedom of movement.

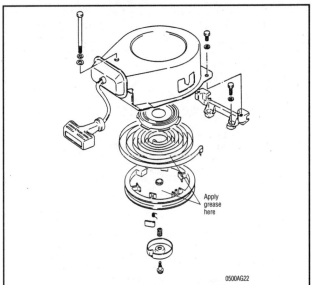

Fig. 7 Lubricate the starter assembly at the following locations— DT2, DT2.2 and DT4

ASSEMBLY

♦ See Figures 8, 9 and 10

1. Position the coil spring in the starter case, feeding the outer portion of the spring into the case first and positioning the remainder in the case gradually.

✽✽✽ CAUTION

During installation, the spring will be placed under high tension. Take all applicable cautions when working with this spring.

2. Place the bent end of the spring into the groove on the drum.

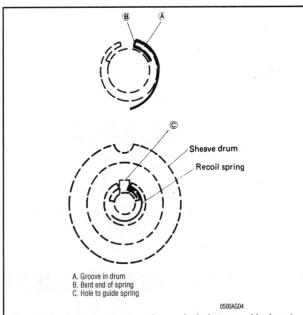

A. Groove in drum
B. Bent end of spring
C. Hole to guide spring

Fig. 8 If the spring is not engaged properly during assembly, insert a rod into the hole on the drum and guide the spring end into engagement with the drum

11-4 HAND REWIND STARTER

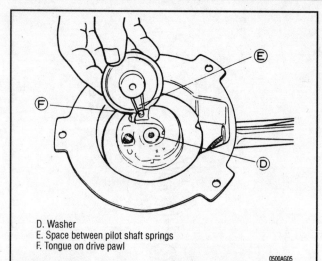

D. Washer
E. Space between pilot shaft springs
F. Tongue on drive pawl

Fig. 9 Ensure the tongue on the drive pawl is properly positioned into the space between the pilot shaft springs on the plate

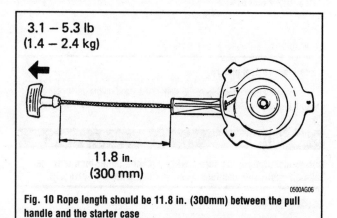

Fig. 10 Rope length should be 11.8 in. (300mm) between the pull handle and the starter case

3. Twist the drum counterclockwise to make sure the spring is positively engaged. If resistance is not felt, the spring may not be engaged properly. If this is the case, insert a rod into the hole on the drum and guide the spring end into engagement with the drum.
4. Install the washer, drive pawl and plate. Ensure the tongue on the drive pawl is properly positioned into the space between the pilot shaft springs on the plate.
5. Secure the drum in place by tightening the bolt securely.
6. Attach the pull rope in to the notch on the drum and turn the drum counterclockwise 6 rotations.
7. Remove the rope from the notch and gently wind it in under the force of the spring to take it up into the drum.
8. Measure the spring resistance using a fish scale. If should be 3.1–5.3 lbs. of pulling force.
9. If resistance is not within specification, adjust the amount of counterclockwise rotations made prior to allowing rope to rewind into the drum.
10. Measure the length of the rope at full extension. Rope length should be 11.8 in. (300mm) between the pull handle and the starter case.

DT6 and DT8

REMOVAL & INSTALLATION

1. Remove the engine cover fasteners and lift off the engine covers.
2. Remove the recoil starter assembly attaching bolts and lift the assembly from the powerhead.

To install:
3. Place the recoil starter assembly on the powerhead and tighten the attaching nuts securely.
4. Install the engine covers and secure them with the cover fasteners.
5. Pull the starter knob several times and check for the proper operation of the ratcheting mechanism.

DISASSEMBLY

♦ See Figure 11

1. Hook the rope on the notch of the drum and gently rotate the drum clockwise to release spring tension.
2. Remove the E-clip and drive plate that covers the drive pawl.
3. Remove the return spring and spacer, then remove the drive pawl.
4. Remove the drum.

➡ It is advisable to wear heavy gloves while removing the spring to prevent your hands from being cut by the sharp spring steel.

※ CAUTION

The starter drum spring is under high tension. If the spring should come loose, it may cause serious damage or personal injury. Take all applicable cautions when working with this spring.

5. Carefully remove the coil spring from the starter case.

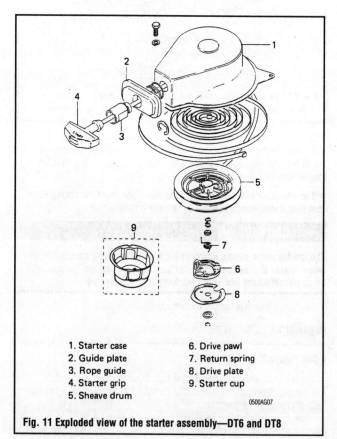

1. Starter case
2. Guide plate
3. Rope guide
4. Starter grip
5. Sheave drum
6. Drive pawl
7. Return spring
8. Drive plate
9. Starter cup

Fig. 11 Exploded view of the starter assembly—DT6 and DT8

CLEANING & INSPECTION

♦ See Figure 12

Clean all components and then blow them dry using compressed air. Remove any trace of corrosion and wipe all metal parts with an oil dampened cloth to prevent future corrosion.
Inspect the rope. Replace the rope if it appears to be weak or frayed. If the rope is frayed, check the holes through which the rope passes for rough edges

HAND REWIND STARTER 11-5

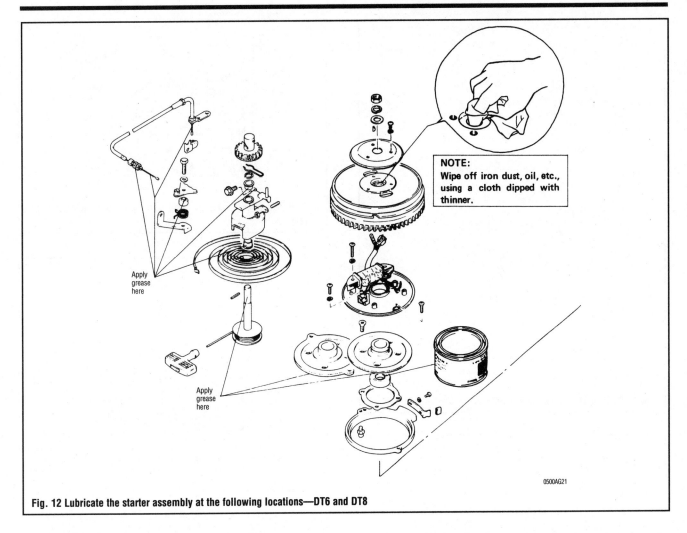

Fig. 12 Lubricate the starter assembly at the following locations—DT6 and DT8

or burrs. Remove the rough edges or burrs with a file and polish the surface until it is smooth.

Inspect the starter return spring end hooks. Replace the spring if it is weak, corroded or cracked. Inspect the inside surface of the starter case and drum for grooves or roughness. Grooves may cause erratic rewinding of the starter rope.

Inspect and lubricate the ratchet mechanism with waterproof grease. Check the mechanism for freedom of movement.

ASSEMBLY

▶ See Figures 13, 14 and 15

1. Position the coil spring in the starter case, feeding the outer portion of the spring into the case first and positioning the remainder in the case gradually.

✲✲✲ CAUTION

During installation, the spring will be placed under high tension. Take all applicable cautions when working with this spring.

2. Place the bent end of the spring into the groove on the drum.
3. Twist the drum counterclockwise to make sure the spring is positively engaged. If resistance is not felt, the spring may not be engaged properly. If this is the case, insert a rod into the hole on the drum and guide the spring end into engagement with the drum.
4. Install the drive pawl, spacer and return spring. Install the return spring first, ensuring the tongue on the spring is first inserted into the hole on the drive plate and then into the hole in the drive pawl.
5. Secure the drum in place by installing the E-clip.
6. Attach the pull rope in to the notch on the drum and turn the drum counterclockwise 6 rotations.

7. Remove the rope from the notch and gently wind it in under the force of the spring to take it up into the drum.
8. Measure the spring resistance using a fish scale. If should be 1.76–3.31 lbs. of pulling force.
9. If resistance is not within specification, adjust the amount of counterclockwise rotations made prior to allowing rope to rewind into the drum.

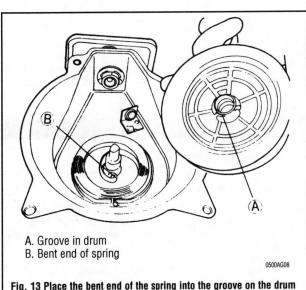

A. Groove in drum
B. Bent end of spring

Fig. 13 Place the bent end of the spring into the groove on the drum

11-6 HAND REWIND STARTER

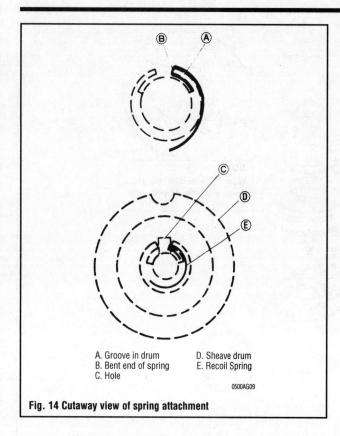

A. Groove in drum
B. Bent end of spring
C. Hole
D. Sheave drum
E. Recoil Spring

Fig. 14 Cutaway view of spring attachment

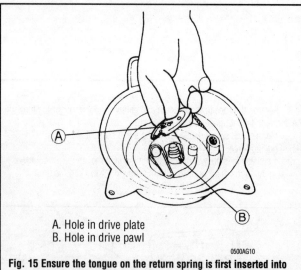

A. Hole in drive plate
B. Hole in drive pawl

Fig. 15 Ensure the tongue on the return spring is first inserted into the hole on the drive plate and then into the hole in the drive pawl

DT9.9 and DT15

REMOVAL & INSTALLATION

▶ See Figures 16 and 17

1. Remove the engine cover fasteners and lift off the engine covers.
2. Disconnect the neutral start interlock cable from the throttle limiter.
3. Remove the recoil starter assembly attaching bolts and lift the assembly from the powerhead.

To install:

4. Place the recoil starter assembly on the powerhead and tighten the attaching nuts securely.

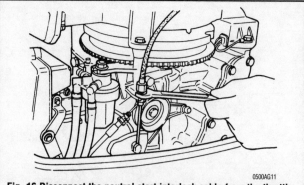

Fig. 16 Disconnect the neutral start interlock cable from the throttle limiter

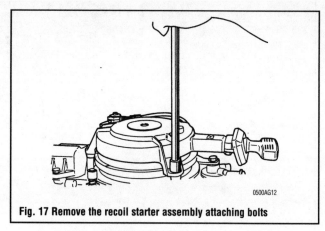

Fig. 17 Remove the recoil starter assembly attaching bolts

5. Install the engine covers and secure them with the cover fasteners.
6. Connect the neutral start cable and adjust it to specification.
7. Pull the starter knob several times and check for the proper operation of the ratcheting mechanism and the neutral starter interlock.

ADJUSTMENT

▶ See Figure 18

1. Place the gear selector lever in the **NEUTRAL** position.
2. Using the adjusting nut to adjust the wire length so that the upper matchmark on the recoil starter aligns with the slit of the stopper arm.

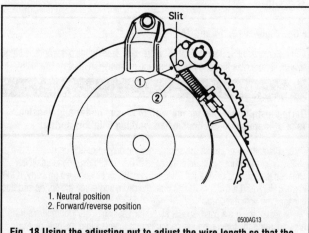

1. Neutral position
2. Forward/reverse position

Fig. 18 Using the adjusting nut to adjust the wire length so that the upper matchmark on the recoil starter aligns with the slit of the stopper arm in the neutral position and the lower matchmark aligns in the forward/reverse position

HAND REWIND STARTER 11-7

3. Place the gear selector lever in the **FORWARD** or **REVERSE** position.
4. The lower matchmark on the recoil starter should align with the slit of the stopper arm.
5. Pull the starter knob several times with the gear selector in the **FORWARD** or **REVERSE** positions and make sure the rope cannot be pulled.

DISASSEMBLY

♦ See Figures 19 and 20

1. Hook the rope on the notch of the drum and gently rotate the drum clockwise to release spring tension.
2. Remove the cotter pin and the washer that secure the stopper arm to the drum stopper.
3. Remove the stopper arm and the drum stopper spring.
4. Remove the bolt, the drive plate and the friction spring.
5. Remove the drive pawl and its spring from the drum.
6. Remove the drum with the starter rope still attached.
7. Remove the drum stopper.

→It is advisable to wear heavy gloves while removing the spring to prevent your hands from being cut by the sharp spring steel.

✹✹ CAUTION

The starter drum spring is under high tension. If the spring should come loose, it may cause serious damage or personal injury. Take all applicable cautions when working with this spring.

8. Carefully remove the coil spring from the starter case.

CLEANING & INSPECTION

♦ See Figure 21

Clean all components and then blow them dry using compressed air. Remove any trace of corrosion and wipe all metal parts with an oil dampened cloth to prevent future corrosion.

Inspect the rope. Replace the rope if it appears to be weak or frayed. If the rope is frayed, check the holes through which the rope passes for rough edges or burrs. Remove the rough edges or burrs with a file and polish the surface until it is smooth.

Inspect the starter return spring end hooks. Replace the spring if it is weak, corroded or cracked. Inspect the inside surface of the starter case and drum for grooves or roughness. Grooves may cause erratic rewinding of the starter rope.

Inspect and lubricate the ratchet mechanism with waterproof grease. Check the mechanism for freedom of movement.

ASSEMBLY

♦ See Figures 22 and 23

1. Position the coil spring in the starter case, feeding the outer portion of the spring into the case first and positioning the remainder in the case gradually.

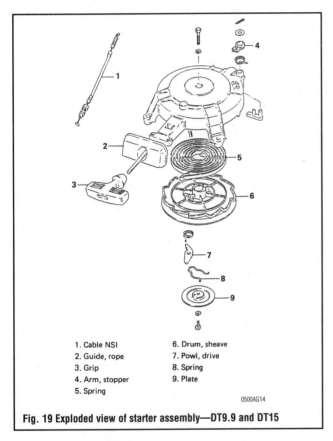

1. Cable NSI
2. Guide, rope
3. Grip
4. Arm, stopper
5. Spring
6. Drum, sheave
7. Powl, drive
8. Spring
9. Plate

Fig. 19 Exploded view of starter assembly—DT9.9 and DT15

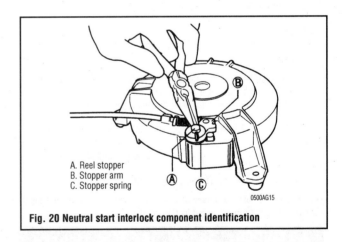

A. Reel stopper
B. Stopper arm
C. Stopper spring

Fig. 20 Neutral start interlock component identification

Fig. 21 Apply waterproof grease to the recoil spring and the bushing

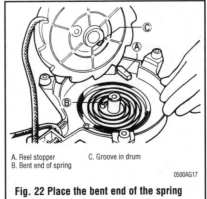

A. Reel stopper
B. Bent end of spring
C. Groove in drum

Fig. 22 Place the bent end of the spring into the groove on the drum

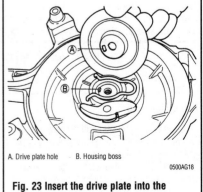

A. Drive plate hole
B. Housing boss

Fig. 23 Insert the drive plate into the housing boss during assembly

11-8 HAND REWIND STARTER

✱✱✱ CAUTION

During installation, the spring will be placed under high tension. Take all applicable cautions when working with this spring.

2. Lubricate the spring and center bushing with waterproof grease.
3. Install the drum stopper.
4. Place the bent end of the spring into the groove on the drum.
5. Twist the drum counterclockwise to make sure the spring is positively engaged. If resistance is not felt, the spring may not be engaged properly.
6. Insert the shorter bent end of the drive pawl spring into the drum hole and then hook the longer bent end on the drive pawl groove. Turn the drive pawl clockwise to ensure the assembly is installed properly.
7. Lubricate and install the drive plate and the friction spring.
8. Install the stopper arm and the drum stopper spring.
9. Install the cotter pin and the washer that secure the stopper arm to the drum stopper.
10. Wind the rope around the drum 2.5 times counterclockwise and hook the rope in to the notch on the drum. Rotate the drum counterclockwise 4 rotations and wind the rope around the drum.

DT20 and DT25

REMOVAL & INSTALLATION

1. Remove the engine cover fasteners and lift off the engine covers.
2. Remove the recoil starter assembly attaching bolts and lift the assembly from the powerhead.

To install:

3. Place the recoil starter assembly on the powerhead and tighten the attaching nuts securely.
4. Install the engine covers and secure them with the cover fasteners.
5. Pull the starter knob several times and check for the proper operation of the ratcheting mechanism.

DISASSEMBLY

♦ See Figure 24

1. Pull the starter rope out as far as it will come and hold the drum with your finger to prevent the rope from rewinding.
2. Hook the rope on the notch of the drum and gently rotate the drum clockwise to release spring tension.
3. With the rope and spring fully stretched and the drum turned all the way, remove the drum securing bolt.
4. Remove the drive plate, pawl, drive pawl springs and drive plate spring.
5. Remove the drum by lifting it gently from the housing.

➡ It is advisable to wear heavy gloves while removing the spring to prevent your hands from being cut by the sharp spring steel.

✱✱✱ CAUTION

The starter drum spring is under high tension. If the spring should come loose, it may cause serious damage or personal injury. Take all applicable cautions when working with this spring.

6. Carefully remove the coil spring from the starter case.

CLEANING & INSPECTION

♦ See Figure 25

Clean all components and then blow them dry using compressed air. Remove any trace of corrosion and wipe all metal parts with an oil dampened cloth to prevent future corrosion.

Inspect the rope. Replace the rope if it appears to be weak or frayed. If the rope is frayed, check the holes through which the rope passes for rough edges or burrs. Remove the rough edges or burrs with a file and polish the surface until it is smooth.

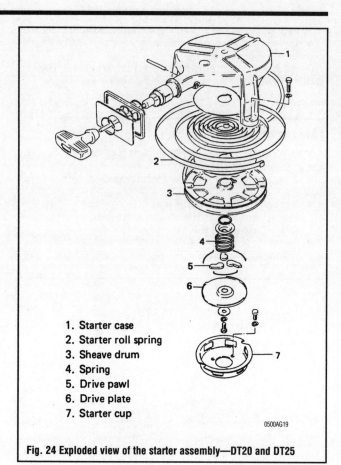

1. Starter case
2. Starter roll spring
3. Sheave drum
4. Spring
5. Drive pawl
6. Drive plate
7. Starter cup

Fig. 24 Exploded view of the starter assembly—DT20 and DT25

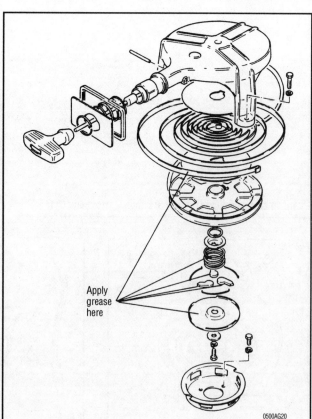

Apply grease here

Fig. 25 Lubricate the starter assembly at the following locations—DT20 and DT25

HAND REWIND STARTER 11-9

Inspect the starter return spring end hooks. Replace the spring if it is weak, corroded or cracked. Inspect the inside surface of the starter case and drum for grooves or roughness. Grooves may cause erratic rewinding of the starter rope.

Inspect and lubricate the ratchet mechanism with waterproof grease. Check the mechanism for freedom of movement.

ASSEMBLY

1. Position the coil spring in the starter case, feeding the outer portion of the spring into the case first and positioning the remainder in the case gradually.

❈❈❈ CAUTION

During installation, the spring will be placed under high tension. Take all applicable cautions when working with this spring.

2. Place the bent end of the spring into the groove on the drum.
3. Twist the drum counterclockwise to make sure the spring is positively engaged. If resistance is not felt, the spring may not be engaged properly. If this is the case, insert a rod into the hole on the drum and guide the spring end into engagement with the drum.
4. Install the washer, drive pawl and plate. Ensure the tongue on the drive pawl is properly positioned into the space between the pilot shaft springs on the plate.
5. Secure the drum in place by tightening the bolt securely.
6. Attach the pull rope in to the notch on the drum and turn the drum counterclockwise 3–4 rotations.
7. Remove the rope from the notch and gently wind it in under the force of the spring to take it up into the drum.

DT25C, DT30C, DT35C and DT40C

REMOVAL & INSTALLATION

1. Remove the engine cover fasteners and lift off the engine covers.
2. Disconnect the neutral start interlock cable from the throttle limiter.
3. Remove the recoil starter assembly attaching bolts and lift the assembly from the powerhead.

To install:

4. Place the recoil starter assembly on the powerhead and tighten the attaching nuts securely.
5. Install the engine covers and secure them with the cover fasteners.
6. Connect the neutral start cable and adjust it to specification.
7. Pull the starter knob several times and check for the proper operation of the ratcheting mechanism and the neutral starter interlock.

ADJUSTMENT

♦ See Figure 26

1. Place the gear selector lever in the **NEUTRAL** position.
2. Using the adjusting nut to adjust the wire length so that the upper matchmark on the recoil starter aligns with the slit of the stopper arm.
3. Place the gear selector lever in the **FORWARD** or **REVERSE** position.
4. The slit on the stopper arm should align between the middle and lower matchmark on the recoil starter.
5. Pull the starter knob several times with the gear selector in the **FORWARD** or **REVERSE** positions and make sure the rope cannot be pulled.

DISASSEMBLY

♦ See Figure 27

1. Invert the starter case and remove the cotter pin, washer, stopper arm, spring and stopper lever.
2. Hook the rope on the notch of the drum and gently rotate the drum clockwise to release spring tension.
3. Remove the drum attaching bolt and drive plate.
4. Remove the drive pawl and its spring from the drum.

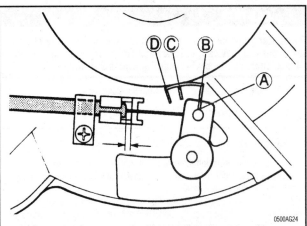

Fig. 26 Use the adjusting nut to adjust the wire length so that the upper matchmark (B) on the recoil starter aligns with the slit (A) of the stopper arm when in neutral and falls between the lower marks (C and D) when in forward or reverse

5. Remove the drum with the starter rope still attached. Make sure the starter spring remains in the starter case.
6. Remove the rope from the drum.

➡It is advisable to wear heavy gloves while removing the spring to prevent your hands from being cut by the sharp spring steel.

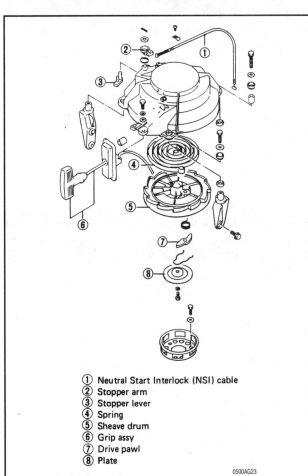

① Neutral Start Interlock (NSI) cable
② Stopper arm
③ Stopper lever
④ Spring
⑤ Sheave drum
⑥ Grip assy
⑦ Drive pawl
⑧ Plate

Fig. 27 Exploded view of the starter assembly— DT25C, DT30C, DT35C and DT40C

11-10 HAND REWIND STARTER

✳ CAUTION

The starter drum spring is under high tension. If the spring should come loose, it may cause serious damage or personal injury. Take all applicable cautions when working with this spring.

7. Carefully remove the coil spring from the starter case.

CLEANING & INSPECTION

◆ See Figure 28

Clean all components and then blow them dry using compressed air. Remove any trace of corrosion and wipe all metal parts with an oil dampened cloth to prevent future corrosion.

Inspect the rope. Replace the rope if it appears to be weak or frayed. If the rope is frayed, check the holes through which the rope passes for rough edges or burrs. Remove the rough edges or burrs with a file and polish the surface until it is smooth.

Inspect the starter return spring end hooks. Replace the spring if it is weak, corroded or cracked. Inspect the inside surface of the starter case and drum for grooves or roughness. Grooves may cause erratic rewinding of the starter rope.

Inspect and lubricate the ratchet mechanism with waterproof grease. Check the mechanism for freedom of movement.

ASSEMBLY

1. Position the coil spring in the starter case, feeding the outer portion of the spring into the case first and positioning the remainder in the case gradually.

✳ CAUTION

During installation, the spring will be placed under high tension. Take all applicable cautions when working with this spring.

2. Lubricate the spring and center bushing with waterproof grease.
3. Place the bent end of the spring into the groove on the drum.
4. Twist the drum counterclockwise to make sure the spring is positively engaged. If resistance is not felt, the spring may not be engaged properly.
5. Insert the shorter bent end of the drive pawl spring into the drum hole and then hook the longer bent end on the drive pawl groove. Turn the drive pawl clockwise to ensure the assembly is installed properly.
6. Install the drive pawl and its spring on the drum.
7. Install the drum drive plate and attaching bolt. Tighten the attaching bolt securely.

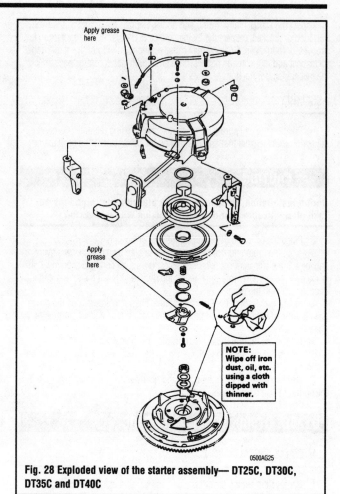

Fig. 28 Exploded view of the starter assembly— DT25C, DT30C, DT35C and DT40C

8. Install the cotter pin, washer, stopper arm, spring and stopper lever on the starter case.
9. Wind the rope around the drum 2.5 times counterclockwise and hook the rope in to the notch on the drum. Then, rotate the drum 4 turns counterclockwise and wind the rope around the drum.

BENDIX TYPE STARTER

DT6 and DT8

REMOVAL & INSTALLATION

1. Pull the starter grip and hold the starter rope fully extended.
2. Untie the knot and remove the starter grip.
3. Remove the engine cover fasteners and lift off the engine cover.
4. Remove the recoil starter assembly attaching bolts and lift the assembly from the powerhead.

To install:

5. Place the recoil starter assembly on the powerhead and tighten the attaching nuts securely.
6. Install the engine cover.
7. Install the starter grip.
8. Pull the starter knob several times and check for the proper operation.

DISASSEMBLY

◆ See Figure 29

1. Drive out the starter pin with a punch to free the starter pinion.
2. Remove the starter clip using snapring pliers.

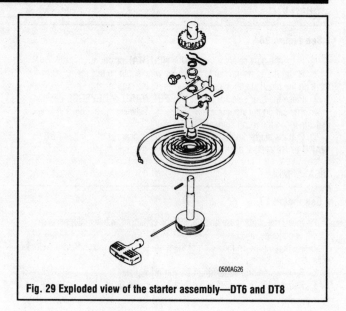

Fig. 29 Exploded view of the starter assembly—DT6 and DT8

HAND REWIND STARTER 11-11

3. Cover the starter drum with a shop rag to prevent the recoil spring from unreeling.
4. Remove the starter drum from the assembly.

➥ It is advisable to wear heavy gloves while removing the spring to prevent your hands from being cut by the sharp spring steel.

✶✶ CAUTION

The recoil spring is under high tension. If the spring should come loose, it may cause serious damage or personal injury. Take all applicable cautions when working with this spring.

5. Remove the recoil spring.

CLEANING & INSPECTION

Clean all components and then blow them dry using compressed air. Remove any trace of corrosion and wipe all metal parts with an oil dampened cloth to prevent future corrosion.

Inspect the rope. Replace the rope if it appears to be weak or frayed. If the rope is frayed, check the holes through which the rope passes for rough edges or burrs. Remove the rough edges or burrs with a file and polish the surface until it is smooth. Inspect the starter return spring end hooks. Replace the spring if it is weak, corroded or cracked. Inspect the inside surface of the starter case and drum for grooves or roughness. Grooves may cause erratic rewinding of the starter rope.

Inspect and lubricate the bendix mechanism and check for freedom of movement.

ASSEMBLY

▸ See Figures 30 and 31

1. Install the recoil spring in the starter housing by winding it clockwise. Insert the tip of the spring into the notch provided in the starter housing.
2. Apply waterproof grease to lubricate the spring and prevent it from rusting.
3. Pull out the starter rope through the notch provided in the recoil starter and wind the rope around two or three times in a clockwise direction. Keep the rope snugly wound up inside the starter reel.
4. Inside the starter housing, match the pawl of the recoil spring to the pawl on the starter reel.
5. Install the starter drum in the housing and secure with the circlip.
6. Secure the pinion in place by driving the starter pin in with a punch.
7. Install the spring on the starter drum shaft.
8. Pull out the starter rope through the notch provided in the starter drum and wind the rope all the way into the starter drum in a clockwise direction.
9. Pass the free end of the rope through the hole provided in the front part of the engine lower cover and attach the starter grip to the rope.
10. Install the recoil starter assembly on the engine, tightening the mounting bolts securely.
11. Pull the starter knob several times and check for the proper operation.

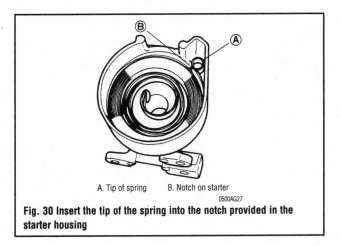

A. Tip of spring   B. Notch on starter

Fig. 30 Insert the tip of the spring into the notch provided in the starter housing

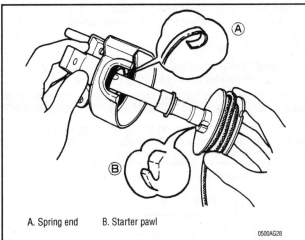

A. Spring end   B. Starter pawl

Fig. 31 Match the pawl of the recoil spring to the pawl on the starter reel

DT20 and DT25

REMOVAL & INSTALLATION

1. Pull the starter grip and hold the starter rope fully extended.
2. Untie the knot and remove the starter grip.
3. Remove the engine cover fasteners and lift off the engine cover.
4. Remove the recoil starter assembly attaching bolts and lift the assembly from the powerhead.

To install:

5. Place the recoil starter assembly on the powerhead and tighten the attaching nuts securely.
6. Install the engine cover.
7. Install the starter grip.
8. Pull the starter knob several times and check for the proper operation.

DISASSEMBLY

▸ See Figure 32

1. Remove the bolt holding the starter pinion gear in place.
2. Drive out the starter pin with a punch to free the starter pinion.

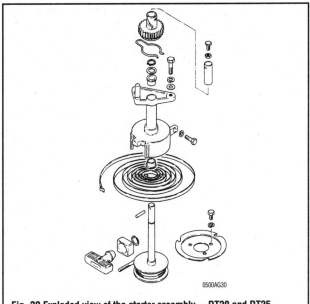

Fig. 32 Exploded view of the starter assembly— DT20 and DT25

3. Remove the starter clip using snapring pliers.
4. Cover the starter drum with a shop rag to prevent the recoil spring from unreeling.
5. Remove the starter drum from the assembly.

➡ It is advisable to wear heavy gloves while removing the spring to prevent your hands from being cut by the sharp spring steel.

❋❋ CAUTION

The recoil spring is under high tension. If the spring should come loose, it may cause serious damage or personal injury. Take all applicable cautions when working with this spring.

6. Remove the recoil spring.

CLEANING & INSPECTION

▶ See Figure 33

Clean all components and then blow them dry using compressed air. Remove any trace of corrosion and wipe all metal parts with an oil dampened cloth to prevent future corrosion.

Inspect the rope. Replace the rope if it appears to be weak or frayed. If the rope is frayed, check the holes through which the rope passes for rough edges or burrs. Remove the rough edges or burrs with a file and polish the surface until it is smooth.

Inspect the starter return spring end hooks. Replace the spring if it is weak, corroded or cracked. Inspect the inside surface of the starter case and drum for grooves or roughness. Grooves may cause erratic rewinding of the starter rope.

Inspect and lubricate the bendix mechanism and check for freedom of movement.

ASSEMBLY

▶ See Figure 34

1. Install the recoil spring in the starter housing by winding it clockwise. Insert the tip of the spring into the notch provided in the starter housing.
2. Apply waterproof grease to lubricate the spring and prevent it from rusting.
3. Pull out the starter rope through the notch provided in the recoil starter and wind the rope around two or three times in a clockwise direction. Keep the rope snugly wound up inside the starter reel.
4. Inside the starter housing, match the pawl of the recoil spring to the pawl on the starter reel.
5. Install the starter drum in the housing and secure with the circlip.
6. Secure the pinion in place by driving the starter pin in with a punch.
7. Install the spring on the starter drum shaft.
8. Pull out the starter rope through the notch provided in the starter drum and wind the rope all the way into the starter drum in a clockwise direction.
9. Pass the free end of the rope through the hole provided in the front part of the engine lower cover and attach the starter grip to the rope.
10. Pull the starter knob several times and check for the proper operation.

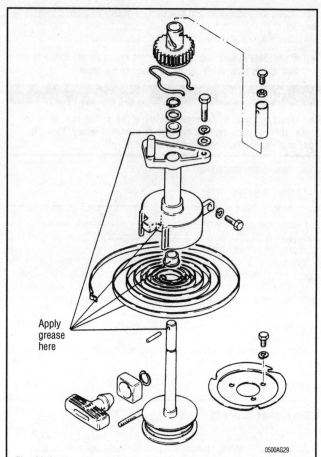

Fig. 33 Apply waterproof grease to the following components—DT20 amd DT25

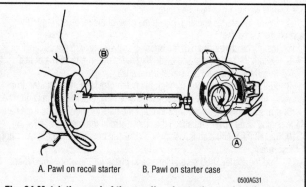

A. Pawl on recoil starter   B. Pawl on starter case

Fig. 34 Match the pawl of the recoil spring to the pawl on the starter reel

GLOSSARY

Understanding your marine mechanic is as important as understanding your outboard. Most boaters know about their boats, but many boaters have difficulty understanding engine terminology. Talking the language of outboards makes it easier to effectively communicate with professional mechanics. It isn't necessary (or recommended) that you diagnose the problem for them, but it will save them time, and you money, if you can accurately describe what is happening. It will also help you to know why your boat does what it is doing, and what repairs were made.

AFTER TOP DEAD CENTER (ATDC): The point after the piston reaches the top of its travel on the compression stroke.

AIR/FUEL RATIO: The ratio of air-to-fuel, by weight, drawn into the engine.

ALTERNATING CURRENT (AC): Electric current that flows first in one direction, then in the opposite direction, continually reversing flow.

ALTERNATOR: A device which produces AC (alternating current) which is converted to DC (direct current) to charge the battery.

AMMETER: An instrument, calibrated in amperes, used to measure the flow of an electrical current in a circuit. Ammeters are always connected in series with the circuit being tested.

AMP/HR. RATING (BATTERY): Measurement of the ability of a battery to deliver a stated amount of current for a stated period of time. The higher the amp/hr. rating, the better the battery.

AMPERE: The rate of flow of electrical current present when one volt of electrical pressure is applied against one ohm of electrical resistance.

ARMATURE: A laminated, soft iron core wrapped by a wire that converts electrical energy to mechanical energy as in a motor or relay. When rotated in a magnetic field, it changes mechanical energy into electrical energy as in a generator.

ATDC: After Top Dead Center.

ATMOSPHERIC PRESSURE: The pressure on the Earth's surface caused by the weight of the air in the atmosphere. At sea level, this pressure is 14.7 psi at 32°F (101 kPa at 0°C).

ATOMIZATION: The breaking down of a liquid into a fine mist that can be suspended in air.

AXIAL PLAY: Movement parallel to a shaft or bearing bore.

BACKFIRE: The sudden combustion of gases in the intake or exhaust system that results in a loud explosion.

BACKLASH: The clearance or play between two parts, such as meshed gears.

BALL BEARING: A bearing made up of hardened inner and outer races between which hardened steel balls roll.

BATTERY: A direct current electrical storage unit, consisting of the basic active materials of lead and sulfuric acid, which converts chemical energy into electrical energy. Used to provide current for the operation of the starter as well as other equipment, such as the radio, lighting, etc.

BEARING: A friction reducing, supportive device usually located between a stationary part and a moving part.

BEFORE TOP DEAD CENTER (BTDC): The point just before the piston reaches the top of its travel on the compression stroke.

BLOCK: See Cylinder Block.

BORE: Diameter of a cylinder.

BTDC: Before Top Dead Center.

BUSHING: A liner, usually removable, for a bearing; an anti-friction liner used in place of a bearing.

CARBON MONOXIDE (CO): A colorless, odorless gas given off as a normal byproduct of combustion. It is poisonous and extremely dangerous in confined areas, building up slowly to toxic levels without warning if adequate ventilation is not available.

CHECK VALVE: Any one-way valve installed to permit the flow of air, fuel or vacuum in one direction only.

CIRCLIP: A split steel snapring that fits into a groove to hold various parts in place.

CIRCUIT BREAKER: A switch which protects an electrical circuit from overload by opening the circuit when the current flow exceeds a pre-determined level. Some circuit breakers must be reset manually, while most reset automatically.

CIRCUIT: Any unbroken path through which an electrical current can flow. Also used to describe fuel flow in some instances.

COMBUSTION CHAMBER: The part of the engine in the cylinder head where combustion takes place.

COMPRESSION CHECK: A test involving cranking the engine with a special high pressure gauge connected to an individual cylinder. Individual cylinder pressure as well as pressure variance across cylinders is used to determine general operating condition of the engine.

COMPRESSION RATIO: The ratio of the volume between the piston and cylinder head when the piston is at the bottom of its stroke (bottom dead center) and when the piston is at the top of its stroke (top dead center).

CONDUCTOR: Any material through which an electrical current can be transmitted easily.

CONNECTING ROD: The connecting link between the crankshaft and piston.

CONTINUITY: Continuous or complete circuit. Can be checked with an ohmmeter.

CRANKCASE: The lower part of an engine in which the crankshaft and related parts operate.

CRANKSHAFT: Engine component (connected to pistons by connecting rods) which converts the reciprocating (up and down) motion of pistons to rotary motion used to turn the driveshaft.

CYLINDER HEAD: The detachable portion of the engine, usually fastened to the top of the cylinder block and containing all or most of the combustion chambers.

CYLINDER: In an engine, the round hole in the engine block in which the piston(s) ride.

DETONATION: An unwanted explosion of the air/fuel mixture in the combustion chamber caused by excess heat and compression, advanced timing, or an overly lean mixture. Also referred to as "ping".

GLOSSARY

DIAPHRAGM: A thin, flexible wall separating two cavities, such as in a vacuum advance unit.

DIGITAL VOLT OHMMETER: An electronic diagnostic tool used to measure voltage, ohms and amps as well as several other functions, with the readings displayed on a digital screen in tenths, hundredths and thousandths.

DIODE: An electrical device that will allow current to flow in one direction only.

DIRECT CURRENT (DC): Electrical current that flows in one direction only.

DISPLACEMENT: The total volume of air that is displaced by all pistons as the engine turns through one complete revolution.

DVOM: Digital volt ohmmeter

ELECTROLYTE: A solution of water and sulfuric acid used to activate the battery. Electrolyte is extremely corrosive.

END-PLAY: The measured amount of axial movement in a shaft.

ENGINE BLOCK: The basic engine casting containing the cylinders, the crankshaft main bearings, as well as machined surfaces for the mounting of other components such as the cylinder head, oil pan, transmission, etc..

FEELER GAUGE: A blade, usually metal, of precisely predetermined thickness, used to measure the clearance between two parts.

FIRING ORDER: The order in which combustion occurs in the cylinders of an engine.

FLAME FRONT: The term used to describe certain aspects of the fuel explosion in the cylinders. The flame front should move in a controlled pattern across the cylinder, rather than simply exploding immediately.

FLAT SPOT: A point during acceleration when the engine seems to lose power for an instant.

FLYWHEEL: A heavy disc of metal attached to the rear of the crankshaft. It smoothes the firing impulses of the engine and keeps the crankshaft turning during periods when no firing takes place. The starter also engages the flywheel to start the engine.

FOOT POUND (ft. lbs. or sometimes, ft. lb.): The amount of energy or work needed to raise an item weighing one pound, a distance of one foot.

FUEL FILTER: A component of the fuel system containing a porous paper element used to prevent any impurities from entering the engine through the fuel system. It usually takes the form of a canister-like housing, mounted in-line with the fuel hose, located anywhere on a vessel between the fuel tank and engine.

FUEL INJECTION: A system that sprays fuel into the cylinder through nozzles. The amount of fuel can be more precisely controlled with fuel injection.

FUSE: A protective device in a circuit which prevents circuit overload by breaking the circuit when a specific amperage is present. The device is constructed around a strip or wire of a lower amperage rating than the circuit it is designed to protect. When an amperage higher than that stamped on the fuse is present in the circuit, the strip or wire melts, opening the circuit.

FUSIBLE LINK: A piece of wire in a wiring harness that performs the same job as a fuse. If overloaded, the fusible link will melt and interrupt the circuit.

HORSEPOWER: A measurement of the amount of work; one horsepower is the amount of work necessary to lift 33,000 lbs. one foot in one minute. Brake horsepower (bhp) is the horsepower delivered by an engine on a dynamometer. Net horsepower is the power remaining (measured at the flywheel of the engine) that can be used to power the vessel after power is consumed through friction and running the engine accessories (water pump, alternator, fan etc.)

HYDROCARBON (HC): Any chemical compound made up of hydrogen and carbon. A major pollutant formed by the engine as a by-product of combustion.

HYDROMETER: An instrument used to measure the specific gravity of a solution.

IMPELLER: The portion of the water pump which provides the propulsion for the coolant to circulate it through the system

INCH POUND (inch lbs.; sometimes in. lb. or in. lbs.): One twelfth of a foot pound.

INJECTOR: A device which receives metered fuel under relatively low pressure and is activated to inject the fuel into the engine under relatively high pressure at a predetermined time.

INTAKE MANIFOLD: A casting of passages or pipes used to conduct air or a fuel/air mixture to the cylinders.

INTAKE SILENCER: An assembly consisting of a housing, and sometimes a filter. The filter element is made up of a porous paper or a wire mesh screening, and is designed to prevent airborne particles from entering the engine. Also see Air Cleaner.

JOURNAL: The bearing surface within which a shaft operates.

JUMPER CABLES: Two heavy duty wires with large alligator clips used to provide power from a charged battery to a discharged battery.

JUMPSTART: Utilizing one sufficiently charged battery to start the engine of another vessel with a discharged battery by the use of jumper cables.

KNOCK: Noise which results from the spontaneous ignition of a portion of the air-fuel mixture in the engine cylinder.

LITHIUM-BASE GREASE: Bearing grease using lithium as a base. Not compatible with sodium-base grease.

LOCK RING: See Circlip or Snapring

MANIFOLD: A casting of passages or set of pipes which connect the cylinders to an inlet or outlet source.

MISFIRE: Condition occurring when the fuel mixture in a cylinder fails to ignite, causing the engine to run roughly.

MULTI-WEIGHT: Type of oil that provides adequate lubrication at both high and low temperatures.

NEEDLE BEARING: A bearing which consists of a number (usually a large number) of long, thin rollers.

NITROGEN OXIDE (NOx): One of the three basic pollutants found in the exhaust emission of an internal combustion engine. The amount of NOx usually varies in an inverse proportion to the amount of HC and CO.

OEM: Original Equipment Manufactured. OEM equipment is that furnished standard by the manufacturer.

GLOSSARY

OHM: The unit used to measure the resistance of conductor-to-electrical flow. One ohm is the amount of resistance that limits current flow to one ampere in a circuit with one volt of pressure.

OHMMETER: An instrument used for measuring the resistance, in ohms, in an electrical circuit.

OXIDES OF NITROGEN: See nitrogen oxide (NOx).

PING: A metallic rattling sound produced by the engine during acceleration. It is usually due to incorrect timing or a poor grade of fuel.

PISTON RING: An open-ended ring which fits into a groove on the outer diameter of the piston. Its chief function is to form a seal between the piston and cylinder wall. Most pistons have three rings: two for compression sealing; one for oil sealing.

POLARITY: Indication (positive or negative) of the two poles of a battery.

PPM: Parts per million; unit used to measure exhaust emissions.

PREIGNITION: Early ignition of fuel in the cylinder, sometimes due to glowing carbon deposits in the combustion chamber.

PRELOAD: A predetermined load placed on a bearing during assembly or by adjustment.

PRESS FIT: The mating of two parts under pressure, due to the inner diameter of one being smaller than the outer diameter of the other, or vice versa; an interference fit.

PSI: Pounds per square inch; a measurement of pressure.

RACE: The surface on the inner or outer ring of a bearing on which the balls, needles or rollers move.

REAR MAIN OIL SEAL: A synthetic or rope-type seal that prevents oil from leaking out of the engine past the rear main crankshaft bearing.

RECTIFIER: A device (used primarily in alternators) that permits electrical current to flow in one direction only.

REGULATOR: A device which maintains the amperage and/or voltage levels of a circuit at predetermined values.

RELAY: A switch which automatically opens and/or closes a circuit.

RESISTANCE: The opposition to the flow of current through a circuit or electrical device, and is measured in ohms. Resistance is equal to the voltage divided by the amperage.

RESISTOR: A device, usually made of wire, which offers a preset amount of resistance in an electrical circuit.

ROCKER ARM: A lever which rotates around a shaft pushing down (opening) the valve with an end when the other end is pushed up by the pushrod. Spring pressure will later close the valve.

ROLLER BEARING: A bearing made up of hardened inner and outer races between which hardened steel rollers move.

RPM: Revolutions per minute (usually indicates engine speed).

SENDING UNIT: A mechanical, electrical, hydraulic or electromagnetic device which transmits information to a gauge.

SENSOR: Any device designed to measure engine operating conditions or ambient pressures and temperatures. Usually electronic in nature and designed to send a voltage signal to an on-board computer, some sensors may operate as a simple on/off switch or they may provide a variable voltage signal (like a potentiometer) as conditions or measured parameters change.

SHIM: Spacers of precise, predetermined thickness used between parts to establish a proper working relationship.

SHORT CIRCUIT: An electrical malfunction where current takes the path of least resistance to ground (usually through damaged insulation). Current flow is excessive from low resistance resulting in a blown fuse.

SLUDGE: Thick, black deposits in engine formed from dirt, oil, water, etc. It is usually formed in engines when oil changes are neglected.

SNAP RING: A circular retaining clip used inside or outside a shaft or part to secure a shaft, such as a floating wrist pin.

SOLENOID: An electrically operated, magnetic switching device.

SPECIFIC GRAVITY (BATTERY): The relative weight of liquid (battery electrolyte) as compared to the weight of an equal volume of water.

SPLINES: Ridges machined or cast onto the outer diameter of a shaft or inner diameter of a bore to enable parts to mate without rotation.

STARTER: A high-torque electric motor used for the purpose of starting the engine, typically through a high ratio geared drive connected to the flywheel ring gear.

STROKE: The distance the piston travels from bottom dead center to top dead center.

TACHOMETER: A device used to measure the rotary speed of an engine, shaft, gear, etc., usually in rotations per minute.

TDC: Top dead center. The exact top of the piston's stroke.

THERMOSTAT: A valve, located in the cooling system of an engine, which is closed when cold and opens gradually in response to engine heating, controlling the temperature of the coolant and rate of coolant flow.

TOP DEAD CENTER (TDC): The point at which the piston reaches the top of its travel on the compression stroke.

TORQUE: Measurement of turning or twisting force, expressed as foot-pounds or inch-pounds.

TUNE-UP: A regular maintenance function, usually associated with the replacement and adjustment of parts and components in the electrical and fuel systems of a engine for the purpose of attaining optimum performance.

VISCOSITY: The ability of a fluid to flow. The lower the viscosity rating, the easier the fluid will flow. 10 weight motor oil will flow much easier than 40 weight motor oil.

VOLT: Unit used to measure the force or pressure of electricity. It is defined as the pressure

VOLTAGE REGULATOR: A device that controls the current output of the alternator or generator.

VOLTMETER: An instrument used for measuring electrical force in units called volts. Voltmeters are always connected parallel with the circuit being tested.

WATER PUMP: Component of the cooling system that mounts on the engine, circulating the coolant under pressure.

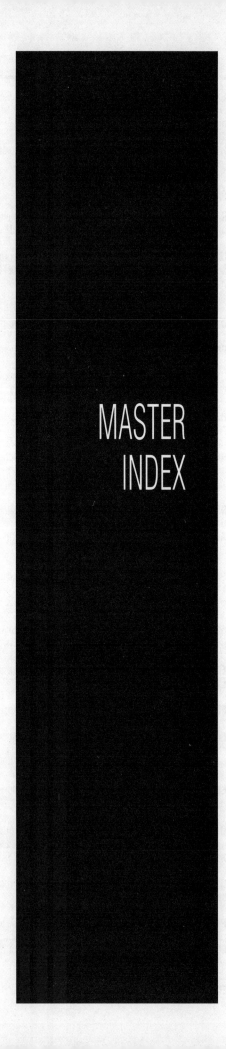

MASTER INDEX

ADDITIONAL INPUTS 4-48
AIR/OIL MIXING VALVE 6-7
    REMOVAL & INSTALLATION 6-7
AIR TEMPERATURE SENSOR 4-39
    DESCRIPTION & OPERATION 4-39
    REMOVAL & INSTALLATION 4-39
    TESTING 4-39
ALL OTHER MODELS 4-25
    INSPECTION & CLEANING 4-26
    REED & REED STOP REPLACEMENT 4-26
    REMOVAL & INSTALLATION 4-26
ALTERNATOR (STATOR) 5-40
    TESTING 5-40
ATMOSPHERIC PRESSURE SENSOR 4-39
    DESCRIPTION & OPERATION 4-39
    REMOVAL & INSTALLATION 4-39
    TESTING 4-39
AVOIDING THE MOST COMMON MISTAKES 1-3
AVOIDING TROUBLE 1-2
BASIC ELECTRICAL THEORY 5-2
    HOW ELECTRICITY WORKS: THE WATER ANALOGY 5-2
    OHM'S LAW 5-2
BATTERY (BOAT MAINTENANCE) 3-9
    BATTERY & CHARGING SAFETY PRECAUTIONS 3-11
    BATTERY CHARGERS 3-11
    BATTERY TERMINALS 3-11
    CHECKING SPECIFIC GRAVITY 3-10
    CLEANING 3-10
    REPLACING BATTERY CABLES 3-12
BATTERY (CHARGING CIRCUIT) 5-41
    BATTERY CABLES 5-45
    BATTERY CHARGERS 5-44
    BATTERY CONSTRUCTION 5-41
    BATTERY LOCATION 5-42
    BATTERY RATINGS 5-42
    BATTERY SERVICE 5-42
    BATTERY STORAGE 5-45
    BATTERY TERMINALS 5-44
    MARINE BATTERIES 5-41
    SAFETY PRECAUTIONS 5-44
BEARINGS 7-44
    GENERAL INFORMATION 7-44
    INSPECTION 7-45
BENDIX TYPE STARTER 11-10
BLEEDING THE OIL INJECTION SYSTEM 6-3
    PROCEDURE 6-3
BOAT MAINTENANCE 3-8
BOATING SAFETY 1-4
BOLTS, NUTS AND OTHER THREADED RETAINERS 2-14
BREAKER POINTS 5-8
    POINT GAP ADJUSTMENT 5-8
    REMOVAL & INSTALLATION 5-9
    TESTING 5-9
BREAKER POINTS IGNITION (MAGNETO IGNITION) 5-7
BUY OR REBUILD? 7-33
CAN YOU DO IT? 1-2
CAPACITIES 3-2
CAPACITOR DISCHARGE IGNITION (CDI) SYSTEM 5-11
CARBURETION 4-3
    BASIC FUNCTIONS 4-5
    CARBURETOR CIRCUITS 4-4
    DUAL-THROAT CARBURETORS 4-6
    GENERAL INFORMATION 4-3
    REMOVING FUEL FROM THE SYSTEM 4-6
CARBURETOR IDLE AIR SCREW SPECIFICATION 3-33
CARBURETOR SERVICE 4-11
CDI UNIT 5-28
    DESCRIPTION & OPERATION 5-28
    REMOVAL & INSTALLATION 5-34
    TESTING 5-28
CHARGING CIRCUIT 5-39

MASTER INDEX 11-17

CHEMICALS 2-3
    CLEANERS 2-4
    LUBRICANTS & PENETRANTS 2-3
    SEALANTS 2-4
CIRCUIT/POSITION 4-7
COMBUSTION 4-3
    ABNORMAL COMBUSTION 4-3
    FACTORS AFFECTING COMBUSTION 4-3
COMBUSTION RELATED PISTON FAILURES 4-10
COMPRESSION CHECK 3-12
    CHECKING COMPRESSION 3-12
    LOW COMPRESSION 3-13
CONDENSER 5-10
    DESCRIPTION & OPERATION 5-10
    REMOVAL & INSTALLATION 5-10
    TESTING 5-10
CONNECTING RODS 7-41
    GENERAL INFORMATION 7-41
    INSPECTION 7-42
CONVERSION FACTORS 2-16
COOLING SYSTEM 6-11
COURTESY MARINE EXAMINATIONS 1-11
CRANKSHAFT 7-43
    GENERAL INFORMATION 7-43
    INSPECTION 7-44
CYLINDER BLOCK AND HEAD 7-35
    GENERAL INFORMATION 7-35
    INSPECTION 7-36
CYLINDER BORES 7-36
    GENERAL INFORMATION 7-36
    INSPECTION 7-37
    REFINISHING 7-37
CYLINDER WALL TEMPERATURE SENSOR 4-36
    DESCRIPTION & OPERATION 4-36
    REMOVAL & INSTALLATION 4-39
    TESTING 4-38
DESCRIPTION AND OPERATION (CAPACITOR DISCHARGE IGNITION (CDI) SYSTEM) 5-11
    SINGLE-CYLINDER IGNITION 5-11
    SUZUKI PEI IGNITION 5-12
DESCRIPTION AND OPERATION (CHARGING CIRCUIT) 5-39
    PRECAUTIONS 5-39
    SINGLE PHASE CHARGING SYSTEM 5-39
    THREE-PHASE CHARGING SYSTEM 5-39
DESCRIPTION AND OPERATION (COOLING SYSTEM) 6-11
    THERMOSTAT 6-12
    WATER PUMP 6-11
DESCRIPTION AND OPERATION (ELECTRONIC FUEL INJECTION) 4-32
    FUEL INJECTION BASICS 4-32
    SUZUKI ELECTRONIC FUEL INJECTION 4-32
DESCRIPTION AND OPERATION (ELECTRONIC IGNITION) 5-38
DESCRIPTION AND OPERATION (GAS ASSISTED TILT) 9-2
DESCRIPTION AND OPERATION (HAND REWIND STARTER) 11-2
DESCRIPTION AND OPERATION (JET DRIVE) 8-51
DESCRIPTION AND OPERATION (MANUAL TILT) 9-2
    SERVICING 9-2
DESCRIPTION AND OPERATION (OIL INJECTION SYSTEM) 6-2
    AIR/OIL MIXING VALVE 6-2
    OIL PUMP 6-2
DESCRIPTION AND OPERATION (OIL INJECTION WARNING SYSTEMS) 6-14
    LOW OIL LEVEL 6-14
    OIL FLOW 6-14
DESCRIPTION AND OPERATION (OVERHEAT WARNING SYSTEM) 6-17
DESCRIPTION AND OPERATION (POWER TILT) 9-3
DESCRIPTION AND OPERATION (POWER TRIM/TILT) 9-6
DESCRIPTION AND OPERATION (REMOTE CONTROL BOX) 10-2

DESCRIPTION AND OPERATION (STARTING CIRCUIT) 5-45
DESCRIPTION AND OPERATION (TILLER HANDLE) 10-7
DETERMINING POWERHEAD CONDITION 7-32
    PRIMARY COMPRESSION TEST 7-32
    SECONDARY COMPRESSION TEST 7-32
DIAPHRAGM TYPE FUEL PUMPS 4-27
    DESCRIPTION & OPERATION 4-27
    OVERHAUL 4-27
    REMOVAL & INSTALLATION 4-27
DIRECTIONS AND LOCATIONS 1-2
DO'S 1-12
DON'TS 1-12
DT2 AND DT2.2 (CARBURETOR SERVICE) 4-11
    ASSEMBLY 4-12
    CLEANING & INSPECTION 4-12
    DISASSEMBLY 4-11
    REMOVAL & INSTALLATION 4-11
DT2 AND DT2.2 (FORWARD ONLY) (LOWER UNIT OVERHAUL) 8-6
    ASSEMBLY 8-8
    CLEANING & INSPECTION 8-7
    DISASSEMBLY 8-6
    SHIMMING PROCEDURE 8-9
DT2 AND DT2.2 (TUNE-UP) 3-18
    IDLE SPEED 3-19
    IGNITION TIMING 3-18
DT2, DT2.2 (REED VALVE SERVICE) 4-25
    REMOVAL & INSTALLATION 4-25
DT2, DT2.2 AND DT4 (OVERHEAD TYPE STARTER) 11-2
    ASSEMBLY 11-3
    CLEANING & INSPECTION 11-3
    DISASSEMBLY 11-2
    REMOVAL & INSTALLATION 11-2
DT4 (CARBURETOR SERVICE) 4-13
    ASSEMBLY 4-14
    CLEANING & INSPECTION 4-13
    DISASSEMBLY 4-13
    REMOVAL & INSTALLATION 4-13
DT4 (LOWER UNIT OVERHAUL) 8-9
    ASSEMBLY 8-10
    CLEANING & INSPECTION 8-9
    DISASSEMBLY 8-9
    SHIMMING PROCEDURE 8-11
DT4 (TUNE-UP) 3-19
    IDLE SPEED 3-19
    IGNITION TIMING 3-19
DT6 AND 1988 DT8 (CARBURETOR SERVICE) 4-14
    ASSEMBLY 4-15
    CLEANING & INSPECTION 4-15
    DISASSEMBLY 4-14
    REMOVAL & INSTALLATION 4-14
DT6 AND DT8 (BENDIX TYPE STARTER) 11-10
    ASSEMBLY 11-11
    CLEANING &INSPECTION 11-11
    DISASSEMBLY 11-10
    REMOVAL & INSTALLATION 11-10
DT6 AND DT8 (LOWER UNIT OVERHAUL) 8-11
    ASSEMBLY 8-14
    DISASSEMBLY 8-11
    INSPECTION & CLEANING 8-13
    SHIMMING PROCEDURE 8-14
DT6 AND DT8 (OVERHEAD TYPE STARTER) 11-4
    ASSEMBLY 11-5
    CLEANING & INSPECTION 11-4
    DISASSEMBLY 11-4
    REMOVAL & INSTALLATION 11-4
DT6 AND DT8 (TUNE-UP) 3-19
    IDLE SPEED 3-20
    IGNITION TIMING 3-19

MASTER INDEX

DT9.9 AND DT15 (LOWER UNIT OVERHAUL) 8-14
    ASSEMBLY 8-16
    CLEANING & INSPECTION 8-15
    DISSASSEMBLY 8-14
    SHIMMING PROCEDURE 8-17
DT9.9 AND DT15 (OVERHEAD TYPE STARTER) 11-6
    ADJUSTMENT 11-6
    ASSEMBLY 11-7
    CLEANING & INSPECTION 11-7
    DISASSEMBLY 11-7
    REMOVAL & INSTALLATION 11-6
DT9.9 AND DT15 (TUNE-UP) 3-21
    IDLE SPEED 3-21
    IGNITION TIMING 3-21
DT9.9, DT15 AND 1989-97 DT8 (CARBURETOR SERVICE) 4-15
    ASSEMBLY 4-17
    CLEANING & INSPECTION 4-16
    DISASSEMBLY 4-16
    REMOVAL & INSTALLATION 4-15
DT20 AND DT25 (BENDIX TYPE STARTER) 11-11
    ASSEMBLY 11-12
    CLEANING &INSPECTION 11-12
    DISASSEMBLY 11-11
    REMOVAL & INSTALLATION 11-11
DT20 AND DT25 (OVERHEAD TYPE STARTER) 11-8
    ASSEMBLY 11-9
    CLEANING & INSPECTION 11-8
    DISASSEMBLY 11-8
    REMOVAL & INSTALLATION 11-8
DT20, DT25 AND DT30 (LOWER UNIT OVERHAUL) 8-19
    ASSEMBLY 8-20
    CLEANING & INSPECTION 8-20
    DISASSEMBLY 8-19
    SHIMMING PROCEDURE 8-22
DT20, DT25 AND DT30 (TUNE-UP) 3-22
    IDLE SPEED 3-22
    IGNITION TIMING 3-22
    THROTTLE LINKAGE ADJUSTMENT 3-22
DT20 TO DT85, DT115 AND DT140 (CARBURETOR SERVICE) 4-17
    ASSEMBLY 4-21
    CLEANING & INSPECTION 4-21
    DISASSEMBLY 4-20
    REMOVAL & INSTALLATION 4-19
DT25C, DT30C, DT35C AND DT40C (OVERHEAD TYPE STARTER) 11-9
    ADJUSTMENT 11-9
    ASSEMBLY 11-10
    CLEANING & INSPECTION 11-10
    DISASSEMBLY 11-9
    REMOVAL & INSTALLATION 11-9
DT35 AND DT40 (LOWER UNIT OVERHAUL) 8-24
    ASSEMBLY 8-24
    CLEANING & INSPECTION 8-24
    DISASSEMBLY 8-24
    SHIMMING PROCEDURE 8-26
DT35 AND DT40 (TUNE-UP) 3-23
    IDLE SPEED 3-25
    IGNITION TIMING 3-23
    THROTTLE LINKAGE 3-26
DT55 AND DT65 (LOWER UNIT OVERHAUL) 8-28
    ASSEMBLY 8-31
    CLEANING & INSPECTION 8-30
    DISASSEMBLY 8-28
    SHIMMING PROCEDURE 8-32
DT55 AND DT65 (TUNE-UP) 3-26
    IDLE SPEED 3-26
    IGNITION TIMING 3-26
    THROTTLE LINKAGE 3-27
DT75 AND DT85 (LOWER UNIT OVERHAUL) 8-35
    ASSEMBLY 8-38
    CLEANING & INSPECTION 8-36
    DISASSEMBLY 8-35
    SHIMMING PROCEDURE 8-39
DT75 AND DT85 (TUNE-UP) 3-29
    CARBURETOR LINKAGE ADJUSTMENT 3-29
    IDLE SPEED 3-29
    IGNITION TIMING 3-29
DT90 AND DT100 (TUNE-UP) 3-30
    CARBURETOR LINKAGE ADJUSTMENT 3-30
    IDLE SPEED 3-30
    IGNITION TIMING 3-30
DT115 AND DT140 (LOWER UNIT OVERHAUL) 8-44
    ASSEMBLY 8-45
    CLEANING & INSPECTION 8-44
    DISASSEMBLY 8-44
    SHIMMING PROCEDURE 8-45
DT115 AND DT140 (TUNE-UP) 3-30
    CARBURETOR LINKAGE 3-31
    IDLE SPEED 3-32
    IGNITION TIMING 3-30
DT150, DT175, DT200 3-32
    CARBURETOR LINKAGE 3-32
    IDLE SPEED 3-32
    IGNITION TIMING 3-32
ELECTRICAL COMPONENTS 5-2
    CONNECTORS 5-4
    GROUND 5-3
    LOAD 5-3
    POWER SOURCE 5-2
    PROTECTIVE DEVICES 5-3
    SWITCHES & RELAYS 5-3
    WIRING & HARNESSES 5-4
ELECTRICAL SYSTEM PRECAUTIONS 5-7
ELECTRONIC FUEL INJECTION 4-32
ELECTRONIC IGNITION 5-38
EMERGENCY STOP SWITCH (REMOTE CONTROL BOX) 10-5
    REMOVAL & INSTALLATION 10-5
    TESTING 10-5
EMERGENCY STOP SWITCH (TILLER HANDLE) 10-10
    REMOVAL & INSTALLATION 10-10
    TESTING 10-10
ENGINE MAINTENANCE 3-2
ENGINE MECHANICAL 7-2
ENGINE REBUILDING SPECIFICATIONS 7-49
ENGINE START SWITCH 10-10
    REMOVAL & INSTALLATION 10-10
    TESTING 10-10
ENGINE STOP SWITCH (REMOTE CONTROL BOX) 10-5
    REMOVAL & INSTALLATION 10-5
    TESTING 10-5
ENGINE STOP SWITCH (TILLER HANDLE) 10-10
    REMOVAL & INSTALLATION 10-10
    TESTING 10-10
EQUIPMENT NOT REQUIRED BUT RECOMMENDED 1-10
    ANCHORS 1-10
    BAILING DEVICES 1-10
    FIRST AID KIT 1-10
    SECOND MEANS OF PROPULSION 1-10
    TOOLS AND SPARE PARTS 1-11
    VHF-FM RADIO 1-11
FASTENERS, MEASUREMENTS AND CONVERSIONS 2-14
FIBERGLASS HULLS 3-8
FLYWHEEL 7-2
    INSPECTION 7-5
    REMOVAL & INSTALLATION 7-2
FUEL 4-2
    ALCOHOL-BLENDED FUELS 4-2

MASTER INDEX 11-19

HIGH ALTITUDE OPERATION 4-2
OCTANE RATING 4-2
RECOMMENDATIONS 4-2
THE BOTTOM LINE WITH FUELS 4-2
VAPOR PRESSURE AND ADDITIVES 4-2
FUEL AND COMBUSTION 4-2
FUEL FILTER 3-4
RELIEVING FUEL SYSTEM PRESSURE 3-5
REMOVAL & INSTALLATION 3-5
FUEL INJECTORS 4-45
DESCRIPTION & OPERATION 4-45
TESTING 4-45
FUEL LINE 4-9
COMMON PROBLEMS 4-10
FUEL PRESSURE REGULATOR 4-46
REMOVAL & INSTALLATION 4-47
TESTING 4-46
FUEL PUMP (FUEL SYSTEM) 4-6
FUEL PUMP (TROUBLESHOOTING) 4-8
FUEL PUMP SERVICE 4-27
FUEL SYSTEM 4-3
FUEL SYSTEM (TROUBLESHOOTING) 4-7
COMMON PROBLEMS 4-8
LOGICAL TROUBLESHOOTING 4-7
FUEL/WATER SEPARATOR 3-6
GAS ASSIST DAMPER 9-2
REMOVAL & INSTALLATION 9-2
TESTING 9-2
GAS ASSISTED TILT 9-2
GEAR COUNTER COIL (ENGINE SPEED SENSOR) 4-45
DESCRIPTION & OPERATION 4-45
TESTING 4-45
GENERAL ENGINE SPECIFICATIONS 3-38
GENERAL INFORMATION (LOWER UNIT) 8-2
HAND REWIND STARTER 11-2
HAND TOOLS 2-5
ELECTRONIC TOOLS 2-10
GAUGES 2-11
HAMMERS 2-9
OTHER COMMON TOOLS 2-10
PLIERS 2-9
SCREWDRIVERS 2-9
SOCKET SETS 2-5
SPECIAL TOOLS 2-10
WRENCHES 2-8
HIGH PRESSURE FUEL PUMP 4-47
TESTING 4-48
HOW TO USE THIS MANUAL 1-2
IGNITION AND ELECTRICAL WIRING DIAGRAMS 5-52
IGNITION COIL (BREAKER POINTS IGNITION) 5-10
DESCRIPTION & OPERATION 5-10
REMOVAL & INSTALLATION 5-11
TESTING 5-11
IGNITION COILS (CAPACITOR DISCHARGE IGNITION SYSTEM) 5-25
DESCRIPTION & OPERATION 5-25
REMOVAL & INSTALLATION 5-26
TESTING 5-25
IGNITION SWITCH 10-6
REMOVAL & INSTALLATION 10-6
TESTING 10-6
IGNITION SYSTEM 3-17
INSIDE THE BOAT 3-8
INTRODUCTION 3-12
JET DRIVE 8-51
JET DRIVE ASSEMBLY 8-51
ADJUSTMENT 8-54
ASSEMBLING 8-57
CLEANING & INSPECTING 8-56
DISASSEMBLY 8-56
REMOVAL & INSTALLATION 8-51
LOWER UNIT 8-2
LOWER UNIT (ENGINE MAINTENANCE) 3-3
DRAINING & FILLING 3-3
OIL RECOMMENDATIONS 3-3
LOWER UNIT OVERHAUL 8-6
LOWER UNIT—NO REVERSE GEAR 8-3
REMOVAL & INSTALLATION 8-3
LOWER UNIT—WITH REVERSE GEAR 8-3
REMOVAL & INSTALLATION 8-3
MAINTENANCE OR REPAIR? 1-2
MANUAL TILT 9-2
MEASURING TOOLS 2-12
DEPTH GAUGES 2-13
DIAL INDICATORS 2-13
MICROMETERS & CALIPERS 2-12
TELESCOPING GAUGES 2-13
MODEL IDENTIFICATION AND SERIAL NUMBERS 8-51
NEUTRAL START SWITCH 10-2
REMOVAL & INSTALLATION 10-2
TESTING 10-2
OIL FLOW SENSOR 6-17
CLEANING & INSPECTION 6-17
REMOVAL & INSTALLATION 6-17
TESTING 6-17
OIL INJECTION SYSTEM 6-2
OIL INJECTION WARNING SYSTEM 6-15
OIL INJECTION WARNING SYSTEMS 6-14
OIL LEVEL SENSOR 6-16
REMOVAL & INSTALLATION 6-16
TESTING 6-16
OIL LINES 6-6
OIL LINE CAUTIONS 6-6
REMOVAL & INSTALLATION 6-6
OIL PUMP 6-5
REMOVAL & INSTALLATION 6-5
OIL PUMP CONTROL ROD 6-8
ADJUSTMENT 6-8
OIL PUMP DISCHARGE RATE (OIL INJECTION) 6-8
OIL PUMP DISCHARGE RATE (OIL INJECTION SYSTEM) 6-7
ADJUSTMENT 6-8
TESTING 6-7
OIL TANK 6-3
CLEANING & INSPECTION 6-4
REMOVAL & INSTALLATION 6-3
OVERHEAD TYPE STARTER 11-2
OVERHEAT SENSOR 6-19
REMOVAL & INSTALLATION 6-19
TESTING 6-19
OVERHEAT WARNING SYSTEM 6-17
OVERHEAT WARNING SYSTEM (OIL INJECTION) 6-18
PISTON PINS 7-39
GENERAL INFORMATION 7-39
INSPECTION 7-39
PISTON RINGS 7-40
GENERAL INFORMATION 7-40
INSPECTION 7-41
PISTONS 7-37
GENERAL INFORMATION 7-37
INSPECTION 7-38
POWER TILT 9-3
POWER TRIM/TILT 9-6
POWERHEAD 7-5
DISASSEMBLY & ASSEMBLY 7-15
REMOVAL & INSTALLATION 7-5
POWERHEAD OVERHAUL TIPS 7-33

11-20 MASTER INDEX

CAUTIONS 7-34
CLEANING 7-34
REPAIRING DAMAGED THREADS 7-34
TOOLS 7-34
POWERHEAD PREPARATION 7-35
POWERHEAD RECONDITIONING 7-32
PROFESSIONAL HELP 1-2
PROPELLER (ENGINE MAINTENANCE) 3-7
PROPELLER (LOWER UNIT) 8-2
REMOVAL & INSTALLATION 8-2
PULSAR/CHARGING/GEAR COUNTER COILS 5-14
DESCRIPTION & OPERATION 5-14
REMOVAL & INSTALLATION 5-20
TESTING 5-15
PULSER COIL 4-45
DESCRIPTION & OPERATION 4-45
TESTING 4-45
PURCHASING PARTS 1-3
RECTIFIER 5-37
DESCRIPTION & OPERATION 5-37
REMOVAL & INSTALLATION 5-38
TESTING 5-37
REED VALVE SERVICE 4-25
REGULATIONS FOR YOUR BOAT 1-4
CAPACITY INFORMATION 1-4
CERTIFICATE OF COMPLIANCE 1-4
DOCUMENTING OF VESSELS 1-4
HULL IDENTIFICATION NUMBER 1-4
LENGTH OF BOATS 1-4
NUMBERING OF VESSELS 1-4
REGISTRATION OF BOATS 1-4
SALES AND TRANSFERS 1-4
VENTILATION 1-5
VENTILATION SYSTEMS 1-5
REGULATOR 5-38
DESCRIPTION & OPERATION 5-38
REMOVAL & INSTALLATION 5-38
TESTING 5-38
REMOTE CONTROL BOX 10-2
REMOTE CONTROL BOX (REMOTE CONTROL BOX) 10-2
REMOVAL & INSTALLATION 10-2
REMOTE CONTROL CABLES 10-2
ADJUSTMENT 10-2
REMOVAL & INSTALLATION 10-2
REQUIRED SAFETY EQUIPMENT 1-5
FIRE EXTINGUISHERS 1-5
PERSONAL FLOTATION DEVICES 1-7
SOUND PRODUCING DEVICES 1-9
TYPES OF FIRES 1-5
VISUAL DISTRESS SIGNALS 1-9
WARNING SYSTEM 1-7
SAFETY IN SERVICE 1-12
SAFETY TOOLS 2-2
EYE AND EAR PROTECTION 2-2
WORK CLOTHES 2-3
WORK GLOVES 2-2
SELF DIAGNOSTIC SYSTEM 4-48
DESCRIPTION & OPERATION 4-48
DIAGNOSIS PROCEDURE 4-49
FAIL SAFE EMERGENCY BACKUP 4-49
SERIAL NUMBER IDENTIFICATION (ENGINE MAINTENANCE) 3-2
SERIAL NUMBER IDENTIFICATION (MAINTENANCE) 3-43
SHIFTING PRINCIPLES 8-2
COUNTERROTATING UNIT 8-2
STANDARD ROTATING UNIT 8-2
SPARK PLUG WIRES 3-17
REMOVAL & INSTALLATION 3-17
TESTING 3-17

SPARK PLUGS 3-13
INSPECTION & GAPPING 3-15
READING SPARK PLUGS 3-14
REMOVAL & INSTALLATION 3-14
SPARK PLUG HEAT RANGE 3-13
SPARK PLUG SERVICE 3-14
SPECIFICATION CHARTS
CAPACITIES 3-2
CARBURETOR IDLE AIR SCREW SPECIFICATION 3-33
CIRCUIT/POSITION 4-7
CONVERSION FACTORS 2-17
ENGINE REBUILDING SPECIFICATIONS 7-49
GENERAL ENGINE SPECIFICATIONS 3-38
OIL PUMP DISCHARGE RATE 6-8
SERIAL NUMBER IDENTIFICATION 3-43
TORQUE SPECIFICATIONS 7-46
TUNE UP SPECIFICATIONS CHARTS 3-34
SPRING COMMISSIONING CHECKLIST 3-35
STANDARD AND METRIC MEASUREMENTS 2-15
STARTER MOTOR 5-46
DESCRIPTION & OPERATION 5-46
OVERHAUL 5-49
REMOVAL & INSTALLATION 5-48
TESTING 5-48
STARTER MOTOR RELAY SWITCH 5-50
DESCRIPTION & OPERATION 5-50
REMOVAL & INSTALLATION 5-51
TESTING 5-51
STARTING CIRCUIT 5-45
SYSTEM TESTING (BREAKER POINTS IGNITION) 5-8
SYSTEM TESTING (CAPACITOR DISCHARGE IGNITION SYSTEM) 5-14
PROCEDURE 5-14
TEST EQUIPMENT 5-4
JUMPER WIRES 5-4
MULTIMETERS 5-5
TEST LIGHTS 5-5
TESTING 5-6
OPEN CIRCUITS 5-7
RESISTANCE 5-6
SHORT CIRCUITS 5-7
VOLTAGE 5-6
VOLTAGE DROP 5-6
THERMOSTAT 6-13
CLEANING & INSPECTION 6-14
REMOVAL & INSTALLATION 6-13
THROTTLE CABLE 10-10
ADJUSTMENT 10-10
REMOVAL & INSTALLATION 10-10
THROTTLE VALVE SENSOR 4-40
ADJUSTMENT 4-42
DESCRIPTION & OPERATION 4-40
REMOVAL & INSTALLATION 4-45
TESTING 4-40
TILLER HANDLE 10-7
TILLER HANDLE (TILLER HANDLE) 10-8
REMOVAL & INSTALLATION 10-8
TILT CYLINDER 9-5
OVERHAUL 9-5
REMOVAL & INSTALLATION 9-5
TILT MOTOR AND PUMP 9-4
BLEEDING THE SYSTEM 9-4
CHECKING FLUID LEVEL 9-4
OVERHAUL 9-5
REMOVAL & INSTALLATION 9-5
TESTING 9-5
TILT SWITCH 9-6
TIMING AND SYNCHRONIZATION 3-17

PREPARATION 3-17
SYNCHRONIZATION 3-17
TIMING 3-17
TOOLS 2-5
TOOLS AND EQUIPMENT 2-2
TORQUE 2-15
TORQUE SPECIFICATIONS 7-46
TRIM TABS, ANODES AND LEAD WIRES 3-8
TRIM/TILT & PIVOT POINTS 3-6
 INSPECTION & LUBRICATION 3-6
TRIM/TILT CYLINDER 9-14
 ASSEMBLY 9-15
 CLEANING & INSPECTION 9-15
 DISASSEMBLY 9-15
 REMOVAL & INSTALLATION 9-14
 TESTING 9-14
TRIM/TILT MOTOR 9-9
 ASSEMBLY 9-13
 CLEANING &INSPECTION 9-12
 DISASSEMBLY 9-11
 REMOVAL & INSTALLATION 9-10
 TESTING 9-9
TRIM/TILT PUMP 9-8
 REMOVAL & INSTALLATION 9-9
 TESTING 9-8
TRIM/TILT RELAY 9-15
 REMOVAL & INSTALLATION 9-15
 TESTING 9-15
TRIM/TILT SWITCH (POWER TRIM/TILT) 9-15
TRIM/TILT SWITCH (REMOTE CONTROL BOX) 10-6
 REMOVAL & INSTALLATION 10-6
 TESTING 10-6
TROUBLESHOOTING 4-7
TROUBLESHOOTING CHARTS
 OIL INJECTION WARNING SYSTEM 6-15
 OVERHEAT WARNING SYSTEM 6-18
TROUBLESHOOTING THE CHARGING SYSTEM 5-40
 OVERCHARGING 5-40
 UNDERCHARGING 5-40
TROUBLESHOOTING THE COOLING SYSTEM 6-12
TROUBLESHOOTING THE ELECTRICAL SYSTEM 5-6
TROUBLESHOOTING THE HAND REWIND STARTER 11-2
TROUBLESHOOTING THE LOWER UNIT 8-2
TROUBLESHOOTING THE OIL INJECTION SYSTEM 6-2
TROUBLESHOOTING THE OIL INJECTION WARNING SYSTEM 6-15
TROUBLESHOOTING THE OVERHEAT WARNING SYSTEM 6-17
TROUBLESHOOTING THE POWER TILT SYSTEM 9-4
TROUBLESHOOTING THE POWER TRIM/TILT SYSTEM 9-7
TROUBLESHOOTING THE REMOTE CONTROLS 10-2
TROUBLESHOOTING THE STARTING SYSTEM 5-46
TROUBLESHOOTING THE TILLER HANDLE 10-8
TUNE UP SPECIFICATIONS CHARTS 3-34
TUNE-UP 3-12
TUNE-UP SEQUENCE 3-12
TWO-STROKE CYCLE 7-2
2-STROKE OIL 3-2
 FILLING 3-2
 OIL RECOMMENDATIONS 3-2
UNDERSTANDING AND TROUBLESHOOTING ELECTRICAL SYSTEMS 5-2
V4 & V6 POWERHEADS 4-22
 ASSEMBLY 4-25
 CLEANING & INSPECTION 4-25
 DISASSEMBLY 4-24
 FUEL LEVEL TEST 4-24
 REMOVAL & INSTALLATION 4-23
V4 AND V6 8-46
 ASSEMBLY 8-47
 CLEANING & INSPECTING 8-47
 DISASSEMBLY 8-46
 SHIMMING PROCEDURE 8-50
WARNING BUZZER 10-7
 REMOVAL & INSTALLATION 10-7
 TESTING 10-7
WARNING SYSTEMS 5-38
 BATTERY VOLTAGE 5-39
 OIL FLOW 5-38
 OIL LEVEL 5-38
 OVER REVOLUTION 5-38
 OVERHEAT 5-38
WATER PUMP 6-12
 CLEANING & INSPECTION 6-13
 REMOVAL & INSTALLATION 6-12
WHERE TO BEGIN 1-2
WINTER STORAGE CHECKLIST 3-35
WIRE AND CONNECTOR REPAIR 5-7

Chilton and Seloc Marine

Chilton and Seloc Marine offer repair and maintenance manuals for nearly everything with an engine. Whether you need to care for a car, truck, sport-utility vehicle, boat, motorcycle, ATV, snowmobile or even a lawn mower, we can provide the information you need to get the most from your vehicles.

With over 230 manuals available in more than 6 lines of Do-It-Yourself books, you're bound to find what you want. Here are just a few types of books we have to offer.

TOTAL CAR CARE

With over 140 titles covering cars, trucks and sport utilities from Acura to Volvo, the Total Car Care series contains everything you need to care for or repair your vehicle. Every manual is based on vehicle teardowns and includes tons of step-by-step photographs, procedures and exploded views. Each manual also contains wiring diagrams, vacuum diagrams and covers every topic from simple maintenance to electronic engine

COLLECTOR'S SERIES HARD-COVER MANUALS

Chilton's Collector's Editions are perfect for enthusiasts of vintage or rare cars. These hard-cover manuals contain repair and maintenance information for all major systems that might not be available elsewhere. Included are repair and overhaul procedures using thousands of illustrations. These manuals offer a range of coverage from as far back as 1940 and as recent as 1997, so you don't need an antique car or truck to be a collector.

SELOC MARINE MANUALS

Chilton Marine offers 40 titles covering outboard engines, stern drives, jet drives, marine diesel and personal watercraft. Written from teardowns, just like our Total Car Care series, each manual is loaded with step-by-step procedures and photography, as well as exploded views, specifications, troubleshooting and wiring diagrams.

GENERAL INTEREST / RECREATIONAL BOOKS

We offer specialty books on a variety of topics including Motorcycles, ATVs, Snowmobiles and automotive subjects like Detailing or Body Repair. Each book from our General Interest line offers a blend of our famous Do-It-Yourself procedures and photography with additional information on enjoying automotive, marine and recreational products. Learn more about the vehicles you use and enjoy while keeping them in top running shape.

TOTAL SERVICE SERIES / SYSTEM SPECIFIC MANUALS

These innovative books offer repair, maintenance and service procedures for automotive related systems. They cover today's complex vehicles in a user-friendly format, which places even the most difficult automotive topic well within the reach of every Do-It-Yourselfer. Each title covers a specific subject from Brakes and Engine Rebuilding to Fuel Injection Systems, Automatic Transmissions and even Engine Trouble

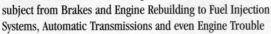

MULTI-VEHICLE SPANISH LANGUAGE MANUALS

Chilton's Spanish language manuals offer some of our most popular titles in Spanish. Each is as complete and easy to use as the English-language counterpart and offers the same maintenance, repair and overhaul information along with specifications charts and tons of illustrations.

Visit your local Chilton® Retailer

For a Catalog, for information, or to order call toll-free: 877-4CHILTON.

1020 Andrew Drive, Suite 200 • West Chester, PA 19380-4291
www.chiltonsonline.com